Biomimetic Radical Chemistry and Applications

Biomimetic Radical Chemistry and Applications

Special Issue Editor
Chryssostomos Chatgilialoglu

MDPI • Basel • Beijing • Wuhan • Barcelona • Belgrade • Manchester • Tokyo • Cluj • Tianjin

Special Issue Editor
Chryssostomos Chatgilialoglu
Consiglio Nazionale delle
Ricerche
Italy

Editorial Office
MDPI
St. Alban-Anlage 66
4052 Basel, Switzerland

This is a reprint of articles from the Special Issue published online in the open access journal *Molecules* (ISSN 1420-3049) (available at: https://www.mdpi.com/journal/molecules/special_issues/biomimetic_radical_chem).

For citation purposes, cite each article independently as indicated on the article page online and as indicated below:

LastName, A.A.; LastName, B.B.; LastName, C.C. Article Title. *Journal Name* **Year**, *Article Number*, Page Range.

ISBN 978-3-03928-392-7 (Pbk)
ISBN 978-3-03928-393-4 (PDF)

© 2020 by the authors. Articles in this book are Open Access and distributed under the Creative Commons Attribution (CC BY) license, which allows users to download, copy and build upon published articles, as long as the author and publisher are properly credited, which ensures maximum dissemination and a wider impact of our publications.

The book as a whole is distributed by MDPI under the terms and conditions of the Creative Commons license CC BY-NC-ND.

Contents

About the Special Issue Editor vii

Preface to "Biomimetic Radical Chemistry and Applications" ix

Chryssostomos Chatgilialoglu, Marios G. Krokidis, Annalisa Masi, Sebastian Barata-Vallejo, Carla Ferreri, Michael A. Terzidis, Tomasz Szreder and Krzysztof Bobrowski
New Insights into the Reaction Paths of Hydroxyl Radicals with Purine Moieties in DNA and Double-Stranded Oligodeoxynucleotides
Reprinted from: *Molecules* **2019**, *24*, 3860, doi:10.3390/molecules24213860 1

Evangelos Balanikas, Akos Banyasz, Gérard Baldacchino and Dimitra Markovitsi
Populations and Dynamics of Guanine Radicals in DNA strands—Direct versus Indirect Generation
Reprinted from: *Molecules* **2019**, *24*, 2347, doi:10.3390/molecules24132347 21

Andrea Peluso, Tonino Caruso, Alessandro Landi, Amedeo Capobianco
The Dynamics of Hole Transfer in DNA
Reprinted from: *Molecules* **2019**, *24*, 4044, doi:10.3390/molecules24224044 37

Zara Molphy, Vickie McKee and Andrew Kellett
Copper *bis*-Dipyridoquinoxaline Is a Potent DNA Intercalator that Induces Superoxide-Mediated Cleavage via the Minor Groove
Reprinted from: *Molecules* **2019**, *24*, 4301, doi:10.3390/molecules24234301 59

Paulina Spisz, Magdalena Zdrowowicz, Samanta Makurat, Witold Kozak, Konrad Skotnicki, Krzysztof Bobrowski and Janusz Rak
Why Does the Type of Halogen Atom Matter for the Radiosensitizing Properties of 5-Halogen Substituted 4-Thio-2′-Deoxyuridines?
Reprinted from: *Molecules* **2019**, *24*, 2819, doi:10.3390/molecules24152819 71

Pawlos S. Tsegay, Yanhao Lai and Yuan Liu
Replication Stress and Consequential Instability of the Genome and Epigenome
Reprinted from: *Molecules* **2019**, *24*, 3870, doi:10.3390/molecules24213870 89

Shinichi Sato and Hiroyuki Nakamura
Protein Chemical Labeling Using Biomimetic Radical Chemistry
Reprinted from: *Molecules* **2019**, *24*, 3980, doi:10.3390/molecules24213980 109

Christian Schöneich
Thiyl Radical Reactions in the Chemical Degradation of Pharmaceutical Proteins
Reprinted from: *Molecules* **2019**, *24*, 4357, doi:10.3390/molecules24234357 127

Balázs Kripli, Bernadett Sólyom, Gábor Speier and József Kaizer
Stability and Catalase-Like Activity of a Mononuclear Non-Heme Oxoiron(IV) Complex in Aqueous Solution
Reprinted from: *Molecules* **2019**, *24*, 3236, doi:10.3390/molecules24183236 143

Elena E. Pohl and Olga Jovanovic
The Role of Phosphatidylethanolamine Adducts in Modification of the Activity of Membrane Proteins under Oxidative Stress
Reprinted from: *Molecules* **2019**, *24*, 4545, doi:10.3390/molecules24244545 155

Giuseppe Maulucci, Ofir Cohen, Bareket Daniel, Carla Ferreri and Shlomo Sasson
The Combination of Whole Cell Lipidomics Analysis and Single Cell Confocal Imaging of Fluidity and Micropolarity Provides Insight into Stress-Induced Lipid Turnover in Subcellular Organelles of Pancreatic Beta Cells
Reprinted from: *Molecules* **2019**, *24*, 3742, doi:10.3390/molecules24203742 171

Anna Vita Larocca, Gianluca Toniolo, Silvia Tortorella, Marios G. Krokidis,
Georgia Menounou, Giuseppe Di Bella, Chryssostomos Chatgilialoglu and Carla Ferreri
The Entrapment of Somatostatin in a Lipid Formulation: Retarded Release and Free Radical Reactivity
Reprinted from: *Molecules* **2019**, *24*, 3085, doi:10.3390/molecules24173085 185

Iulia Matei, Cristina Maria Buta, Ioana Maria Turcu, Daniela Culita, Cornel Munteanu and Gabriela Ionita
Formation and Stabilization of Gold Nanoparticles in Bovine Serum Albumin Solution
Reprinted from: *Molecules* **2019**, *24*, 3395, doi:10.3390/molecules24183395 197

Katarzyna Taras-Goslinska, Fabrizio Vetica, Sebastián Barata-Vallejo,
Virginia Triantakostanti, Bronisław Marciniak and Chryssostomos Chatgilialoglu
Converging Fate of the Oxidation and Reduction of 8-Thioguanosine
Reprinted from: *Molecules* **2019**, *24*, 3143, doi:10.3390/molecules24173143 213

Konrad Skotnicki, Katarzyna Taras-Goslinska, Ireneusz Janik and Krzysztof Bobrowski
Radiation Induced One-Electron Oxidation of 2-Thiouracil in Aqueous Solutions
Reprinted from: *Molecules* **2019**, *24*, 4402, doi:10.3390/molecules24234402 231

Aggeliki Stathi, Michael Mamais, Evangelia D. Chrysina and Thanasis Gimisis
Anomeric Spironucleosides of β-D-Glucopyranosyl Uracil as Potential Inhibitors of Glycogen Phosphorylase
Reprinted from: *Molecules* **2019**, *24*, 2327, doi:10.3390/molecules24122327 259

Nicolas Millius, Guillaume Lapointe and Philippe Renaud
Two-Step Azidoalkenylation of Terminal Alkenes Using Iodomethyl Sulfones
Reprinted from: *Molecules* **2019**, *24*, 4184, doi:10.3390/molecules24224184 271

Liye Fu, Antonina Simakova, Sangwoo Park, Yi Wang, Marco Fantin and Krzysztof Matyjaszewski
Axially Ligated Mesohemins as Bio-Mimicking Catalysts for Atom Transfer Radical Polymerization
Reprinted from: *Molecules* **2019**, *24*, 3969, doi:10.3390/molecules24213969 281

About the Special Issue Editor

Chryssostomos Chatgilialoglu (Dr.) Research Director at the Italian National Research Council (CNR) in Bologna and Visiting Professor at the Center for Advanced Technologies, Adam Mickiewicz University, in Poznań (Poland). He is also Co-Founder and President of the company Lipinutragen. He received a doctorate degree in Industrial Chemistry from Bologna University in 1976 and completed his postdoctoral studies at York University (UK) and the National Research Council of Canada, Ottawa. From March 2014 to May 2016 he was appointed as the Director of the Institute of Nanoscience and Nanotechnology at the NCSR "Demokritos" in Athens (Greece). He was Chairman of the COST Actions Free Radicals in Chemical Biology (2007–2011) and Biomimetic Radical Chemistry (2013–2016).

He is the author or co-author of more than 290 publications in peer-reviewed journals and 36 book chapters. He is the author and editor of several books, including Organosilanes in Radical Chemistry, Wiley 2004; Encyclopedia of Radicals in Chemistry, Biology and Materials, Wiley 2012; and Membrane Lipidomics for Personalized Health, Wiley 2015. He has been invited as a Guest Editor for Special Issues in several scientific journals and was an invited speaker over 260 times at congresses and institutions.

His research group is active in the field of free radical chemistry addressed to applications in life sciences. In recent years, he developed biomimetic chemistry of radical stress and related biomarkers. The discovery of the endogenous formation of trans-lipids, research on 5′,8-cyclopurine DNA lesions, and fatty acid-based lipidomics attracted worldwide attention. He is responsible for introducing tris(trimethylsilyl)silane as a radical-based reducing agent, and for this he was the winner of the Fluka Prize "Reagent of the Year 1990".

Preface to "Biomimetic Radical Chemistry and Applications"

Free radicals have attracted considerable attention in various research areas, including organic synthesis, material science, atmospheric chemistry, radiation chemistry, pharmacology, biology, and medicine. The "free radical theory of aging", the role of enzymes like superoxide dismutase (SOD) or ribonucleotide reductase (RNR), the antioxidant network, and the role of vitamins are now milestones of the life sciences. Free radicals are generated in the biological environment as a result of normal intracellular metabolism and function as physiological signaling molecules that participate in the modulation of apoptosis, stress responses, and proliferation. During inflammatory response, their concentrations can increase up to 100-fold excess, and if not properly quenched, they can have a negative effect by causing damages to biomolecules. Therefore, the estimation of the type and extent of damages, as well as the mechanisms and efficiency of protective and repair systems, are important subjects in the life sciences.

When studying free-radical-based chemical mechanisms, biomimetic chemistry and the design of established biomimetic models come into play to perform experiments in a controlled environment, suitably designed to be in strict connection with cellular conditions. In this Special Issue, the biomimetic approach is presented with new insights and reviews of the current knowledge in the field of radical-based processes relevant to health, such as biomolecular damages and repair, signaling and biomarkers, biotechnological applications, and novel synthetic approaches. Several subjects are presented, with 12 articles and 6 reviews written by specialists in the fields.

In the area of DNA, the Special Issue reports on: (i) new insights into the reaction of hydroxyl radicals with genetic material and the role of purine lesions as biomarkers; (ii) guanine radicals generated in single, double, and G-quadruplex oligonucleotides studied by nanosecond transient absorption spectroscopy; (iii) an overview of work on the dynamics of hope transfer in DNA; (iv) the mechanism of copper artificial metallo-nuclease to induce superoxide-mediated cleavage via the minor groove; (v) the influence of the type of halogen atom in the radiosensitizing properties of substituted uridines; (vi) an overview of replication stress and the consequential instability of the genome and epigenome.

In the area of proteins, two important reviews deal with the chemical labelling of proteins using biomimetic radical chemistry and the role of thiyl radical reactions in the degradation of protein pharmaceuticals. New insights on mechanistic studies of a mononuclear non-heme oxoiron(IV) complex in aqueous solution as catalase-like activity are also described in an original article. In the area of membrane lipids, two reviews provide thorough descriptions of the role of phosphatidylethanolamine-derived adducts as mediators of reactive aldehydes and effects on membrane properties, and the combination of various technologies to study stress-induced lipid turnover in subcellular organelles of pancreatic beta cells. Biotechnological applications in this Special Issue concern two fields: (i) the entrapment of somatostatin in lipid formulation with the discovery of a new radical reactivity arising from the sulfur-containing peptides and its nano-delivery formulation; (ii) the formation and stabilization of gold nanoparticles in bovine serum albumin solution.

Mechanistic studies of radical processes in pharmacological applications, which also inspire biological mechanisms, are represented by various articles: the oxidation of 8-thioguanosine and 2-thiouracil by photolytic and radiolytic conditions; bioinspired synthetic strategies radical-based toward anomeric spironucleosides as potential inhibitors of glycogen phosphorylase and for the preparation of azido-derivatives via a radical azidoalkylation of alkenes; synthesis of two new iron-porphyrin-based catalysts inspired by naturally occurring proteins such as horseradish peroxidase, hemoglobin, and cytochrome P450 tested for atom transfer radical polymerization (ATRP), obtaining polymers with specific properties.

This Special Issue gives the reader a wide overview of biomimetic radical chemistry, where molecular mechanisms have been defined and molecular libraries of products are developed to also be used as traces for the discovery of some relevant biological processes. The biomimetic approach is a convenient tool, since the achievements in free radical mechanisms can be easily transferred to a better comprehension of the radical-based biological pathways in living organisms, to foster advancements in health and diseases. In addition, the identification of modified biomolecules paves the way for molecular libraries and the evaluation of in vivo damage through biomarkers.

Chryssostomos Chatgilialoglu
Special Issue Editor

Article

New Insights into the Reaction Paths of Hydroxyl Radicals with Purine Moieties in DNA and Double-Stranded Oligodeoxynucleotides

Chryssostomos Chatgilialoglu [1,2,*], Marios G. Krokidis [1,3], Annalisa Masi [1], Sebastian Barata-Vallejo [1,4], Carla Ferreri [1], Michael A. Terzidis [1,5], Tomasz Szreder [5] and Krzysztof Bobrowski [5,*]

1. Istituto per la Sintesi Organica e la Fotoreattività, Consiglio Nazionale delle Ricerche, 40129 Bologna, Italy; m.krokidis@inn.demokritos.gr (M.G.K.); annalisa.masi@isof.cnr.it (A.M.); sebastian.barata@isof.cnr.it (S.B.-V.); carla.ferreri@isof.cnr.it (C.F.); mterzidi@gmail.com (M.A.T.)
2. Center for Advanced Technologies, Adam Mickiewicz University, 61-614 Poznań, Poland
3. Institute of Nanoscience and Nanotechnology, N.C.S.R. "Demokritos", 15310 Agia Paraskevi Attikis, Greece
4. Departamento de Quimíca Organíca, Facultad de Farmacia y Bioquímica, Universidad de Buenos Aires, Junin 954, Buenos Aires CP 1113, Argentina
5. Centre of Radiation Research and Technology, Institute of Nuclear Chemistry and Technology, Dorodna 16, 03-195 Warsaw, Poland; t.szreder@ichtj.waw.pl
* Correspondence: chrys@isof.cnr.it (C.C.); kris@ichtj.pl (K.B.); Tel.: +39-051-6398309 (C.C.)

Academic Editor: Michael Smietana
Received: 24 September 2019; Accepted: 23 October 2019; Published: 26 October 2019

Abstract: The reaction of hydroxyl radical (HO•) with DNA produces many primary reactive species and many lesions as final products. In this study, we have examined the optical spectra of intermediate species derived from the reaction of HO• with a variety of single- and double-stranded oligodeoxynucleotides and ct-DNA in the range of 1 µs to 1 ms by pulse radiolysis using an Intensified Charged Coupled Device (ICCD) camera. Moreover, we applied our published analytical protocol based on an LC-MS/MS system with isotopomeric internal standards to enable accurate and precise measurements of purine lesion formation. In particular, the simultaneous measurement of the four purine 5′,8-cyclo-2′-deoxynucleosides (cPu) and two 8-oxo-7,8-dihydro-2′-deoxypurine (8-oxo-Pu) was obtained upon reaction of genetic material with HO• radicals generated either by γ-radiolysis or Fenton-type reactions. Our results contributed to the debate in the literature regarding absolute level of lesions, method of HO• radical generation, 5′R/5′S diastereomeric ratio in cPu, and relative abundance between cPu and 8-oxo-Pu.

Keywords: DNA damage; 5′,8-cyclopurines; 8-oxo-dG; free radicals; pulse radiolysis; gamma radiolysis; Fenton reaction; oligonucleotides

---

1. Introduction

Hydroxyl radicals (HO•) are highly reactive with many compounds and DNA is not an exception. Indeed, HO• radicals are known for their reactivity and ability to cause chemical modifications to DNA, the site of attack being both the base moieties (85–90%) and the 2-deoxyribose units [1,2]. The attack at H5′ of DNA by HO• radicals is estimated to be 55% of all possible sugar positions and the resulting C5′ radical in the purine nucleotide moieties leads to the formation of purine 5′,8-cyclo-2′-deoxynucleosides (cPu) as final products (Figure 1A) [3,4]. The 5′,8-cyclo-2′-deoxyadenosine (cdA) and 5′,8-cyclo-2′-deoxyguanosine (cdG) exist in 5′R and 5′S diastereoisomeric forms (Figure 1B). These tandem-type lesions, generated by the attack of HO• radicals or direct irradiation damage [5], have been identified in mammalian cellular DNA in vivo [6–10] and are substrates of nucleotide excision repair (NER) [4,11,12]. On the other hand, the addition of HO• radicals to the guanine and adenine moieties affords a variety of products including the well-known 8-oxo-7,8-dihydro-2′-deoxyadenosine (8-oxo-dA) and 8-oxo-7,8-dihydro-2′-deoxyguanosine (8-oxo-dG) lesions (Figure 1C) [13]. Like HO• radicals, other oxidizing species such as H_2O_2, singlet oxygen or $ONOO^-$ are able to generate 8-oxo-Pu lesions that are removed by the base excision repair (BER) system [14].

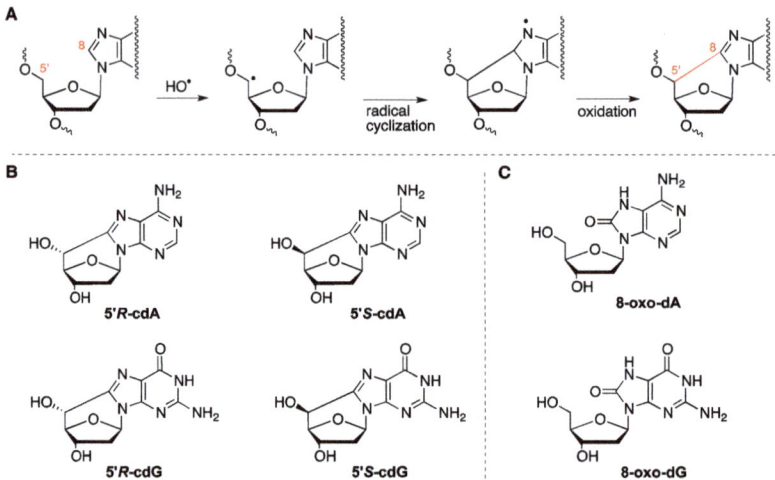

Figure 1. (**A**) purine 2′-deoxynucleoside reacts with hydroxyl radical (HO•) yielding the purine 5′,8-cyclo-2′-deoxynucleoside (cPu) via cyclization of C5′ radical followed by oxidation; (**B**) structures of 5′,8-cyclo-2′-deoxyadenosine (cdA) and 5′,8-cyclo-2′-deoxyguanosine (cdG) in their 5′R and 5′S diastereomeric forms; (**C**) structures of 8-oxo-2′-deoxyadenosine (8-oxo-dA) and 8-oxo-2′-deoxyguanosine (8-oxo-dG).

Attempts to accurately determine the level of the four cPu lesions in DNA are numerous [4,15,16]. A detailed protocol for the simultaneous quantification of the four cPu lesions and two 8-oxo-Pu of DNA has also been provided by some of us [4,16]. Liquid chromatography-tandem mass spectrometry (LC-ESI-MS/MS) analysis, following a top–down approach starting from the genetic material and going down to a single nucleoside level, establishes accurate quantification of these lesions. The use of isotopically labeled reference compounds for the lesions further enhances the reliability of the process, increasing to a great extent the reproducibility and the recovery of the quantification protocol.

Pulse radiolysis studies on the reaction of HO• radicals with DNA and its model systems (oligonucleotides) are limited due to the cost of starting material. In the present work, we explored the Intensified Charge-Coupled Device (ICCD) as an alternative of photomultiplier (PMT) for transient spectra measurements [17]. The possibility to record the transient spectra using the ICCD camera has the advantage of reduced sample size. Such an approach has been already successfully applied for pulse radiolysis studies on the reaction of HO• radicals with the calcium-saturated forms of wild-type calmodulin and its Met-deficient mutant [18].

The objective of this work is dual: (i) to gain information on the optical absorption of transient spectra of the reaction of HO• radical with DNA and its model systems by pulse radiolysis, and (ii) to apply our protocol based on the stable isotope-dilution tandem mass spectrometry technique for the quantification of HO• radical induced cPu and 8-oxo-Pu lesions within DNA and its model systems in gamma-irradiated aqueous solutions or Fenton-type reactions. In principle, the two techniques employed (pulse radiolysis of transient spectra in the range of 1 µs to 1 ms and LC-ESI-MS/MS for product identification) complement each other, that is, acquired information of the intermediate reactive species that lead to the observed stable products on the same material when exposed to HO• radicals. In order to achieve our aims, we examined the reactivity of HO• radical with calf-thymus DNA (ct-DNA) and a variety of single-stranded (ss) or doubled-stranded (ds) oligodeoxynucleotides (ODNs), which are the simplest biomimetic models of DNA that respect the biological characteristics (Table 1).

Table 1. The sequences of the single stranded (ss) oligodeoxynucleotides (ODN) used in this study.

Strands	Sequence (5′-3′)	Length
ODN1	CGT ATG GTA TCG	12
ODN2	CGA TAC CAT ACG	12
ODN3	CGA TGG GGT ACG	12
ODN4	CGT ACC CCA TCG	12
ODN5	GGG (TTA GGG)$_3$	21
ODN6	CCC (TAA CCC)$_3$	21

2. Results and Discussion

2.1. Radiolytic Production of Transients

Radiolysis of neutral water leads to the reactive species e_{aq}^-, HO• and H•, as shown in Reaction 1, together with H$^+$ and H$_2$O$_2$. The values in parentheses represent the radiation chemical yields (G) in units of μmol J^{-1}. In N$_2$O-saturated solution (~0.02 M of N$_2$O), e_{aq}^- are converted into HO• radical via Reaction 2 (k_2 = 9.1 × 10^9 M^{-1} s^{-1}), with G(HO•) = 0.55 μmol J^{-1}, i.e., HO• radicals and H• atoms account for 90 and 10%, respectively, of the reactive species [19,20]. The rate constants for the reactions of HO• radicals and H• atoms with DNA (Reactions 3 and 4) have been reported to be ca. 2.5 × 10^8 M^{-1} s^{-1} and 6 × 10^7 M^{-1} s^{-1}, respectively [19,20]:

$$H_2O + \gamma\text{-}irr/e\text{-}beam \rightarrow e_{aq}^-(0.27), HO^\bullet(0.28), H^\bullet(0.062), \qquad (1)$$

$$e_{aq}^- + N_2O + H_2O \rightarrow HO^\bullet + N_2 + HO^-, \qquad (2)$$

$$HO^\bullet + DNA \text{ (or ODN)} \rightarrow \text{radical product}, \qquad (3)$$

$$H^\bullet + DNA \text{ (or ODN)} \rightarrow \text{radical product}. \qquad (4)$$

2.2. Pulse Radiolysis in Aqueous Solutions

Pulse radiolysis is a time-resolved technique that gives an opportunity to look into very short time domains. Therefore, it allows detection and spectral/kinetic characterization of very short-lived transients like radicals, radical-ions, and excited states. In a typical experiment, the UV-Vis spectral changes obtained from the pulse irradiation of N$_2$O-saturated solution containing ca. 1 mM of nucleoside are monitored. The possibility to record the transient spectra using the ICCD camera has a great advantage over PMT since it allows to work with extremely valuable micro-volume liquid samples.

2.2.1. Comparison of PMT and ICCD Detection Methods Using Nucleosides

In order to check first reliability of the ICCD camera, the transient spectra resulted from the reactions of HO• radicals with single nucleosides (dC, dG, T, and dA) were recorded by the PMT and ICCD camera, and then the two spectra obtained were compared.

The spectral changes obtained by the two detection systems after pulse irradiation of a N$_2$O-saturated sodium phosphate 50 mM, pH 7, solution of 1 mM 2′-deoxycytidine (representing pyrimidine derivative) superimpose more than satisfactorily and are shown in Figure 2 (left panel). The optical absorption spectra taken 2 μs after the pulse are characterized by two distinctive absorption bands with λ_{max} ~350 and 440 nm. The present results are in agreement with those from the literature [21,22]. This spectrum is mainly due to 5-OH-6-yl radical **1** formed by addition at C5 with ~87% yield, 6-OH-5-yl radicals **2** formed by addition at C6 (Figure 3) and with some minor contribution of H•-adduct radicals. The H-atom abstraction from the sugar moiety can also be considered; however, the resulting radicals do not absorb significantly in the wavelength region of interest.

Figure 2. Transient absorption spectra recorded using PMT (○) and ICCD (−); (Left) 2 µs after electron pulse in N$_2$O-saturated phosphate buffered (50 mM) aqueous solution containing 1 mM 2′-deoxycytidine (dC); (Right) 1 µs after electron pulse in N$_2$O-saturated phosphate buffered (50 mM) aqueous solution containing 1 mM 2′-deoxyguanosine (dG), at pH 7.

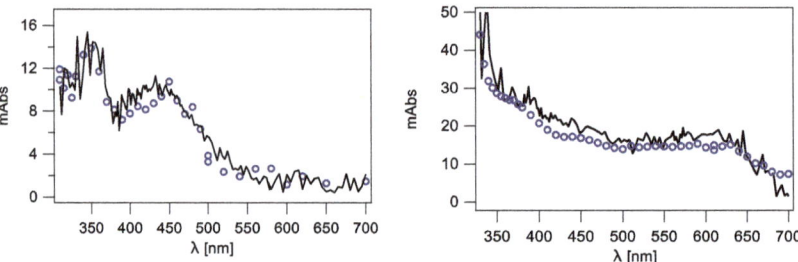

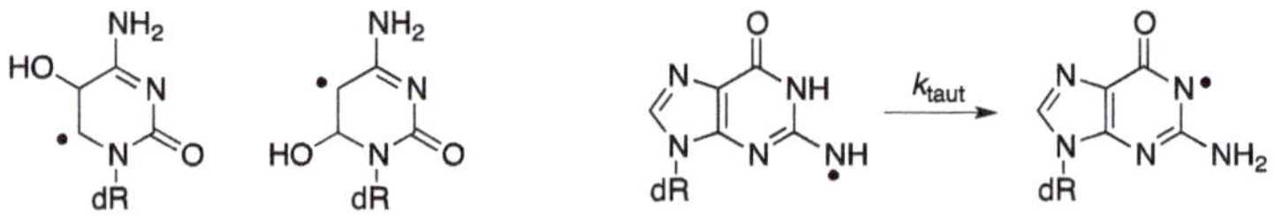

Figure 3. The main species generated by the reaction of HO• radical with dC and dG, respectively.

Similarly, the spectral changes obtained by the two detection systems for 2′-deoxyguanosine (purine representative) superimpose satisfactorily and are shown in Figure 2 (right panel). The optical absorption spectra taken 1 µs after the pulse are characterized by a broad absorption band with a weakly marked maximum at λ_{max} ~610 nm. This absorption band was earlier assigned to a guanyl radical **3** (Figure 3) formed by hydrogen abstraction from the exocyclic NH$_2$ with ~65% yields, which undergoes further a water-assisted tautomerization to the most stable isomer **4** with a $k_{taut} = 2.3 \times 10^4$ s^{-1} [23–25]. Moreover, the absorption spectra obtained are in agreement with those reported in the literature. The spectrum in the range 400–600 nm is flat without a clearly pronounced maximum. Contribution of 8-hydroxyl radical adduct to the absorption spectrum in the short wavelength range (<400 nm) also has to be taken into account [23,25].

2.2.2. Mixture of dC and dG Nucleosides

With this information in hand, the transient absorption spectra resulted from the reaction of HO• radicals with the mixture of dG and dC present in an aqueous solution in a concentration ratio of 1:1 were recorded only by an ICCD camera on the time domain between 1 µs to 1 ms (Figure 4). The optical absorption spectrum recorded 1 µs after the pulse is characterized by a broad absorption band in the region of 600–650 nm, a distinctive shoulder in the region of 350–400 nm and a sharp absorption band with λ_{max} ~310 nm. These spectral characteristics are consistent with the presence of transients derived from dG and dC. Moreover, absorption intensity of the 610 nm band is nearly half of the absorption intensity measured in the solution containing only 2′-deoxyguanosine (Figure 2, right panel). This observation is not surprising taking into account equal concentrations of dG and dC nucleosides and their respective rate constants with HO• radicals which are very similar, and equal to $(5.7 \pm 0.4) \times 10^9$ M^{-1}s^{-1} [24], and $(6.0 \pm 1.5) \times 10^9$ M^{-1}s^{-1} [21], respectively.

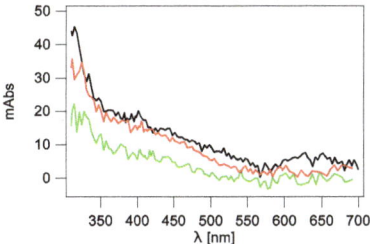

Figure 4. Transient absorption spectra recorded using ICCD (−) 1 μs, (−) 50 μs, and (−) 1 ms after electron pulse in N$_2$O-saturated phosphate buffered (50 mM) aqueous solutions containing mixture of 2′-deoxyguanosine (dG, 0.5 mM) and 2′-deoxycytidine (dC, 0.5 mM) with concentration ratio 1:1 at pH 7.

2.2.3. Mixture of dC, dG, T and dA Nucleosides vs. Calf-Thymus DNA (ct-DNA)

The subsequent chemical system subjected to irradiation was a phosphate buffered (50 mM) aqueous solution containing a mixture of four nucleosides: dC, dG, T (thymidine) and dA (2′-deoxyadenosine) in the following concentration ratios. The first pair of nucleosides (dC and dG) and the second pair of nucleosides (T and dA) were present in a concentration ratio 1:1. In turn, the two respective pairs of nucleosides were present in a concentration ratio 2:3, which mimics the ratio of these nucleosides present in ct-DNA.

The optical absorption spectrum recorded 1 μs after the pulse in an aqueous buffered solution containing a mixture of four nucleosides (dC, dG, T and dA) (Figure 5, left panel) is similar to that observed in solution with a mixture of two nucleosides (dC and dG, see Figure 4) except two features: the absorption band in the region >600 nm is absent while two distinctive shoulders in the region 460–500 nm and 400–420 nm appear. The first feature can be rationalized by taking into account the concentration ratio of nucleosides present in the solution as 2:2:3:3 and their respective rate constants with HO• radicals which are nearly equal [2]. Taking a simple competition kinetics of these four nucleosides for HO• radicals, one can easily calculate that at most 20% of all available HO• radicals can react with dG and give rise to the guanyl-type radicals. In turn, the second feature can be explained by the spectral characteristics of radicals derived from dA [26], formed in the reaction of 30% of HO• radicals. On the other hand, the optical spectrum recorded 1 μs after the pulse in an aqueous solution containing ct-DNA (Figure 5, right panel) is not very different from that recorded in a solution containing a mixture of four nucleosides, except for the fact that the absorption intensity is two-fold weaker. Interestingly, the time evolution of the radicals formed in both systems is different (clearly seen by comparison of the absorption spectra in the time domain between 50 μs and 1 ms) showing higher stability of radicals in ct-DNA.

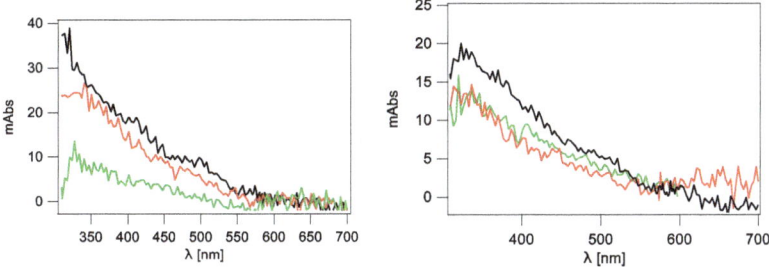

Figure 5. Transient absorption spectra recorded using ICCD in N$_2$O-saturated phosphate buffered (50 mM) aqueous solutions containing (left panel) a mixture of four nucleosides (see text) and (right panel) ct-DNA, at natural pH: (−) 1 μs, (−) 50 μs, and (−) 1 ms after electron pulse.

2.2.4. Single Stranded 12-Mer Oligodeoxynucleotides

The subsequent chemical systems subjected to irradiation were the aqueous solutions containing one of four single stranded 12-mer oligonucleotides (cf. Table 1): ODN1, ODN2, ODN3, or ODN4.

Figure 6 shows the optical absorption spectra recorded 1 μs after the pulse for ODN1 and ODN3, which are very similar to the absorption spectrum recorded 1 μs after the pulse in aqueous solutions containing ct-DNA (Figure 5, right panel). Similar spectra are obtained for ODN2 and ODN4 (see Figure S1 in Supporting Information). Three of them (except ODN1) are characterized by the absorption band which can be assigned to the guanyl-type radical **3** (cf. Figure 3). Surprisingly, the spectrum recorded in aqueous solution of ODN1 (oligonucleotide containing four dG nucleotides) does not show optical features which can be assigned to this radical (Figure 6, left panel). It seems that the number of dG present in these single stranded 12-mer oligonucleotides is not the only factor determining the efficiency of the intermediate **3** formation. Perhaps, the peculiar conformation arrangement of ODN1 excludes the access of HO• radical to the NH$_2$ moiety for H-atom abstraction.

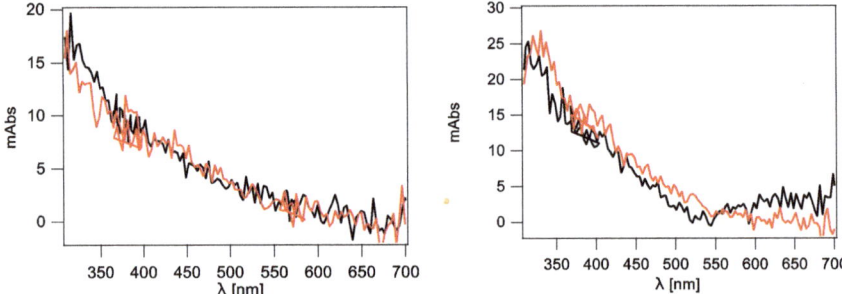

Figure 6. Transient absorption spectra recorded using ICCD in N$_2$O-saturated phosphate buffered (50 mM) aqueous solutions containing 5′-CGT ATG GTA TCG-3′ (ODN1) (left panel) and 5′-CGA TGG GGT ACG-3′ (ODN3) (right panel) at natural pH: (−) 1 μs and (−) 50 μs after electron pulse.

It is worth mentioning that the formation of the radical intermediate **3** and its tautomerization process (**3**→**4**) taking place on the ms scale is an important process, shown to occur in G-quadruplex through oxidation followed by deprotonation step [27–29].

2.2.5. Double-Stranded 12-Mer Oligodeoxynucleotides

The last two chemical systems subjected to irradiation were the aqueous solutions containing double-stranded (ds) 12-mer oligonucleotides: ODN1/ODN2 or ODN3/ODN4. In our previous studies, we used the same ds-oligonucleotide sequences for investigating the oxidation potential upon increasing the number of consecutive Gs [30]. The optical absorption spectra recorded 1 μs after the pulse in aqueous solutions containing one of two ds-oligonucleotides are very similar to each other (Figure 7) and resemble the spectrum recorded 1 μs after the pulse in aqueous solutions containing ct-DNA (Figure 5, right panel). Moreover, the time evolution of spectra recorded at 1 μs and 50 μs shows clearly that the formed radicals are stable within this time domain. The lack of the absorption band >600 nm indicates absence of guanyl-type radicals which might result from the specific structure of the double-stranded DNA (Figure 7). It is reasonable to assume that the HO• radical is not able to reach the NH$_2$ moiety for H-atom abstraction due to the steric encumbrance in ds-ODNs or the tautomerization process (**3**→**4**) is very fast with respect to our time-scale experiments.

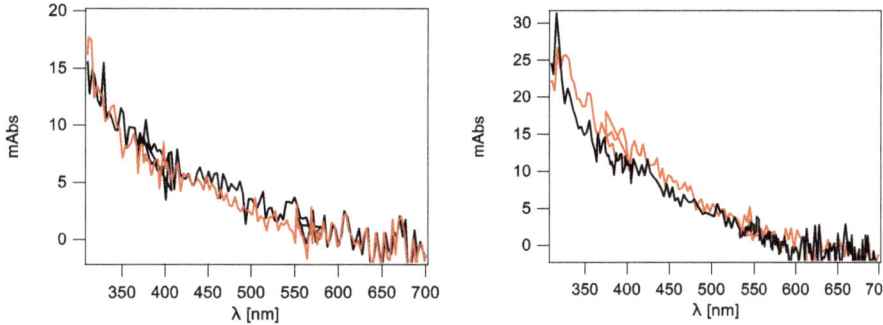

Figure 7. Transient absorption spectra recorded using ICCD in N$_2$O-saturated phosphate buffered (50 mM) aqueous solutions containing (left panel) ds-(ODN1/ODN2) and (right panel) ds-(ODN3/ODN4) at natural pH: (−) 1 µs and (−) 50 µs after electron pulse.

2.3. γ-Radiolysis

2.3.1. Hydroxyl Radical–Induced Formation of Purine Lesions in ct-DNA

The reactions of HO$^\bullet$ radicals with DNA were carried out using ct-DNA. Two preparation procedures of ct-DNA solutions for irradiation were used:

(i) the commercial ct-DNA solution containing 1 mM Tris-HCl, pH 7.5, with 1 mM NaCl and 1 mM EDTA was firstly lyophilized and then 200 µL of a N$_2$O saturated ct-DNA aqueous solutions (0.5 mg/mL) at natural pH were prepared;

(ii) the commercial ct-DNA solution was desalted by ethanol precipitation (removal of the additives Tris-HCl, NaCl and EDTA) and then 200 µL of a N$_2$O saturated ct-DNA (0.5 mg/mL) were prepared in 50 mM phosphate buffer, pH 7.2

All samples were irradiated (in triplicate) under steady-state conditions at different doses (0, 10, 20, 35, and 50 Gy). The quantification of the lesions was executed in two independent steps. Firstly, the sample was analyzed via an HPLC-UV system coupled with a sample collector. According to this first clean-up step, the quantification of the unmodified nucleosides took place, based on their absorbance at 260 nm, whereas, at their time-windows when the lesions are eluted, fractions were collected and pooled. The concentrated samples containing the modified DNA bases were injected subsequently to LC-MS/MS to be analyzed and quantified independently [10,16,31,32]. In the absence of the unmodified nucleosides, which cause solubility problems to arise, the sample can be concentrated prior to LC-MS/MS analysis increasing the overall sensitivity of the quantification method. The use of isotopically labeled lesions (Figure S2) maintains the reproducibility and the recovery of the quantification protocol within the levels that are generally accepted for reliability. The calibration curves for the quantification of the lesions and the list of MRM transitions employed for the quantification are reported in Figure S3 and Table S1, respectively.

The radiation induced formation of 8-oxo-dG, 5′S-cdG, 5′R-cdG, 8-oxo-dA, 5′S-cdA and 5′R-cdA in ct-DNA applying Procedure (i) is shown in Figure 8 (for data, see Table S2). As expected, the number of the lesions studied increases with the increment of the dose. The reaction product profile obtained employing Procedure (ii) (see Figure S4 and Table S3), where ct-DNA is treated for removing additives such as Tris-HCl, NaCl and EDTA, is comparable with the results gathered with Procedure (i); this indicates that the additives originally present in the commercial ct-DNA sample have no significant interference in the reaction outcome.

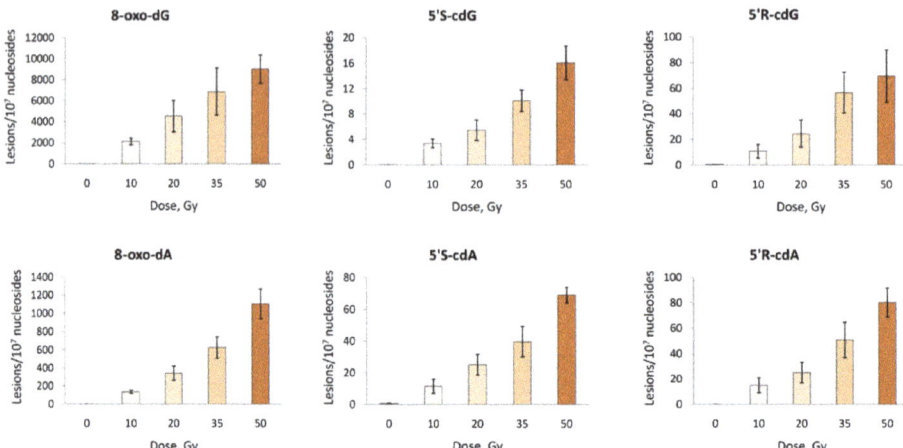

Figure 8. Radiation induced formation of 8-oxo-dG, 5′S-cdG, 5′R-cdG, 8-oxo-dA, 5′S-cdA and 5′R-cdA in ct-DNA Procedure (i). Each sample was exposed to 0, 10, 20, 35, and 50 Gy dose. The values represent the mean ± SD of n = 3 independent experiments.

From Figure 8, it becomes evident that the main reaction products detected are 8-oxo-dG and 8-oxo-dA, which are also formed in a ca. 10:1 ratio. The intramolecular cyclization products 5′S-cdG, 5′R-cdG, 5′S-cdA and 5′R-cdA are also formed, albeit in lower yields. We have further analyzed the data obtained employing Procedures (i) and (ii) (Tables S2 and S3) by plotting the number of each lesion formed vs. the radiation dose; the slope of the lines obtained (Figure 9 for Procedure (i) and Figure S5 for Procedure (ii)) represents the number of lesions formed per Gy which are reported in Table S3.

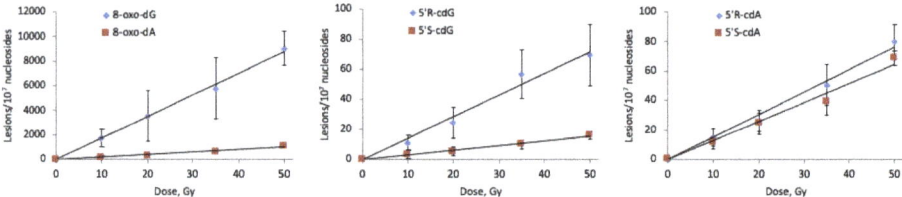

Figure 9. Radiation induced formation of 8-oxo-dG, 8-oxo-dA, 5′R-cdG, 5′S-cdG, 5′R-cdA, and 5′S-cdA in ct-DNA Procedure (i). Each sample was exposed to 0, 10, 20, 35, and 50 Gy dose in N_2O-saturated aqueous solutions; the values represent the mean ± SD of n = 3 independent experiments.

After having verified that Procedure (i) is indeed equivalent to (ii) (Table S4), the data obtained using both procedures at each dose were gathered and treated as a unique experiment, affording the data reported in column 2 of Table 2. The 5′R diastereomer is formed predominantly leading to 5′R/5′S ratio of 4.5 for cdG and 1.2 for cdA (column 3). It is very gratifying to observe how our data perfectly match previous results obtained and published by our group (columns 4 and 5) [33], considering different batches of ct-DNA, the labor-intensive enzymatic digestion/prepurification/enrichment of cPu lesions protocol, and the use of different analytical instrumentation. The level of total 8-oxo-Pu was found to be ~40-fold excess of total cPu lesions and the 8-oxo-dG/8-oxo-dA ratio of 7.7. The yields of four cPu lesions were found to be similar. It is worth recalling that 5′R/5′S ratios of 8.3 for cdG and 6 for cdA were obtained in water upon irradiation of free nucleosides [34,35], indicating that the diastereomer ratio is dependent on the molecular complexity. It is also worth mentioning that, in earlier work [36] on similar experiments, the level of lesions was reported to be much higher than in the present work for cdG and for cdA, (these data are reported in the last two columns of Table 2) and later the quantification protocol used was strongly criticized, being inappropriate for these measurements [16].

Table 2. The levels (lesions/10^7 nu/Gy) of 8-oxo-dG, 8-oxo-dA, 5'R-cdG, 5'S-cdG, 5'R-cdA and 5'S-cdA from the irradiation of N_2O saturated ct-DNA (0.5 mg/mL) aqueous solutions.

Lesion	Lesions/10^7 nu/Gy [a]	5'R/5'S [a]	Lesions/10^7 nu/Gy [b]	5'R/5'S [b]	Lesions/10^7 nu/Gy [c]	5'R/5'S [c]
8-oxo-dG	171.8 ± 13.0		200.1 ± 3.03		780	
8-oxo-dA	22.22 ± 1.25		28.04 ± 0.46		72	
5'R-cdG	1.40 ± 0.12	4.5	2.98 ± 0.10	4.7	151 [d]	~3
5'S-cdG	0.31 ± 0.02		0.64 ± 0.06		50 [d]	
5'R-cdA	1.55 ± 0.08	1.2	1.47 ± 0.14	1.5	114 [d]	~4
5'S-cdA	1.30 ± 0.06		0.95 ± 0.07		28 [d]	

[a] This work; Procedure (i)+(ii); [b] From Ref. [33]; The original data are plotted in Supporting information (Figure S6). In the Ref. [33] only cdG and cdA together with 5'R/5'S ratio were reported and the diastereoisomeric ratio of cdG was erroneously reported to be 7 instead of 4.7; [c] From Ref. [36], where only cdG and cdA together with 5'R/5'S ratios were given; [d] The values of each diastereoisomer were calculated from the cdG (201 lesions/10^7 nu/Gy) or cdA (142 lesions/10^7 nu/Gy) taking into consideration the 5'R/5'S ratio [36].

2.3.2. Hydroxyl Radical–Induced Formation of Purine Lesions in Double Stranded 21-Mer Oligonucleotides

The reaction of HO• radicals with double-stranded 21-mer oligonucleotide ODN5/ODN6 (see Table 1 for the ODNs sequences and Table S5, Figures S7–S9 for characterization) was studied under standard radiolytic conditions. For this purpose, 200 µL of N_2O-saturated aqueous solutions containing ds-(ODN5/ODN6) (0.5 mg/mL) at natural pH were irradiated under steady-state conditions with a dose rate of 2.5 Gy min^{-1} at room temperature, followed by our optimized routine enzymatic oligonucleoside digestion and LC-MS/MS analysis. As expected, both 5'R and 5'S diastereomers of cdA and cdG as well as 8-oxo-dA and 8-oxo-dG were generated, and the number of lesions studied increased proportionally with the increment of the dose (Figure 10 and Table S6).

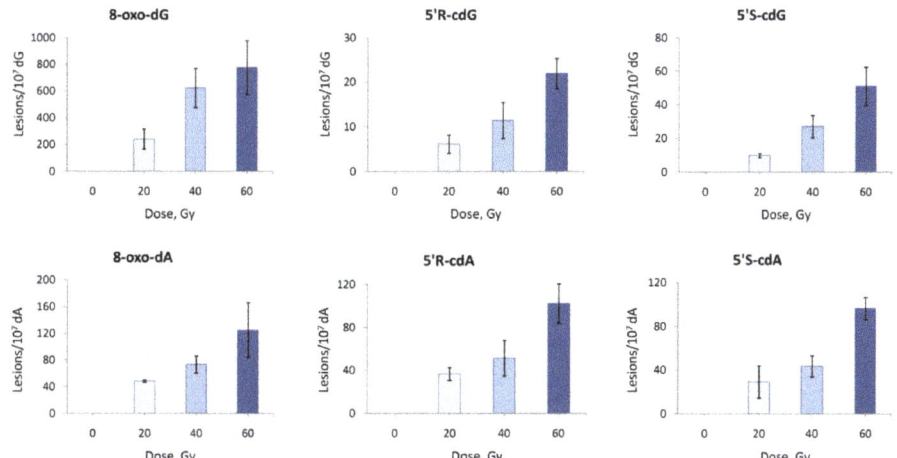

Figure 10. Radiation induced formation of 8-oxo-dG, 5'S-cdG, 5'R-cdG, 8-oxo-dA, 5'S-cdA and 5'R-cdA in double-stranded 21-mer oligonucleotides; Each sample was exposed to 0, 20, 40 and 60 Gy dose in N_2O-saturated aqueous solutions. The values represent the mean ± SD of $n = 3$ independent experiments.

From the analysis of the data reported in Figure 10, it turns out that the main reaction products detected are 8-oxo-dG and 8-oxo-dA, formed in an approximately 6.5:1 ratio. Intramolecular cyclization products 5′S-cdG, 5′R-cdG, 5′S-cdA and 5′R-cdA are also formed in the same fashion but in lower yields (Figure 10). Further analysis of the data reported in Figure 10 by plotting the number of each lesion detected vs. the radiation dose shows a linear correlation. The slope of the lines obtained (Figure 11) represents the number of lesions formed per Gy and these data are reported in Table 3. Analysis of the data shown in Table 3 proves that, in our experiments, the formation 8-oxo-dG is 6.4 times greater than 8-oxo-dA; regarding the cPu lesions detected, cdA lesions are higher than cdG and in 5′R/5′S ratios of ca 1.16 and 0.48, respectively. Although the diastereoisomeric ratio in cdA is similar to that observed in ct-DNA, in cdG, it is 10-fold smaller (0.48 vs. 4.5). It is also worth mentioning our previous work [37], where single-stranded and G-quadruplex of Tel22 d[AGGG(TTAGGG)$_3$] and mutated Tel24 d[TTGGG(TTAGGG)$_3$A] were exposed to HO$^\bullet$ radicals in similarity with ODN5 of ds-(ODN5/ODN6) in the present study. Indeed, for Tel22, it was reported 0.4 and 0.9 lesion/10^7 dG/Gy for 5′R-cdG and 5′S-cdG, respectively, with a 5′R/5′S ratios of 0.44. All these data confirm that the diastereomer ratio is dependent on the molecular complexity and detailed theoretical calculations on the transition states are needed for a better understanding of C5′ radical cyclization in ds-ODNs.

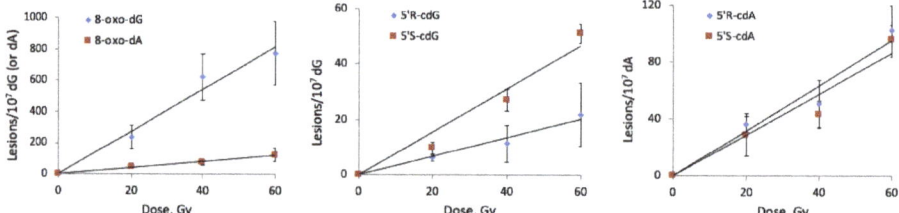

Figure 11. Radiation induced formation of 8-oxo-dG, 8-oxo-dA, 5′R-cdG, 5′S-cdG, 5′R-cdA, and 5′S-cdA in double-stranded 21-mer oligonucleotides; Each sample was exposed to 0, 20, 40 and 60 Gy dose in N$_2$O-saturated aqueous solutions. The values represent the mean ± SD of n = 3 independent experiments.

Table 3. The levels (lesions/10^7 nu/Gy) of 8-oxo-dG, 8-oxo-dA, 5′R-cdA, 5′S-cdA, 5′R-cdG and 5′S-cdG in irradiated ds-ODNs.

	ds-(ODN5/ODN6)	
Lesion	Lesions/10^7 dG/Gy	Lesions/10^7 dA/Gy
8-oxo-dG	13.49 ± 1.87	
8-oxo-dA		2.11 ± 0.30
5′R-cdG	0.32 ± 0.04	
5′S-cdG	0.67 ± 0.18	
5′R-cdA		1.60 ± 0.29
5′S-cdA		1.38 ± 0.27

2.4. Hydroxyl Radical Generated by Fenton Reactions and Formation of Purine Lesions in Double Stranded 21-Mer Oligonucleotides

Hydrogen peroxide (H$_2$O$_2$) reacts with the reduced-state transition metal ions, like Fe^{2+} or Cu^{1+}, to give HO$^\bullet$ radicals (reaction 5) [38,39]:

$$H_2O_2 + Fe^{2+} (Cu^{1+}) \rightarrow HO^- + HO^\bullet + Fe^{3+} (Cu^{2+}). \tag{5}$$

One-electron reduction of hydrogen peroxide occurs with a reduction potential of +0.38 V (H$_2$O$_2$,H$^+$/H$_2$O,HO$^\bullet$, pH 7, vs. NHE). Thus, the relatively long-lived oxidant H$_2$O$_2$ upon reduction generates a potent and indiscriminant oxidant, like HO$^\bullet$ radical.

This reaction, referred to as the Fenton reaction, has been reported to be responsible for some of the toxicity associated with H_2O_2 in vivo. The reduction of H_2O_2 in biological systems can occur via reaction with the reduced forms of several redox active metals such as the ferrous ion (Fe^{2+}) or cuprous ion (Cu^{1+}). H_2O_2 toxicity is highly dependent on the presence/location of reactive forms of Fe^{2+} or Cu^{1+} ions [40].

In this section, the role of HO• radicals generated by Fenton reaction (5) with ds-oligodeoxynucleotides ds-(ODN5/ODN6) was investigated in some details. The measurements of cPu lesions in ct-DNA by Fenton-type reagents was reported by Wang and coworkers [41]. They used Cu(II) or Fe(II) 12.5 µM, H_2O_2 100 µM and ascorbate 1 mM with proportional increments up to 1600 µM of H_2O_2 in 250 µL solution containing 75 µg of ct-DNA. For our studies, we used the same concentration ratio of Fenton-type reagents and followed an analogous approach.

The reaction was explored employing three different systems that consisted of 50 µg of ds-(ODN5/ODN6) in 200 µL solution and proportional increment of $CuCl_2$, H_2O_2 and ascorbate: 5 µM/40 µM/0.4 mM, 10 µM/80 µM/0.8 mM, and 15 µM/120 µM/1.2 mM, respectively (Table S7). Observing Figure 12 and Table S7, it becomes clear that, in all experiments, 5'R-cdG, 5'S-cdG, 5'R-cdA, 5'S-cdA, 8-oxo-dG, and 8-oxo-dA lesions are formed. It is also gratifying to observe an increment in the lesions number with the increased concentration of the reagents employed in the Fenton system (Figure 12 and Table S8). The main lesion detected in the three systems studied is 8-oxo-dG followed by 8-oxo-dA. It is interesting to point out that the ratio between 8-oxo-dG and 8-oxo-dA increases as the concentration of the reagents employed increases; that is, 8-oxo-dG/8-oxo-dA ratio 3.4:1, 5.6:1, 6.4:1 for $CuCl_2$/H_2O_2/ascorbate: 5 µM/40 µM/0.4 mM, 10 µM/80 µM/0.8 mM, and 15 µM/120 µM/1.2 mM, respectively (Figure 12B). This result is in agreement with the fact that the oxidation of dG proceeds faster than the oxidation of dA [42]. Regarding the formation of the 5',8-cyclopurines, the four assessed lesions (5'R-cdG, 5'S-cdG, 5'R-CdA and 5'S-cdA) increase with the increment of the reagents concentration employed in the Fenton reaction (Figure 12A); it should be mentioned that the 5'R/5'S ratio for both the cdG and the cdA lesions remains almost constant in the three reaction conditions studied being ~0.76 for cdG and ~1.56 for cdA (Table S9).

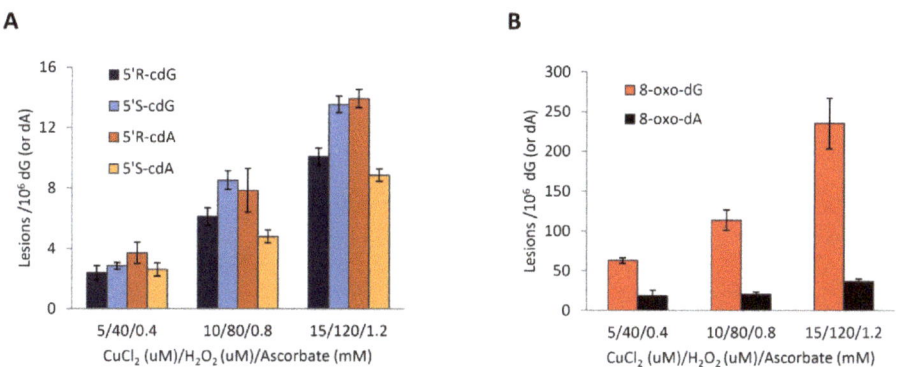

Figure 12. $CuCl_2$/H_2O_2/Ascorbate-induced formation of (**A**) 5'R-cdG, 5'S-cdG, 5'R-cdA and 5'S-cdA and (**B**) 8-oxo-dG and 8-oxo-dA in ds-ODNs. The numbers represent the mean value (±standard deviation) of $n = 3$ independent experiments.

After analyzing the reaction outcome at various proportional increments of $CuCl_2$, H_2O_2, and ascorbate, it was deemed proper to study the reaction evolution with time. For doing so, the most reactive reaction conditions previously studied were employed, that is, $CuCl_2$, H_2O_2 and ascorbate 15 µM/120 µM/1.2 mM, respectively. As expected, in all the experiments, an increment of the reaction products was observed with the increase of the reaction time (Tables S10 and S11). Figure 13 (A and B) shows the plots of the mean value of each purine lesion studied at different reaction times; analogously,

Figure 14 (left side) shows the plots for 8-oxo-dA and 8-oxo-dG. The linear ds-ODNs lesion rates could just reflect that the reaction half-life is much longer than the experimental time-frame, $t_{1/2} \gg 2$ h. The slope of the lines can be interpreted as the number of lesions produced in 1 min of reaction (Table 4). The data reported in Table 4 (column 2) show that the main lesions detected are 8-oxo-dG (3.88 lesions per minute) and 8-oxo-dA (0.60 lesions per minute) in a 6.5:1 ratio, respectively. The 5′R-cdG, 5′S-cdG, 5′R-cdA and 5′S-cdA lesions are also formed, albeit in lower yields ranging between 0.14 and 0.20 lesions per minute.

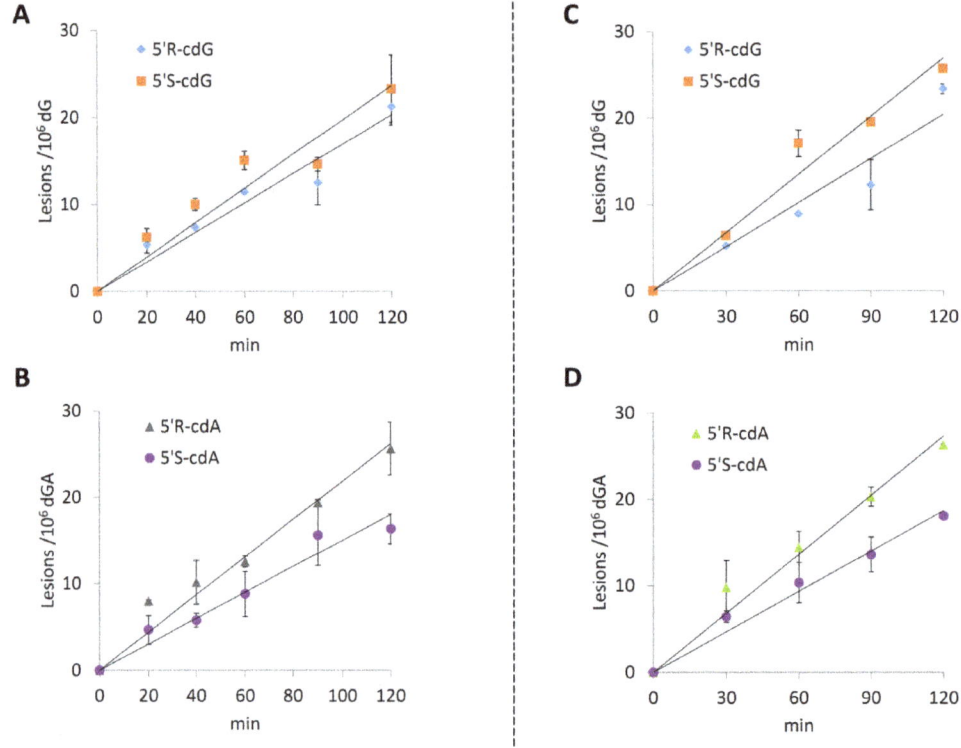

Figure 13. Kinetic study by Fenton reaction. (A and B): $CuCl_2/H_2O_2$/Ascorbate-induced formation of 5′R-cdG, 5′S-cdG, 5′R-cdA and 5′S-cdA at 0, 20, 40, 60, 90 and 120 min. ds-(ODN5/ODN6) treated with $CuCl_2$ (15 µM), H_2O_2 (120 µM), Ascorbate (1.2 mM); (C and D): Fe^{2+}/H_2O_2/Ascorbate-induced formation of 5′R-cdG, 5′S-cdG, 5′R-cdA and 5′S-cdA at 0, 30, 60, 90 and 120 min. ds-(ODN5/ODN6) treated with Fe^{2+} (15 µM), H_2O_2 (120 µM), Ascorbate (1.2 mM). All points are the mean value (± standard deviation) of $n = 2$ independent experiments.

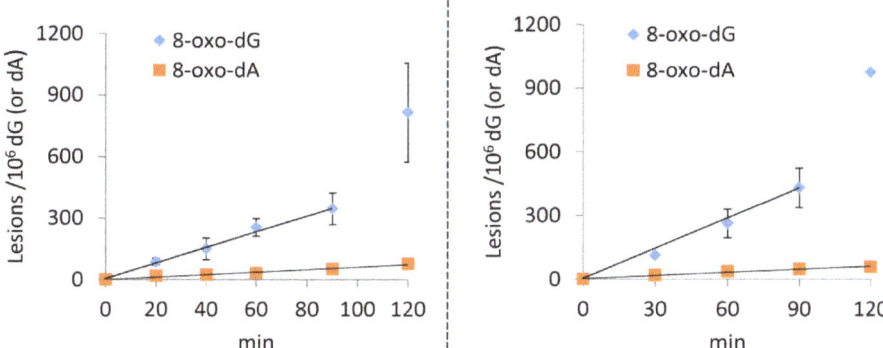

Figure 14. Kinetic study by Fenton reaction. (Left side): CuCl2/H2O2/Ascorbate-induced formation of 8-oxo-dG and 8-oxo-dA at 0, 20, 40, 60, 90 and 120 min. ds-(ODN5/ODN6) treated with $CuCl_2$ (15 µM), H_2O_2 (120 µM); (Right side): Fe^{2+}/H_2O_2/Ascorbate-induced formation of 8-oxo-dG and 8-oxo-dA at 0, 30, 60, 90 and 120 min. ds-(ODN5/ODN6) treated with Fe^{2+} (15 µM), H_2O_2 (120 µM), Ascorbate (1.2 mM). All points are the mean value (± standard deviation) of $n = 2$ independent experiments.

Table 4. The line slope for each lesion obtained from the plot of the mean value of each purine lesion studied at different reaction times as reported in Figure 13 for cPu and Figure 14 for 8-oxo-Pu.

Lesions	From Cu^{1+}	From Fe^{2+}
5′R-cdG	0.16	0.18
5′S-cdG	0.17	0.21
5′R-cdA	0.20	0.21
5′S-cdA	0.14	0.14
8-oxo-dG	3.88	4.81
8-oxo-dA	0.60	0.49

We continued our investigations on ds-(ODN5/ODN6) lesion formation by replacing Cu^+ with Fe^{2+} in the Fenton system, using a preestablished concentration of Fe(II)/H_2O_2/ascorbate: 15 µM/120 µM/1.2, which corresponds to the most efficient ratio of reagents towards ds-ODNs lesion formation when using Cu^+ as the metal source. As expected, the Fenton system employing iron also induced a time-dependent formation of the assessed purine lesions (Tables S13 and S14). Figure 13 (C and D) shows the plots of the mean value of each purine lesion studied at different reaction times; analogously, Figure 14 (left side) shows the plots for 8-oxo-dA and 8-oxo-dG. The data reported in Table 4 (column 3) show the number of lesions produced in 1 min, showing that by replacing Cu^+ with Fe^{2+} in these Fenton-type of reactions the outcome is similar. It is also gratifying to see that the 5′R/5′S ratio in the two Fenton systems are similar, i.e., after 120 min for the cuprous ion the ratios are 0.91 for cdG and 1.57 for cdA, whereas, for the ferrous ion, the ratios are 0.91 for cdG and 1.45 for cdA (cf. Table S12 with Table S15).

Figure 15 summarizes the reaction mechanism we conceived for the above described Fenton-type reactions. The Fenton reaction produces HO• radical and oxidation of copper and iron (Cu^{2+} and Fe^{3+}). The role of ascorbate is to maintain copper and iron in the reduced state (Cu^+ and Fe^{2+}) with the formation of ascorbyl radical anion (Asc•−). There is a plethora of paths for the reaction of HO• radical with DNA or ds-ODNs [1,2]. It can be estimated that 2–3% only occur by hydrogen abstraction at C5′ position in the purine nucleotide moieties leading to C5′ radical. After the intramolecular addition to C8–N7 double bond, the formation of an heteroaromatic aminyl radical results. This radical can be easily oxidized by Cu^{2+} or Fe^{3+} to afford the final lesion after deprotonation. At the nucleoside level, we reported a rate constant of 8.3×10^8 $M^{-1}s^{-1}$ for the aminyl radical with $Fe(CN)_6^{3-}$ [43,44].

Alternatively, if the [Asc•−] is built-up in a fairly high steady-state concentration, there will be the possibility of radical disproportionation with the regeneration of ascorbate and the cPu lesion formation within the biomacromolecule. It is worth mentioning that, after 60 min, the reactions were quenched by adding 0.4 mg L-methionine as described in the protocol of Wang [41], more likely in order to eliminate remaining oxidizing species. However, control experiments without L-methionine gave similar results (results not shown).

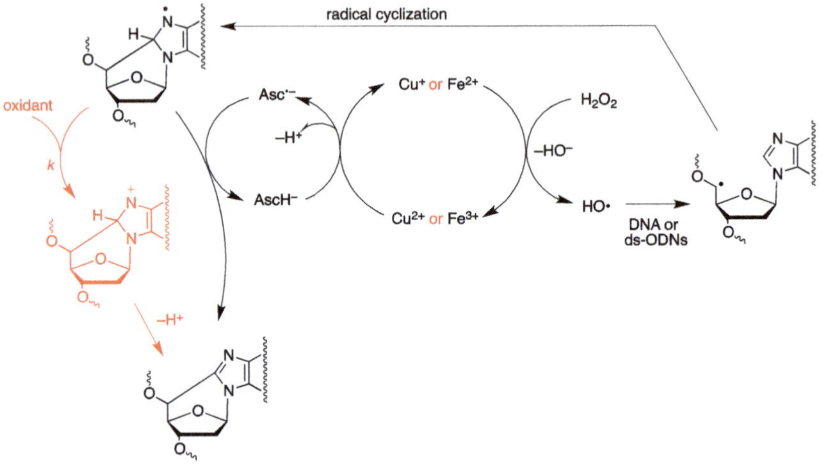

Figure 15. The conceived reaction mechanism for the formation of cPu lesion in ds-ODNs by Fenton-type reactions.

3. Materials and Methods

3.1. Chemicals, Reagents and Enzymes

Nuclease P1 from Penicillium citrinum, phosphodiesterase I and II, alkaline phosphatase from bovine intestinal mucosa, DNase I and DNase II, benzonase 99%, BHT, deferoxamine mesylate and pentostatin were purchased from Sigma-Aldrich (Steinheim, Germany). RNase T1 was from Thermo Fisher Scientific (Waltham, MA, USA) and RNase A from Roche Diagnostic GmbH, (Mannheim, Germany). The 3 kDa cut-off filters were obtained from Millipore (Bedford, OH, USA). Chemicals for the synthesis of oligonucleotides were purchased from Sigma Aldrich, Fluka and Link Technologies. $CuCl_2$, L-methionine, L-ascorbic acid and alkaline phosphatase were purchased from Sigma-Aldrich. Hydrogen peroxide (30%) and solvents (HPLC-grade) were purchased from Fisher Scientific. 2′-deoxyadenosine monohydrate and 2′-deoxyguanosine were purchased from Berry & Associates Inc. (Dexter, USA). Isotopic labeled internal standards of 5′R-cdA, 5′S-cdA, 5′R-cdG, 5′S-cdG, 8-oxo-dG and 8-oxo-dA (see Supporting Information) were prepared according to the previously reported procedures [31]. Ultrapure water (18.3 MΩcm) distilled and deionized water (Milli-Q water) were purified by a Milli-Q system (Merck-Millipore, Bedford, OH, USA).

3.2. Oligodeoxynucleotides (ODNs) Synthesis and Purification

ODNs were prepared by automated synthesis using the DMT- and β-(cyanoethyl)-phosphoramidite method, on CPG supports (500 Å), with an Expedite 8900 DNA synthesizer (Applied Biosystems, Foster City, CA, USA) at 1 µmol scale. Following their synthesis, the DMTr-on ODNs were cleaved from the solid support and deprotected by the method of two syringes using an AMA reagent (NH_4OH (30%)/CH_3NH_2 (40%) 1:1) for 10 min at room temperature. The AMA solution containing the cleaved ODN was placed in a sealed vial and heated for 15 min at 55 °C. The solvent was then removed in a Speedvac. The crude 5'-DMT-on oligomers were purified and detritylated on-column by RP-HPLC (Grace Vydac C18 column, 5 μm, 50 × 22 mm). The ODNs were further purified by SAX HPLC (preparative DNA Pac PA-100 column, 5 μm, 22 × 250 mm). TRIS.HCl 25 mM, pH = 8 (buffer A) and TRIS. HCl 25 mM, NaClO$_4$ 0.5 M, pH 8.0 (buffer B) were used at a flow rate of 9 mL/min eluting with 2–30% B in 30 min, 30% B for 10 min, then 30–45% B in 5 min monitoring at 254 nm. The purified fractions were concentrated, desalted on Water SepPakTM-C18-cartridges (Milford, MA, USA) and lyophilized again. The final DNA yield was estimated by UV absorption in aqueous solution measured at 254 nm on a Cary 100 UV/Vis Spectrometer (Agilent, Cernusco sul Naviglio, Italy) following standard procedures. Electrospray Ionization (ESI) was used to characterize the purified ODNs. Maldi-TOF mass spectrometry (ODN1 to ODN4) [30] and Electrospray Ionization (ESI) (ODN5 and ODN6) were used to characterize the purified ODNs (see Table S5 and Figure S7).

3.3. Preparation of Double Stranded Oligonucleotide Substrates

The oligonucleotide strands were annealed to the complementary strands in equimolar concentrations in buffer solution containing 10 mM sodium phosphate, 100 mM NaCl, 0.1 mM EDTA, pH 7.2. The substrates were constructed by heating the two strands of the substrates at 90 °C for 10 min and subsequently allowing the temperature to slowly drop down to the room temperature (25 °C). Melting temperatures (Tm) of the substrates were measured with a Cary 100 UV/Vis spectrometer (Agilent, Cernusco sul Naviglio, Italy) using a 1 mL quartz cuvette with a 1 cm path length. This allowed monitoring of the absorbance of the solutions at 260 nm as a function of the temperature. The temperature cycles were recorded from 20 to 80 °C per strand with a temperature controller at a heating rate of 0.3 °C/min. UV melting curves of 21-mer duplexes ODN5/ODN6 (See Figure S8).

CD spectra were recorded on a Jasco *J-710* spectropolarimeter (Cremella, Italy) using a quartz cuvette (0.1 cm optical path length) at a scanning speed of 50 nm/min with 1 s response time. Measurements at the range of 200–360 nm were the average of four accumulations at 295 K and smoothed with Origin, Version 8.00 program (OriginLab Corporation, Northampton, MA, USA). The 21-mer duplexes ODN5/ODN6 contained an aqueous solution of 50 mM sodium phosphate, pH 7.2 and 50 μM double stranded oligonucleotide substrates (Figure S9. The reported spectrum was obtained by subtracting the spectrum of blank (aqueous solution of sodium phosphate buffer).

3.4. Pulse Radiolysis

Pulse radiolysis experiments with time-resolved UV-vis optical absorption detection were carried out at the Institute of Nuclear Chemistry and Technology in Warsaw, Poland. The linear electron accelerator (LAE 10) delivering 10 ns pulses with electron energy about 10 MeV was applied as a source of irradiation. The 150 W xenon arc lamp E7536 (Hamamatsu, Shizuoka, Japan) was used as a monitoring light source. The respective wavelengths were selected by MSH 301 (Lot Oriel Gruppe) motorized monochromator/spectrograph with two optical output ports. The time dependent intensity of the analyzing light was measured by means of photomultiplier (PMT) R955 (Hamamatsu, Shizuoka, Japan). A signal from detector was digitized using a WaveSurfer 104MXs-B (1 GHz, 10 GS/s, LeCroy) oscilloscope. Alternatively, iSTAR Intensified Charge-Coupled Device (ICCD) (A-DH720-18F-03) detector with W-type photocathode and 18 mm Multi-Channel Plate (MCP) image intensifier was used for transient spectra measurements. Minimum optical gate width of this detector was <5 ns with a spectral range of 180–850 nm. In order to avoid photodecomposition and/or photobleaching effects in the samples, the UV or VIS cut-off filters were used. However, no evidence of such effects was found within the time domains monitored. Water filter was used to eliminate near IR wavelengths. Optical path of microcells was 1 cm with a total volume of irradiated solution about 300 μl. All experiments were carried out at the ambient temperature ~22 °C. The spectral range which can be covered with the existing pulse radiolysis set-up is comprised between 300–700 nm.

The total dose per electron pulse was determined before each series of experiments by a thiocyanate dosimeter (N$_2$O-saturated aqueous solution containing 10 mM KSCN) using $G \times \varepsilon = 5.048 \times 10^{-3}$ mol J^{-1} M^{-1} cm^{-1} for the (SCN)$_2$$^{\bullet-}$ radical anion at λ = 472 nm.

3.5. γ-Radiolysis Experiments

Each sample of ds-DNA (50 µg) dissolved in 200 µL of phosphate buffer 50 mM was placed in a 2 mL glass vial. Irradiations were performed at room temperature (22 ± 2 °C) using a ^{60}Co-Gammacell at different doses (dose rates: 2.5 Gy/min). The exact absorbed radiation dose was determined with the Fricke chemical dosimeter, by taking G(Fe^{3+}) 1.61 µmol J^{-1}. In particular, the irradiation doses used were 20, 40, and 60 Gy, and the solutions were saturated by N$_2$O. All the irradiation experiments were performed in triplicates.

3.6. Fenton-Type Reagent Treatments of ds-ODNs

3.6.1. CuCl$_2$ with L-Methionine

Aliquots of ds-ODNs (50 µg) were incubated with CuCl$_2$ (5–15 µM), H$_2$O$_2$ (40–120 µM), Ascorbate (0.40–1.2 mM) in a 200 µL solution containing 25 mM NaCl and 50 mM phosphate (pH 7.2) at room temperature under aerobic conditions for 60 min. Ascorbate was added to maintain copper in the reduced state (Cu$^+$), so that it could participate in the Fenton reaction. Chemicals used in the Fenton-type reagent treatment of ds-ODNs were freshly prepared in doubly distilled water. After 60 min, the reactions were terminated by adding 0.4 mg L-methionine (control treatments were performed also without L-methionine), and the ODN samples were desalted by ethanol precipitation (10% volume of 3 M sodium acetate, pH 5.2 and 3 volumes of 100% Ethanol). The samples were mixed and frozen overnight at −20 °C. In the morning, the samples were centrifuged at 13,000 RPM at 4 degrees for 45 min. The supernatant was decanted. The pellet was washed and centrifuged again, for only 15 min, with 80% EtOH. The supernatant was decanted and the pellet air dried.

3.6.2. Kinetic Study by Cu^{2+}/H$_2$O$_2$ of ds-ODNs with L-Methionine

Aliquots of ds-ODNs (16 µg) were incubated with CuCl$_2$ (15 µM), H$_2$O$_2$ (120 µM), Ascorbate (1.2 mM) in a 64 µL solution containing 25 mM NaCl and 50 mM phosphate (pH 7.2) at room temperature under aerobic conditions. After 20–40–60–90–120 min, the reactions were terminated by adding 115 µg of L-methionine, and the DNA samples were desalted by ethanol precipitation (10% volume of 3M sodium acetate, pH 5.2 and 3 volumes of 100% Ethanol). The samples were mixed and frozen overnight at −20 °C. In the morning, the samples were centrifuged at 13,000 RPM at 4 degrees for 45 min. The supernatant was decanted. The pellet was washed and centrifuged again, for only 15 min, with 80% EtOH. The supernatant was decanted and the pellet air dried.

3.6.3. Kinetic Study by Fe^{2+}/H$_2$O$_2$) of ds-ODNs with L-Methionine

Aliquots of ds-ODNs (50 µg) were incubated with Fe(NH$_4$)$_2$(SO$_4$)$_2$·6H$_2$O (15 µM), H$_2$O$_2$ (120 µM), Ascorbate (1.2 mM) in a 200 µL solution containing 25 mM NaCl and 50 mM phosphate (pH 7.2) at room temperature under aerobic conditions. After 30–60–90–120 min, the reactions were terminated by adding 336 µg of L-methionine, and the DNA samples were desalted by ethanol precipitation (10% volume of 3 M sodium acetate, pH 5.2 and 3 volumes of 100% Ethanol). The samples were mixed and freeze overnight at −20 °C. In the morning, the samples were centrifuged at 13,000 RPM at 4 degrees for 45 min. The supernatant was decanted. The pellet was washed and centrifuged again, for only 15 min, with 80% EtOH. The supernatant was decanted and the pellet air dried. Fenton-type reagent treatment (Fe^{2+}/H$_2$O$_2$) of ds-ODNs was performed also without L-methionine to elucidate the role of L-methionine in the progress of the reaction.

3.7. Enzymatic Digestion of the ct-DNA and ds-ODNs

In addition, 50 µg of ct-DNA or ds-ODN were dissolved in 100 µL of Ar flushed 10 mM Tris-HCl (pH 7.9), containing 10 mM $MgCl_2$, 50 mM NaCl, 0.2 mM pentostatin, 5 µM BHT and 3 mM deferoxamine and the internal standards were added ([$^{15}N_5$]-5′S-cdA, [$^{15}N_5$]-5′R-cdA, [$^{15}N_5$]-5′S-cdG, [$^{15}N_5$]-5′R-cdG, [$^{15}N_5$]-8-oxo-dG and [$^{15}N_5$]-8-oxo-dA) as previously described [31]. Furthermore, 3 U of benzonase (in 20 mM Tris-HCl pH 8.0, 2 mM $MgCl_2$ and 20 mM NaCl), 4 mU phosphodiesterase I, 3 U DNAse I, 2 mU of phosphodiesterase II and 2 U of alkaline phosphatase were added and the mixture was incubated at 37 °C. After 21 h, 35 µL of Ar flushed buffer containing 0.3 M AcONa (pH 5.6) and 10 mM $ZnCl_2$ were added along with 0.5 U of Nuclease P1 (in 30 mM AcONa pH 5.3, 5 mM $ZnCl_2$ and 50 mM NaCl), 4 mU PDE II and 125 mU of DNAse II and the mixture was further incubated at 37°C for extra 21 h. A step-quenching with 1% formic acid solution (final pH ~ 7) was followed, the digestion mixture was placed in a microspin filter (3 kDa) and the enzymes were filtered off by centrifugation at 14,000× g (4 °C) for 20 min. Subsequently, the filtrate was freeze-dried before HPLC analysis, clean-up, and enrichment.

3.8. HPLC Analysis and Quantification of Modified Nucleosides by Stable Isotope LC-MS/MS

The quantification of the modified nucleosides (in lesions/10^6 nucleosides units) in the enzymatically digested samples (spiked with the ^{15}N-labeled nucleosides) was based on the parallel quantification of the unmodified nucleosides after HPLC clean-up and sample enrichment and the quantification of the single lesions by stable isotope dilution LC-MS/MS analysis [16]. HPLC-UV clean-up and enrichment of the enzyme free samples were performed using a gradient program (2 mM ammonium formate, acetonitrile and methanol) while the fractions containing the lesions were collected, freeze-dried, pooled, freeze-dried again, and redissolved in Milli-Q water before been injected for LC-MS/MS analysis. Detection was performed in multiple reaction monitoring mode (MRM) using the two most intense and characteristic precursor/product ion transitions for each DNA lesion [32,33].

4. Conclusions

Pulse radiolysis in a series of 12 mer as ss-ODNs or ds-ODNs and ct-DNA gives only limited information. In particular, (i) the time evolution of spectra recorded at 1 µs and 50 µs shows that the formed radicals are stable within this time domain, and, (ii) in three ss-ODNs cases, it shows the absorption band > 600 nm that can be assigned to guanyl-type radicals and its decay due to the tautomerization.

The use of cPu lesions as a candidate marker of DNA damage is increasingly appreciated [4]. They offer, together with 8-oxo-Pu lesions, a profile of purine lesions with different properties: cPu as markers of HO• radical damage in aging and diseases (repaired by NER), whereas 8-oxo-Pu are the results of various oxidizing species including HO• radical (repaired by BER). In the present work, the simultaneous measurement of the four cPu and two 8-oxo-Pu upon reaction of genetic material with HO• radicals, generated either by γ-radiolysis or Fenton-type reaction, contribute to greater knowledge on an absolute level of lesions according to the method of HO• radical generation, relative abundance between cPu and 8-oxo-Pu, and the diastereomer ratio in cPu, with the latter one being associated with molecular complexity. The robustness of analytical protocol, which does not produce artifactual oxidations, was also provided by dose curve dependence and comparison of different methods of HO• radical generation, thus rendering our results useful to shed light on disagreements in the literature regarding the formation of DNA purine lesions [16].

Supplementary Materials: The following are available online at http://www.mdpi.com/1420-3049/24/21/3860/s1, Figure S1: Transient absorption spectra, Figure S2: ^{15}N isotopic labeled compounds, Figure S3: Calibration curves for the quantification of the lesions, Figure S4 and S5: Radiation induced formation of cPu and 8-oxo-Pu in ct-DNA Procedure (ii), Figure S6: Radiation induced formation of cPu and 8-oxo-Pu in ds-ODNs, Figure S7: ESI spectra of ODN5 and ODN6, Figure S8: UV melting curves of 21-mer duplexes, Figure S9: CD spectra of 21-mer duplexes,

Table S1: A list of MRM transitions, Table S2: Total amount of cPu and 8-oxo-Pu in DNA applying Procedure (i), Table S3: Total amount of cPu and 8-oxo-Pu in DNA applying Procedure (ii), Table S4: The levels cPu and 8-oxo-Pu in ct-DNA DNA, Table S5: Sequences and molecular masses of the synthesized ODNs, Table S6: Total amount of cPu and 8-oxo-Pu in double-stranded 21-mer oligonucleotides, Table S7: The levels of cPu and 8-oxo-Pu upon Fenton-type reagent treatment of ds-ODNs, Table S8: Total amount of 8-oxo-purine lesions and cPu lesions upon Fenton-type reagent treatment of ds-ODNs, Table S9. The diastereoisomeric ratios ($5'R/5'S$) for both cdG and cdA upon Fenton-type reagent treatment of ds-ODNs, Table S10: The levels of cPu and 8-oxo-Pu upon treatment of ds-ODNs with Fenton-type reagents (copper), Table S11: Total amount of 8-oxo-purine lesions and cPu lesions formation upon treatment of ds-ODNs with Fenton-type reagents (copper), Table S12: The diastereoisomeric ratios ($5'R/5'S$) for both cdG and cdA upon treatment of ds-ODNs with Fenton-type reagents (copper), Table S13: The levels of cPu and 8-oxo-Pu upon treatment of ds-ODNs with Fenton-type reagents (iron), Table S14. Total amount of 8-oxo-purine lesions and cPu lesions formation upon treatment of ds-ODNs with Fenton-type reagents (iron), Table S15: The diastereoisomeric ratios ($5'R/5'S$) for both cdG and cdA upon treatment of ds-ODNs with Fenton-type reagents (iron).

Author Contributions: Conceptualization, C.C.; Coordinated the work, C.C. and K.B.; Synthesis and characterization of Oligonucleotides A.M.; Experiments in pulse radiolysis M.A.T., T.S., and K.B; conducted experiments in Gammacell and collected the worked-up samples C.C., S.B.-V., A.M., and C.F.; performed the LC-MS/MS analyses, M.G.K.; Data Analysis, C.C., M.G.K., and K.B.; writing—original draft preparation, C.C., M.G.K., A.M., C.F., and K.B.; all authors contributed to the figures; and all authors reviewed the manuscript.

Funding: This research was funded by the Marie Skłodowska-Curie European Training Network (ETN) ClickGene: Click Chemistry for Future Gene Therapies to Benefit Citizens, Researchers and Industry [H2020-MSCAETN-2014-642023].

Acknowledgments: The support given by the EU COST Action CM1201 "Biomimetic Radical Chemistry" is kindly acknowledged. C.C. is grateful to Vincent W. Bowry for helpful discussions.

Conflicts of Interest: The authors declare no conflict of interest.

References

1. Greenberg, M.M. (Ed.) *Radical and Radical Ion Reactivity in Nucleic Acid Chemistry*; John Wiley & Sons: Honoken, NJ, USA, 2009.
2. von Sonntag, C. *Free-Radical-Induced DNA Damage and Its Repair. A Chemical Perspective*; Springer Science: Berlin/Heidelberg, Germany, 2006.
3. Chatgilialoglu, C.; Ferreri, C.; Terzidis, M.A. Purine 5′,8-cyclonucleoside lesions: Chemistry and biology. *Chem. Soc. Rev.* **2011**, *40*, 1368–1382. [CrossRef] [PubMed]
4. Chatgilialoglu, C.; Ferreri, C.; Geacintov, N.E.; Krokidis, M.G.; Liu, Y.; Masi, A.; Shafirovich, V.; Terzidis, M.A.; Tsegay, P.S. 5′,8-Cyclopurine Lesions in DNA Damage: Chemical, Analytical, Biological and Diagnostic Significance. *Cells* **2019**, *8*, 513. [CrossRef] [PubMed]
5. Adhikary, A.; Becker, D.; Palmer, B.J.; Heizer, A.N.; Sevilla, M.D. Direct formation of the C5′-radical in the sugar-phosphate backbone of DNA by high energy radiation. *J. Phys. Chem. B* **2012**, *116*, 5900–5906. [CrossRef] [PubMed]
6. Mitra, D.; Luo, X.; Morgan, A.; Wang, J.; Hoang, M.P.; Lo, J.; Guerrero, C.R.; Lennerz, J.K.; Mihm, M.C.; Wargo, J.A.; et al. An ultraviolet-radiation-independent pathway to melanoma carcinogenesis in the red hair/fair skin background. *Nature* **2012**, *491*, 449–453. [CrossRef] [PubMed]
7. Wang, J.; Clauson, C.L.; Robbins, P.D.; Niedernhofer, L.J.; Wang, Y. The oxidative DNA lesions 8,5′-cyclopurines accumulate with aging in a tissue-specific manner. *Aging Cell* **2012**, *11*, 714–716. [CrossRef] [PubMed]
8. Yu, Y.; Guerrero, C.R.; Liu, S.; Amato, N.J.; Sharma, Y.; Gupta, S.; Wang, Y. Comprehensive assessment of oxidatively induced modifications of DNA in a rat model of human Wilson's disease. *Mol. Cell. Proteom.* **2016**, *15*, 810–817. [CrossRef] [PubMed]
9. Robinson, A.R.; Yousefzadeh, M.J.; Rozgaja, T.A.; Wang, J.; Li, X.; Tilstra, J.S.; Feldman, C.H.; Gregg, S.Q.; Johnson, C.H.; Skoda, E.M.; et al. Spontaneous DNA damage to the nuclear genome promotes senescence, redox imbalance and aging. *Redox Biol.* **2018**, *17*, 259–273. [CrossRef]
10. Krokidis, M.; Louka, M.; Efthimiadou, E.; Zervou, S.-K.; Papadopoulos, K.; Hiskia, A.; Ferreri, C.; Chatgilialoglu, C. Membrane lipidome reorganization and accumulation of tissue DNA lesions in tumor-bearing mice: An exploratory study. *Cancers* **2019**, *11*, 480. [CrossRef]

11. Kropachev, K.; Ding, S.; Terzidis, M.A.; Masi, A.; Liu, Z.; Cai, Y.; Kolbanovskiy, M.; Chatgilialoglu, C.; Broyde, S.; Geancitov, N.E.; et al. Structural basis for the recognition of diastereomeric 5′,8-cyclo-2′-deoxypurine lesions by the human nucleotide excision repair system. *Nucleic. Acids Res.* **2014**, *42*, 5020–5032. [CrossRef]
12. Shafirovich, V.; Kolbanovskiy, M.; Kropachev, K.; Liu, Z.; Cai, Y.; Terzidis, M.A.; Masi, A.; Chatgilialoglu, C.; Amin, S.; Dadali, A.; et al. Nucleotide excision repair and impact of site-specific 5′,8-cyclopurine and bulky DNA lesions on the physical properties of nucleosomes. *Biochemistry* **2019**, *58*, 561–574. [CrossRef]
13. Yu, Y.; Wang, P.; Cui, Y.; Wang, Y. Chemical analysis of DNA damage. *Anal. Chem.* **2018**, *90*, 556–576. [CrossRef] [PubMed]
14. Cui, L.; Ye, W.; Prestwich, E.G.; Wishnok, J.S.; Taghizadeh, K.; Dedon, P.C.; Tannenbaum, S.R. Comparative analysis of four oxidized guanine lesions from reactions of DNA with peroxynitrite, single oxygen, and γ-radiation. *Chem Res. Toxicol.* **2013**, *26*, 195–202. [CrossRef] [PubMed]
15. Dizdaroglu, M.; Coskun, E.; Jaruga, P. Measurement of oxidatively induced DNA damage and its repair, by mass spectrometric techniques. *Free Radic. Res.* **2015**, *49*, 525–548. [CrossRef] [PubMed]
16. Chatgilialoglu, C. Cuclopurine (cPu) lesions: What, how and why. *Free Radic. Res.* **2019**, *53*, 941–943. [CrossRef]
17. Baldacchino, G.; Hickel, B.A. method to improve the nonrepetitive acquisition of transient absorption spectra with an intensified charge-coupled device camera. *Rev. Sci. Instrum.* **1998**, *69*, 1605–1609. [CrossRef]
18. Nauser, T.; Jacoby, M.; Koppenol, W.H.; Squier, T.C.; Schöneich, C. Calmoduline methionine residues are targets for one electron oxidation by hydroxyl radicals: Formation of S∴N three-electron bonded radical complexes. *Chem. Commun.* **2005**, 587–589. [CrossRef]
19. Buxton, G.V.; Greenstock, C.L.; Helman, W.P.; Ross, A.B. Critical review of rate constants for hydrated electrons, hydrogen atoms and hydroxyl radicals (OH/O$^-$) in aqueous solution. *J. Phys. Chem. Ref. Data* **1988**, *17*, 513–886. [CrossRef]
20. Ross, A.B.; Mallard, W.G.; Helman, W.P.; Buxton, G.V.; Huie, R.E.; Neta, P. *NDRLNIST Solution Kinetic Database-Ver. 3*; Notre Dame Radiation Laboratory, Notre Dame, IN and NIST Standard Reference Data: Gaithersburg, MD, USA, 1998.
21. Hissung, A.; von Sonntag, C. Radiolysis of Cytosine, 5-Methyl Cytosine and 2′-Deoxycytidine in Deoxygenated Aqueous Solution. A Pulse Spectroscopic and Pulse Conductometric Study of the OH Adduct. *Z. Nat.* **1978**, *33*, 321–328. [CrossRef]
22. Aravandikumar, C.T.; Schuchmann, M.N.; Rao, B.S.M.; von Sonntag, J.; von Sonntag, C. The reactions of cytidine and 2′deoxycytidine with SO$_4^{\bullet-}$ revisited. Pulse radiolysis and product studies. *Org. Biomol. Chem.* **2003**, *1*, 401–408. [CrossRef]
23. Chatgilialoglu, C.; D'Angelantonio, M.; Guerra, M.; Kaloudis, P.; Mulazzani, Q.G. A reevaluation of the ambident reactivity of guanine moiety towards hydroxyl radicals. *Angew. Chem. Int. Ed.* **2009**, *48*, 2214–2217. [CrossRef]
24. Chatgilialoglu, C.; Caminal, C.; Guerra, M.; Mulazzani, Q.G. Tautomers of one-electron-oxidized guanosine. *Angew. Chem. Int. Ed.* **2005**, *44*, 6030–6032. [CrossRef] [PubMed]
25. Chatgilialoglu, C.; D'Angelantonio, M.; Kciuk, G.; Bobrowski, K. New Insights into the Reaction Paths of Hydroxyl Radicals with 2′-Deoxyguanosine. *Chem. Res. Toxicol.* **2011**, *24*, 2200–2206. [CrossRef] [PubMed]
26. Vieira, A.J.S.C.; Steenken, S. Pattern of OH Radical Reaction with Adenine and Its Nucleosides and Nucleotides. Characterization of Two Types of Isomeric OH Adduct and Their Unimolecular Transformation Reactions. *J. Am. Chem. Soc.* **1990**, *112*, 6986–6994. [CrossRef]
27. Wu, L.D.; Liu, K.H.; Jie, J.L.; Song, D.; Su, H.M. Direct Observation of Guanine Radical Cation Deprotonation in G-Quadruplex DNA. *J. Am. Chem. Soc.* **2015**, *137*, 259–266. [CrossRef] [PubMed]
28. Banyasz, A.; Martinez-Fernandez, L.; Balty, C.; Perron, M.; Douki, T.; Improta, R.; Markovitsi, D. Absorption of Low-Energy UV Radiation by Human Telomere G-Quadruplexes Generates Long-Lived Guanine Radical Cations. *J. Am. Chem. Soc.* **2017**, *139*, 10561–10568. [CrossRef]
29. Balanikas, E.; Banyasz, A.; Baldacchino, G.; Markovitsi, D. Populations and Dynamics of Guanine Radicals in DNA strands—Direct versus Indirect Generation. *Molecules* **2019**, *24*, 2347. [CrossRef]
30. Capobianco, A.; Caruso, T.; D'Urci, A.M.; Fusco, S.; Masi, A.; Scrima, M.; Chatgilialoglu, C.; Peluso, A. Delocalized hole domains in guanine rich DNA oligonucleotides. *J. Phys. Chem. B* **2015**, *119*, 5462–5466. [CrossRef]

31. Terzidis, M.A.; Chatgilialoglu, C. An ameliorative protocol for the quantification of purine 5′,8-cyclo-2′-deoxynucleosides in oxidized DNA. *Front. Chem.* **2015**, *3*, 47. [CrossRef]
32. Krokidis, M.G.; Terzidis, M.A.; Efthimiadou, E.; Zervou, S.K.; Kordas, G.; Papadopoulos, K.; Hiskia, A.; Kletsas, D.; Chatgilialoglu, C. Purine 5′,8-cyclo-2′-deoxynucleoside lesions: Formation by radical stress and repair in human breast epithelial cancer cells. *Free Radic. Res.* **2017**, *51*, 470–482. [CrossRef]
33. Terzidis, M.A.; Ferreri, C.; Chatgilialoglu, C. Radiation-induced formation of purine lesions in single and double stranded DNA: Revised quantification. *Front. Chem.* **2015**, *3*, 18. [CrossRef]
34. Chatgilialoglu, C.; Bazzanini, R.; Jimenez, L.B.; Miranda, M.A. (5′S)- and (5′R)-5′,8-cyclo-2′-deoxyguanosine: Mechanistic insights on the 2′-deoxyguanosin-5′-yl radical cyclization. *Chem. Res. Toxicol.* **2007**, *20*, 1820–1824. [CrossRef] [PubMed]
35. Boussicault, F.; Kaloudis, P.; Caminal, C.; Mulazzani, Q.G.; Chatgilialoglu, C. The fate of C5′ radicals of purine nucleosides under oxidative conditions. *J. Am. Chem. Soc.* **2008**, *130*, 8377–8385. [CrossRef] [PubMed]
36. Belmadoui, N.; Boussicault, F.; Guerra, M.; Ravanat, J.-L.; Chatgilialoglu, C.; Cadet, J. Radiation-induced formation of purine 5′,8-cyclonucleosides in isolated and cellular DNA: High stereospecificity and modulating effect of oxygen. *Org. Biomol. Chem.* **2010**, *8*, 3211–3219. [CrossRef] [PubMed]
37. Terzidis, M.A.; Prisecaru, A.; Molphy, Z.; Barron, N.; Randazzo, A.; Dumont, E.; Krokidis, M.G.; Kellett, A.; Chatgilialoglu, C. Radical-induced purine lesion formation is dependent on DNA helical topology. *Free Radic. Res.* **2016**, *50*, S91–S101. [CrossRef]
38. Miller, D.M.; Buettner, G.R.; Aust, S.D. Transition metals as catalysts of "autoxidation" reactions. *Free Radic. Biol. Med.* **1990**, *8*, 95–108. [CrossRef]
39. Stohs, S.J.; Bagchi, D. Oxidative mechanisms in the toxicity of metal ions. *Free Radic. Biol. Med.* **1995**, *18*, 321–336. [CrossRef]
40. Haliwell, B.; Gutteridge, J.M.C. *Free Radicals in Biology and Medicine*, 5th ed.; Oxford University Press: Oxford, UK, 2015.
41. Guerrero, C.R.; Wang, J.; Wang, Y. Induction of 8,5′-cyclo-2′-deoxyadenosine and 8,5′-cyclo-2′-deoxyguanosine in isolated DNA by Fenton-type reagents. *Chem. Res. Toxicol.* **2013**, *26*, 1361–1366. [CrossRef]
42. Bergeron, F.; Auvré, F.; Radicella, J.P.; Ravanat, J.-L. HO• radicals induce an unexpected high proportion of tandem base lesions refractory to repair by DNA glycosylases. *Proc. Natl. Acad. Sci. USA* **2010**, *107*, 5528–5533. [CrossRef]
43. Flyunt, R.; Bazzanini, R.; Chatgilialoglu, C.; Mulazzani, Q.G. Fate of the 2′-Deoxyadenosin-5′-yl Radical under Anaerobic Conditions*J. Am. Chem. Soc.* **2000**, *122*, 4225–4226. [CrossRef]
44. Chatgilialoglu, C.; Guerra, M.; Mulazzani, Q.G. Model studies of DNA C5′ radicals. Selective generation and reactivity of 2′-deoxyadenosin-5′-yl radical. *J. Am. Chem. Soc.* **2003**, *125*, 3839–3848. [CrossRef]

Sample Availability: Samples of the compounds are not available from the authors.

© 2019 by the authors. Licensee MDPI, Basel, Switzerland. This article is an open access article distributed under the terms and conditions of the Creative Commons Attribution (CC BY) license (http://creativecommons.org/licenses/by/4.0/).

Article

Populations and Dynamics of Guanine Radicals in DNA strands—Direct versus Indirect Generation

Evangelos Balanikas [1], Akos Banyasz [1,2], Gérard Baldacchino [1] and Dimitra Markovitsi [1,*]

[1] LIDYL, CEA, CNRS, Université Paris-Saclay, F-91191 Gif-sur-Yvette, France; vangelis.balanikas@cea.fr (E.B.); akos.banyasz@ens-lyon.fr (A.B.); gerard.baldacchino@cea.fr (G.B.)
[2] Univ Lyon, ENS de Lyon, CNRS UMR 5182, Université Claude Bernard Lyon 1, Laboratoire de Chimie, F-69342 Lyon, France
[*] Correspondence: dimitra.markovitsi@cea.fr; Tel.: +33-169084656

Academic Editor: Chryssostomos Chatgilialoglu
Received: 22 May 2019; Accepted: 18 June 2019; Published: 26 June 2019

Abstract: Guanine radicals, known to be involved in the damage of the genetic code and aging, are studied by nanosecond transient absorption spectroscopy. They are generated in single, double and four-stranded structures (G-quadruplexes) by one and two-photon ionization at 266 nm, corresponding to a photon energy lower than the ionization potential of nucleobases. The quantum yield of the one-photon process determined for telomeric G-quadruplexes (**TEL25/Na$^+$**) is $(5.2 \pm 0.3) \times 10^{-3}$, significantly higher than that found for duplexes containing in their structure GGG and GG sequences, $(2.1 \pm 0.4) \times 10^{-3}$. The radical population is quantified in respect of the ejected electrons. Deprotonation of radical cations gives rise to (G-H1)$^{\bullet}$ and (G-H2)$^{\bullet}$ radicals for duplexes and **G**-quadruplexes, respectively. The lifetimes of deprotonated radicals determined for a given secondary structure strongly depend on the base sequence. The multiscale non-exponential dynamics of these radicals are discussed in terms of inhomogeneity of the reaction space and continuous conformational motions. The deviation from classical kinetic models developed for homogeneous reaction conditions could also be one reason for discrepancies between the results obtained by photoionization and indirect oxidation, involving a bi-molecular reaction between an oxidant and the nucleic acid.

Keywords: DNA; guanine quadruplexes; radicals; electron holes; oxidative damage; photo-ionization; time-resolved spectroscopy; inhomogeneous reactions

1. Introduction

Guanine (**G**) radicals are major actors in the oxidatively generated damage to the genetic code [1]. The reason is that **G** is the nucleobase with the lowest oxidation potential [2]. Therefore, electron holes (radical cations) created on other nucleobases of a DNA helix, may reach **G** sites following a charge transfer process and, subsequently, undergo irreversible chemical reactions [3–7]. Various reaction mechanisms have been determined [8,9]. Some of them, such as formation of the well-known oxidation marker 8-oxo-7,8-dihydro-2′-deoxyguanosine (8-oxodGuo), involve directly the radical cation (**G**)$^{\bullet+}$. However, this charged species is prone to loss of a proton, giving rise to deprotonated radicals, labeled (G-H1)$^{\bullet}$ [10–14] and (G-H2)$^{\bullet}$ [15–19] (Figure 1), depending on the position from which the proton is lost (Figure 1). Further reactions implicate deprotonated radicals [8]. Accordingly, the fraction of (**G**)$^{\bullet+}$ that undergoes deprotonation, as well as the lifetime of the various radicals are expected to play a pivotal role in the relative yields of the final reaction products.

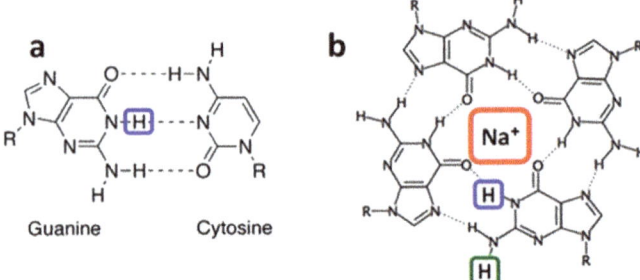

Figure 1. In double helices guanine is paired to cytosine (**a**). Guanines may also self-associate forming a tetrad (**b**) which is the building block of **G**-quadruplexes. (**G-H1**)• and (**G-H2**)• radicals correspond to the transfer of the protons encased in blue and green, respectively, toward the aqueous solvent. Na⁺ encased in red represents a sodium ion located in the central cavity of the G-quadruplex.

Two different approaches, based on time-resolved techniques, have been used in order to characterize the dynamics of **G** radicals. On the one hand, (**G**)•⁺ are formed directly by photoionization [10,18–26]. On the other, they are created in an indirect way via a charge transfer reaction requiring mediation of an external oxidant. In turn, the latter may be generated either by laser [17,27–29] or electron pulses [11,13,15,16,30]. During the past few years, important discrepancies started to appear in the reported lifetimes of the **G** radicals. For example, indirect oxidation using sulfate ions ($SO_4^{•-}$) reported that base-pairing induces a faster decay of (**G-H1**)• on the ms time-scale [28]. In contrast, the lifetimes found for (**G-H1**)• by direct photoionization increase in the following order: single strand, double strand, four-stranded structure (G-quadruplex) [18,24,26]. More surprisingly, while one indirect study of **G**-quadruplexes reported that radical cations decay with a lifetime of 0.1 ms giving rise to (**G-H1**)• radicals [29], another study, using exactly the same oxidant, showed that (**G**)•⁺ deprotonation in **G**-quadruplexes, occurring on the µs time-scale, gives rise to (**G-H2**)• [17]. This was explained by the participation of the hydrogen in position 1 to a hydrogen bond (Figure 1) [17]. The latter conclusion was supported by our direct photoionization studies, which, in addition, found that (**G-H2**)• → (**G-H1**)• tautomerisation takes place on the ms time-scale [18,19].

The above mentioned discrepancies appear by comparing results obtained for the same secondary structure but different base sequences. However, it is also reported that the base sequence may affect radical dynamics. This is the case of (**G**)•⁺ in four-stranded structures [17–19] and of deprotonated adenine radicals in duplexes [24]. Therefore, it is important to explore if the two approaches used for the study of radical dynamics agree when experiments are performed for exactly the same system. This is one objective of the present work.

Our study was performed by nanosecond laser photolysis and used the direct photoionization approach with excitation at 266 nm. We focused on three different types of DNA structures whose study by the indirect approach is well described [28,29]:

- two single strands composed of 30 bases **S1**: 5'-CGTACTCTTTGGTGGGTCGGTTCTTTCTAT-3', and **S2**: 3'-GCATGAGAAACCACCCAGCCAAGAAAGATA-5',
- the duplex **D** formed by hybridization of **S1** with its complementary strand **S2**, and
- the monomolecular **G**-quadruplex formed by folding of the human telomeric sequence 5'-TAGGG(TTAGGG)₃TT-3' in the presence of Na⁺ ions, abbreviated as **TEL25/Na⁺**.

The second objective of our study is to examine the extent to which the dynamics of **G** radicals are affected by the base sequence within a given secondary structure (single-, double- or four-stranded). To this end, the present results were compared with those obtained by us previously following the same methodology for a single strand corresponding to the human telomer repeat 5'-TTAGGG-3' [18], a duplex composed of the guanine-cytosine pairs in alternating sequence **GC$_5$** [26], another human telomer **G**-quadruplex formed by a somewhat shorter sequence, 5'-GGG(TTAGGG)$_3$-3' in the presence of Na$^+$ ions (**TEL21/Na$^+$**) [18] and a tetramolecular G-quadruplex formed by association of four TGGGGT strands (**TG4T)$_4$/Na$^+$** [19].

For **D** and **TEL25/Na$^+$**, the probability that **G** radicals generated upon direct absorption of single photons with energy lower than the **G** ionization potential was also examined. This unexpected mono-photonic ionization at long wavelengths, suggested by a few authors [31–33], has been evidenced recently by concomitant quantification of ejected electrons and generated radicals [18,19,24,26]. It was further supported by the detection of the well-known oxidation marker 8-oxo-7,8-dihydro-2'-deoxyguanosine (8-oxodGuo) in solutions of purified genomic DNA [34] and telomeric **G**-quadruplexes [18] irradiated by continuous light sources at wavelengths ranging from 254 to 295 nm.

2. Results and Discussion

2.1. Methodology: Advantages and Limitations

A key point in our methodology is that the DNA solution does not contain any additive besides the phosphate buffer. In addition, electrons are ejected at zero-time in respect of the time resolution of the setup which is ~30 ns. At this time, the ejected electrons have been already hydrated [35]. Under the latter configuration, they exhibit a broad absorption band peaking at 720 nm with a molar absorption coefficient ε of 19,700 mol^{-1} L cm^{-1} [36]. With respect to this property, they can be quantified. For better precision, their decay was fitted with a mono-exponential function $A_0 + A_1 \exp(-t/\tau_1)$ (Figure 2). Subsequently, the A_1 value, associated to ε, provides the initial concentration of the hydrated ejected electrons $[e_{hyd}^-]_0$. In such an experiment, electrons may originate not only from DNA photoionization, but also from two-photon ionization of water. In order to avoid the latter process, which precludes quantitative correlation between ejected electrons and generated radicals, weak excitation intensities ($\leq 2 \times 10^6$ W cm^{-2}) were used. Under these conditions, no hydrated electrons were detected for the aqueous solvent alone (Figure 2). Moreover, electrons may react with nucleic acids [37]. However, this unwanted effect is prevented because the hydrated electrons are scavenged by the phosphate groups of the buffer [38], which are present in much higher concentrations than the DNA multimers.

An important drawback of radical generation by direct photoionization is that, in the same time, a series of photoproducts, possibly involving reaction intermediates, are formed [39]. The spectra of such species may overlap with those of radicals, determined after 2 μs, when the hydrated electrons have disappeared. This is, in particular, the case of pyrimidine (6-4) pyrimidone photoproducts (64PPs) formed following reactions between two pyrimidines [40,41], as well as adenine-adenine [42] and adenine-thymine dimers [43–47] and their reaction intermediates [25,48]. All these compounds absorb in the 300 to 400 nm range, exactly where the absorption of **G** radicals is particularly intense. Fortunately, **G** radicals exhibit additional characteristic peaks in the visible spectral domain, thus allowing their identification and quantification.

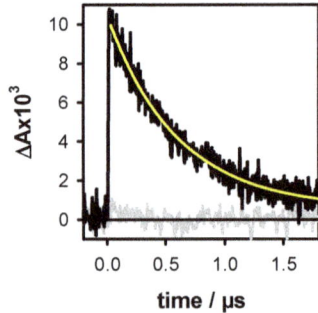

Figure 2. The transient absorption signals recorded at 700 nm for the duplex **D** (black) and the buffer alone (grey) with an excitation intensity of 2×10^6 W cm^{-2}. The yellow line represents the fit with a mono-exponential function $A_0 + A_1\exp(-t/\tau_1)$. Within the precision of our measurements, the intensity of the signals at 720 nm and 700 nm are same. As the latter are less noisy, the electron concentration was systematically determined at this wavelength.

Finally, a key condition in our methodology is to avoid exciting DNA multimers that have been altered as a result of either photoionization or other photochemical reactions. This is achieved by using a large quantity of solution, which makes such experiments both slow and expensive. Typically, 40 mL of **D** or **TEL25/Na$^+$** solutions are needed for recording a transient absorption spectrum over a single time scale. Considerably larger quantities are required in the case of **S1** and **S2** because the yield of dimeric photoproducts is much higher in single strands [49–51]. Therefore, the study of single strands was limited to radical dynamics.

More details on the experimental protocols are given in the Materials and Methods Section.

2.2. One- and Two-Photon Ionization

Electron ejection upon 266 nm laser excitation of nucleic acids, provoked by two-photon ionization has been exploited to study oxidative damage to DNA [52,53]. In this study, we tried to keep as low as possible the contribution of the two-photon process but without completely eliminating it, otherwise the transient absorption signals stemming from both the hydrated electrons and the radicals become too weak to be observed. In order to disentangle between one and two-photon effects, the laser intensity was varied and at each step, we determined $[e_{hyd}^-]_0$. Subsequently, the ionization curve was obtained by plotting $[e_{hyd}^-]_0/[h\nu]$ as a function $[h\nu]$. The latter quantity represents the concentration of absorbed photons per pulse in the probed volume of the studied solution. The experimental points are fitted with the linear model function $[e_{hyd}^-]_0/[h\nu] = \varphi_1 + \alpha[h\nu]$. The intercept on the ordinate provides the one-photon ionization quantum yield φ_1, while the slope is proportional to the two-photon ionization yield φ_2, which depends on the laser intensity, $\varphi_2 \propto \alpha[h\nu]$.

The ionization curves obtained for **D** and **TEL25/Na$^+$** are shown in Figure 3. The φ_1 determined for the duplex is $(2.1 \pm 0.4) \times 10^{-3}$, while a much higher value, $(5.2 \pm 0.3) \times 10^{-3}$, is found for the G-quadruplex. The higher propensity of G-quadruplexes to undergo electron detachment upon absorption of single photons at 266 nm is in line with previously reported results [18,19,24–26]. However, in addition, the present work brings to light some subtle differences.

In the case of duplexes, electron detachment is facilitated by the occurrence of one GG and one GGG sequences, for a total of thirty base pairs, composing **D**. As a matter of fact, a φ_1 value of $(1.2 \pm 0.2) \times 10^{-3}$, was found for the duplex **GC$_5$** [26] while those determined for alternating and homopolymeric AT duplexes amount to $(1.3 \pm 0.2) \times 10^{-3}$ and $(1.5 \pm 0.3) \times 10^{-3}$, respectively [24,25]. This is in line with previous findings that the oxidation potential of G is decreased upon stacking, rendering GG and GGG triplets traps [54,55] for hole transfer [7,56–58] and preferential sites for redox reactions [59].

Considering the above base sequence effect found for duplexes, it is understandable that telomeric G-quadruplexes, composed of four interconnected GGG stacks, exhibit more efficient one-photon ionization. However, our results show that not only GGG stacks play a role in this process. The φ_1 value determined for **TEL25/Na$^+$** is slightly higher compared to that of **TEL21/Na$^+$** $(4.5 \pm 0.6) \times 10^{-3}$. The difference in the base sequence of these systems is the presence of two flanking groups TT and TA in **TEL25/Na$^+$**. These flanking groups do not participate neither to tetrad nor to loop formation.

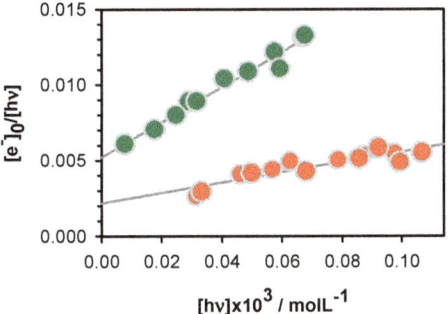

Figure 3. The ionization curves obtained for the duplex **D** (red) and the G-quadruplex **TEL25/Na$^+$** (green); [e$_{hyd}^-$]$_0$ and [hv] denote, respectively, the zero-time concentration of hydrated ejected electrons and the concentration of absorbed photons per laser pulse. Experimental points (circles) are fitted with the linear model function [e$_{hyd}^-$]$_0$/[hv] = φ_1 + α[hv] (grey).

2.3. Radicals in Single and Double Strands

The transient absorption spectrum obtained for **D** at 5 µs (Figure 4) resembles closely that of the deprotonated (G-H1)$^\bullet$ radicals [21]. As discussed in the literature [28,60], deprotonation of guanine radical cations in duplexes may proceed by the transfer of a hydrogen atom to either the cytosine or the aqueous solvent. The spectra of these two deprotonated guanine radicals were computed by quantum chemistry methods for a short duplex composed of two guanine-cytosine pairs in alternating sequences (Figure 6b in reference [26]). It appeared that only the transfer of the proton to the aqueous solvent induces a long red tail in the radical absorption spectrum. Quite recent calculations performed for a guanine-cytosine pair using a larger basis set [61] showed the existence of a weak intensity band between 600 and 650 nm for (G-H1)$^\bullet$ radical. This feature can be distinguished in the **D** spectrum at 5 µs.

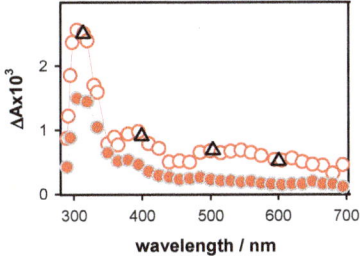

Figure 4. The differential absorption spectra determined for the duplex **D** at 5 µs (empty circles; average ΔA from 3 to 7 µs) and 10 ms (full circles; average ΔA from 8 to 12 ms). The triangles denote relative intensities of the 5 µs spectrum obtained using oxidation by SO$_4^{\bullet-}$ [28].

The radical concentration at 5 µs, determined from the differential absorption at 500 nm and the molar absorption coefficient reported for the corresponding monomeric radical (1500 mol^{-1} L cm^{-1}) [10], is 4.8 × 10^{-7} mol L^{-1}. This value is quite close to the initial electron concentration [e$_{hyd}^-$]$_0$, determined for the same excitation energy (5.1 × 10^{-7} mol L^{-1}). The 12% difference falls in the experimental error bar, so that it cannot be excluded that the somewhat lower concentration of radical is due to a reaction taking place at shorter times.

At longer times, the relative intensity of the UV band, in respect to the absorption in the visible spectral domain increases, suggesting contribution of photoproducts appearing on the ms time scale. For example, thymine 64PPs are formed within 4 ms [41]. The coexistence of radicals and photoproducts is also reflected in the dependence of the decays recorded on the ms time scale as a function of the laser intensity (Figure 5). Those at 500 nm remain unchanged (Figure 5a), showing that the dynamics of radicals formed by one- or two-photon ionization is the same. However, dimers are generated by one-photon processes, thus their relative concentration is higher at low excitation intensities. **S1** and **S2** exhibit similar behavior in this respect but, as single strands are more prone to dimerization reactions compared to duplexes [49–51], the effect on the 305 nm decay is much stronger. An example is given in Figure S1.

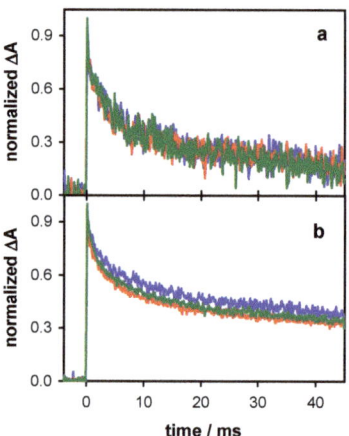

Figure 5. The normalized transient absorption signals recorded for the duplex **D** at 500 (**a**) and 305 nm (**b**) for excitation energies of 4 mJ (blue), 6 mJ (green) and 7 mJ (red), corresponding to decreasing φ_1/φ_2 ratios.

The decays recorded at 500 nm over two time-scales for **S1**, **S2** and **D** are shown in Figure 6. They have been fitted with exponential functions and the absorbance at 2 µs has been normalized to 1. For all three systems, an absorbance loss of about 20% was observed within the first 150 µs while at 45 ms only 8% of the initial absorbance persists for **S1** and **S2** and 12% for **D**. The time needed for the signal to decrease by a factor of 2 ($t_{1/2}$) is 1.8 ms and 2.2 ms, respectively, for **S1** and **S2** and significantly longer (4 ms) for **D**. A lengthening of $t_{1/2}$ from 1 to 4 ms was also found upon base-pairing of adenine tracts [24].

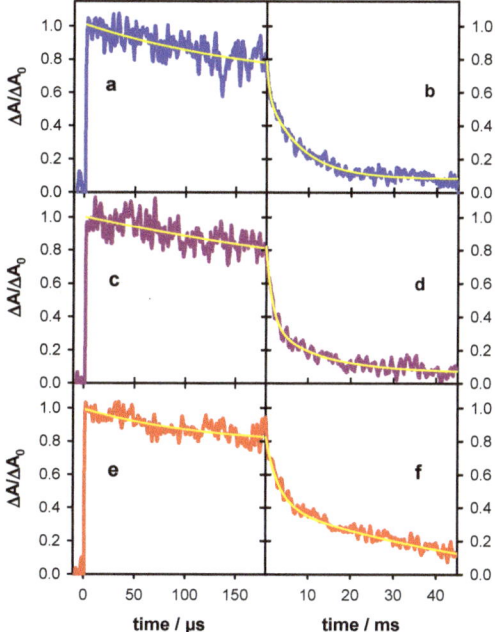

Figure 6. Transient absorption traces recorded for the single strand **S1** (blue; (**a**,**b**)), the single strand **S2** (violet; (**c**,**d**)) and the duplex **D** (red; (**e**,**f**)) at 500 nm. Yellow lines correspond to fits with mono-exponential (**a**,**c**,**e**) and bi-exponential (**b**,**d**,**f**) functions. For all signals, the absorbance at 2 μs (ΔA_0) was normalized to 1.

It is interesting to compare the dynamics of guanine radicals determined in the present work with those of two other systems studied previously by the same methodology: TTAGGG [18] and **GC₅** [26]. This is illustrated in Figure 7, where the dynamics at 500 nm between 0.15 and 15 ms are shown. For clarity, only the fitted functions are presented. It is noted that those of TTAGGG and **GC₅** remain constant between 2 μs and 0.15 ms. It appears that, although for all systems the most important part of the absorbance decays within this time-scale, the decay patterns are specific to each system.

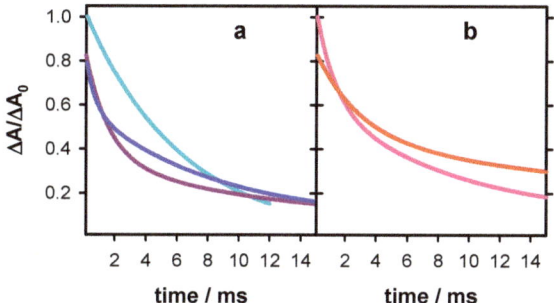

Figure 7. Dynamics of deprotonated guanine radicals observed at 500 nm in single (**a**) and double (**b**) strands. **S1** (blue), **S2** (violet), TTAGGG (cyan; data from reference [18]), **D** (red) and GC₅ (pink; data from reference [26]). For clarity, only the fitted functions of the transient absorption signals are shown. For all signals, the absorbance at 2 μs (ΔA_0) was set equal to 1.

2.4. Radicals in G-Quadruplexes

The differential absorption spectra determined for **TEL25/Na$^+$** exhibit important variations as a function of time on the visible spectral domain, where **G** radicals are expected to absorb (Figure 8). The spectrum at 3 μs it is characterized by a very broad absorption band, indicating the presence of at least two species. At 0.5 ms, the differential absorbance has decreased between 400 and 600 nm while it has been hardly altered at longer wavelengths. A rather symmetrical band peaking at 600 nm was observed. The latter resembles that of monomeric (G-H2)$^•$ radicals [10,16]. As found by Su and coll. [17] and confirmed by us, for both monomolecular (**TEL21/Na$^+$**) [18] and tetramolecular **(TG4T)$_4$/Na$^+$** [19] G-quadruplexes deprotonation of radical cations gives rise to (G-H2)$^•$ radicals because H1 protons participate in Hoogsteen hydrogen bonds (Figure 1b). Moreover, these studies evidenced that deprotonation is much slower compared to other DNA systems, for which it occurs on the ns time-scale [11,13]. Accordingly, the broad absorption band present in the 3 μs spectrum was attributed to a mixture of the **G** radical cation and the (G-H2)$^•$ radical. The peak at 600 nm is still present at 10 ms (Figure 8b; see also normalized spectra in Figure S2). This contrasts with the behavior of the two previously studied G-quadruplexes **TEL21/Na$^+$** and **(TG4T)$_4$/Na$^+$**, for which complete (G-H2)$^•$ → (G-H1)$^•$ tautomerisation has already occurred at this time. However, it cannot be ruled out that a small population of (G-H1)$^•$ radicals is also present. The problem is that the spectrum below 500 nm is dominated by an unknown photoproduct, which does not stem from radicals, as attested by the dependence of the decays on the excitation intensity (Figure S4). Its fingerprint is also present in the steady-state differential absorption spectra recorded before and after irradiation (Figure S3).

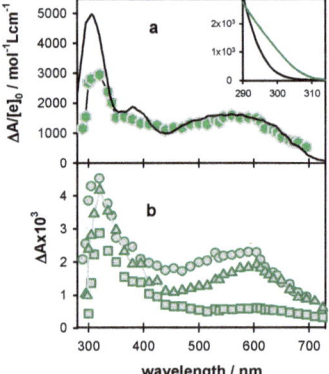

Figure 8. Differential absorption spectra determined for **TEL25/Na$^+$** at 3 μs ((**a**); hexagons; average ΔA from 2 to 4 μs), 5 μs ((**b**); circles; average ΔA from 3 to 7 μs), 0.5 ms ((**b**); triangles; average ΔA from 0.3 to 0.7 ms) and 10 ms ((**b**); squares; average ΔA from 8 to 12 ms). The black line in (**a**) is a linear combination of the spectra corresponding to the radical cation (45%) [10] and the (G-H2)$^•$ radical (55%) [16] of monomeric guanosine, considered with their ε values. In the inset, the steady-state absorption spectra of dGMP (black) [62] and **TEL25/Na$^+$** (green; see also Figure S5) are shown. The ε is given per base.

For a quantitative description of the radical population, we determined the concentration of hydrated ejected electrons [e$_{hyd}^-$]$_0$ produced by the same excitation intensity as that used for recording the transient spectra in Figure 8 (15.6 × 10^{-7} mol L^{-1}). Subsequently, we represented the transient spectrum recorded at 3 μs on ΔA/[e$^-$]$_0$ scale, (Figure 8a) and reconstructed the broad absorption band in the visible spectral range by linear combinations of the (G)$^{•+}$ [10] and (G-H2)$^•$ [16] spectra, reported for monomeric guanosines. The best agreement in the 450–700 nm area is obtained for combinations 45 (±2)% of (G)$^{•+}$ with 55 (±2)% for (G-H2)$^•$. The lower intensity found for the G-quadruplex spectrum around 400 nm is explained by the fact that the radical cation in G-quadruplexes absorbs less than the mono-nucleotide dGMP, while at 500 nm the molar absorption coefficient is practically the same [18]. Moreover, the differential absorbance of **TEL25/Na$^+$** of the UV band is lower because its ground state absorption is stronger than that of dGMP, as shown in the inset of Figure 8 (see also Figure S5). The radical cation population surviving at 3 µs (45%) is quite close to what was found for **TEL21/Na$^+$** (50%) [18] but significantly higher compared to **(TG4T)$_4$/Na$^+$** (25%) [19].

Based on the spectrum at 0.5 ms (Figure 8b) and using a molar absorption coefficient of 2100 mol^{-1} L cm^{-1} at 600 nm, determined for monomeric (G-H2)$^\bullet$ radicals [10,16], we found that their concentration corresponds to 60 ± 2% of the initial radical concentration. This means that ~40% of the radical cations reacted between 0.5 µs and 0.5 ms, through a process other than deprotonation to (G-H2)$^\bullet$. This is also reflected in the transient absorption signals at 500 nm, dominated by the radical cation and 605 nm dominated by the (G-H2)$^\bullet$ radical (Figure 9). The former shows a sharp decrease described by a time-constant of 6 µs (Figure 9a), while a concomitant rise cannot be distinguished on the latter (Figure 9d). As expected, the decays on the ms time-scale are wavelength dependent, $t_{1/2}$ being 2.4 and 3.1 ms, at 500 and 605 nm, respectively.

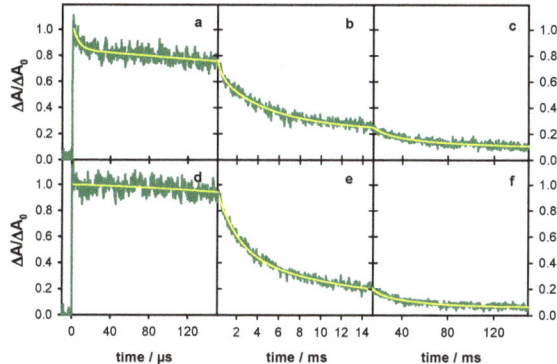

Figure 9. The transient absorption traces recorded for **TEL25/Na$^+$** (green) at 500 nm (a,b,c) and 605 nm (d,e,f). The yellow lines correspond to fits with the bi-exponential or tri-exponential functions. For all signals, the absorbance at 2 µs (ΔA_0) was set equal to 1.

The spectral evolution in Figure 8 and the associated dynamics in Figure 9 greatly differ from those reported previously for **TEL21/Na$^+$** [18]. For the G-quadruplex structure formed by the shorter sequence, 50% of the radical cation population deprotonates with a time constant of 1.2 ms instead of 6 µs for the longer one. Moreover, the disappearance of the (G-H2)$^\bullet$ radical in **TEL21/Na$^+$** is concomitant with that of the radical cation, giving rise to (G-H1)$^\bullet$ radical whose population at 5 ms amounts to 50% of the initial radical population.

3. Reaction Schemes of Nucleic Acids

In general, reactions involving radicals of nucleobases are likely to be bi-molecular. For example, the formation of 8-oxodGuo involves a hydration step requiring addition of a water molecule to G$^{+\bullet}$, while that of guanine-thymine adducts requires the attack of thymine to the (G)$^{\bullet+}$ [8,9]. The probability that the reactants come close to each other is not homogeneous over the three-dimensional space but it is determined by the conformation of the nucleic acid, which, in turn, structures the local environment, including the water network [63]. Thus, conformational motions, occurring on the same timescale as the reaction, may have two effects—on the one hand, it may differentiate the behavior of various reaction sites and, on the other, it may modify the behavior of a given site in the course of the observation. As a result, important deviations appear from the classical models widely used to describe kinetics of chemical reactions in homogenous solutions. The underlying assumption in such models is that of a well-stirred chemical reactor, which means that at the time scale of the observation, there is an internal averaging of all the reaction sites, randomly distributed in three dimensions. The lack of these conditions leads to multiscale decay patterns and renders the notion of rate constant inappropriate, with the reaction rate being time dependent. A good illustration of such multiscale dynamics in DNA is provided by the relaxation of the electronic excited states in helical structures, involving interactions among nucleobases, which spans over, at least, five decades of time [64–66]. In general, the description of inhomogeneous dynamical processes necessitates specific theoretical treatments and/or simulations, developed in various fields such as photocatalysis, charge and energy transport in restricted geometries, polymerization reactions, reactions in biological cells, etc. (see for example references [67–71]).

Given the above considerations, the fits of the transient absorption signals with multi-exponential decays presented in Figures 6 and 9 are, in principle, devoid of physical meaning. We simply used the fitted functions for a quantitative description of the decays, allowing easier comparison among the dynamics of the various systems (Figure 7, Table 1). However, in the case of **TEL25/Na$^+$**, we refer to a time constant of 6 µs (Figure 9a). Although the exponential nature of this decay is certainly an approximation, its association with changes observed in the time-resolved spectra (Figure 8) and the quantification of the radical population allowed the assigning of this characteristic time to the reaction of approximately 45% of the population of initially created radical cations, while the other 55% reacted much faster.

Coming to the comparison of our results with those reported for the same systems using oxidation by sulfate radical ions, there is one common point. Our transient spectra recorded for **D** at 5 µs are in agreement with those reported in reference [28], as attested by a few comparative points also shown in Figure 4. However, the spectra of **TEL25/Na$^+$** obtained by the two methods are in stark contrast—we detected a spectral evolution which is correlated in a quantitative way to (G)$^{\bullet+}$ and (G-H2)$^\bullet$ radicals. The study performed via the indirect approach did not reveal any time-dependence of the spectra, which were attributed to (G-H1)$^\bullet$ radicals [29]. Most dramatic divergences appear in the dynamics. In Table 1, the half times determined for the studied systems by the two approaches are shown. In all cases, the $t_{1/2}$ values found from direct photoinonization are significantly shorter. The largest difference is encountered for the single strands for which the $t_{1/2}$ values reported in indirect oxidation studies are more than one order of magnitude larger than those found by photoionization.

Table 1. Half-times (in ms) at which the intensity of the transient absorbance signals is decreased by a factor of 2.

Method	S1	S2	D	TEL25/Na$^+$
direct photoionization	1.8 [1]	2.2 [1]	4 [1]	2.4 [1]/3.1 [2]
indirect oxidation	120 [3] [28]	40 [3] [28]	7.5 [3] [28]	8 [3] [29]

[1] 500 nm; [2] 605 nm; [3] 510 nm.

The discrepancies between the results obtained by the two methods could be explained by the non-classical reaction schemes discussed above, involved in radical generation.

In direct photoionization, radicals are formed in zero time in respect to our time resolution. At the earliest time that the spectra of radicals can be recorded (2–3 µs), we found that their concentration equals that of the observed ejected electrons. Thus, we were able to follow the fate of the entire radical population, even if part of the deprotonation process was missed.

In the indirect approach, the laser induced reaction occurring at zero time is the production of sulfate radicals ($Na_2S_2O_8 \rightarrow SO_4^{\bullet-}$), while the charge transfer reaction with DNA is a diffusion controlled bi-molecular process [72]. As the nucleobase undergoing the oxidation is not necessarily a guanine [72], there are potentially 30 electron donating sites per single strand, 60 per duplex and 25 per G-quadruplex. The occurrence of many spatially correlated electron donors, renders the reaction scheme highly inhomogeneous. In addition, nucleic acids are negatively charged electrolytes making the approach of a negatively charged donor particularly selective. Thus, it would not be surprising that the formation of radicals is not limited in a few µs, on which the corresponding transient absorption exhibits a clear rise (Figure 3 in reference [29]). This fast rise, which has been correlated to a reaction rate, may simply concern only part of the sulfate radicals located in positions favoring the reaction, while other sulfate radicals react on longer times. A simple estimation of the sulfate radical concentration produced under the described experimental conditions (13×10^{-6} mol L^{-1}) shows that it is indeed twice as high as **G** radicals (6.7×10^{-6} mol L^{-1}). Details are given in the SI. However, it is also possible that non-homogenous diffusion controlled reactions is not the only reason for the longer transient absorption decays found via indirect oxidation. As a matter of fact, the 1 s spectrum reported in Figure 2 in reference [28] clearly differs from those at 5 µs and 10 ms corresponding to **G** radicals. It could be due, for example, to species resulting from reactions of **G** radicals with impurities present in Na$_2$S$_2$O$_8$. Considering that impurities in analytical grade chemicals may reach 1–2%, their concentration in the studied solutions could be two orders of magnitude higher than that of **G** radicals (see SI).

In photoionization experiments, the DNA solutions contain no additives which may react with radicals and shorten their lifetimes. In order to check if the phosphate buffer, which scavenges the produced hydrated electrons, gave rise to secondary reactions, we: (i) diluted it by a factor of 10; and (ii) replaced it by a NaCl solution with the same ionic strength, but none of these modifications altered the dynamics. Along the same line, it was found that, within the precision of our measurements, neither the population nor the decays of deprotonated radicals are affected by oxygen, air equilibrated and argon saturated solutions giving the same signals. These observations suggest that the formation of the final reaction products stemming from (G-H1)$^{\bullet}$ and (G-H2)$^{\bullet}$ radicals involve just water molecules and/or parts of the nucleic acid itself (other nucleobases, 2-deoxyribose moieties).

4. Materials and Methods

4.1. Spectroscopic Measurements

Steady-state absorption spectra were recorded using a Lambda 850 (Perkin-Elmer) spectrophotometer. The transient absorption setup used as an excitation source for the fourth harmonic of a Nd:YAG laser (Spectra-Physics, Quanta Ray). The excited area at the surface of the sample was 0.6×1.0 cm^2. The analyzing beam, orthogonal to the exciting beam, was provided by a 150 W Xe-arc lamp (Applied Photophysics, OSRAM XBO). Its optical path length through the sample was 1 cm while its thickness was limited to 0.1 cm in order to use the most homogeneous part of the light. It was dispersed in a Jobin-Yvon SPEX 270M monochromator, detected by a Hamamatsu R928 photomultiplier and recorded by a Lecroy Waverunner oscilloscope (6084). For measurements on the sub µs-scale, the Xe-arc lamp was intensified via an electric discharge. Transient absorption spectra were recorded using a wavelength-by-wavelength approach. Fast shutters were placed in the path of both laser and lamp beams, thus, the excitation rate was decreased from 10 Hz to 0.2 Hz. The incident pulse energy at the surface of the sample was measured using a NIST traceable pyroelectric sensor (OPHIR Nova2/PE25). Potential variations during a measurement were monitored by detecting a fraction of the exciting beam by a photodiode. In addition, the absorbance of the naphthalene triplet state, whose quantum yield in cyclohexane is 0.75 [73], served as actinometer.

4.2. Sample Preparation and Handling

Lyophilized oligonucleotides, purified by reversed phase HPLC and tested by MALDI-TOF, were purchased from Eurogentec Europe. They were dissolved in a phosphate buffer (0.15 mol L^{-1} NaH$_2$PO$_4$, 0.15 mol L^{-1} Na$_2$HPO$_4$), prepared using ultrapure water delivered by a MILLIPORE (Milli-Q Integral) system. The pH, measured by a HANNA Instr. Apparatus (pH 210), was adjusted to 7 by the addition of a concentrated NaOH solution. A dry bath (Eppendorf-ThermoStatplus) was used for thermal treatment. For the formation of double and four-stranded structures, an appropriate mother solution (2 mL) was heated to 96 °C during 5 min, cooled to the melting point of the corresponding system (cooling time: 1 h), where the temperature was maintained for 10 min. Subsequently, the solution was cooled to 4 °C (cooling time: 2 h), where it was incubated overnight. Representative melting curves are shown in Figure S6. The melting points, found for **D** and **TEL25/Na⁺** are, respectively, 76 °C and 62 °C.

Oligonucleotide solutions were kept at −20 °C. Prior to time-resolved experiments, the sample (2 mL contained in a 1 cm × 1 cm QZ cell) was mildly stirred and its temperature was maintained at 23 ± 0.5 °C. We checked that the stirring did not artificially shorten the decays by cutting it off during each measurement. The absorbance on the excitation side was 0.25 ± 0.02 over 0.1 cm, corresponding to concentrations of approximately 1×10^{-5} mol L^{-1}, 5×10^{-6} mol L^{-1} and 1.2×10^{-5} mol L^{-1}, respectively for single, double and four-stranded systems. These values are at least one order of magnitude higher than the concentration of ejected hydrated electrons. At each wavelength, a series of three successive signals, resulting from 20–50 laser shots each, were recorded. If judged to be reproducible, they were averaged to reduce the signal-to-noise ratio.

5. Conclusions

The present study on **G** radicals, formed by direct photoionization of nucleic acids using low intensity laser pulses, brought new insights and raised new questions regarding radical generation and reactivity in nucleic acids. Below, we focus on a few points which deserve attention in respect to future developments.

In line with previous studies, it was found that G-quadruplexes exhibit a larger propensity than duplexes to photoeject an electron upon absorption of low energy photons. One hypothesis suggested previously is that electron ejection occurs after population of excited charge transfer states involving different bases, followed by charge separation [74]. According to such a scenario, the guanine core should behave as a deep trap for the positive charge, the negative charge remaining on an external base (adenine or thymine). This could explain our observation that, going from **TEL21/Na⁺** to **TEL25/Na⁺** by the addition of TA and TT steps at the two ends of the telomeric sequence, the quantum yield of one photon ionization at 266 nm increases from 4.5×10^{-3} to 5.2×10^{-3}. A systematic study of G-quadruplexes with carefully chosen flanking groups and loops could contribute to check the validity of the above mentioned mechanism.

Our methodology, allowing the determination of populations of the various types of radicals in respect to the ejected electrons, showed that the most important part of radical cations undergoes deprotonation. The lifetime of deprotonated radicals is independent of external conditions (phosphate buffer, oxygen, excitation intensity) but it does depend on the base sequence forming a given secondary structure. This behavior suggests that guanine radicals react internally, with other parts of the nucleic acid and/or water molecules participating in the local structure [63,75]. However, we found no indication in the literature about DNA lesions issued from internal reactions of **G** deprotonated radicals or for any reaction involving (G-H2)• radicals. Molecular modeling [76] will certainly help understanding such radical reactions.

For all types of radicals, the dynamics deviates from classical reaction kinetics describing monomolecular and bimolecular reactions that take place in homogenous three-dimensional environment. This deviation may also interfere in studies of **G** radicals formed indirectly by the mediation of an oxidant. When the oxidation step involves diffusion of the reactants, the long-time behavior of radicals may be blurred by delayed oxidation due to non-homogeneous, and, therefore, multiscale reactions. However, such indirect studies, if they are limited to early times may bring precious information. This is the case, for example, of the work by Su et al. [17], which managed to grasp important features of radical cations in G-quadruplexes.

Supplementary Materials: The following are available online, Figure S1: Dependence of the radical decays in **S1** on the excitation intensity, Figure S2: Post-irradiation steady-state differential spectra of **TEL25/Na⁺**, Figure S3: Normalized transient absorption spectra of **TEL25/Na⁺**, Figure S4: Dependence of the radical decays in **TEL25/Na⁺** on the excitation intensity, Figure S5: Steady-state absorption spectra of **D** and **TEL25/Na⁺**, Figure S6: melting curves of **D** and **TEL25/Na⁺**, Estimation of radical concentrations in reference [28].

Author Contributions: Conceptualization, D.M.; methodology, D.M. and A.B.; software, A.B.; validation, E.B., A.B., G.B. and D.M.; formal analysis, E.B. and D.M.; investigation, E.B. and A.B.; writing—original draft preparation, D.M.; review and editing, E.B., A.B., G.B. and D.M; supervision, D.M. and G.B.; project administration, D.M.; funding acquisition, D.M.

Funding: This work was supported by the European Program H2020 MSCA ITN [grant No. 765266 – LightDyNAmics project].

Acknowledgments: In this section you can acknowledge any support given which is not covered by the author contribution or funding sections. This may include administrative and technical support, or donations in kind (e.g., materials used for experiments).

Conflicts of Interest: The authors declare no conflicts of interest.

References

1. Cadet, J.; Davies, K.J.A. Oxidative DNA damage & repair: An introduction. *Free Radic. Biol. Med.* **2017**, *107*, 2–12. [PubMed]
2. Palecek, E.; Bartosik, M. Electrochemistry of Nucleic Acids. *Chem. Rev.* **2012**, *112*, 3427–3481. [CrossRef] [PubMed]
3. Lewis, F.D.; Letsinger, R.L.; Wasielewski, M.R. Dynamics of photoinduced charge transfer and hole transport in synthetic DNA hairpins. *Acc. Chem. Res.* **2001**, *34*, 159–170. [CrossRef] [PubMed]
4. Kawai, K.; Majima, T. Hole Transfer Kinetics of DNA. *Acc. Chem. Res.* **2013**, *46*, 2616–2625. [CrossRef] [PubMed]
5. Genereux, J.C.; Barton, J.K. Mechanisms for DNA Charge Transport. *Chem. Rev.* **2010**, *110*, 1642–1662. [CrossRef] [PubMed]
6. Kanvah, S.; Joseph, J.; Schuster, G.B.; Barnett, R.N.; Cleveland, C.L.; Landman, U. Oxidation of DNA: Damage to Nucleobases. *Acc. Chem. Res.* **2010**, *43*, 280–287. [CrossRef]
7. Giese, B.; Amaudrut, J.; Köhler, A.-K.; Spormann, M.; Wessely, S. Direct observation of hole transfer through DNA by hopping between adenine bases and by tunneling. *Nature* **2001**, *412*, 318–320. [CrossRef]
8. Cadet, J.; Douki, T.; Ravanat, J.L. Oxidatively generated damage to the guanine moiety of DNA: Mechanistic aspects and formation in cells. *Acc. Chem. Res.* **2008**, *41*, 1075–1083. [CrossRef]
9. Di Mascio, P.; Martinez, G.R.; Miyamoto, S.; Ronsein, G.E.; Medeiros, M.H.G.; Cadet, J. Singlet Molecular Oxygen Reactions with Nucleic Acids, Lipids, and Proteins. *Chem. Rev.* **2019**, *119*, 2043–2086. [CrossRef]
10. Candeias, L.P.; Steenken, S. Stucture and acid-base properties of one-electron-oxidized deoxyguanosine, guanosine, and 1-methylguanosine. *J. Am. Chem. Soc.* **1989**, *111*, 1094–1099. [CrossRef]
11. Kobayashi, K.; Tagawa, S. Direct observation of guanine radical cation deprotonation in duplex DNA using pulse radiolysis. *J. Am. Chem. Soc.* **2003**, *125*, 10213–10218. [CrossRef] [PubMed]
12. Adhikary, A.; Kumar, A.; Becker, D.; Sevilla, M.D. The guanine cation radical: Investigation of deprotonation states by ESR and DFT. *J. Phys. Chem. B* **2006**, *110*, 24171–24180. [CrossRef] [PubMed]
13. Kobayashi, K.; Yamagami, R.; Tagawa, S. Effect of base sequence and deprotonation of guanine cation radical in DNA. *J. Phys. Chem. B* **2008**, *112*, 10752–10757. [CrossRef] [PubMed]
14. Adhikary, A.; Khanduri, D.; Sevilla, M.D. Direct Observation of the Hole Protonation State and Hole Localization Site in DNA-Oligomers. *J. Am. Chem. Soc.* **2009**, *131*, 8614–8619. [CrossRef] [PubMed]
15. Chatgilialoglu, C.; Caminal, C.; Guerra, M.; Mulazzani, Q.G. Tautomers of one-electron-oxidized guanosine. *Angew. Chem. Int. Ed.* **2005**, *44*, 6030–6032. [CrossRef] [PubMed]
16. Chatgilialoglu, C.; Caminal, C.; Altieri, A.; Vougioukalakis, G.C.; Mulazzani, Q.G.; Gimisis, T.; Guerra, M. Tautomerism in the guanyl radical. *J. Am. Chem. Soc.* **2006**, *128*, 13796–13805. [CrossRef]
17. Wu, L.D.; Liu, K.H.; Jie, J.L.; Song, D.; Su, H.M. Direct Observation of Guanine Radical Cation Deprotonation in G-Quadruplex DNA. *J. Am. Chem. Soc.* **2015**, *137*, 259–266. [CrossRef]

18. Banyasz, A.; Martinez-Fernandez, L.; Balty, C.; Perron, M.; Douki, T.; Improta, R.; Markovitsi, D. Absorption of Low-Energy UV Radiation by Human Telomere G-Quadruplexes Generates Long-Lived Guanine Radical Cations. *J. Am. Chem. Soc.* **2017**, *139*, 10561–10568. [CrossRef]
19. Banyasz, A.; Balanikas, E.; Martinez-Fernadez, L.; Baldacchino, G.; Douki, T.; Improta, R.; Markovitsi, D. Radicals generated in guanine nanostructures by photo-ionization: Spectral and dynamical features. *J. Phys. Chem. B* **2019**, *123*, 4950–4957. [CrossRef]
20. Wala, M.; Bothe, E.; Görner, H.; Schulte-Frohlinde, D. Quantum yields for the generation of hydrated electrons and single strand breaks in poly(C), poly(A) and single-stranded DNA in aqueous solution on 20 ns laser excitation at 248 nm. *J. Photochem. Photobiol. A Chem.* **1990**, *53*, 87–108. [CrossRef]
21. Candeias, L.P.; Steenken, S. Ionization of purine nucleosides and nucleotides and their components by 193-nm laser photolysis in aqueous solution: Model studies for oxidative damage of DNA. *J. Am. Chem. Soc.* **1992**, *114*, 699–704. [CrossRef]
22. Candeias, L.P.; O'Neill, P.; Jones, G.D.D.; Steenken, S. Ionization of polynucleotides and DNA in aqueous solution by 193 nm pulsed laser light: Identification of base derived radicals. *Int. J. Radiat. Biol.* **1992**, *61*, 15–20. [CrossRef] [PubMed]
23. Kuimova, M.K.; Cowan, A.J.; Matousek, P.; Parker, A.W.; Sun, X.Z.; Towrie, M.; George, M.W. Monitoring the direct and indirect damage of DNA bases and polynucleotides by using time-resolved infrared spectroscopy. *Proc. Natl. Acad. Sci. USA* **2006**, *103*, 2150–2153. [CrossRef] [PubMed]
24. Banyasz, A.; Ketola, T.; Muñoz-Losa, A.; Rishi, S.; Adhikary, A.; Sevilla, M.D.; Martinez-Fernandez, L.; Improrta, R.; Markovitsi, D. UV-induced Adenine Radicals Induced in DNA A-tracts: Spectral and Dynamical Characterization. *J. Phys. Chem. Lett.* **2016**, *7*, 3949–3953. [CrossRef] [PubMed]
25. Banyasz, A.; Ketola, T.; Martinez-Fernandez, L.; Improta, R.; Markovitsi, D. Adenine radicals generated in alternating AT duplexes by direct absorption of low-energy UV radiation. *Faraday Discuss.* **2018**, *207*, 181–197. [CrossRef]
26. Banyasz, A.; Martinez-Fernandez, L.; Improta, R.; Ketola, T.M.; Balty, C.; Markovitsi, D. Radicals generated in alternating guanine-cytosine duplexes by direct absorption of low-energy UV radiation. *Phys. Chem. Chem. Phys.* **2018**, *20*, 21381–21389. [CrossRef] [PubMed]
27. Rokhlenko, Y.; Geacintov, N.E.; Shafirovich, V. Lifetimes and Reaction Pathways of Guanine Radical Cations and Neutral Guanine Radicals in an Oligonucleotide in Aqueous Solutions. *J. Am. Chem. Soc.* **2012**, *134*, 4955–4962. [CrossRef]
28. Rokhlenko, Y.; Cadet, J.; Geacintov, N.E.; Shafirovich, V. Mechanistic Aspects of Hydration of Guanine Radical Cations in DNA. *J. Am. Chem. Soc.* **2014**, *136*, 5956–5962. [CrossRef]
29. Merta, T.J.; Geacintov, N.E.; Shafirovich, V. Generation of 8-oxo-7,8-dihydroguanine in G-Quadruplexes Models of Human Telomere Sequences by One-electron Oxidation. *Photochem. Photobiol.* **2019**, *95*, 244–251. [CrossRef]
30. Latus, A.; Alam, M.S.; Mostafavi, M.; Marignier, J.L.; Maisonhaute, E. Guanosine radical reactivity explored by pulse radiolysis coupled with transient electrochemistry. *Chem. Commun.* **2015**, *51*, 9089–9092. [CrossRef]
31. Crespo-Hernandez, C.E.; Arce, R. Near threshhold photo-oxidation of dinucleotides containing purines upon 266 nm nanosecond laser excitation. The role of base stacking, conformation and sequence. *J. Phys. Chem. B* **2003**, *107*, 1062–1070. [CrossRef]
32. Gabelica, V.; Rosu, F.; Tabarin, T.; Kinet, C.; Antoine, R.; Broyer, M.; De Pauw, E.; Dugourd, P. Base-dependent electron photodetachment from negatively charged DNA strands upon 260-nm laser irradiation. *J. Am. Chem. Soc.* **2007**, *129*, 4706–4713. [CrossRef] [PubMed]
33. Marguet, S.; Markovitsi, D.; Talbot, F. One and two photon ionization of DNA single and double helices studied by laser flash photolysis at 266 nm. *J. Phys. Chem. B* **2006**, *110*, 11037–11039. [CrossRef] [PubMed]
34. Gomez-Mendoza, M.; Banyasz, A.; Douki, T.; Markovitsi, D.; Ravanat, J.L. Direct Oxidative Damage of Naked DNA Generated upon Absorption of UV Radiation by Nucleobases. *J. Phys. Chem. Lett.* **2016**, *7*, 3945–3948. [CrossRef] [PubMed]
35. Gauduel, Y.; Migus, A.; Chambaret, J.P.; Antonetti, A. Femtosecond Reactivity of Electron in Aqueous Solutions. *Rev. Phys. Appl.* **1987**, *22*, 1755–1759. [CrossRef]
36. Torche, F.; Marignier, J.L. Direct Evaluation of the Molar Absorption Coefficient of Hydrated Electron by the Isosbestic Point Method. *J. Phys. Chem. B* **2016**, *120*, 7201–7206. [CrossRef] [PubMed]

37. Ma, J.; Wang, F.; Denisov, S.A.; Adhikary, A.; Mostafavi, M. Reactivity of prehydrated electrons toward nucleobases and nucleotides in aqueous solution. *Sci. Adv.* **2017**, *3*, e1701669. [CrossRef]
38. Buxton, G.V.; Greenstock, C.L.; Helman, W.P.; Ross, A.B. Critical review of rate constants for reactions of hydrated electrons, hydrogen atoms and hydroxyl radicals (·OH/O⁻) in aqueous solution. *J. Phys. Chem. Ref. Data* **1988**, *17*, 513–886. [CrossRef]
39. Cadet, J.; Grand, A.; Douki, T. Solar UV radiation-induced DNA Bipyrimidine photoproducts: Formation and mechanistic insights. *Top. Curr. Chem.* **2015**, *356*, 249–275.
40. Douki, T.; Voituriez, L.; Cadet, J. Measurement of pyrimidine (6-4) photoproducts in DNA by a mild acidic hydrolysis-HPLC fluorescence detection assay. *Chem. Res. Toxicol.* **1995**, *8*, 244–253. [CrossRef]
41. Marguet, S.; Markovitsi, D. Time-resolved study of thymine dimer formation. *J. Am. Chem. Soc.* **2005**, *127*, 5780–5781. [CrossRef] [PubMed]
42. Porschke, D. Analysis of a Specific Photoreaction in Oligodoxyadenylic and Polydeoxyadenylic acids. *J. Am. Chem. Soc.* **1973**, *95*, 8440–8446. [CrossRef] [PubMed]
43. Bose, S.N.; Davies, R.J.H.; Sethi, S.K.; McCloskey, J.A. Formation of an adenine-thymine photoadduct in the deoxydinucleosides monophosphate d(TpA) and in DNA. *Science* **1983**, *220*, 723–725. [CrossRef] [PubMed]
44. Bose, S.N.; Kumar, S.; Davies, R.J.H.; Sethi, S.K.; McCloskey, J.A. The Photochemistry of d(T-A) in Aqueous Solution and Ice. *Nucleic Acids Res.* **1984**, *12*, 7929–7947. [CrossRef] [PubMed]
45. Kumar, S.; Sharma, N.D.; Davies, R.J.H.; Phillipson, D.W.; McCloskey, J.A. The isolation and and characterization of a new type of dimeric adenine photoproduct in UV-irradiated deoxyadenylates. *Nucleic Acids Res.* **1987**, *15*, 1199–1216. [CrossRef] [PubMed]
46. Kumar, S.; Joshi, P.C.; Sharma, N.D.; Bose, S.N.; Davies, R.J.H.; Takeda, N.; McCloskey, J.A. Adenine Photodimerization in Deoxyadenylate Sequences—Elucidation of the Mechanism through Structural Studies of a Major d(ApA) Photoproduct. *Nucleic Acids Res.* **1991**, *19*, 2841–2847. [CrossRef] [PubMed]
47. Zhao, X.D.; Nadji, S.; Kao, J.L.F.; Taylor, J.S. The structure of d(TpA)*, the major photoproduct of thymidylyl-(3'-5')-deoxyadenosine. *Nucleic Acids Res.* **1996**, *24*, 1554–1560. [CrossRef] [PubMed]
48. Banyasz, A.; Martinez-Fernandez, L.; Ketola, T.; Muñoz-Losa, A.; Esposito, L.; Markovitsi, D.; Improta, R. Excited State Pathways Leading to Formation of Adenine Dimers. *J. Phys. Chem. Lett.* **2016**, *7*, 2020–2023. [CrossRef]
49. Clingen, P.H.; Davies, R.J.H. Quantum yields of adenine photodimerization in poly(deoxyadenylic acid) and DNA. *J. Photochem. Photobiol. B Biol.* **1997**, *38*, 81–87. [CrossRef]
50. Douki, T. Effect of denaturation on the photochemistry of pyrimidine bases in isolated DNA. *J. Photochem. Photobiol. B* **2006**, *82*, 45–52. [CrossRef]
51. McCullagh, M.; Lewis, F.; Markovitsi, D.; Douki, T.; Schatz, G.C. Conformational control of TT dimerization in DNA conjugates. A molecular dynamics study. *J. Phys. Chem. B* **2010**, *114*, 5215–5221. [CrossRef] [PubMed]
52. Görner, H. Photochemistry of DNA and related biomolecules: Quantum yields and consequences of photoionization. *J. Photochem. Photobiol. B Biol.* **1994**, *26*, 117–139. [CrossRef]
53. Cadet, J.; Wagner, J.R.; Angelov, D. Biphotonic Ionization of DNA: From Model Studies to Cell. *Photochem. Photobiol.* **2019**, *95*, 59–72. [CrossRef] [PubMed]
54. Meggers, E.; Michel-Beyerle, M.E.; Giese, B. Sequence dependent long range hole transport in DNA. *J. Am. Chem. Soc.* **1998**, *120*, 12950–12955. [CrossRef]
55. Yoshioka, Y.; Kitagawa, Y.; Takano, Y.; Yamaguchi, K.; Nakamura, T.; Saito, I. Experimental and theoretical studies on the selectivity of GGG triplets toward one-electron oxidation in B-form DNA. *J. Am. Chem. Soc.* **1999**, *121*, 8712–8719. [CrossRef]
56. Barnett, R.N.; Cleveland, C.L.; Joy, A.; Landman, U.; Schuster, G.B. Charge migration in DNA: Ion-gated transport. *Science* **2001**, *294*, 567–571. [CrossRef]
57. Takada, T.; Kawai, K.; Fujitsuka, M.; Majima, T. Direct observation of hole transfer through double-helical DNA over 100 A. *Proc. Natl. Acad. Sci. USA* **2004**, *101*, 14002–14006. [CrossRef] [PubMed]
58. Renaud, N.; Harris, M.A.; Singh, A.P.N.; Berlin, Y.A.; Ratner, M.A.; Wasielewski, M.R.; Lewis, F.D.; Grozema, F.C. Deep-hole transfer leads to ultrafast charge migration in DNA hairpins. *Nat. Chem.* **2016**, *8*, 1015–1021. [CrossRef]
59. Stemp, E.D.A.; Arkin, M.R.; Barton, J.K. Oxidation of guanine in DNA by Ru(phen)$_2$(dppz)$^{3+}$ using the flash-quench technique. *J. Am. Chem. Soc.* **1997**, *119*, 2921–2925. [CrossRef]

60. Steenken, S. Purine-Bases, Nucleosides and Nucleotides—Aqueous-Solution Redox Chemistry and transformation Reactions of their Radical Cations and e- and OH Adducts. *Chem. Rev.* **1989**, *89*, 503–520. [CrossRef]
61. Kumar, A.; Sevilla, M.D. Excited States of One-Electron Oxidized Guanine-Cytosine Base Pair Radicals: A Time Dependent Density Functional Theory Study. *J. Phys. Chem. A* **2019**, *123*, 3098–3108. [CrossRef] [PubMed]
62. Onidas, D.; Markovitsi, D.; Marguet, S.; Sharonov, A.; Gustavsson, T. Fluorescence properties of DNA nucleosides and nucleotides: A refined steady-state and femtosecond investigation. *J. Phys. Chem. B* **2002**, *106*, 11367–11374. [CrossRef]
63. Laage, D.; Elsaesser, T.; Hynes, J.T. Water Dynamics in the Hydration Shells of Biomolecules. *Chem. Rev.* **2017**, *117*, 10694–10725. [CrossRef] [PubMed]
64. Borrego-Varillas, R.; Cerullo, G.; Markovitsi, D. Exciton Trapping Dynamics in DNA Multimers. *J. Phys. Chem. Lett.* **2019**, *10*, 1639–1643. [CrossRef] [PubMed]
65. Banyasz, A.; Gustavsson, T.; Onidas, D.; Changenet-Barret, P.; Markovitsi, D.; Importa, R. Multi-Pathway Excited State Relaxation of Adenine Oligomers in Aqueous Solution: A Joint Theoretical and Experimental Study. *Chem. Eur. J.* **2013**, *19*, 3762–3774. [CrossRef] [PubMed]
66. Markovitsi, D.; Talbot, F.; Gustavsson, T.; Onidas, D.; Lazzarotto, E.; Marguet, S. Complexity of excited state dynamics in DNA. *Nature* **2006**, *441*, E7. [CrossRef] [PubMed]
67. Blumen, A.; Klafter, J.; Zumofen, G. Models for reaction dynamics in glasses. In *Optical Spectroscopy of Glasses*; Zschokke, I., Ed.; Springer: Dordrecht, the Netherlands, 1986; pp. 199–265.
68. Markovitsi, D.; Germain, A.; Millie, P.; Lécuyer, I.; Gallos, L.; Argyrakis, P.; Bengs, H.; Ringsdorf, H. Triphenylene columnar liquid crystals: Excited states and energy transfer. *J. Phys. Chem.* **1995**, *99*, 1005–1017. [CrossRef]
69. Emelianova, E.V.; Athanasopoulos, S.; Silbey, R.J.; Beljonne, D. 2D Excitons as Primary Energy Carriers in Organic Crystals: The Case of Oligoacenes. *Phys. Rev. Lett.* **2010**, *104*, 206405. [CrossRef]
70. Benichou, O.; Chevalier, C.; Klafter, J.; Meyer, B.; Voituriez, R. Geometry-controlled kinetics. *Nat. Chem.* **2010**, *2*, 472–477. [CrossRef]
71. Dolgushev, M.; Guerin, T.; Blumen, A.; Benichou, O.; Voituriez, R. Contact Kinetics in Fractal Macromolecules. *Phys. Rev. Lett.* **2015**, *115*, 208301. [CrossRef]
72. Candeias, L.P.; Steenken, S. Electron transfer in di(deoxy)nucleoside phosphates in aqueous solution: Rapid migration of oxidative damage (via adenine) to guanine. *J. Am. Chem. Soc.* **1993**, *115*, 2437–2440. [CrossRef]
73. Amand, B.; Bensasson, R. Determination of triplet quantum yields by laser flash absorption spectroscopy. *Chem. Phys. Lett.* **1975**, *34*, 44–48. [CrossRef]
74. Bhat, V.; Cogdell, R.; Crespo-Hernández, C.E.; Datta, A.; De, A.; Haacke, S.; Helliwell, J.; Improta, R.; Joseph, J.; Karsili, T.; et al. Photocrosslinking between nucleic acids and proteins: General discussion. *Faraday Discuss.* **2018**, *207*, 283–306. [CrossRef] [PubMed]
75. Gervasio, F.L.; Laio, A.; Iannuzzi, M.; Parrinello, M. Influence of DNA structure on the reactivity of the guanine radical cation. *Chem. Eur. J.* **2004**, *10*, 4846–4852. [CrossRef] [PubMed]
76. Dumont, E.; Monari, A. Understanding DNA under oxidative stress and sensitization: The role of molecular modeling. *Front. Chem.* **2015**, *3*, 43. [CrossRef]

Sample Availability: Samples not available.

© 2019 by the authors. Licensee MDPI, Basel, Switzerland. This article is an open access article distributed under the terms and conditions of the Creative Commons Attribution (CC BY) license (http://creativecommons.org/licenses/by/4.0/).

Review

The Dynamics of Hole Transfer in DNA

Andrea Peluso, Tonino Caruso, Alessandro Landi and Amedeo Capobianco *

Dipartimento di Chimica e Biologia "A. Zambelli", Università di Salerno, via Giovanni Paolo II, 132, I-84084 Fisciano (SA), Italy; apeluso@unisa.it (A.P.); tcaruso@unisa.it (T.C.); alelandi1@unisa.it (A.L.)
* Correspondence: acapobianco@unisa.it

Received: 28 September 2019; Accepted: 2 November 2019; Published: 7 November 2019

Abstract: High-energy radiation and oxidizing agents can ionize DNA. One electron oxidation gives rise to a radical cation whose charge (hole) can migrate through DNA covering several hundreds of Å, eventually leading to irreversible oxidative damage and consequent disease. Understanding the thermodynamic, kinetic and chemical aspects of the hole transport in DNA is important not only for its biological consequences, but also for assessing the properties of DNA in redox sensing or labeling. Furthermore, due to hole migration, DNA could potentially play an important role in nanoelectronics, by acting as both a template and active component. Herein, we review our work on the dynamics of hole transfer in DNA carried out in the last decade. After retrieving the thermodynamic parameters needed to address the dynamics of hole transfer by voltammetric and spectroscopic experiments and quantum chemical computations, we develop a theoretical methodology which allows for a faithful interpretation of the kinetics of the hole transport in DNA and is also capable of taking into account sequence-specific effects.

Keywords: DNA oxidation; DNA hole transfer; DNA; quantum dynamics; Electron transfer; charge transfer

1. Introduction

Eley and Spivey first envisioned DNA as a possible conduit for conveying electrical charges, via the π system of stacked nucleobases [1]. However, long-range charge transport in DNA was discovered only in the 1990s, by Barton and coworkers [2].

Since then, a large body of experimental evidence has accumulated, showing that one electron oxidation on a DNA donor site (D) produces a hole that can migrate through the double helix, covering long distances (up to several hundreds of Å) until an irreversible oxidative damage takes place at an acceptor site (A) (see Figure 1) [3–5].

Living cells are continuously exposed to endogenously generated as well as external agents that can oxidize DNA [6]. That may result in the corruption of genetic information with potentially serious consequences, including mutagenesis and cancer [7].

Aside from its enormous biologic relevance, long-range hole transport in DNA has attracted much interest because: (*i*) it is useful for detecting structural changes in DNA resulting from alterations of the regular π-π stacking [8,9]; and (*ii*) it enables a potential use of DNA as a dielectric material in field-effect transistors and organic light-emitting diodes, hopefully leading to biosustainable devices [10–15].

The stability and the conformational flexibility of double stranded DNA largely result from the interaction of water and counterions with the charged sugar-phosphate backbone [16]. Dielectric effects strongly affect the oxidation of nucleobases inside DNA, even if the nucleobases experience a strongly hydrophobic environment as opposite to the charged phosphate backbone. Indeed, the oxidation potentials of oligonucleotides and short B-DNA sequences inferred from voltammetric measurements were found to be strongly dependent on pH (due to possible pH mediated proton transfer) and counterion concentration [17–23]. Moreover, molecular dynamics simulations and quantum chemical computations revealed that structural fluctuations causing the redistribution of Na$^+$ counterions and their associated water molecules strongly affect the energy of the HOMO levels of the nucleobase units inside DNA [24].

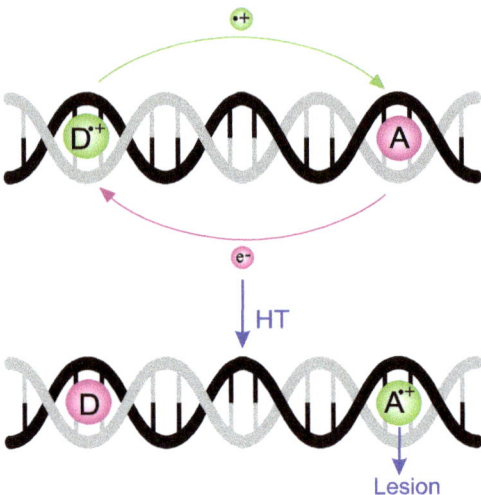

Figure 1. Schematic representation of the hole transport in DNA. The first ionization takes place at the donor site (D), where a hole is generated. The hole migrates through the double helix of DNA, reaching the acceptor site (A), where the final irreversible oxidative damage (lesion) occurs.

Aside from B-DNA, a very efficient hole transport has been observed also in guanine quadruplex stacked sequences adsorbed on a mica substrates [25]. Indeed, in recent years, growing attention has been given to the oxidation of G-quadruplexes, for they occupy telomeric regions of chromosomes, often found in oncogene promoter sequences [26].

HT is known to be strongly dependent also on the specific sequence of nucleosides and on DNA conformation. In view of all previous considerations, it is clear that addressing the dynamics of hole transport (HT) from a theoretical viewpoint constitutes a very difficult task. Several approaches have been used to study the kinetics of hole transfer in double stranded B-DNA and different conclusions about its underlying mechanism have been reached so far [27–40]. Nevertheless, a few firm points have been established. The observed products of the one-electron oxidation are rather insensitive to the process by which DNA is oxidized [4]. For DNA sequences containing guanine (G), the final radical cation usually localizes at G sites [6,7,41], because guanine is the most readily oxidized natural occurring nucleobase. Runs of two or more adjacent Gs are often used as thermodynamic traps to localize the hole, which is then detected as an alkali-labile lesion, because DNA steps composed of consecutive guanines experience a further lowering of the oxidation potential [42–47]. Furthermore, it is now well assessed that steps composed of adjacent adenine (A) nucleobases greatly facilitate the hole transport [18,21,48–50], while consecutive stacked thymines (T) and cytosines (C) act as barrier sites, strongly attenuating hole transfer efficiency, due to their higher ionization energy [4].

Herein, we review the work carried out by our research group in the last decade on the dynamics of hole transfer in DNA, focusing particularly on electrochemical measurements and showing how their outcomes have led to building up a quite general kinetics model for treating hole transfer in DNA and other molecular wires.

2. Hole Site Energy

The dynamics of HT in DNA is modulated by two quantities: (*i*) the hole energies of the nucleobases, the actual redox sites in DNA [51]; and (*ii*) the electronic couplings between adjacent sites. Both the observed oxidation free energies of nucleobases, nucleosides and oligonucleotides in aqueous environment and the hole trapping efficiencies of DNA sequences originate from those quantities.

Although the importance of dielectric effects has been fully recognized, until recently, the majority of studies concerning long range hole transfer in DNA employed hole site energies inferred from the gas phase [29,32–34,52–55]. That is far from being satisfactory because the actual ionization energy of a nucleobase in hydrated DNA is strongly affected by dielectric effects, hydrogen bonding of complementary bases, and, above all, intrastrand π–π stacking interactions [18,56,57].

The redox properties of nucleobases, nucleosides, (oligo)nucleotides and DNA sequences have been deeply investigated by voltammetric techniques [17,58,59]. Although cyclic voltammetry measurements show the presence of collateral reactions leading to irreversible processes [60,61], the hole site energy spacing inferred from electrochemistry closely matches the one obtained by liquid-jet photoelectron spectroscopy (PES) not suffering from the above problem [56].

A selection of hole site energies of DNA constituents obtained by different techniques is reported in Table 1.

Table 1. Oxidation free energies (ΔG^{ox}) and adiabatic ionization energies (*I*) of DNA constituents relative to guanine or its derivatives. N, nucleobases; Ns, nucleosides; Nt, nucleotides. All data are expressed in eV.

	$I(N)$ [a]	$\Delta G^{ox}(Ns)$ [b]	$\Delta G^{ox}(N)$ [c]	$\Delta G^{ox}(Nt)$ [c]	$\Delta G^{ox}(Ns)$ [d]	$\Delta G^{ox}(Ns)$ [e]	$I_g(N)$ [f]
A	+0.41	+0.5	+0.27	+0.30	+0.47	+0.15	+0.49
C	+0.74	+0.8	+0.61	+0.57	+0.65	+0.38	+0.91
T	+0.75	+0.8	+0.45	+0.52	+0.62	+0.31	+1.10

[a] Ref. [62], density functional theory (DFT) computations including aqueous environment, via'the polarizable continuum model (PCM) [63]. [b] Ref. [64], PES measurements in water integrated with ab initio computations. [c] Ref. [17], voltammetry, water pH 7. [d] Ref. [58], voltammetry, acetonitrile. [e] Ref. [65], nanosecond spectroscopy, data adjusted as in Ref. [64]. [f] Ref. [66], gas phase photoionization mass-spectrometry.

With the exception of spectroscopic measurements (sixth column), whose reliability for pyrimidine derivatives is quite modest [64,65], a substantially good agreement is found for the data of Table 1 referring to solvated environment. PES, voltammetry, and quantum chemical predictions find guanine as the most easily oxidizable nucleobase; the hole energy of adenine is ≈0.4 eV higher than guanine, while pyrimidine derivatives are oxidized at a potential higher by 0.6–0.8 eV than G.

A graphical comparison of the hole energies for the aqueous environment (first column of Table 1) with those referred to gas phase (last column of Table 1) is presented in Figure 2.

Solvation acts by somewhat leveling the hole energies of DNA constituents. While the hole energy of adenine (relative to guanine) is scarcely affected by solvation, cytosine, and, above all, thymine become comparably easier to ionize in solution. Indeed, oxidative damages are often observed at T sites in oligonucleotides which lack G [67–70].

The data in Table 1 do not include the effects of the H-bonded complementary base on ionization energies of nucleobases. The lowering of the oxidation potential of G due to the base pairing with C had been predicted by theoretical computations [71], and experimentally estimated by the increase of the oxidation rate of guanosine (Guo) upon cytidine (Cyd) pairing [57,72]. However, a direct measurement of the above quantity was not available until 2005, when an electrochemical study carried out in our laboratories settled the question [73].

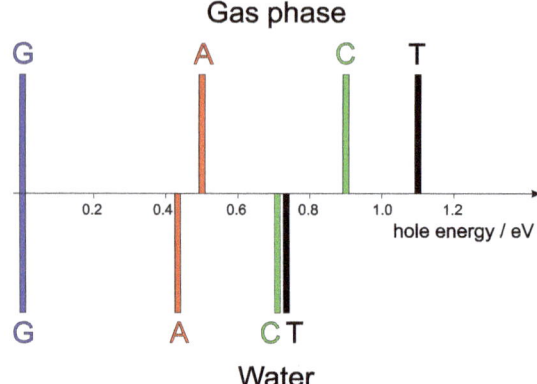

Figure 2. Relative (to guanine) hole energies of DNA nucleobases in the gas phase (top) and in aqueous environment (bottom).

Guanosine and deoxycytidine (dCyd) were properly functionalized to make them soluble in chloroform, a solvent in which the association constant for the formation of the Watson–Crick Guo:dCyd H-bonded complex is sufficiently high to permit its detection [74,75]. Then, voltammetric measurements of solutions containing Guo, dCyd, and their mixtures were carried out. The same procedure was later adopted for the adenosine (Ado) deoxythymidine (dThd) pair [76]. The main results of our investigations are summarized in Figure 3. The differential pulse voltammogram of the solution containing an equimolar amount of Guo and dCyd (Figure 3a) shows two well-resolved peaks, one occurring at the same potential observed for solutions containing only the Guo nucleoside, which can therefore be assigned to the fraction of free Guo in solution, and the other occurring at a potential lower by 0.34 V is assigned to the Guo:dThd Watson–Crick complex. Ado:dThd voltammograms exhibit a similar behavior and identical conclusions were inferred for the Ado:dThd hydrogen bond complex (Figure 3b).

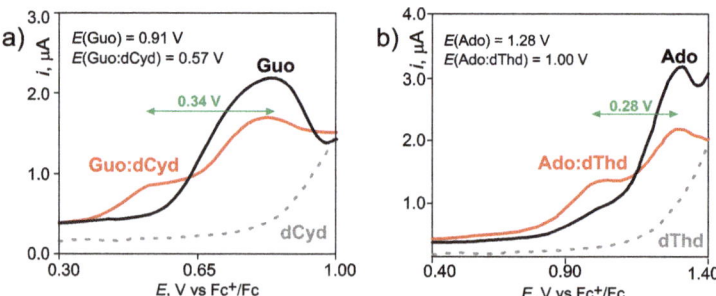

Figure 3. Differential pulse voltammograms of nucleoside derivatives in CHCl$_3$ at 298 K on glassy carbon electrode. (**a**) Solutions containing only Guo 2.0 mM (black line); only dCyd 2.0 mM (dashed gray line); and Guo 2.0 mM and dCyd 2.0 mM (red line). (**b**) Solutions containing only Ado 2.0 mM (black line); only dThd 2.0 mM (dashed gray line); and Ado 2.0 mM and dThd 20.0 mM (red line). Scan rate, 100 mV/s. Supporting electrolyte Bu$_4$NClO$_4$. Oxidation potentials are referred to the Ferrocenium/Ferrocene (Fc$^+$/Fc) redox couple. Green arrows indicate the lowering of the oxidation potential of purine nucleosides upon pairing via H-bond with their complementary pyrimidine nucleosides. Adapted with permission from *J. Am. Chem. Soc.* **2005**, *127*, 15040–15041 and *J. Am. Chem. Soc.* **2007**, *129*, 15347–15353. Copyright (2005, 2007) American Chemical Society.

H-bond association with complementary nucleosides lowers the oxidation free energy of Guo and Ado by ca. 0.3 eV, because ionization causes a substantial increase of the binding energy in the oxidized Watson–Crick complex with respect to its neutral counterpart [77].

The voltammograms in Figure 3 show no anodic signal for pyrimidine nucleosides in chloroform. Indeed, oxidizing pyrimidine derivatives in solution is a very difficult experimental task [58,64,78]. Nevertheless, the first estimate of the energy of a low lying excited state of DNA with the hole localized on cytidine in the Watson–Crick complex with guanosine was inferred by spectroelectrochemistry measurements [79].

NIR spectra of solutions containing Guo:dCyd mixtures were recorded in an electrochemical cell equipped with an optically transparent thin-layer electrode kept at +0.57 V versus Fc^+/Fc in $CHCl_3$ and CH_2Cl_2. At that potential, solutions containing only Guo or only dCyd are not oxidized, whereas solutions containing both species exhibit a well-resolved anodic peak (Figure 3). A positive broad band (Figure 4), not observed during the oxidation of solutions containing only Guo or only dCyd, was recorded in the difference spectrum, at approximately 10,600 cm^{-1} in CH_2Cl_2 and at 10,200 cm^{-1} in $CHCl_3$. Upon replacing cytidine with 5-methylcytidine, whose ionization energy is expected to be lower than that of cytidine by ca. 1400 cm^{-1} [66,79], that band was red shifted to 9100 cm^{-1} in CH_2Cl_2 and to 8700 cm^{-1} in $CHCl_3$. On the basis of the above evidence and with the support of time dependent DFT (TDDFT) computations, that signal was assigned to the charge-transfer (CT) localizing the hole on cytidine (Figure 4).

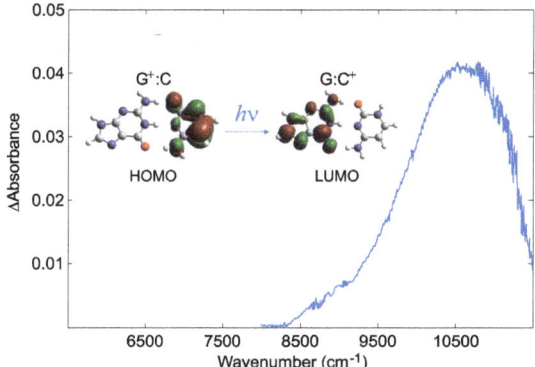

Figure 4. The charge transfer band of the [Guo:dCyd]$^+$ complex recorded in dichloromethane at a controlled potential of +0.57 V vs. Fc^+/Fc. It corresponds to the transition from the HOMO, a Kohn–Sham π orbital localized on cytosine, to the LUMO, a π^* orbital of the guanine moiety. Adapted with permission from *Angew. Chem. Int. Ed.* **2009**, *48*, 9526–9528. Copyright (2009) Wiley-VCH Verlag.

3. Electronic Couplings

Stacking interactions are by far the most important inter-base interactions for hole transfer because they provide the electronic couplings for long-range hole transfer. The effect of stacking interactions can be addressed by a simple two state quantum model, according to which the hole energy levels of two stacked nucleobases (X and Y) are shifted up and down with respect to those of unstacked ones by a quantity related to the difference between the hole energies of the two nucleobases ε_X, ε_Y, and to the strength of the stacking interactions, J_{XY}.

In the case of identical nucleobases X = Y (Figure 5, left), the hole energy shift upon pairing of nucleobases is just the electronic coupling element. Instead (Figure 5, right), if $\varepsilon_X \gg \varepsilon_Y$, and $\varepsilon_X - \varepsilon_Y \gg 2J_{XY}$, then the J_{XY} coupling term is not effective in lowering the hole energy of the stack, so that the lowest eigenvalue of the two state model Hamiltonian, $\mathcal{H}$, is nearly coincident with ε_Y.

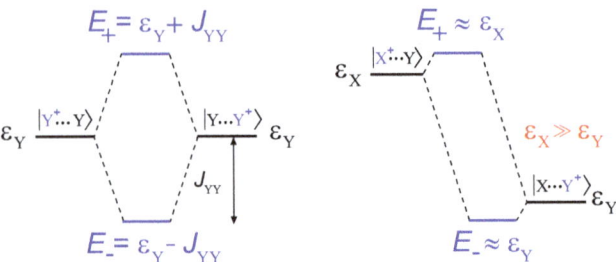

Figure 5. The two limiting cases predicted by the two state model for the ionization energy of two stacked nucleobases. (**Left**) The remotion of an electron from two identical unstacked Y nucleobases pair gives rise to two diabatic states, $|Y^+ \cdots Y\rangle$ and $|Y \cdots Y^+\rangle$ with hole energy ε_Y; upon formation of a π stacking interaction, the two states are coupled each other, and the energy levels of the electron hole (E_+, E_-) are shifted up and down with respect to those of unstacked pair by the quantity J_{YY}, representing the strength of the π stacking interaction. (**Right**) If the hole energy of X is far larger than that of Y, then the J_{XY} coupling term is not effective in lowering the ionization energy of the stack, so that the lowest eigenvalue of $\mathcal{H}$, E_-, is nearly coincident with ε_Y, and the highest eigenvalue of $\mathcal{H}$, E_+, is nearly coincident with ε_X.

Provided that two identical Y nucleobases assume a regular conformation, i.e. they are efficiently stacked, the J_{YY} coupling term can be estimated as the lowering of the ionization energy of the YY stacked sequence, with respect to that of a strand containing only one Y oxidizable nucleobases, on the assumption that all other nucleobases have higher hole site energies.

The reliability of the two state model has been verified by voltammetric experiments [47,80]. Figure 6 reports the differential pulse voltammograms recorded in water for the 5′-ACCCCA-3′ and 5′-AACCAA-3′ single stranded DNA oligonucleotides. A lowering of the oxidation potential amounting to 0.31 V is observed for the sequence containing two consecutive adenines. Measurements carried out for sequences containing an increasing number of adjacent adenines end capped by thymine nucleobases confirmed that result; anodic peaks for the first oxidation were detected at 0.97, 0.90 and 0.82 V vs. Ag/AgCl for 5′-TTAATT-3′, 5′-TTAAAT-3′, and 5′-TAAAAT-3′, single strands, respectively [80].

If oligonucleotides possessing adjacent adenines assume conformations in which nucleobases are well stacked altogether, as is indeed the case for A-rich tracts [80–86], which are known to confer structural rigidity to DNA [87,88], the two state model holds. Therefore, disregarding the coupling between A and C as a first approximation, J_{AA} was estimated to amount to $\approx$0.3 eV by the voltammograms of Figure 6 [80].

The observed progressive lowering of the oxidation potential upon increasing the number of consecutive stacked adenines is well predicted by PCM/DFT calculations carried out for the same single stranded DNA sequences used in experiments. Indeed, the computed ionization potential shifts are in very good agreement with the observed oxidation potentials (see Figure 7). Furthermore, the analysis of spin distributions (Figure 7) indicates that the observed oxidation potential shifts can be assigned to orbital mixing effects among stacked nucleobases, which lead to the formation of delocalized polarons [85]. Notably, there has been a vivid debate about the role of delocalized polarons in the charge transport in DNA [18,21,39,89–101]. Aside from voltammetric evidence [80], the formation of the $A_n^{\bullet+}$ polaron has been later on observed also by time dependent spectroscopy measurements carried out for oxidized DNA hairpins possessing two or more intervening A-T steps [50].

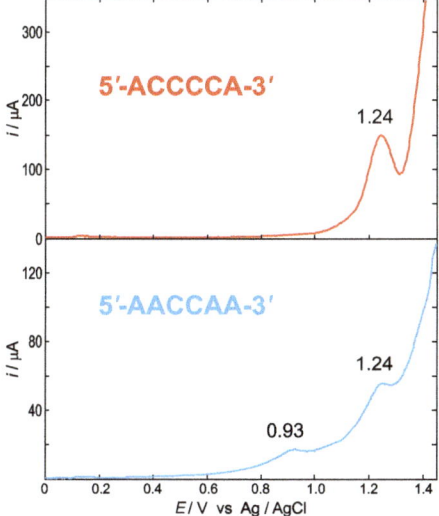

Figure 6. Differential pulse voltammetry of 2.0 mM single stranded 5′-ACCCCA-3′ (top, red) and 0.50 mM 5′-AACCAA-3′ (bottom, blue) in 50 mM phosphate buffer solution. Internal reference electrode Ag/AgCl (3.0 M KCl). A lowering of the first anodic signal amounting to 0.31 V is observed in passing from the sequence not containing stacked adenines to the one holding two stacked A's. Adapted with permission from *J. Phys. Chem. B* **2013**, *117*, 8947–8953. Copyright (2013) American Chemical Society.

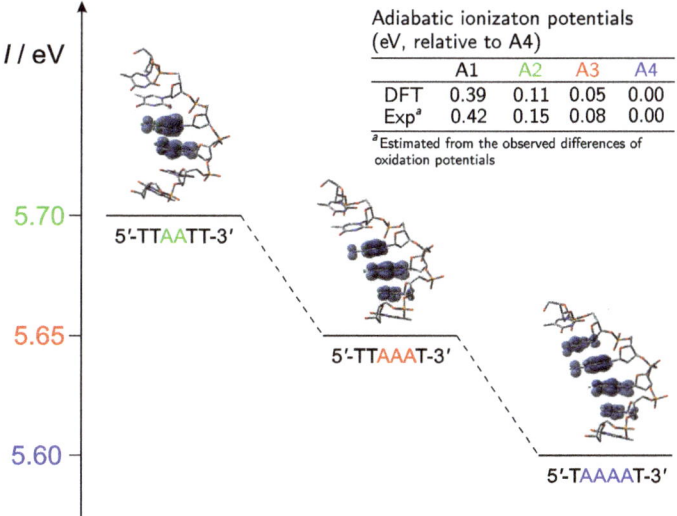

Figure 7. Predicted adiabatic ionization potential and spin densities for the 5′-TTAATT-3′ (A2), 5′-TTAAAT-3′ (A3), and 5′-TAAAAT-3′ (A4) single stranded ionized oligonucleotides. Inset: Comparison between relative computed ionization energies and observed oxidation free energies. The lowering of the oxidation potential observed for sequences lacking G upon increasing the number of stacked adenines has been rationalized in terms of resonance effects: the increasing stability of the hole is due to its delocalization over the entire adenine bridge. Adapted from Ref. [85] with permission from the PCCP Owner Societies.

A progressing lowering of the oxidation potential was also observed in single and double stranded oligonucleotides possessing an increasing number of consecutive guanines, showing that hole site energies can be lowered up to ≈0.3 eV for sites composed of six consecutive guanines [47]. The observed shift of the oxidation potential leads to $J_{GG} \approx 0.1$ V, an electron coupling significantly lower than that observed for adenine. That result stems from the lower extent of hole delocalization on G steps. Theoretical studies have concluded that the positive charge is almost entirely localized on ionized 5'-G in GG steps, possibly due to strong heteroatom-π electrostatic interactions [33,44,100,102–108]. Indeed, for DNA sequences containing up to two consecutive guanines, the oxidative damage is not equally distributed over G sites, but preferentially occurs at the 5' G of the GG step [42,102,109]. However, DNA cleavage efficiency does not depend only on hole trapping efficiencies, but it also relies on the kinetic mechanism by which DNA damage occurs, therefore no unambiguous conclusion can be drawn for the preferred cleavage site based only on the preferred ionization site [67,68,106,110].

The extension of the set of electronic couplings to pyrimidine nucleobases is an experimentally very challenging task because of the well-known difficulties of oxidizing pyrimidine derivatives in solution, especially by voltammetric techniques [73,76,111]. Therefore, to retrieve the whole set of coupling parameters also including pyrimidine derivatives, we resorted to PCM/DFT computations, which for purine bases had proved to provide faithful descriptions of the available experimental data [85].

To evaluate the effects of π–π stacking on ionization energies separately from hydrogen bonding, we had to resort to single stranded DNA sequences. In detail, we considered the two sets of tetrameric single stranded sequences 5'-XXYX-3' and 5'-XYZX-3', in which Y and Z are natural occurring DNA nucleobases and X is a nucleoside analogue possessing an ionization energy much higher than DNA nucleobases. Single stranded tetramers are the simplest sequences in which both nucleobases of the YZ tract occupy an internal position inside the strand. That permits minimizing inconsistencies arising from different exposure to the solvent. Figure 5 (right) shows that any coupling of X with DNA nucleobases is ineffective on the oxidation potential of nucleobases. In that case, according to the two state model, the in situ hole site energy of a DNA nucleobase nearly coincides with the ionization energy of 5'-XXYX-3', whereas J_{YZ} intrastrand coupling terms are given by:

$$J_{YZ} = \sqrt{\Delta I \left(\Delta I - \Delta \varepsilon \right)}, \tag{1}$$

where:

$$\Delta \varepsilon = \varepsilon_Y - \varepsilon_Z \approx I_{XXYX} - I_{XXZX}, \tag{2}$$

$$\Delta I = E_{YZ}^- - \varepsilon_Z \approx I_{XYZX} - I_{XXZX}, \tag{3}$$

in which I denotes ionization potential. The assumptions underlying Equations (1)–(3) are illustrated in Figure 8 [62].

Hole site energies and coupling terms inferred by Equations (1)–(3), with X = 6-azauracil (a nucleobase analogue with very high ionization energy due to the electron withdrawing effect of nitrogen [112], employed as a growth inhibitor of microorganisms which is known to incorporate into nucleic acids [113–115]), are reported in Table 2.

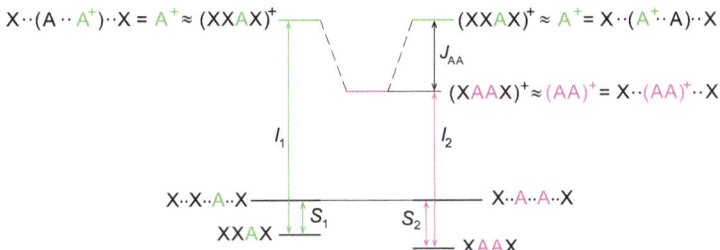

Figure 8. Adenine embedded in the 5'-XXAX-3' and 5'-XAAX-3' tetrameric single strands, where X is a nucleobase with an ionization energy much larger than that of A. Dotted lines denote non-interacting nucleobases, S denote stabilization energies due to stacking interactions in neutral strands and I are ionization potentials. Because the ionization energy of X is much larger than that of A, the effect of coupling terms is negligible in (XXAX)$^+$, so that the hole is fully localized on A. Since A in XXAX and AA in XAAX experience almost identical environments, the difference of the stacking energy in XXAX and XAAX neutral strands is expected to be small ($S_1 - S_2 \approx 0$); therefore, the difference of hole energies of XXAX and XAAX oligonucleotides is well approximated by the difference of ionization energies, $J_{AA} \approx I_1 - I_2$. Reproduced from Ref. [62] with permission from the PCCP Owner Societies.

Table 2. Hole site energies (ε_Y, eV, relative to G$^+$) and electronic coupling parameters for stacked 5'-YZ-3' base pairs (J_{YZ}, eV). Y and Z denote native DNA nucleobases. All values are taken from Ref. [62].

Y	ε_Y	J_{YG}	J_{YA}	J_{YC}	J_{YT}
G	0.00	0.09	0.15	0.23	0.14
A	0.43	0.15	0.24	0.16	0.08
C	0.68	0.23	0.16	0.12	0.12
T	0.70	0.14	0.08	0.12	0.12

Hole transport properties, oxidation potentials, trapping efficiencies and several other properties of oxidized DNA can be addressed by the tight binding (TB) Hamiltonian commonly used for the dynamics of charge transport [34,54,116–123]:

$$\mathcal{H} = \varepsilon_L |L\rangle\langle L| + \sum_{n=1}^{L-1} \varepsilon_n |n\rangle\langle n| + \left(J_{n,n+1} |n\rangle\langle n+1| + \text{H.c.} \right). \qquad (4)$$

In $\mathcal{H}$, only the interactions between nearest neighbor sites are considered. L is the number of nucleobases, $|n\rangle$ constitutes a set of orthogonal diabatic states with the charge fully localized on the nth nucleobase, ε_n is the hole energy of $|n\rangle$ assumed to be independent of nucleobase sequence, and $J_{n,m}$ are the electronic coupling elements between $|n\rangle$ and $|m\rangle$, i.e., the interaction energies due to π–π stacking. Hole site energies and electronic couplings are the parameters to be employed in the model Hamiltonian.

In Figure 9, the ionization energies of several tetrameric single stranded DNA sequences predicted by DFT computations are compared with those obtained by diagonalizing the TB Hamiltonian of Equation (4) using the parameters inferred by Equations (1)–(3), (Table 2). Predictions by the TB Hamiltonian are in excellent agreement with the outcomes of DFT computations and also with experimental evidence. Ionization energies of TTAAAT and TTAATT single strands are found by TB to be higher by +0.05 and +0.13 eV, respectively, than that of TAAAAT, to be compared with the corresponding oxidation potentials inferred by voltammetric measurements: +0.08 and +0.15 V (see Figure 7).

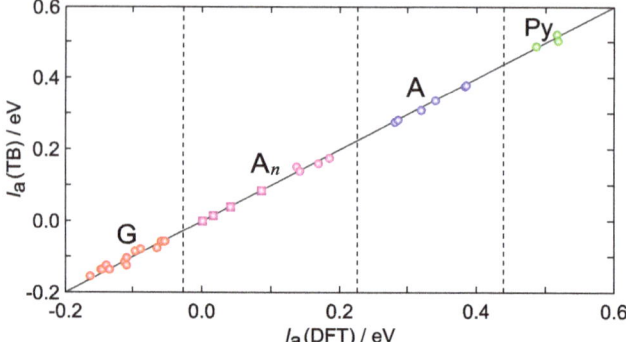

Figure 9. Comparison of adiabatic ionization energies (relative to XXGX, X being 6-azauracil) predicted by DFT (abscissa) and TB (ordinate) computations for tetrameric single stranded DNA oligonucleotides (circles). Hole energies for TA$_n$T (n = 3–6) sequences (squares) have been computed only at the TB level. Full line has null intercept and unitary slope. Dashed lines separate the regions of potential corresponding to the oxidation of G (red), A$_n$ tracts (violet), A (blue) and pyrimidines (green). Very similar oxidation patterns were found by differential pulse voltammograms recorded in buffered aqueous solutions for single and double stranded DNA oligonucleotides and DNA itself (see Refs. [47,59,124,125]). Reproduced from Ref. [62] with permission from the PCCP Owner Societies.

4. The Dynamics of Hole Transfer in DNA

The kinetics of HT in DNA is known to exhibit two regimes: for shorter distance between the hole donor and the hole acceptor nucleobases the rate exponentially depends on distance, whereas a much weaker distance dependence is observed for longer donor–acceptor distances [39,45]. That behavior is exemplified by Giese experiments on double-stranded 3'-G(T)$_n$GGG-5' oligonucleotides, in which a hole is injected onto the single G site via photoexcitation of a suitably modified nucleotide, and the yields of oxidation products formed at the initial site (P_G) and at the trap site (P_{GGG}) are measured [45]. Giese and coworkers observed that for shorter sequences, up to $n = 3$, the product ratio P_{GGG}/P_G drops by a factor of ca. 8 for each additional adenine:thymine (A:T) step. In longer sequences, for $n = 4–7$, the P_{GGG}/P_G ratio exhibits a much weaker distance dependence, whereas, for $n = 7–16$, no substantial change in P_{GGG}/P_G was detected. Experimental results were interpreted by admitting a switch of HT mechanism from coherent superexchange in the short range regime to a thermally induced multistep or multirange hopping for the long range regime [28,36,39,122,126,127]. Nevertheless, theoretical models aimed at describing both the long- and short-range regimes in the framework of a single mechanism have also been proposed [122,128].

It is worth noting that the weak dependence of HT in the long range regime is consistent with different underlying mechanisms based on very different underlying physics [129]. Therefore, the detection of the mechanism for HT in DNA cannot rely solely on experimental observations. Theoretical modeling and simulations thus appear to be essential tools for solving such a high complex problem.

Following to some extent the recent work of Parson [130–133], we recently presented a novel methodology especially suited for addressing the dynamics of charge transport in molecular wires [134], which has been applied to the DNA oligomers investigated by Giese, whose work is particularly appealing, for it provides the yield ratios of the water trapping products of hole transfer for a large number of DNA oligomers [45].

Our approach relies on the multi-step kinetic model illustrated in Figure 10, in which D$^+$(Bridge)A and D(Bridge)A$^+$ denote the initial state with the charge localized on the donor and the final CT state, respectively; [D$^+$(Bridge)A]* and [D(Bridge)A$^+$]* denote the ensembles of structures in which the hole donor and acceptor are in vibronic resonance with each other; and P_D and P_A denote the products of oxidative damage occurring at donor and acceptor sites, respectively. The HT mechanism starts with an activation step, which brings the donor and the acceptor groups into electronic degeneracy (Step 1); Step 2 represents the elementary electron transfer between resonant donor and acceptor groups, followed by relaxation of all the non equilibrium species to their minimum energy structures (including solvent) and formation of hole transfer products (Step 3).

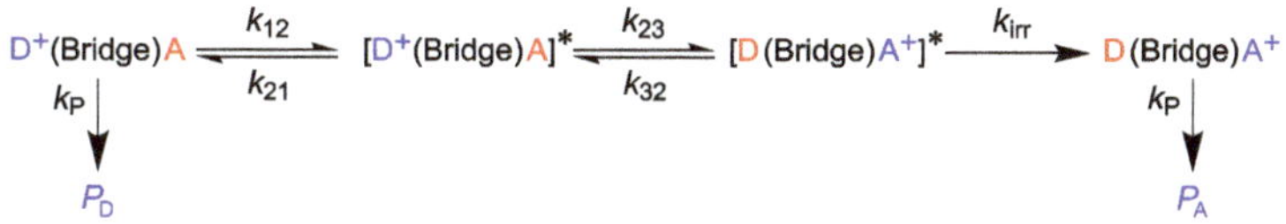

$$D^+(Bridge)A \underset{k_{21}}{\overset{k_{12}}{\rightleftharpoons}} [D^+(Bridge)A]^* \underset{k_{32}}{\overset{k_{23}}{\rightleftharpoons}} [D(Bridge)A^+]^* \xrightarrow{k_{irr}} D(Bridge)A^+$$

$$\downarrow k_P \qquad\qquad\qquad\qquad\qquad\qquad\qquad\qquad\qquad\qquad \downarrow k_P$$

$$P_D \qquad\qquad\qquad\qquad\qquad\qquad\qquad\qquad\qquad\qquad\qquad P_A$$

Figure 10. The kinetic scheme adopted to model HT in DNA. $D^+(Bridge)A$ and $D(Bridge)A^+$ indicate the initial and the finale state, possibly giving rise to oxidation product P_D and P_A. $[D^+(Bridge)A]^*$ and $[D(Bridge)A^+]^*$ denote the ensembles of structures in which the hole donor and acceptor are in vibronic resonance with each other, so that $k_{23} = k_{32}$. $k_{21} = k_{irr}$ and k_P have been taken from experimental data; k_{23} has been computed by resolving the time dependent Schrödinger equation; k_{12} is an adjustable parameter to be inferred by comparing computed and experimental product yield ratios.

Step 3 and the reverse of Step 1 take into account the solvent response to a nonequilibrium charge distribution of the solute; pump–probe experiments in water solutions showed that solvent relaxation occurs in a few tens of femtoseconds, thus we set $k_{21} = k_{irr} = 10^{13}$ s^{-1} [135,136]. k_P has been set to 10^7 s^{-1}, taken from the rate of deprotonation of Guo$^{\cdot+}$ [137,138], likely the first step of formation of the products of oxidative damage [139–141].

Because the DNA oligomers studied by Giese contain several consecutive rigid A:T steps, it is possible to assume that Step 1 is governed only by solvent motion rather than by backbone reorganization. However, no experimental information about k_{12} is available, therefore k_{12} has been taken as an adjustable parameter to be inferred from experimental results.

Step 2 is the hole transfer, which is mainly governed by the nuclear motion of nucleobases [142,143]. Because of the local rigidity of the DNA backbone due to the $(A:T)_n$ tract [88], we can make the reasonable assumption that the elements of the ensemble of the activated HT reactants differ from each other only for solvent configurations, which are irrelevant for calculation of the rates of the elementary HT step. Therefore, quantum dynamics computations can be carried out for only one of the typical equilibrium configurations of the different oligonucleotides. In our approach, the kinetic constant for the charge transfer step has been determined as $k_{23} = 1/\tau_{23}$, where τ_{23} are transition times taken at the complete population of the final state, i.e., when the hole is fully localized on the GGG site. To compute transition times, we numerically solved the time dependent Schrödinger equation:

$$i\hbar \frac{\partial |\psi(t)\rangle}{\partial t} = \mathcal{H}|\psi(t)\rangle. \tag{5}$$

Upon introducing the vibronic nature of the diabatic states in Equation (4), $|n\rangle \to |n\rangle \otimes |\nu\rangle \equiv |n,\nu\rangle$, where $|\nu\rangle$ denotes the manifold of the harmonic vibrational states of the nth nucleobase, the TB Hamiltonian can be cast in the form:

$$\mathcal{H} = \sum_{n,\nu}(\varepsilon_n + E_\nu)|n,\nu\rangle\langle n,\nu| + \sum_{n,m,\nu,\mu} J_{nm}\langle\mu|\nu\rangle|n\rangle\langle m| + \sum_{n,m,\nu,\mu,i} \frac{\partial J_{nm}}{\partial Q_i}\langle\mu|Q_i|\nu\rangle|n\rangle\langle m|, \tag{6}$$

in which the Born–Oppenheimer approximation has been used. E_ν denote vibrational energies, and the last summation has been introduced to take into account the possible fluctuations of the electronic couplings J_{nm} due to interbase oscillations along the ith normal coordinate, causing dynamical deformation of regular double stranded DNA [52,91,144,145].

The $\partial J_{nm}/\partial Q_i$ factor has been kept fixed at 0.1 eV/Å, ε and J parameters have been taken from Table 2, and the G/A inter-strand coupling term has been set to 0.012 eV. Consistent with the Hamiltonian of Equation (6), the time dependent wave function is expanded over the set of Born–Oppenheimer products of time independent basis functions:

$$|\psi(t)\rangle = \sum_{n,\nu} C_{n,\nu}(t) |n,\nu\rangle. \tag{7}$$

To make calculations possible, the computational load connected with the huge number of integrals has to be strongly reduced. That goal is achieved by partitioning the Hilbert space into a set of subspaces with a fixed number of vibrations that are allowed to be simultaneously excited. Only the normal modes that are effectively coupled to hole transfer vibrations, i.e., the modes which are allowed to change their quantum number during the transition, are included in computations. Active modes are selected according to Duschinsky's transformation, as the ones giving rise to the largest displacement of geometrical coordinates upon electronic excitation [121,134].

We have considered the possibility that in Giese sequences HT may occur either intrastrand, via T units, or interstrand, mediated by the adenine nucleobases of the complementary helix. The results of quantum dynamics simulations are reported in Figure 11, where the logarithm plot of the predicted kinetic constants of Step 2 against the distance between G and GGG nucleobases is reported. Computations predict that, for shorter oligonucleotides, $n = 1$–3 the charge transfer goes intrastrand; the computed hole transfer times being in excellent agreement with their experimental counterpart: simulations yields a straight line with a slope $\beta = 0.63$ Å as the distance parameter of the Marcus–Levich–Jortner equation [146], to be compared with the experimental value $\beta \approx 0.6$ Å [45].

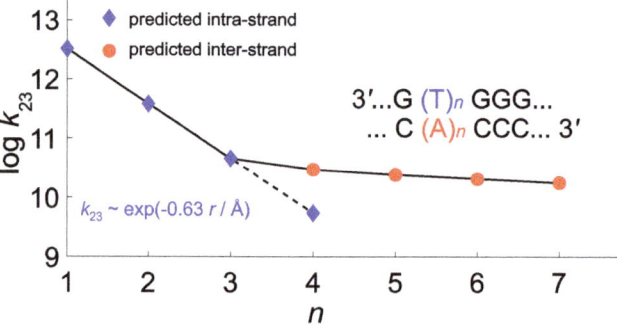

Figure 11. Predicted rate constants k_{23} for the HT step in $3'$-G(T)$_n$GGG-$5'$ DNA sequences, as a function of the number of bases separating the donor (G) and the acceptor (GGG) sites; blue diamonds, Intrastrand; red circles, Interstrand HT via the (A)$_n$ bridge of the complementary helix. The predicted parameter of the Marcus–Levich–Jortner equation (in blue) for the short distance regime ($n = 1$–3) is in excellent agreement with its experimental counterpart. For $n \geq 4$, HT is predicted to occur interstrand, mediated by the adenine bridge. Adapted with permission from *J. Phys. Chem. Lett.* **2019**, *10*, 1845–1851. Copyright (2019) American Chemical Society.

Interstrand HT becomes about 1 order of magnitude faster than intrastrand HT for $n = 4$. For $n \geq 4$, our computations predict that HT rates are almost distance-independent, in good agreement with experimental results. In all cases, hole transfer is predicted to occur via superexchange, since negligible populations have been found on adenine or thymine bridge. According to our simulations, the different distance dependence of HT rates in the short and long distance regimes results from the balance of two contrasting effects: On the one hand, as the number of bridging adenines increases, the energy barrier becomes comparatively lower due to the formation of delocalized polarons, thus favoring hole tunneling along the strand. On the other hand, upon increasing the length of the adenine bridge, the tunneling distance also increases, thus lessening the efficiency of HT, as illustrated in Figure 12.

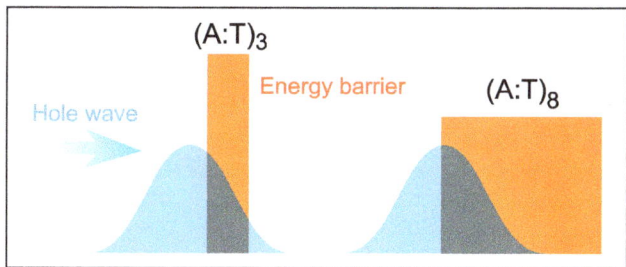

Figure 12. The contrasting effects of the bridge in the HT for the sequences studied by Giese: In short sequences (**left**), superexchange is favored by the short tunneling distance and disfavored by high barriers. In longer sequences (**right**), barrier height decreases due to formation of delocalized $A_n^{\bullet+}$ polarons thus favoring the HT, but tunneling distance increases, thus attenuating the efficiency of HT at the same time.

Yield ratios P_G/P_{GGG} have then been obtained by numerically solving the set of ordinary differential equations (ODEs) of the kinetic model of Figure 10. The experimental yield ratios are compatible with the adopted kinetics model only if k_{12} is set to values of the order of 10^{10} s^{-1}. By taking fixed that value for all the cases analyzed by Giese, thus assuming that the rate of the activation process (Step 1) is independent of the bridge length, computations yield the P_{GGG}/P_G ratios reported in Figure 13. Our predictions are in excellent agreement with experimental results, both in the short and in the long range regimes. In particular, a very weak dependence on n is predicted for $n = 4$–7.

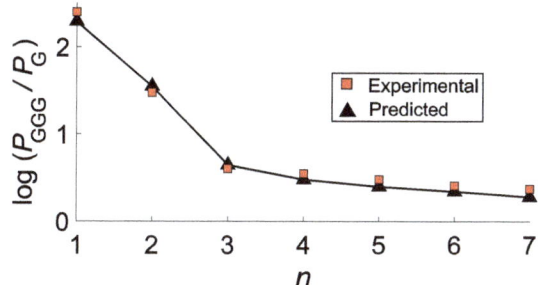

Figure 13. Predicted (black triangles) and experimental (red squares) P_{GGG}/P_G yield ratios for the HT in d-G(T)$_n$GGG as a function of the number of bases separating the donor and the acceptor G sites. Computed values have been obtained by setting $k_{12} = 10^{10}$ s^{-1} for all the investigated sequences, obtained by solving the ODEs equations for the kinetic scheme of Figure 10. Adapted with permission from *J. Phys. Chem. Lett.* **2019**, *10*, 1845–1851. Copyright (2019) American Chemical Society.

The computational approach and the set of parameters used for treating Giese sequences are broadly applicable to other DNA oligomers; preliminary results which will appear in a forthcoming paper show that our treatment correctly predicts the yields of DNA oxidative damage in several oligonucleotides studied by Schuster [19].

5. Discussion

Hole transfer in DNA is a complex process in which many chemico-physical factors play a role, among which hole site energies, electronic couplings among nucleobases, solvent relaxation, and backbone reorganization times are the most relevant. Hole site energies in DNA are significantly sequence dependent, but that dependence can be reasonably handled using a tight binding approximation in which only interactions among nearest neighbor nucleobases are considered, on condition that reliable electronic couplings are also used. Here, we show that electrochemical well tailored experiments can provide very reliable values of those quantities. Although nucleobase oxidations are usually irreversible processes and prevent the obtainment of standard redox potentials, since voltammetric peaks do not refer to equilibrium conditions, hole site energies and electronic couplings inferred from electrochemical measurements provide a very robust set of parameters for predicting the yields of oxidative DNA damages of several oligonucleotides.

Electrochemical measurements have also shed light on one among the most debated issues on long range hole transfer in DNA: the establishment of charge delocalized domains. Voltammetric measurements have shown a progressive lowering of the first anodic peak potential as the number of adjacent homo-bases (guanine or adenine) increases in DNA sequences. That result is particularly important for adenine: the formation of an AA step considerably lowers the hole site energy, by ca. 0.3 eV, whereas, for GG step, the effect is lower, amounting to about 0.1 eV. That observation is an unambiguous evidence of the establishment of delocalized hole domains in DNA oligonucleotides.

Delocalized domains play a key role in long range hole transfer in DNA, enabling a distance independent regime in adenine rich oligonucleotides, a peculiar property of DNA which should be more deeply explored in organic electronics.

Our developed kinetic model based on hole site energies and electronic couplings inferred from voltammetric measurements is able to reproduce both the yield ratios of damaged products and the correct time scale of HT for the DNA sequences investigated by Giese. Present simulations only include the vibronic ground state localized at the single G as the initial state. However, that approximation should not constitute a severe drawback, inasmuch as the vibrational frequencies of the normal coordinates of single nucleobases that are effectively coupled to hole motion exceed thermal energy at room temperature. Instead, thermal populations of vibrational states possibly affecting hole transfer rates could arise from interbase oscillation modes. Indeed, the electronic coupling for the hole transport in oxidized DNA is expected to depend to a larger extent on variations of the rise coordinate and to a lesser extent on variations of the twist coordinate, interbase motions originating from low frequencies vibrations of DNA backbone [34,103]. We have averaged such thermal effects through the last term of Equation (6). Nevertheless, a more systematic treatment could rely on the thermo field dynamics theory, in which temperature effects are already included in the Hilbert space [147].

As a final remark, we note that our approach could be easily extended also to the hole transport in G4-quadruplexes, provided reliable hole energies and electronic couplings are available for such complex systems [148]. Work is in progress along that line.

Author Contributions: Conceptualization, A.P. and A.C.; funding acquisition, A.P., T.C. and A.C.; investigation, A.C., T.C. and A.L.; methodology, A.C., A.L. and A.P.; project administration and resources, A.P., A.C. and T.C.; software, A.C. and A.L.; and writing, A.C., A.P. and A.L.

Funding: This research was funded by Ministero of Università e Ricerca MIUR, project PRIN 2009K3RH7N and PON2007-2014 (Relight project), by the Università di Salerno, grant FARB RBF12CLQD-002, and COST Action, grant CM1201. We acknowledge the CINECA (HP10CK2XJU and HP10CYW18T awards) under the ISCRA initiative for the availability of high-performance computing resources.

Conflicts of Interest: The authors declare no conflict of interest.

References

1. Eley, D.D.; Spivey, D.I. Semiconductivity of Organic Substances. Part 9. Nucleic Acid in the Dry State. *Trans. Faraday Soc.* **1962**, *58*, 411–415. [CrossRef]
2. Murphy, C.J.; Arkin, M.R.; Jenkins, Y.; Ghatlia, N.D.; Bossmann, S.H.; Turro, N.J.; Barton, J.K. Long-Range Photoinduced Electron Transfer through a DNA Helix. *Science* **1993**, *262*, 1025–1029. [CrossRef] [PubMed]
3. Genereux, J.C.; Barton, J.K. Mechanisms for DNA Charge Transport. *Chem. Rev.* **2010**, *110*, 1642–1662. [CrossRef] [PubMed]
4. Kanvah, S.; Joseph, J.; Schuster, G.B.; Barnett, R.N.; Cleveland, C.L.; Landman, U. Oxidation of DNA: Damage to Nucleobases. *Acc. Chem. Res.* **2010**, *43*, 280–287. [CrossRef]
5. Kawai, K.; Majima, T. Hole Transfer Kinetics of DNA. *Acc. Chem. Res.* **2013**, *46*, 2616–2625. [CrossRef]

6. Cadet, J.; Douki, T.; Ravanat, J.L. Oxidatively Generated Damage to the Guanine Moiety of DNA: Mechanistic Aspects and Formation in Cells. *Acc. Chem. Res.* **2008**, *41*, 1075–1083. [CrossRef]
7. Cooke, M.S.; Evans, M.D.; Dizdaroglu, M.; Lunec, J. Oxidative DNA Damage: Mechanisms, Mutation, and Disease. *FASEB J.* **2003**, *17*, 1195–1214. [CrossRef]
8. Sontz, P.A.; Muren, N.B.; Barton, J.K. DNA Charge Transport for Sensing and Signaling. *Acc. Chem. Res.* **2012**, *45*, 1792–1800. [CrossRef]
9. Zwang, T.J.; Tse, E.C.M.; Barton, J.K. Sensing DNA through DNA Charge Transport. *ACS Chem. Biol.* **2018**, *13*, 1799–1809. [CrossRef]
10. Endres, R.G.; Cox, D.L.; Singh, R.R.P. The Quest for High-Conductance DNA. *Rev. Mod. Phys.* **2004**, *76*, 195–214. [CrossRef]
11. Singh, B.; Sariciftci, N.S.; Grote, J.G.; Hopkins, F.K. Bio-Organic-Semiconductor-Field-Effect-Transistor Based on Deoxyribonucleic Acid Gate Dielectric. *J. Appl. Phys.* **2006**, *100*, 024514. [CrossRef]
12. Zalar, P.; Kamkar, D.; Naik, R.; Ouchen, F.; Grote, J.G.; Bazan, G.C.; Nguyen, T.Q. DNA Electron Injection Interlayers for Polymer Light-Emitting Diodes. *J. Am. Chem. Soc.* **2011**, *133*, 11010–11013. [CrossRef] [PubMed]
13. Zhang, Y.; Zalar, P.; Kim, C.; Collins, S.; Bazan, G.C.; Nguyen, T.Q. DNA Interlayers Enhance Charge Injection in Organic Field-Effect Transistors. *Adv. Mater.* **2012**, *24*, 4255–4260. [CrossRef] [PubMed]
14. Shi, W.; Han, S.; Huang, W.; Yu, J. High Mobility Organic Field-Effect Transistor Based on Water-Soluble Deoxyribonucleic Acid via Spray Coating. *Appl. Phys. Lett.* **2015**, *106*, 043303. [CrossRef]
15. Gomez, E.F.; Venkatraman, V.; Grote, J.G.; Steckl, A.J. Exploring the Potential of Nucleic Acid Bases in Organic Light Emitting Diodes. *Adv. Mater.* **2015**, *27*, 7552–7562. [CrossRef]
16. Makarov, V.; Pettitt, B.M.; Feig, M. Solvation and Hydration of Proteins and Nucleic Acids: A Theoretical View of Simulation and Experiment. *Acc. Chem. Res.* **2002**, *35*, 376–384. [CrossRef]
17. Oliveira-Brett, A.M.; Piedade, J.A.P.; Silva, L.A.; Diculescu, V.C. Voltammetric Determination of All DNA Nucleotides. *Anal. Biochem.* **2004**, *332*, 321–329. [CrossRef]
18. O'Neill, M.A.; Barton, J.K. DNA Charge Transport: Conformationally Gated Hopping through Stacked Domains. *J. Am. Chem. Soc.* **2004**, *126*, 11471–11483. [CrossRef]
19. Joseph, J.; Schuster, G.B. Emergent Functionality of Nucleobase Radical Cations in Duplex DNA: Prediction of Reactivity Using Qualitative Potential Energy Landscapes. *J. Am. Chem. Soc.* **2006**, *128*, 6070–6074. [CrossRef]
20. Basko, D.M.; Conwell, E.M. Effect of Solvation on Hole Motion in DNA. *Phys. Rev. Lett.* **2002**, *88*, 098102. [CrossRef]
21. Conwell, E.M.; Bloch, S.M.; McLaughlin, P.M.; Basko, D.M. Duplex Polarons in DNA. *J. Am. Chem. Soc.* **2007**, *129*, 9175–9181. [CrossRef] [PubMed]
22. Capobianco, A.; Caruso, T.; Celentano, M.; La Rocca, M.V.; Peluso, A. Proton Transfer in Oxidized Adenosine Self-Aggregates. *J. Chem. Phys.* **2013**, *139*, 145101–4. [CrossRef] [PubMed]
23. Kumar, A.; Sevilla, M.D. Proton-Coupled Electron Transfer in DNA on Formation of Radiation-Produced Ion Radicals. *Chem. Rev.* **2010**, *110*, 7002–7023. [CrossRef] [PubMed]
24. Barnett, R.N.; Cleveland, C.L.; Joy, A.; Landman, U.; Schuster, G.B. Charge Migration in DNA: Ion-Gated Transport. *Science* **2001**, *294*, 567–571. [CrossRef] [PubMed]
25. Livshits, G.I.; Stern, A.; Rotem, D.; Borovok, N.; Eidelshtein, G.; Migliore, A.; Penzo, E.; Wind, S.J.; Di Felice, R.; Skourtis, S.S.; Cuevas, J.C.; Gurevich, L.; Kotlyar, A.B.; Porath, D. Long-Range Charge Transport in Single G-Quadruplex DNA Molecules. *Nat. Nanotechnol.* **2014**, *9*, 1040–1046. [CrossRef]
26. Neidle, S. Quadruplex Nucleic Acids as Novel Therapeutic Targets. *J. Med. Chem.* **2016**, *59*, 5987–6011. [CrossRef]
27. Jortner, J.; Bixon, M.; Langenbacher, T.; Michel-Beyerle, M.E. Charge Transfer and Transport in DNA. *Proc. Natl. Acad. Sci. USA* **1998**, *95*, 12759. [CrossRef]
28. Bixon, M.; Giese, B.; Wessely, S.; Langenbacher, T.; Michel-Beyerle, M.E.; Jortner, J. Long-Range Charge Hopping in DNA. *Proc. Natl. Acad. Sci. USA* **1999**, *96*, 11713–11716. [CrossRef]
29. Voityuk, A.A.; Jortner, J.; Bixon, M.; Rösch, N. Energetic of Hole Transfer in DNA. *Chem. Phys. Lett.* **2000**, *324*, 430–434. [CrossRef]
30. Berlin, Y.A.; Burin, A.L.; Ratner, M.A. Charge Hopping in DNA. *J. Am. Chem. Soc.* **2001**, *123*, 260–268. [CrossRef]

31. Voityuk, A.A.; Jortner, J.; Bixon, M.; Rösch, N. Electronic Coupling between Watson-Crick Pairs for Hole Transfer and Transport in Desoxyribonucleic Acid. *J. Chem. Phys.* **2001**, *114*, 5614–5620. [CrossRef]
32. Troisi, A.; Orlandi, G. Hole Migration in DNA: A Theoretical Analysis of the Role of Structural Fluctuations. *J. Phys. Chem. B* **2002**, *106*, 2093–2101. [CrossRef]
33. Senthilkumar, K.; Grozema, F.C.; Fonseca Guerra, C.; Bickelhaupt, F.M.; Siebbeles, L.D.A. Mapping the Sites for Selective Oxidation of Guanines in DNA. *J. Am. Chem. Soc.* **2003**, *125*, 13658–13659. [CrossRef] [PubMed]
34. Senthilkumar, K.; Grozema, F.C.; Fonseca Guerra, C.; Bickelhaupt, F.M.; Lewis, F.D.; Berlin, Y.A.; Ratner, M.A.; Siebbeles, L.D.A. Absolute Rates of Hole Transfer in DNA. *J. Am. Chem. Soc.* **2005**, *127*, 14894–14903. [CrossRef]
35. Berlin, Y.A.; Kurnikov, I.V.; Beratan, D.; Ratner, M.A.; Burin, A.L. DNA Electron Transfer Processes: Some Theoretical Notions. In *Long-Range Charge Transfer in DNA II*; Schuster, G.B., Ed.; Springer: Berlin/Heidelberg, Germany, 2004; Volume 237, pp. 1–36.
36. Grozema, F.C.; Tonzani, S.; Berlin, Y.A.; Schatz, G.C.; Siebbeles, L.D.A.; Ratner, M.A. Effect of Structural Dynamics on Charge Transfer in DNA Hairpins. *J. Am. Chem. Soc.* **2008**, *130*, 5157–5166. [CrossRef]
37. Grozema, F.C.; Tonzani, S.; Berlin, Y.A.; Schatz, G.C.; Siebbeles, L.D.A.; Ratner, M.A. Effect of GC Base Pairs on Charge Transfer through DNA Hairpins: The Importance of Electrostatic Interactions. *J. Am. Chem. Soc.* **2009**, *131*, 14204–14205. [CrossRef]
38. Gollub, C.; Avdoshenko, S.; Gutierrez, R.; Berlin, Y.; Cuniberti, G. Charge Migration in Organic materials: Can Propagating Charges Affect the Key Physical Quantities Controlling Their Motion? *Isr. J. Chem.* **2012**, *52*, 452–460. [CrossRef]
39. Renaud, N.; Berlin, Y.A.; Lewis, F.D.; Ratner, M.A. Between Superexchange and Hopping: An Intermediate Charge-Transfer Mechanism in polyA-polyT DNA Hairpins. *J. Am. Chem. Soc.* **2013**, *135*, 3953–3963. [CrossRef]
40. Kubař, T.; Gutièrrez, R.; Kleinekathöfer, U.; Cuniberti, G.; Elstner, M. Modeling charge transport in DNA Using multi-scale Methods. *Phys. Status Solidi B* **2013**, *250*, 2277–2287. [CrossRef]
41. Lewis, F.D.; Liu, X.; Liu, J.; Hayes, R.T.; Wasielewski, M.R. Dynamics and Equilibria for Oxidation of G, GG, and GGG Sequences in DNA Hairpins. *J. Am. Chem. Soc.* **2000**, *122*, 12037–12038. [CrossRef]
42. Saito, I.; Takayama, M.; Kawanishi, S. Photoactivatable DNA-Cleaving Amino Acids: Highly Sequence-Selective DNA Photocleavage by Novel L-Lysine Derivatives. *J. Am. Chem. Soc.* **1995**, *117*, 5590–5591. [CrossRef]
43. Hall, D.B.; Holmlin, R.E.; Barton, J.K. Oxidative DNA Damage through Long-Range Electron Transfer. *Nature* **1996**, *382*, 731–735. [CrossRef] [PubMed]
44. Yoshioka, Y.; Kitagawa, Y.; Takano, Y.; Yamaguchi, K.; Nakamura, T.; Saito, I. Experimental and Theoretical Studies on the Selectivity of GGG Triplets toward One-Electron Oxidation in B-Form DNA. *J. Am. Chem. Soc.* **1999**, *121*, 8712–8719. [CrossRef]
45. Giese, B.; Amaudrut, J.; Köhler, A.K.; Spormann, M.; Wessely, S. Direct Observation of Hole Transfer through DNA by Hopping between Adenine Bases and by Tunneling. *Nature* **2001**, *412*, 318–320. [CrossRef] [PubMed]
46. Lee, Y.A.; Durandin, A.; Dedon, P.C.; Geacintov, N.E.; Shafirovich, V. Oxidation of Guanine in G, GG, and GGG Sequence Contexts by Aromatic Pyrenyl Radical Cations and Carbonate Radical Anions: Relationship between Kinetics and Distribution of Alkali-Labile Lesions. *J. Phys. Chem. B* **2008**, *112*, 1834–1844. [CrossRef] [PubMed]
47. Capobianco, A.; Caruso, T.; D'Ursi, A.M.; Fusco, S.; Masi, A.; Scrima, M.; Chatgilialoglu, C.; Peluso, A. Delocalized Hole Domains in Guanine-Rich DNA Oligonucleotides. *J. Phys. Chem. B* **2015**, *119*, 5462–5466. [CrossRef]
48. Genereux, J.C.; Wuerth, S.M.; Barton, J.K. Single-Step Charge Transport through DNA over Long Distances. *J. Am. Chem. Soc.* **2011**, *133*, 3863–3868. [CrossRef]
49. Muren, N.B.; Olmon, E.D.; Barton, J.K. Solution, Surface, and Single Molecule Platforms for the Study of DNA-Mediated Charge Transport. *Phys. Chem. Chem. Phys.* **2012**, *14*, 13754–13771. [CrossRef]
50. Harris, M.A.; Mishra, A.K.; Young, R.M.; Brown, K.E.; Wasielewski, M.R.; Lewis, F.D. Direct Observation of the Hole Carriers in DNA Photoinduced Charge Transport. *J. Am. Chem. Soc.* **2016**, *138*, 5491–5494. [CrossRef]

51. Paleček, E.; Bartošík, M. Electrochemistry of Nucleic Acids. *Chem. Rev.* **2012**, *112*, 3427–3481. [CrossRef]
52. Troisi, A.; Orlandi, G. The Hole Transfer in DNA: Calculation of Electron Coupling between Close Bases. *Chem. Phys. Lett.* **2001**, *344*, 509–518. [CrossRef]
53. Cramer, T.; Krapf, S.; Koslowski, T. DNA Charge Transfer: An Atomistic Model. *J. Phys. Chem. B* **2004**, *108*, 11812–11819. [CrossRef]
54. Kubař, T.; Woiczikowski, P.B.; Cuniberti, G.; Elstner, M. Efficient Calculation of Charge-Transfer Matrix Elements for Hole Transfer in DNA. *J. Phys. Chem. B* **2008**, *112*, 7937–7947. [CrossRef] [PubMed]
55. Kitoh-Nishioka, H.; Ando, K. Charge-Transfer Matrix Elements by FMO-LCMO Approach: Hole Transfer in DNA with Parameter Tuned Range-Separated DFT. *Chem. Phys. Lett.* **2015**, *621*, 96–101. [CrossRef]
56. Pluharová, E.; Slavíček, P.; Jungwirth, P. Modeling Photoionization of Aqueous DNA and Its Components. *Acc. Chem. Res.* **2015**, *48*, 1209–1217. [CrossRef] [PubMed]
57. Kawai, K.; Wata, Y.; Ichinose, N.; Majima, T. Selective Enhancement of the One-Electron Oxidation of Guanine by Base Pairing with Cytosine. *Angew. Chem. Int. Ed.* **2000**, *39*, 4327–4329. [CrossRef]
58. Seidel, C.A.M.; Schulz, A.; Sauer, M.H.M. Nucleobase-Specific Quenching of Fluorescent Dyes. I. Nucleobase One-Electron Redox Potentials and Their Correlation with Static and Dynamic Quenching Efficiencies. *J. Phys. Chem.* **1996**, *100*, 5541–5553. [CrossRef]
59. Brotons, A.; Mas, L.A.; Metters, J.P.; Banks, C.E.; Iniesta, J. Voltammetric Behaviour of Free DNA Bases, Methylcytosine and Oligonucleotides at Disposable Screen Printed Graphite Electrode Platforms. *Analyst* **2013**, *138*, 5239–5249. [CrossRef]
60. Dryhurst, G.; Elving, P.J. Electrochemical Oxidation of Adenine: Reaction Products and Mechanisms. *J. Electrochem. Soc.* **1968**, *115*, 1014–1020. [CrossRef]
61. Faraggi, M.; Broitman, F.; Trent, J.B.; Klapper, M.H. One-Electron Oxidation Reactions of Some Purine and Pyrimidine Bases in Aqueous Solutions. Electrochemical and Pulse Radiolysis Studies. *J. Phys. Chem.* **1996**, *100*, 14751–14761. [CrossRef]
62. Capobianco, A.; Landi, A.; Peluso, A. Modeling DNA Oxidation in Water. *Phys. Chem. Chem. Phys.* **2017**, *19*, 13571–13578. [CrossRef] [PubMed]
63. Tomasi, J.; Mennucci, B.; Cammi, R. Quantum Mechanical Continuum Solvation Models. *Chem. Rev.* **2005**, *105*, 2999–3094. [CrossRef] [PubMed]
64. Schroeder, C.A.; Pluhařová, E.; Seidel, R.; Schroeder, W.P.; Faubel, M.; Slavíček, P.; Winter, B.; Jungwirth, P.; Bradforth, S.E. Oxidation Half-Reaction of Aqueous Nucleosides and Nucleotides via Photoelectron Spectroscopy Augmented by ab Initio Calculations. *J. Am. Chem. Soc.* **2015**, *137*, 201–209. [CrossRef] [PubMed]
65. Steenken, S.; Jovanovic, S.V. How Easily Oxidizable Is DNA? One-Electron Reduction Potentials of Adenosine and Guanosine Radicals in Aqueous Solution. *J. Am. Chem. Soc.* **1997**, *119*, 617–618. [CrossRef]
66. Orlov, V.M.; Smirnov, A.N.; Varshavsky, Y.M. Ionization Potentials and Electron-Donor Ability of Nucleic Acid Bases and Their Analogues. *Tetrahedron Lett.* **1976**, *48*, 4377–4378. [CrossRef]
67. Abraham, J.; Gosh, A.K.; Schuster, G.B. One-Electron Oxidation of DNA Oligomers That Lack Guanine: Reaction and Strand Cleavage at Remote Thymines by Long-Distance Radical Cation Hopping. *J. Am. Chem. Soc.* **2006**, *128*, 5346–5347.
68. Ghosh, A.; Joy, A.; Schuster, G.B.; Douki, T.; Cadet, J. Selective One-Electron Oxidation of Duplex DNA Oligomers: Reaction at Thymines. *Org. Biomol. Chem.* **2008**, *6*, 916–928. [CrossRef]
69. Joseph, J.; Schuster, G.B. One-Eelectron Oxidation of DNA: Reaction at Thymine. *Chem. Commun.* **2010**, *46*, 7872–7878. [CrossRef]
70. Barnett, R.N.; Joseph, J.; Landman, U.; Schuster, G.B. Oxidative Thymine Mutation in DNA: Water-Wire-Mediated Proton-Coupled Electron Transfer. *J. Am. Chem. Soc.* **2013**, *135*, 3904–3914. [CrossRef]
71. Colson, A.O.; Besler, B.; Sevilla, M.D. Ab Initio Molecular Orbital Calculation of DNA Base Pair Radical Ions: Effects of Base Pairing on Proton Transfer Energies, Electron Affinities and Ionization Potentials. *J. Phys. Chem.* **1992**, *96*, 9787–9794. [CrossRef]
72. Kawai, K.; Wata, Y.; Hara, M.; Toyo, S.; Majima, T. Regulation of One-Electron Oxidation Rate of Guanine by Base Pairing with Cytosine Derivatives. *J. Am. Chem. Soc.* **2002**, *124*, 3586–3590. [CrossRef] [PubMed]
73. Caruso, T.; Carotenuto, M.; Vasca, E.; Peluso, A. Direct Experimental Observation of the Effect of the Base Pairing on the Oxidation Potential of Guanine. *J. Am. Chem. Soc.* **2005**, *127*, 15040–15041. [CrossRef] [PubMed]

74. Kyogoku, Y.; Lord, R.C.; Alexander, R. An infrared study of the hydrogen-bonding specificity of hypoxanthine and other nucleic acid derivatives. *Biochim. Biophys. Acta* **1969**, *179*, 10–17. [CrossRef]
75. Williams, L.D.; Chawla, B.; Shaw, B.R. The hydrogen bonding of cytosine with guanine: Calorimetric and 1H-NMR analysis of the molecular interactions of nucleic acid bases. *Biopolymers* **1987**, *26*, 591–603. [CrossRef] [PubMed]
76. Caruso, T.; Capobianco, A.; Peluso, A. The Oxidation Potential of Adenosine and Adenosine-Thymidine Base-Pair in Chloroform Solution. *J. Am. Chem. Soc.* **2007**, *129*, 15347–15353. [CrossRef] [PubMed]
77. Capobianco, A.; Caruso, T.; Fusco, S.; Terzidis, M.A.; Masi, A.; Chatgilialoglu, C.; Peluso, A. The Association Constant of 5′,8-cyclo-2′-Deoxyguanosine with Cytidine. *Front. Chem.* **2015**, *3*, 22. [CrossRef]
78. Psciuk, B.T.; Lord, R.L.; Munk, B.H.; Schlegel, H.B. Theoretical Determination of One-Electron Oxidation Potentials for Nucleic Acid Bases. *J. Chem. Theory Comput.* **2012**, *12*, 5107–5123. [CrossRef]
79. Capobianco, A.; Carotenuto, M.; Caruso, T.; Peluso, A. The Charge-Transfer Band of an Oxidized Watson-Crick Guanosine-Cytidine Complex. *Angew. Chem. Int. Ed.* **2009**, *48*, 9526–9528. [CrossRef]
80. Capobianco, A.; Caruso, T.; Celentano, M.; D'Ursi, A.M.; Scrima, M.; Peluso, A. Stacking Interactions between Adenines in Oxidized Oligonucleotides. *J. Phys. Chem. B* **2013**, *117*, 8947–8953. [CrossRef]
81. Isaksson, J.; Acharya, S.; Barman, J.; Cheruku, P.; Chattopadhyaya, J. Single-Stranded Adenine-Rich DNA and RNA Retain Structural Characteristics of Their Respective Double-Stranded Conformations and Show Directional Differences in Stacking Pattern. *Biochemistry* **2004**, *43*, 15996–16010. [CrossRef]
82. Zubatiuk, T.A.; Shishkin, O.V.; Gorb, L.; Hovorun, D.M.; Leszczynski, J. B-DNA Characteristics Are Preserved in Double stranded $d(A)_3 \cdot d(T)_3$ and $d(G)_3 \cdot d(C)_3$ Mini-Helixes: Conclusions from DFT/M06-2X Study. *Phys. Chem. Chem. Phys.* **2013**, *15*, 18155–18166. [CrossRef] [PubMed]
83. Capobianco, A.; Peluso, A. The Oxidization Potential of AA Steps in Single Strand DNA Oligomers. *RSC Adv.* **2014**, *4*, 47887–47893. [CrossRef]
84. Zubatiuk, T.; Kukuev, M.A.; Korolyova, A.S.; Gorb, L.; Nyporko, A.; Hovorun, D.; Leszczynski, J. Structure and Binding Energy of Double-Stranded A-DNA Mini-helices: Quantum-Chemical Study. *J. Phys. Chem. B* **2015**, *119*, 12741–12749. [CrossRef] [PubMed]
85. Capobianco, A.; Caruso, T.; Peluso, A. Hole Delocalization over Adenine Tracts in Single Stranded DNA Oligonucleotides. *Phys. Chem. Chem. Phys.* **2015**, *17*, 4750–4756. [CrossRef]
86. Capobianco, A.; Velardo, A.; Peluso, A. Single-Stranded DNA Oligonucleotides Retain Rise Coordinates Characteristic of Double Helices. *J. Phys. Chem. B* **2018**, *122*, 7978–7989. [CrossRef]
87. El Hassan, M.A.; Calladine, C.R. Conformational Characteristics of DNA: Empirical Classifications and a Hypothesis for the Conformational Behaviour of Dinucleotide Steps. *Philos. Trans. R. Soc. A* **1997**, *355*, 43–100. [CrossRef]
88. Calladine, C.R.; Drew, H.R.; Luisi, B.F.; Travers, A.A. *Understanding DNA*, 3rd ed.; Elsevier Academic Press: Oxford, UK, 2004; Chapter 3.
89. Dandliker, P.J.; Holmlin, R.E.; Barton, J.K. Oxidative Thymine Dimer Repair in the DNA Helix. *Science* **1997**, *275*, 1465–1468. [CrossRef]
90. Henderson, P.T.; Jones, D.; Hampikian, G.; Kan, Y.; Schuster, G.B. Long-distance Charge Transport in Duplex DNA: The Phonon-Assisted Polaron-like Hopping Mechanism. *Proc. Natl. Acad. Sci. USA* **1999**, *96*, 8353–8358. [CrossRef]
91. Conwell, E.M.; Rakhamanova, S.V. Polarons in DNA. *Proc. Natl. Acad. Sci. USA* **2000**, *97*, 4556–4560. [CrossRef]
92. Schuster, G.B.; Landman, U. Long-Range Charge Transfer in DNA. II. *Top. Curr. Chem.* **2004**, *236*, 139.
93. Shao, F.; O'Neill, M.A.; Barton, J.K. Long Range Oxidative Damage to Cytosine in Duplex DNA. *Proc. Natl. Acad. Sci. USA* **2004**, *101*, 17914–17919. [CrossRef] [PubMed]
94. Takada, T.; Kawai, K.; Fujitsuka, M.; Majima, T. Rapid Long-Distance Hole Transfer through Consecutive Adenine Sequence. *J. Am. Chem. Soc.* **2006**, *128*, 11012–11013. [CrossRef] [PubMed]
95. Lewis, F.D.; Zhu, H.; Daublain, P.; Cohen, B.; Wasielewski, M.R. Hole Mobility in DNA A Tracts. *Angew. Chem. Int. Ed.* **2006**, *45*, 7982–7985. [CrossRef] [PubMed]
96. Zeidan, T.A.; Carmieli, R.; Kelley, R.F.; Wilson, T.M.; Lewis, F.D.; Wasielewski, M.R. Charge-Transfer and Spin Dynamics in DNA Hairpin Conjugates with Perylenediimide as a Base-Pair Surrogate. *J. Am. Chem. Soc.* **2008**, *130*, 13945–13955. [CrossRef] [PubMed]

97. Vura-Weis, J.; Wasielewski, M.R.; Thazhathveetil, A.K.; Lewis, F.D. Efficient Charge Transport in DNA Diblock Oligomers. *J. Am. Chem. Soc.* **2009**, *131*, 9722–9727. [CrossRef]
98. Blaustein, G.S.; Lewis, F.D.; Burin, A.L. Kinetics of Charge Separation in Poly(A)–Poly(T) DNA Hairpins. *J. Phys. Chem. B* **2010**, *114*, 6732–6739. [CrossRef]
99. Kravec, S.M.; Kinz-Thompson, C.D.; Conwell, E.M. Localization of a Hole on an Adenine-Thymine Radical Cation in B-Form DNA in Water. *J. Phys. Chem. B* **2011**, *115*, 6166–6171. [CrossRef]
100. Kumar, A.; Sevilla, M.D. Density Functional Theory Studies of the Extent of Hole Delocalization in One-Electron Oxidized Adenine And Guanine Base Stacks. *J. Phys. Chem. B* **2011**, *115*, 4990–5000. [CrossRef]
101. Rooman, M.; Wintjens, R. Sequence and Conformation Effects on Ionization Potential and Charge Distribution of Homo-Nucleobase Stacks Using M06-2X Hybrid Density Functional Theory Calculations. *J. Biomol. Struct. Dyn.* **2014**, *32*, 532–545. [CrossRef]
102. Saito, I.; Takayama, M.; Sugiyama, H.; Nakatani, K.; Tsuchida, A.; Yamamoto, M. Photoinduced DNA Cleavage via Electron Transfer: Demonstration that Guanine Residues Located 5' to Guanine Are the Most Electron-Donating Sites. *J. Am. Chem. Soc.* **1995**, *117*, 6406–6407. [CrossRef]
103. Sugiyama, H.; Saito, I. Theoretical Studies of GG-Specific Photocleavage of DNA via Electron Transfer: Significant Lowering of Ionization Potential and 5'-Localization of HOMO of Stacked GG Bases in B-form DNA. *J. Am. Chem. Soc.* **1996**, *118*, 7063–7068. [CrossRef]
104. Sies, H.; Schulz, W.A.; Steenken, S. Adjacent Guanines as Preferred Sites for Strand Breaks in Plasmid DNA Irradiated with 193 nm and 248 nm UV Laser Light. *J. Photochem. Photobiol. B Biol.* **1996**, *32*, 97–102. [CrossRef]
105. Prat, F.; Houk, K.N.; Foote, C.S. Effect of Guanine Stacking on the Oxidation of 8-Oxoguanine in B-DNA. *J. Am. Chem. Soc.* **1998**, *120*, 845–846. [CrossRef]
106. Saito, I.; Nakamura, T.; Nakatani, K. Mapping of the Hot Spots for DNA Damage by One-Electron Oxidation: Efficacy of GG Doublets and GGG Triplets as a Trap in Long-Range Hole Migration. *J. Am. Chem. Soc.* **1998**, *120*, 12686–12687. [CrossRef]
107. Voityuk, A.A. Are Radical Cation States Delocalized over GG and GGG Hole Traps in DNA? *J. Phys. Chem. B* **2005**, *109*, 10793–10796. [CrossRef] [PubMed]
108. Kumar, A.; Sevilla, M.D. Photoexcitation of Dinucleoside Radical Cations: A Time-Dependent Density Functional Study. *J. Phys. Chem. B* **2006**, *110*, 24181–24188. [CrossRef] [PubMed]
109. Ito, K.; Inoue, S.; Yamamoto, K.; Kawanishi, S. 8-Hydroxydeoxyguanosine Formation at the 5' Site of 5'-GG-3' Sequences in Double-Stranded DNA by UV Radiation with Riboflavin. *J. Biol. Chem.* **1993**, *268*, 13221–13227.
110. Steinbrecher, T.; Koslowski, T.; Case, D.A. Direct Simulation of Electron Transfer Reactions in DNA Radical Cations. *J. Phys. Chem. B* **2008**, *112*, 16935–16944. [CrossRef]
111. Pitterl, F.; Chervet, J.P.; Oberacher, H. Electrochemical Simulation of Oxidation Processes Involving Nucleic Acids Monitored with Electrospray Ionization-Mass Spectrometry. *Anal. Bioanal. Chem.* **2010**, *397*, 1203–1215. [CrossRef]
112. Centore, R.; Fusco, S.; Peluso, A.; Capobianco, A.; Stolte, M.; Archetti, G.; Kuball, H.G. Push-Pull Azo-Chromophores Containing Two Fused Pentatomic Heterocycles and Their Nonlinear Optical Properties. *Eur. J. Org. Chem.* **2009**, 3535–3543. [CrossRef]
113. Anano, S.; Kurashina, Y.; Anraku, Y.; Mizuno, D. A Possible Recognition of Ribonucleotides by DNA Dependent RNA Polymerase of *E. coli*. *J. Biochem.* **1971**, *70*, 9–20. [CrossRef]
114. Exinger, F.; Lacroute, F. 6-Azauracil Inhibition of GTP Biosynthesis in Saccharomyces Cerevisiae. *Curr. Genet.* **1992**, *22*, 9–11. [CrossRef]
115. Oyelere, A.K.; Strobel, S.A. Site Specific Incorporation of 6-Azauridine into the Genomic HDV Ribozyme Active Site. *Nucleosides Nucleotides Nucleic Acids* **2001**, *20*, 1851–1858. [CrossRef]
116. Pope, M.; Swenberg, C.E. *Electronic Processes in Organic Crystals and Polymers*; Oxford University Press: Oxford, UK, 1999.
117. Peluso, A.; Del Re, G. On the Occurrence of an Electron-Transfer Step in Aromatic Nitration. *J. Phys. Chem.* **1996**, *100*, 5303–5309. [CrossRef]
118. Senthilkumar, K.; Grozema, F.C.; Bickelhaupt, F.M.; Siebbeles, L.D.A. Charge Transport in Columnar Stacked Triphenylenes: Effects of Conformational Fluctuations on Charge Transfer Integrals and Site Energies. *J. Chem. Phys.* **2003**, *119*, 9809–9817. [CrossRef]

119. Borrelli, R.; Di Donato, M.; Peluso, A. Quantum Dynamics of Electron Transfer from Bacteriochlorophyll to Pheophytin in Bacterial Reaction Centers. *J. Chem. Theory Comput.* **2007**, *3*, 673–680. [CrossRef]
120. Brisker-Klaiman, D.; Peskin, U. Coherent Elastic Transport Contribution to Currents through Ordered DNA Molecular Junctions. *J. Phys. Chem. C* **2010**, *114*, 19077–19082. [CrossRef]
121. Borrelli, R.; Capobianco, A.; Landi, A.; Peluso, A. Vibronic Couplings and Coherent Electron Transfer in Bridged Systems. *Phys. Chem. Chem. Phys.* **2015**, *17*, 30937–30945. [CrossRef]
122. Levine, A.D.; Iv, M.; Peskin, U. Length-Independent Transport Rates in Biomolecules by Quantum Mechanical Unfurling. *Chem. Sci.* **2016**, *7*, 1535–1542. [CrossRef]
123. Borrelli, R.; Peluso, A. Elementary Electron Transfer Reactions: From Basic Concepts to Recent Computational Advances. *WIREs Comput. Mol. Sci.* **2013**, *3*, 542–559. [CrossRef]
124. Zhou, M.; Zhai, Y.; Dong, S. Electrochemical Sensing and Biosensing Platform Based on Chemically Reduced Graphene Oxide. *Anal. Chem.* **2009**, *81*, 5603–5613. [CrossRef]
125. Brotons, A.; Vidal-Iglesias, F.J.; Solla-Gullón, J.; Iniesta, J. Carbon Materials for the Electrooxidation of Nucleobases, Nucleosides and Nucleotides toward Cytosine Methylation Detection: A Review. *Anal. Methods* **2016**, *8*, 702–715. [CrossRef]
126. Renger, T.; Marcus, R.A. Variable Range Hopping Electron Transfer Through Disordered Bridge States: Application to DNA. *J. Phys. Chem. A* **2003**, *107*, 8404–8419. [CrossRef]
127. Bixon, M.; Jortner, J. Incoherent Charge Hopping and Conduction in DNA and Long Molecular Chains. *Chem. Phys.* **2005**, *319*, 273–282. [CrossRef]
128. Zhang, Y.; Liu, C.; Balaeff, A.; Skourtis, S.S.; Beratan, D.N. Biological Charge Transfer via Flickering Resonance. *Proc. Natl. Acad. Sci. USA* **2014**, *111*, 10049–10054. [CrossRef]
129. Levine, A.D.; Iv, M.; Peskin, U. Formulation of Long-Range Transport Rates through Molecular Bridges: From Unfurling to Hopping. *J. Phys. Chem. Lett.* **2018**, *9*, 4139–4145. [CrossRef]
130. Parson, W.W. Vibrational Relaxations and Dephasing in Electron-Transfer Reactions. *J. Phys. Chem. B* **2016**, *120*, 11412–11418. [CrossRef]
131. Parson, W.W. Effects of Free Energy and Solvent on Rates of Intramolecular Electron Transfer in Organic Radical Anions. *J. Phys. Chem. A* **2017**, *121*, 7297–7306. [CrossRef]
132. Parson, W.W. Electron-Transfer Dynamics in a Zn-Porphyrin-Quinone Cyclophane: Effects of Solvent, Vibrational Relaxations, and Conical Intersections. *J. Phys. Chem. B* **2018**, *122*, 854–3863. [CrossRef]
133. Parson, W.W. Temperature Dependence of the Rate of Intramolecular Electron Transfer. *J. Phys. Chem. B* **2018**, *122*, 8824–8833. [CrossRef]
134. Landi, A.; Borrelli, R.; Capobianco, A.; Peluso, A. Transient and Enduring Electronic Resonances Drive Coherent Long Distance Charge Transport in Molecular Wires. *J. Phys. Chem. Lett.* **2019**, *10*, 1845–1851. [CrossRef]
135. Jimenez, R.; Fleming, G.R.; Kumar, P.V.; Maroncelli, M. Femtosecond Solvation Dynamics of Water. *Nature* **1994**, *369*, 471–473. [CrossRef]
136. Fleming, G.R.; Cho, M. Chromophore-Solvent Dynamics. *Annu. Rev. Phys. Chem.* **1996**, *47*, 109–134. [CrossRef]
137. Kobayashi, K.; Tagawa, S. Direct Observation of Guanine Radical Cation Deprotonation in Duplex DNA Using Pulse Radiolysis. *J. Am. Chem. Soc.* **2003**, *125*, 10213–10218. [CrossRef]
138. Rokhlenko, Y.; Cadet, J.; Geacintov, N.E.; Shafirovich, V. Mechanistic Aspects of Hydration of Guanine Radical Cations in DNA. *J. Am. Chem. Soc.* **2014**, *136*, 5956–5962. [CrossRef]
139. Candeias, L.P.; Steenken, S. Structure and Acid-Base Properties of One-Electron-Oxidized Deoxyguanosine, Guanosine, and 1-Methylguanosine. *J. Am. Chem. Soc.* **1989**, *111*, 1094–1099. [CrossRef]
140. Chatgilialoglu, C.; Caminal, C.; Guerra, M.; Mulazzani, Q.G. Tautomers of One-Electron-Oxidized Guanosine. *Angew. Chem. Int. Ed.* **2005**, *44*, 6030–6032. [CrossRef]
141. Chatgilialoglu, C.; Caminal, C.; Altieri, A.; Vougioukalakis, G.C.; Mulazzani, Q.G.; Gimisis, T.; Guerra, M. Tautomerism in the Guanyl Radical. *J. Am. Chem. Soc.* **2006**, *128*, 13796–13805. [CrossRef]
142. Borrelli, R.; Capobianco, A.; Peluso, A. Hole Hopping Rates in Single Strand Oligonucleotides. *Chem. Phys.* **2014**, *440*, 25–30. [CrossRef]
143. Velardo, A.; Borrelli, R.; Capobianco, A.; La Rocca, M.V.; Peluso, A. First Principle Analysis of Charge Dissociation and Charge Recombination Processes in Organic Solar Cells. *J. Phys. Chem. C* **2015**, *119*, 18870–18876. [CrossRef]

144. Macía, E. Electrical Conductance in Duplex DNA: Helical Effects and Low-Frequency Vibrational Coupling. *Phys. Rev. B* **2007**, *76*, 245123. [CrossRef]
145. Li, G.; Govind, N.; Ratner, M.A.; Cramer, C.J.; Gagliardi, L. Influence of Coherent Tunneling and Incoherent Hopping on the Charge Transfer Mechanism in Linear Donor-Bridge-Acceptor Systems. *J. Phys. Chem. Lett.* **2015**, *6*, 4889–4897. [CrossRef]
146. Marcus, R.A. Electron Transfer Reactions in Chemistry. Theory and Experiment. *Rev. Mod. Phys.* **1993**, *65*, 599–610. [CrossRef]
147. Borrelli, R.; Gelin, M.F. Quantum Eelectron-Vibrational Dynamics at Finite Temperature: Thermo Field Dynamics Approach. *J. Chem. Phys.* **2016**, *145*, 224101. [CrossRef]
148. Wu, J.; Meng, Z.; Lu, Y.; Shao, F. Efficient Long-Range Hole Transport Through G-Quadruplexes. *Chem. Eur. J.* **2017**, *23*, 13980–13985. [CrossRef]

© 2019 by the authors. Licensee MDPI, Basel, Switzerland. This article is an open access article distributed under the terms and conditions of the Creative Commons Attribution (CC BY) license (http://creativecommons.org/licenses/by/4.0/).

Article

Copper *bis*-Dipyridoquinoxaline Is a Potent DNA Intercalator that Induces Superoxide-Mediated Cleavage via the Minor Groove

Zara Molphy [1,2]**, Vickie McKee** [1,3] **and Andrew Kellett** [1,2,]*

1. School of Chemical Sciences, National Institute for Cellular Biotechnology and Nano Research Facility, Dublin City University, Glasnevin, Dublin 9, Ireland; zara.molphy@dcu.ie (Z.M.); vickie.mckee@dcu.ie (V.M.)
2. Synthesis and Solid-State Pharmaceutical Centre, School of Chemical Sciences, Dublin City University, Glasnevin, Dublin 9, Ireland
3. Department of Physics, Chemistry and Pharmacy, University of Southern Denmark, Campusvej 55, 5230 Odense M, Denmark
* Correspondence: andrew.kellett@dcu.ie; Tel.: +353-1-7005461

Academic Editor: Chryssostomos Chatgilialoglu
Received: 31 October 2019; Accepted: 20 November 2019; Published: 26 November 2019

Abstract: Herein, we report the synthesis, characterisation, X-ray crystallography, and oxidative DNA binding interactions of the copper artificial metallo-nuclease [Cu(DPQ)$_2$(NO$_3$)](NO$_3$), where DPQ = dipyrido[3,2-*f*:2′,3′-*h*]quinoxaline. The cation [Cu(DPQ)$_2$]$^{2+}$ (Cu-DPQ), is a high-affinity binder of duplex DNA and presents an intercalative profile in topoisomerase unwinding and viscosity experiments. Artificial metallo-nuclease activity occurs in the absence of exogenous reductant but is greatly enhanced by the presence of the reductant Na-*L*-ascorbate. Mechanistically, oxidative DNA damage occurs in the minor groove, is mediated aerobically by the Cu(I) complex and is dependent on both superoxide and hydroxyl radical generation. To corroborate cleavage at the minor groove, DNA oxidation of a cytosine–guanine (5′-CCGG-3′)-rich oligomer was examined in tandem with a 5-methylcytosine (5′-C5mCGG-3′) derivative where 5mC served to sterically block the major groove and direct damage to the minor groove. Overall, both the DNA binding affinity and cleavage mechanism of Cu-DPQ depart from Sigman's reagent [Cu(1,10-phenanthroline)$_2$]$^{2+}$; however, both complexes are potent oxidants of the minor groove.

Keywords: DNA damage; copper; chemical nuclease; intercalation; free radical oxidation

1. Introduction

Since the structural elucidation of duplex DNA, the construction of small molecules that recognise and react at specific sites to modify DNA structure, reactivity and biological repair processes has been an area of considerable research interest. The discovery of the synthetic chemical nuclease [Cu(1,10-phenanthroline)$_2$]$^{2+}$ (Cu-Phen) in 1979, sparked efforts toward the development of new artificial metallonucleases with DNA cleavage mediated through the generation of reactive oxygen species (ROS) at the DNA interface [1]. A kinetic study revealed that the nuclease activity of the Cu(II) complex firstly involves reduction to the activated Cu(I) species, which binds reversibly to double-stranded DNA, while the second step involves oxidation of the non-covalently bound Cu(I) species by H$_2$O$_2$ (or O$_2$) to generate reactive oxygen species (ROS) responsible for strand scission [2]. Cu-Phen degrades DNA in a mechanism involving hydrogen atom abstraction from the C1′ deoxyribose position, resulting in the production of the C1′ radical as a major oxidative product. The C1′ site is only accessible by minor groove binding agents oriented toward H-1′, such as Cu-Phen and its derivatives [3,4]. Other targets of oxidative attack via the minor groove include C-H bonds located at C4′ and C5′

positions that are well characterised sites of attack by Mn(TMPyP)/KHSO$_5$, iron bleomycin and the enediyne antibiotics [5–10]. Cu-Phen and several derivatives have shown promising antitumoral [11,12], antifungal [13], antimicrobial [14,15] and protein engineering properties [16]. However, limitations associated with the Cu-Phen scaffold include (*i*) weak DNA binding constant; (*ii*) binds both DNA and protein biomolecules without specificity; (*iii*) a high dissociation constant of the second coordinated 1,10 phenanthroline ligand; and (*iv*) a reliance on an exogenous reductant (ascorbate or thiols) and oxidant to initiate ROS production to mediate C–H bond activation and strand scission [7]. Accordingly, modulation of the Cu-Phen scaffold represents an interesting developmental challenge. To circumvent dissociation of the second Phen group, the Clip-Phen ligand was developed to link both Phen ligands via their C2- or C3-carbons by a short flexible arm (Figure 1) [17,18]. Another significant property of this ligand is that its primary amine can be functionalised with sequence-directing groups including minor groove binders netropsin and distamycin [19–23], intercalating acridine conjugates [24], or oligonucleotides [25] to localise nuclease activity.

In recent years, we have focused on introducing modifications to the Cu-Phen scaffold to enhance DNA recognition, oxidative damage and cytotoxicity [11]. Approaches including inner sphere modifications using coordinated carboxylate groups [13]; dicarboxylate groups (e.g., o-phthalate or octanedioate) [26–31]; designer phenazine intercalators [32,33]; tris-(2-pyridylmethyl)amine (TPMA) caging ligands [34,35]; and the introduction of a binuclear *di*-cation have been undertaken [26,30,36]. Previously, we reported a series of phenazine-functionalized Cu(II) phenanthroline complexes—with general formula [Cu(N,N')(Phen)]$^{2+}$ (where N,N' = dipyridoquinoxaline (DPQ) or dipyridophenazine (DPPZ))—that offered a significant enhancement of DNA binding affinity relative to Cu–Phen [32]. A microfluidic "on-chip"-based method identified Cu-DPQ-Phen as the most active chemical nuclease in the series and later, the hydroxyl radical was identified as the predominant species responsible for the oxidative cleavage event [33]. We also probed the chemical nuclease efficacy on restricted plasmids with varying GC content. Significant differences were noted in the chemical nuclease efficacy of the Cu-N,N'-Phen complex series compared to Cu-Phen. Cu-Phen displayed higher chemical nuclease activity on pUC19 (51% GC), while each of the Cu-N,N'-Phen complexes had enhanced activity on the pBC4 plasmid (59% GC), leading us to hypothesise that this general class of compound may oxidatively target cytosine–phosphate–guanine (CpG) islands.

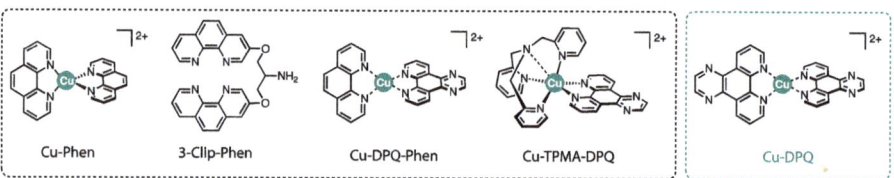

Figure 1. Structures of Cu-Phen, 3-Clip-Phen ligand and ternary copper(II) complexes incorporating 1,10-phenanthroline (Phen) and dipyridoquinoxaline (DPQ) Cu-DPQ-Phen and Cu-TPMA-DPQ developed in our laboratory [32]. The current study focuses on the preparation and chemical nuclease activity of Cu-DPQ (assuming dissociation of nitrato ligand in solution).

Our interest in developing new chemical nucleases with high efficacy and enhanced targeting properties prompted this study, where we report the synthesis and X-ray structural characterisation of [Cu(dipyrido[3,2-*f*:2',3'-*h*]quinoxaline)$_2$(NO$_3$)](NO$_3$). The aim of this study was to investigate what influence two DPQ ligands have within the core Cu-(N,N')$_2$ complex when applied as an artificial chemical nuclease. Additionally, the DNA binding and oxidative DNA damage profile of Cu–DPQ was compared to the well-studied Cu-Phen and Cu–DPQ-Phen complexes previously developed in our laboratory [32,33].

2. Experimental

2.1. Materials and Methods

Chemicals and reagents were sourced from Sigma-Aldrich (St. Louis, MO, USA) or Tokyo Chemical Industry (TCI, Oxford, UK Ltd.) and were used without further purification. High-performance liquid chromatography (HPLC)-grade chloroform ($CHCl_3$), methanol (MeOH) and acetonitrile (CH_3CN) were used as received. ^1H and ^{13}C-NMR spectra were obtained using a Bruker AC 400 and 600 MHz NMR spectrometer. pH was monitored by a Mettler Toledo InLab Expert Pro-ISM pH probe (Columbus, OH, USA). Electrospray ionization mass spectra (ESI-MS) were recorded using a Thermo Fisher Exactive Orbitrap (Waltham, MA, USA) mass spectrometer coupled to an Advion TriVersa Nanomate (Ithaca, NY, USA) injection system with samples prepared in 100% HPLC-grade CH_3CN prior to analysis. UV-visible spectrometry studies were carried out on a Shimadzu UV-2600. FT-IR spectra were conducted using a Perkin Elmer Spectrum Two Spectrometer (Waltham, MA, USA). For biological studies, complexes were prepared in molecular-biology-grade DMF and further diluted in 80 mM HEPES buffer (Waltham, MA, USA). DNA plasmids and enzymes were purchased from New England BioLabs (NEB, Ipswich, MA, USA). Solutions of the title complex (Cu-DPQ) for biological studies were carried out in DMF with further dilutions made in 80 mM HEPES.

2.2. Synthesis of [Cu(DPQ)$_2$NO$_3$](NO$_3$)

[Cu(DPQ)$_2$NO$_3$](NO$_3$) was prepared by treating 2 equivalents of DPQ (0.0451g, 0.194 mmol) with 1 equivalent of Cu(NO$_3$)$_2$·3H$_2$O (0.0234 g, 0.097 mmol) under ethanolic reflux for 3 h. A green solid was isolated and washed with ice cold EtOH and dried by desiccation (yield: 0.050 g, 78%). ATR-FTIR (cm^{-1}): 596, 643, 724, 820, 1016, 1082, 1128, 1213, 1290, 1370, 1392, 1439, 1491, 1582, 3033, 3040. Elemental analysis calculated for: $C_{28}H_{16}CuN_{10}O_6$: C, 51.58; H, 2.47; N, 21.48; Cu, 9.75. %Found: C, 51.09; H, 2.28; N, 21.13; Cu, 9.95. ESI-MS: m/z calculated: 263.5 $[M]^{2+}$ found: 263.6. Solubility: DMSO, DMF.

2.3. X-Ray Crystallography

The data were collected at 150(2) K on a Bruker-Nonius Apex II CCD diffractometer using MoK$_\alpha$ radiation (λ = 0.71073Å) and were corrected for Lorentz and polarisation effects. Data were processed as a 2-component twin and corrected for absorption. The structure was solved by dual-space methods (SHELXT) [37] and refined on F^2 using all the reflections (SHELXL) [38]. All the non-hydrogen atoms were refined using anisotropic atomic displacement parameters and hydrogen atoms bonded to carbon were inserted at calculated positions using a riding model, hydrogen atoms of solvate water were not located or included in the refinement since partial-occupancy and disorder meant no single H-bonding network could be identified. Details for data collection and refinement are summarised in Supplementary S2. CCDC 1959974 contains the supplementary crystallographic data for this paper. These data can be obtained free of charge from The Cambridge Crystallographic Data Centre via www.ccdc.cam.ac.uk/data_request/cif.

Crystal Data for [Cu(DPQ)$_2$NO$_3$](NO$_3$)·2H$_2$O, $C_{28}H_{20}N_{10}O_8Cu$ (M = 688.08 g/mol): monoclinic, space group $P2_1$ (no. 4), a = 7.3882(12) Å, b = 29.205(5) Å, c = 13.363(2) Å, β= 103.686(3)°, V = 2801.4(8) Å^3, Z = 4, T = 150 (2) K, μ(MoKα) = 0.852 mm^{-1}, Dcalc = 1.631 g/cm^3, 5048 reflections measured (4.198° $\leq 2\theta \leq$ 50.054°), 5048 unique, which were used in all calculations. R_{int} = 0.044 (for HKLF 4 data containing 39,211 reflections, with I > 3sig(I), R_{sigma} = 0.0677 The final R1 was 0.0566 ($I > 2\sigma(I)$) and wR2 was 0.1379 (all data).

2.4. DNA Binding Studies

EtBr displacement, viscosity, and topoisomerase-I-mediated DNA relaxation assay were conducted according to methods previously reported by Kellett et al. [39,40]

2.5. DNA Damage Studies

Oxidative DNA damage studies including covalent recognition elements and free radical scavengers on pUC19 were carried out according to published methods [33].

3. CpG Sequence Studies

3.1. PCR Primer Design

Forward and reverse primers for the pBR322 vector (4361 bp) were designed in order to generate a 798 bp sequence (54% GC) by PCR. The transcript, which contains four individual CCGG sequences (CpG islands), was produced using PCR (35 cycles) with 1 ng pBR322 plasmid (NEB, N3033L) using 2× MyTaq Red Mix (Bioline) at suitable annealing temperatures for the primer pair. The sequence length was verified using a 1 Kb DNA ladder. Further details can be found in Supplementary S5.

3.2. Restriction Enzyme Studies

The 798 bp amplicon was exposed to a range of enzymes including HpaII (NEB, R0171S), MspI (NEB, R0106S) and HpaII-MT (NEB, M0214S). A quantity of 400 ng of the amplified sequence was treated with each enzyme for 1 h at 37 °C as per the manufactures' guidelines. Endonuclease recognition at each 5'-C⇓CGG-3' site was confirmed by agarose gel electrophoresis with HpaII and MspI (isoschizomers). Successful methylation of internal cytosine residues (5mC) of each CCGG island by HpaII methyltransferase was confirmed when HpaII was observed to have no restriction effect due to its sensitivity to CpG methylation. Electrophoresis experiments were performed at 60 V for 60 min on a 2% agarose gel.

3.3. DNA Damage

A quantity of 400 ng of the 798 bp amplicon was treated with increasing concentrations of Cu-DPQ in the absence and presence of 1 mM of exogenous reductant (Na-L-ascorbate). A further experiment was carried out involving pre-treatment with HpaII methyltransferase and subsequent exposure of the complex in the absence and presence of 1 mM reductant.

4. Results and Discussion

4.1. Preparation and Characterisation of Cu-DPQ

The dipyrido[3,2-*f*:2',3'-*h*]quinoxaline ligand (DPQ) was prepared using a previously reported method involving Schiff-base condensation of 1,10-phenanthroline-5,6-dione with ethylenediamine [32]. [Cu(DPQ)$_2$(NO$_3$)](NO$_3$) was generated by treating 2 equivalents of DPQ with 1 equivalent of Cu(NO$_3$)$_2$·3H$_2$O under ethanolic reflux and was isolated by vacuum filtration as a green solid. The complex was characterised by elemental analysis, ESI-MS and attenuated total reflectance (ATR) Fourier transform infrared (FTIR) spectroscopies. [Cu(DPQ)$_2$(NO$_3$)](NO$_3$) was isolated in good yield with elemental analysis results for carbon, hydrogen, nitrogen, and copper, in agreement with theoretical values. ESI-MS analysis contained the expected parent ion peak for the cationic complex [Cu(DPQ)$_2$]$^{2+}$ (lane1). Characteristic bands for C=C, C=N and C-N stretching of the DPQ ligand were monitored and shifts in both C=N and C-N were noted when the complex was formed (Figure S2). The electronic spectrum of the complex was measured in the UV-visible range, monitored over time and compared to Cu-Phen and Cu-DPQ-Phen. No notable differences were observed in the *d-d* transition region (600–900 cm^{-1}) upon prolonged incubation; however, the appearance of a new band ~445 cm^{-1} in the spectrum of all three complexes indicates potentially significant rearrangement in DMF solution (Figure S3).

Single crystal X-ray structure analysis of the dihydrated complex [Cu(DPQ)$_2$(NO$_3$)](NO$_3$)·2H$_2$O revealed that the asymmetric unit contains two independent, but similar, [Cu(DPQ)$_2$(NO$_3$)]$^+$ cations. The copper ions are 5-coordinate and their geometry is close to trigonal bipyrimidal with the nitrate anions in the trigonal plane (Figure 2); the τ values [41] are 0.75 and 0.77 for Cu1 and Cu2, respectively. The mean planes of the DPQ ligands are inclined at 46.35(9)° about Cu1 and 43.37(9)° at Cu2. The cations are π-stacked into zig-zag sheets perpendicular to the *c* axis (Figure S4), the spaces between sheets are occupied by the uncoordinated nitrate counter anions and solvate water molecules, which make up a hydrogen-bonded network (incorporating some disorder of both anion and water). The Cambridge Structural Database [42] includes several other examples of structures containing the $[Cu(DPQ)_2X]^{n+}$ or $[Cu(Phen)_2X]^{n+}$ cations (X = water or anionic ligand, n = 1 or 2). The structures of these cations are broadly similar to that reported here provided the ligand X is a weak donor [43,44] or if there is no fifth donor [45]. However, if X is thiocyanate, azide or, chloride square planar or tetragonal geometries were observed [46].

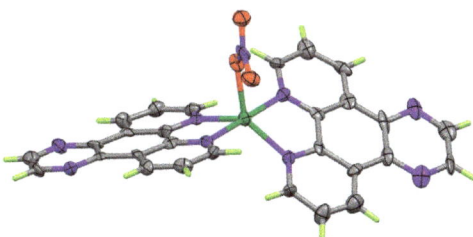

Figure 2. X-ray structure of one of the two independent $[Cu(DPQ)_2NO_3]^+$ cations showing 50% probability ellipsoids (colour scheme: copper, green; nitrogen, purple; carbon, grey; and oxygen, red).

4.2. DNA Binding Experiments

To establish how Cu-DPQ interacts with duplex DNA a range of binding studies were conducted. Firstly, the apparent binding constant (K_{app}) of the complex to calf thymus DNA was determined using a high-throughput 96-well plate ethidium bromide (EtBr) displacement assay [39]. DNA was treated with an excess of EtBr, which becomes highly fluorescent when bound to DNA. Solutions were then treated with increasing concentrations of Cu-DPQ in order to reduce the fluorescence intensity, which is indicative of the ejection of bound EtBr from DNA. Cu-DPQ displayed high binding affinity in the range of ~10^7 M(bp)$^{-1}$. This binding value is superior to that of Cu-Phen (~10^5 M(bp)$^{-1}$) and broadly in line with high-affinity complexes Cu-DPQ-Phen, Cu-DPPZ-Phen and actinomycin D (Figure 3A) [32]. To probe the DNA binding mode of Cu-DPQ further, viscosity analysis with salmon testes dsDNA fibres was investigated. Here, hydrodynamic changes revealed an increasing trend in viscosity (relative to complex loading) indicative of extension and unwinding effects associated with intercalation. The extent of unwinding elicited by Cu-DPQ surpassed that of Cu-Phen and EtBr (Figure 3B).

The topoisomerase-I-mediated DNA relaxation assay was next applied to study the intercalative effects of Cu-DPQ on supercoiled (SC) plasmid DNA (Figure 3C). This experiment was monitored by gel electrophoresis as DNA isoforms migrate at different rates through agarose due to their size and charge. Generally, intercalation results in the gradual relaxation of negatively (-) supercoiled plasmid to relaxed open circular (0) plasmid and upon further drug loading, the plasmid becomes unwound in the opposite direction resulting in positively (+) coiled SC DNA [47]. Cu-DPQ was observed to relax (-) SC plasmid to its OC form at ~2.5 μM, beyond which (+) SC topoisomers were identified. At high complex concentrations of 20 and 50 μM (lanes 14 and 15), DNA nicking was identified whereby a fixed band remains in line with the OC isoform. These results are broadly equivalent to those observed for the ternary Cu-DPQ-Phen complex (Figure S6A) and represent an ~8-fold enhancement in the unwinding effects of the semi-intercalator Cu-Phen (Figure S6B).

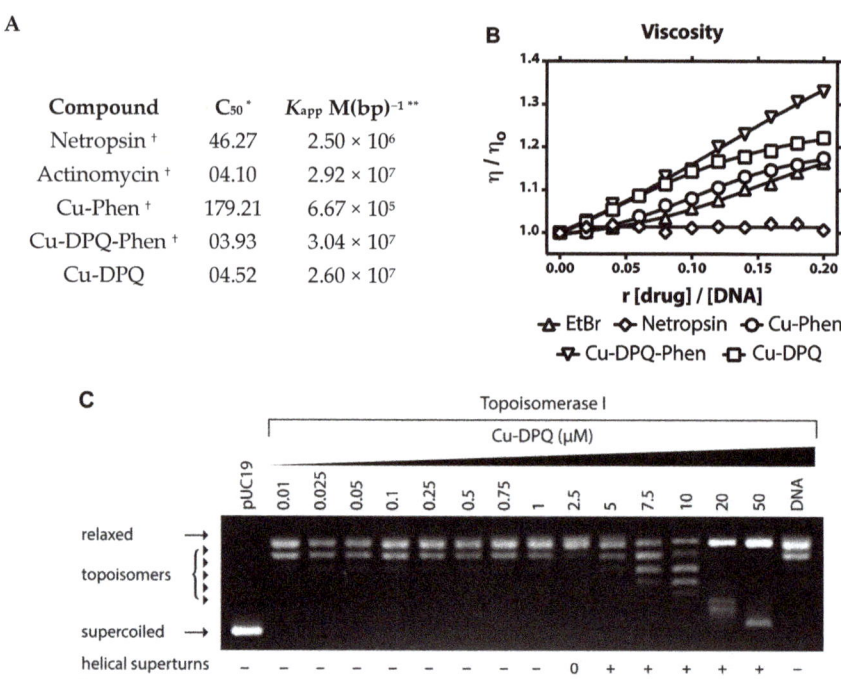

Figure 3. (**A**) Apparent DNA binding constants (K_{app}) of Cu-DPQ to EtBr-saturated solutions of dsDNA (* C_{50} = concentration required to reduce fluorescence by 50%; ** $K_{app} = K_e \times 12.6/C_{50}$ where $K_e = 9.5 \times 10^6$ M(bp)$^{-1}$; † previously reported). (**B**) Relative changes in viscosity profiles of stDNA in the presence of Cu-DPQ, Cu-DPQ-Phen, Cu-Phen, netropsin and EtBr-treated salmon testes dsDNA. (**C**) Topoisomerase-I-mediated DNA relaxation assay.

4.3. DNA Damage Studies on SC pUC19

The cleavage efficiency of Cu-DPQ was monitored by agarose gel electrophoresis over a narrow concentration range of 0.25 µM, 0.50 µM, 1.0 µM, and 2.5 µM (Figure 4A, lanes 2–5). Cu-DPQ was initially reduced to the active Cu(I) state with 1 mM exogenous reductant (Na-L-ascorbate) prior to the introduction of plasmid DNA. Cu-DPQ showed concentration-dependent relaxation of SC to OC with almost complete degradation upon treatment with 2.5 µM of the complex. The cleavage efficiency of Cu-DPQ on pUC19 is greater than Cu-DPQ-Phen and Cu-Phen complexes, illustrating the influence of the bis-ligated DPQ ligand on oxidative DNA damage [33]. Interestingly, in the absence of reductant, Cu-DPQ is capable of 'self-activation' since the emergence of OC DNA is visible in the presence of 2.5 µM of this complex with higher concentrations emerging at 10 µM (Figure 4B, lane 7).

To determine how the complex cleaves plasmid DNA, pUC19 was pre-treated with an excess of non-covalent recognition agents including methyl green (major groove binder), netropsin (minor groove binder) and cobalt(III) hexamine chloride (electrostatic binder). Pre-incubation with methyl green (Figure 4A lanes 6–9) resulted in enhanced nuclease activity compared to methyl green free experiments (lanes 2–5). However, priming the plasmid with netropsin (lanes 10–13) afforded notable protection to the minor groove resulting in a reduction in oxidative DNA damage. Finally, pre-treatment with the electrostatic binding agent had negligible effects (lanes 14–17). These results are in agreement with those previously reported with Cu-Phen, Cu-Phen-DPQ and Cu-TPMA-N,N' complexes where the presence of methyl green results in enhanced chemical nuclease activity due to its minor groove priming effects [1,33,34].

To probe the DNA damage profile of Cu-DPQ, a range of radical scavengers were introduced to the nuclease experiment to investigate the nature of ROS species involved in DNA oxidation (Figure 4C). This study identified the superoxide radical ($O_2^{\bullet-}$) as a key radical involved in the scission process as pre-treatment with tiron (4,5-dihydroxy-1,3-benzenedisulfonic acid disodium salt) significantly impeded oxidative damage to the plasmid (Figure 4D, lane 12). Moderate protection was afforded by potassium iodide (KI, Figure 4D, lanes 1–4) and dimethyl sulfoxide (DMSO, Figure 4D, lanes 13–16), which scavenge H_2O_2 and $^{\bullet}OH$ species, respectively. Singlet oxygen ($^1O_2$), however, does not appear to play a significant role in the oxidative cleavage mechanism of Cu-DPQ. The overall trend in damage inhibition follows $O_2^{\bullet-} > H_2O_2 \simeq {}^{\bullet}OH > {}^1O_2$ and departs substantially from Cu-Phen and Cu-DPQ-Phen where hydroxyl radical generation predominates [33]. Interestingly, we previously identified that $O_2^{\bullet-}$ plays a major role in oxidative profile of [Cu(o-phthalate)(1,10-phenanthroline)] and the Cu-TPMA-N,N′ series [31,34]. Finally, copper chelating agents ethylenediaminetetraacetic acid (EDTA) and neocuprione were shown to afford full protection to the SC substrate (Figure S7).

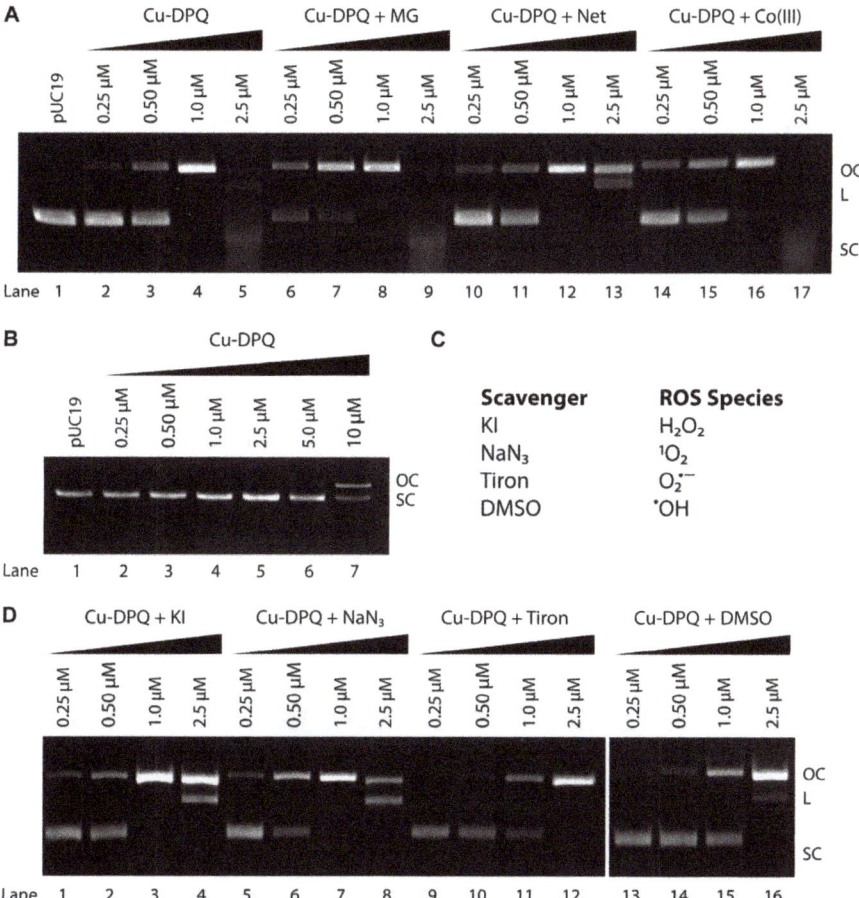

Figure 4. (**A**) DNA cleavage reactions on pUC19 in the presence of non-covalent recognition elements methyl green (MG), netropsin (Net) and cobalt(III) hexamine chloride (Co(III)). (**B**) Nuclease in the absence of reductant. (**C**) ROS scavengers employed in this study. (**D**) DNA cleavage reactions in the presence of selected ROS scavengers.

5. Studies with CpG and Methylated CpG Islands

To probe the DNA oxidation profile of Cu-DPQ further, cleavage experiments involving a 798 bp dsDNA sequence (amplified from pBR322) containing four distinct cytosine–phosphate–guanine (CpG) islands (5′-CCGG-3′) were undertaken. The internal cytosine residues of this transcript were methylated using HpaII methyltransferase (HpaII-MT) to generate an equivalent amplicon with four 5′-C5mCGG-3′ sites (Figure 5AII and Figure 5B lane 3). To verify the presence of four CpG islands, the non-methylated sequence was treated with HpaII and isoschizomer MspI. Both enzymes recognise and cleave the 5′-C⇓CGG-3′ tract to produce the desired fragmentation pattern (Figure 5AI and Figure 5B lanes 4 and 5; supplementary S5). HpaII is sensitive to CpG methylation and was inactive against the 5mC sequence (Figures 5AIII and 5B lane 7), while isoschizomer MspI is insensitive to DNA methylation and cleaved all four methylation sites (Figure 5AIV and Figure 5B lane 6).

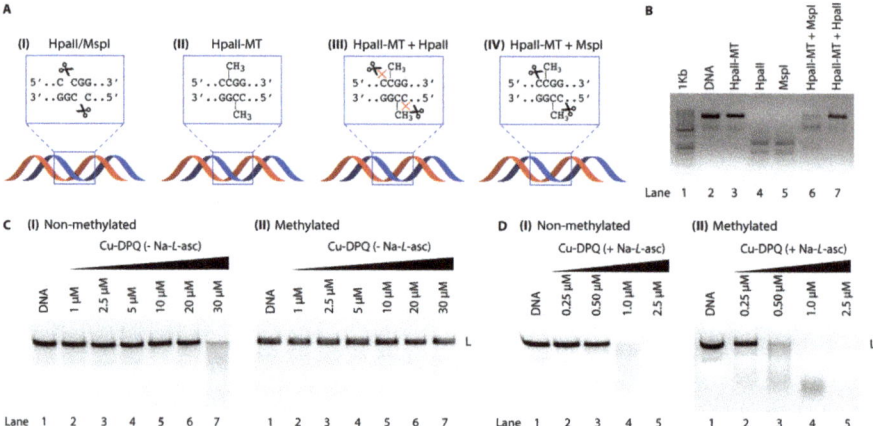

Figure 5. (**A**) Cartoon representation of enzyme restriction sites. (**B**) Control experiment with isoschizomers HpaII and MspI in the presence and absence of HpaII-MT. (**C**) A quantity of 400 ng of 798 bp linear sequence (**I** non-methylated and **II** methylated) treated with Cu-DPQ in the absence of reductant. (**D**) 400 ng of 798 bp linear sequence (**I** non-methylated and **II** methylated) treated with Cu-DPQ in the presence of reductant Na-L-asc.

Cleavage experiments with Cu-DPQ in the absence of reductant were then examined using non-methylated and methylated sequences (Figure 5CI and II). At the highest tested concentration (30 µM), the non-methylated sequence was extensively sheared while the 5mC amplicon remained intact. Similar experiments were then conducted in the presence of reductant (Figure 5DI and II). Here, the activated Cu-DPQ complex degraded the non-methylated sequence at 1.0 µM, while shearing of the methylated transcript started at 0.25 µM with almost complete ablation at 0.5 µM. These results indicate that under 'self-activation' conditions (i.e., without reductant), DNA cleavage may occur in the major groove of CpG islands. When these islands contain 5mC residues, the methyl group serves to sterically block the major groove, thereby preventing Cu-DPQ-mediated oxidation. In its reduced Cu(I) form, Cu-DPQ appears to be a potent oxidant of the minor groove since limiting access to the major groove of CpG islands enhances oxidation. Therefore, 5mC serves to direct the complex to the minor groove—in much the same way as methyl green—and this residency serves to increase the oxidative cleavage effects of Cu-DPQ in its reduced form.

6. Conclusions

The Cu-DPQ complex was generated in high purity and its structure determined by single-crystal X-ray crystallography. In the solid state, the crystal of [Cu(DPQ)$_2$(NO$_3$)](NO$_3$)·2H$_2$O revealed the asymmetric unit contains two independent, but similar, [Cu(DPQ)$_2$(NO$_3$)]$^+$ cations. The five-coordinated copper(II) ions [CuN$_4$O] adopt a geometry close to trigonal bipyrimidal with the nitrate anions in the trigonal plane. Cu-DPQ is a high-affinity DNA binder with potent intercalative properties compared to Cu-Phen, as evidenced by the DNA fluorescence displacement assay, viscosity and topoisomerase-I DNA unwinding studies. The complex is capable of 'self-activation' by inducing DNA damage in the absence of exogenous reductant but is greatly enhanced in its presence. The oxidative DNA damage profile of Cu-DPQ was studied in the presence of free radical scavenging species, with results demonstrating the complex to be an efficient oxidiser of pUC19 plasmid DNA mediating oxidative DNA damage predominately through the generation of the superoxide radical species (O$_2$•$^-$) with involvement from the hydroxyl radical (•OH). By pre-exposing plasmid DNA with non-covalent steric blocking agents of methyl green (major groove) or netropsin (minor groove), the minor groove was identified as the preferred DNA oxidation site. To help confirm this effect, we introduced 5-methylcytosine (5mC) into a 798 bp construct at four individual CpG islands. Here, 5mC served to sterically block the major groove and enhance chemical nuclease activity, thereby supporting preferential oxidative cleavage at the minor groove. This mono-nuclear copper(II) DNA damaging agent represents an interesting therapeutic lead for the treatment of human cancer and warrants future in vitro evaluation.

Supplementary Materials: Supplementary materials are available online.

Author Contributions: Conceptualisation, A.K., Z.M.; Methodology, A.K., Z.M.; Investigation, Z.M., V.M.; Writing—Original Draft, A.K., Z.M., V.M.; Writing—Reviewing & Editing, A.K., Z.M., V.M.; Visualization, A.K., Z.M.; Supervision, A.K.

Funding: This publication has emanated from research supported in part by a research grant from Science Foundation Ireland (SFI) and is co-funded under the European Regional Development Fund under Grant Number 12/RC/2275_P2. Mass spectrometry and UV analysis was carried out at the Nano Research Facility in Dublin City University which was funded under the Programme for Research in Third Level Institutions (PRTLI) Cycle 5. The PRTLI is co-funded through the European Regional Development Fund (ERDF), part of the European Union Structural Funds Programme 2011-2015. Support from the SFI Career Development Award 15/CDA/3648, the Synthesis and Solid-State Pharmaceutical Centre (SSPC) and SFI under grant number 12/RC/2275 is gratefully acknowledged. This work was also supported by the Marie Skłodowska-Curie Innovative Training Network (ITN) ClickGene (H2020-MSCA-ITN-2014-642023).

Acknowledgments: A.K. thanks Deepak Chandran for assistance with the crystallisation of the title complex.

Conflicts of Interest: The authors declare no conflict of interest.

References

1. Sigman, D.S.; Graham, D.R.; D'Aurora, V.; Stern, A.M. Oxygen-dependent cleavage of DNA by the 1,10-phenathroline cuporous complex. Inhibition of Escherichia coli DNA polymerase I. *J. Biol. Chem.* **1979**, *254*, 12269–12272. [PubMed]
2. Thederahn, T.B.; Kuwabara, M.D.; Larsen, T.A.; Sigman, D.S. Nuclease activity of 1,10-phenanthroline-copper: Kinetic mechanism. *J. Am. Chem. Soc.* **1989**, *111*, 4941–4946. [CrossRef]
3. Sigman, D.S. Nuclease Activity of 1,10-Phenanthroline-Copperion. *Acc. Chem. Res.* **1986**, *19*, 180–186. [CrossRef]
4. Goyne, T.E.; Sigman, D.S. Nuclease activity of 1,10-phenanthroline-copper ion. Chemistry of deoxyribose oxidation. *J. Am. Chem. Soc.* **1987**, *109*, 2846–2848. [CrossRef]
5. Sigman, D.S. Chemical Nucleases. *Biochemistry* **1990**, *29*, 9097–9105. [CrossRef]
6. Sigman, D.S.; Mazumder, A.; Perrin, D.M. Chemical Nucleases. *Chem. Rev.* **1993**, *93*, 2295–2316. [CrossRef]
7. Pratviel, G.; Bernadou, J.; Meunier, B. Carbon—Hydrogen Bonds of DNA Sugar Units as Targets for Chemical Nucleases and Drugs. *Angew. Chem. Int. Ed.* **1995**, *34*, 746–769. [CrossRef]
8. Chen, T.; Greenberg, M.M. Model Studies Indicate That Copper Phenanthroline Induces Direct Strand Breaks via β-Elimination of the 2′-Deoxyribonolactone Intermediate Observed in Enediyne Mediated DNA Damage. *J. Am. Chem. Soc.* **1998**, *120*, 3815–3816. [CrossRef]
9. Pitié, M.; Boldron, C.; Pratviel, G. DNA Oxidation by Copper and Manganese Complexes. *Adv. Inorg. Chem.* **2006**, *58*, 77–130.

10. Pitié, M.; Pratviel, G. Activation of DNA Carbon–Hydrogen Bonds by Metal Complexes. *Chem. Rev.* **2010**, *110*, 1018–1059. [CrossRef]
11. Kellett, A.; Molphy, Z.; Slator, C.; McKee, V. Recent advances in anticancer copper compounds. In *Metal-Based Anticancer Agents*; Casini, A., Vessières, A., Meier-Menches, S.M., Eds.; RSC Metallobiology; Royal Society of Chemistry: Cambridge UK, 2019; pp. 91–119.
12. McGivern, T.J.P.; Slator, C.; Kellett, A.; Marmion, C.J. Innovative DNA-Targeted Metallo-prodrug Strategy Combining Histone Deacetylase Inhibition with Oxidative Stress. *Mol. Pharmaceutics* **2018**, *15*, 5058–5071. [CrossRef] [PubMed]
13. Prisecaru, A.; McKee, V.; Howe, O.; Rochford, G.; McCann, M.; Colleran, J.; Pour, M.; Barron, N.; Gathergood, N.; Kellett, A. Regulating Bioactivity of Cu^{2+} Bis-1,10-phenanthroline Artificial Metallonucleases with Sterically Functionalized Pendant Carboxylates. *J. Med. Chem.* **2013**, *56*, 8599–8615. [CrossRef] [PubMed]
14. McCann, M.; Kellett, A.; Kavanagh, K.; Devereux, M.; Santos, A.L.S. Deciphering the Antimicrobial Activity of Phenanthroline Chelators. *Curr. Med. Chem.* **2012**, *19*, 2703–2714. [CrossRef] [PubMed]
15. McCann, M.; Santos, A.L.S.; da Silva, B.A.; Romanos, M.T.V.; Pyrrho, A.S.; Devereux, M.; Kavanagh, K.; Fichtner, I.; Kellett, A. In vitro and in vivo studies into the biological activities of 1,10-phenanthroline, 1,10-phenanthroline-5,6-dione and its copper(II) and silver(I) complexes. *Toxicol. Res.* **2012**, *1*, 47–54. [CrossRef]
16. Larragy, R.; Fitzgerald, J.; Prisecaru, A.; McKee, V.; Leonard, P.; Kellett, A. Protein engineering with artificial chemical nucleases. *Chem. Commun.* **2015**, *51*, 12908–12911. [CrossRef]
17. Pitié, M.; Donnadieu, B.; Meunier, B. Preparation of the New Bis(phenanthroline) Ligand "Clip-Phen" and Evaluation of the Nuclease Activity of the Corresponding Copper Complex. *Inorg. Chem.* **1998**, *37*, 3486–3489.
18. Pitié, M.; Sudres, B.; Meunier, B. Dramatic increase of the DNA cleavage activity of Cu(Clip-phen) by fixing the bridging linker on the C3 position of the phenanthroline units. *Chem. Commun.* **1998**, *0*, 2597–2598. [CrossRef]
19. Bailly, C.; Henichart, J.P. DNA recognition by intercalator-minor-groove binder hybrid molecules. *Bioconjugate Chem.* **1991**, *2*, 379–393. [CrossRef]
20. Bailly, C.; Chaires, J.B. Sequence-Specific DNA Minor Groove Binders. Design and Synthesis of Netropsin and Distamycin Analogues. *Bioconjugate Chem.* **1998**, *9*, 513–538. [CrossRef]
21. Pitié, M.; Burrows, C.J.; Meunier, B. Mechanisms of DNA cleavage by copper complexes of 3-Clip-Phen and of its conjugate with a distamycin analogue. *Nucleic Acids Res.* **2000**, *28*, 4856–4864. [CrossRef]
22. Pitié, M.; Van Horn, J.D.; Brion, D.; Burrows, C.J.; Meunier, B. Targeting the DNA Cleavage Activity of Copper Phenanthroline and Clip-Phen to A·T Tracts via Linkage to a Poly-N-methylpyrrole. *Bioconjugate Chem.* **2000**, *11*, 892–900. [CrossRef] [PubMed]
23. Bales, B.C.; Kodama, T.; Weledji, Y.N.; Pitié, M.; Meunier, B.; Greenberg, M.M. Mechanistic studies on DNA damage by minor groove binding copper–phenanthroline conjugates. *Nucleic Acids Res.* **2005**, *33*, 5371–5379. [CrossRef] [PubMed]
24. Boldron, C.; Ross, S.A.; Pitié, M.; Meunier, B. Acridine Conjugates of 3-Clip-Phen: Influence of the Linker on the Synthesis and the DNA Cleavage Activity of Their Copper Complexes. *Bioconjugate Chem.* **2002**, *13*, 1013–1020. [CrossRef] [PubMed]
25. Laurent, A.; Wright, M.; Laayoun, A.; Meunier, B.; Tissot, L.; Pitie, M. Clip-Phen Conjugates for the Specific Cleavage of Nucleic Acids. *Nucleosides Nucleotides Nucleic Acids* **2007**, *26*, 927–930. [CrossRef]
26. Kellett, A.; O'Connor, M.; McCann, M.; McNamara, M.; Lynch, P.; Rosair, G.; McKee, V.; Creaven, B.; Walsh, M.; McClean, S.; et al. Bis-phenanthroline copper(II) phthalate complexes are potent in vitro antitumour agents with 'self-activating' metallo-nuclease and DNA binding properties. *Dalton Trans.* **2011**, *40*, 1024–1027. [CrossRef]
27. Kellett, A.; O'Connor, M.; McCann, M.; Howe, O.; Casey, A.; McCarron, P.; Kavanagh, K.; McNamara, M.; Kennedy, S.; May, D.D.; et al. Water-soluble bis(1,10-phenanthroline) octanedioate Cu2+ and Mn2+ complexes with unprecedented nano and picomolar in vitro cytotoxicity: Promising leads for chemotherapeutic drug development. *MedChemComm* **2011**, *2*, 579–584. [CrossRef]
28. O'Connor, M.; Kellett, A.; McCann, M.; Rosair, G.; McNamara, M.; Howe, O.; Creaven, B.S.; McClean, S.; Kia, A.F.-A.; O'Shea, D.; et al. Copper(II) complexes of salicylic acid combining superoxide dismutase mimetic properties with DNA binding and cleaving capabilities display promising chemotherapeutic potential with fast acting in vitro cytotoxicity against cisplatin sensitive and resista. *J. Med. Chem.* **2012**, *55*, 1957–1968. [CrossRef]

29. Kellett, A.; Howe, O.; O'Connor, M.; McCann, M.; Creaven, B.S.; McClean, S.; Foltyn-Arfa Kia, A.; Casey, A.; Devereux, M. Radical-induced DNA damage by cytotoxic square-planar copper(II) complexes incorporating o-phthalate and 1,10-phenanthroline or 2,2′-dipyridyl. *Free Radic. Biol. Med.* **2012**, *53*, 564–576. [CrossRef]
30. Prisecaru, A.; Devereux, M.; Barron, N.; McCann, M.; Colleran, J.; Casey, A.; McKee, V.; Kellett, A. Potent oxidative DNA cleavage by the di-copper cytotoxin: [Cu2(µ-terephthalate)(1,10-phen)4]2+. *Chem. Commun.* **2012**, *48*, 6906–6908. [CrossRef]
31. Slator, C.; Barron, N.; Howe, O.; Kellett, A. [Cu(o-phthalate)(phenanthroline)] Exhibits Unique Superoxide-Mediated NCI-60 Chemotherapeutic Action through Genomic DNA Damage and Mitochondrial Dysfunction. *ACS Chem. Biol.* **2016**, *11*, 159–171. [CrossRef]
32. Molphy, Z.; Prisecaru, A.; Slator, C.; Barron, N.; McCann, M.; Colleran, J.; Chandran, D.; Gathergood, N.; Kellett, A. Copper Phenanthrene Oxidative Chemical Nucleases. *Inorg. Chem.* **2014**, *53*, 5392–5404. [CrossRef] [PubMed]
33. Molphy, Z.; Slator, C.; Chatgilialoglu, C.; Kellett, A. DNA oxidation profiles of copper phenanthrene chemical nucleases. *Front. Chem.* **2015**, *3*, 1–9. [CrossRef] [PubMed]
34. Zuin Fantoni, N.; Molphy, Z.; Slator, C.; Menounou, G.; Toniolo, G.; Mitrikas, G.; McKee, V.; Chatgilialoglu, C.; Kellett, A. Polypyridyl-Based Copper Phenanthrene Complexes: A New Type of Stabilized Artificial Chemical Nuclease. *Chem. Eur. J.* **2019**, *25*, 221–237. [CrossRef] [PubMed]
35. Toniolo, G.; Louka, M.; Menounou, G.; Fantoni, N.Z.; Mitrikas, G.; Efthimiadou, E.K.; Masi, A.; Bortolotti, M.; Polito, L.; Bolognesi, A.; et al. [Cu(TPMA)(Phen)](ClO4)2: Metallodrug Nanocontainer Delivery and Membrane Lipidomics of a Neuroblastoma Cell Line Coupled with a Liposome Biomimetic Model Focusing on Fatty Acid Reactivity. *ACS Omega* **2018**, *3*, 15952–15965. [CrossRef] [PubMed]
36. Slator, C.; Molphy, Z.; McKee, V.; Long, C.; Brown, T.; Kellett, A. Di-copper metallodrugs promote NCI-60 chemotherapy via singlet oxygen and superoxide production with tandem TA/TA and AT/AT oligonucleotide discrimination. *Nucleic Acids Res.* **2018**, *46*, 2733–2750. [CrossRef] [PubMed]
37. Sheldrick, G.M. SHELXT. *Acta Crystallogr.* **2015**, *A71*, 3–8. [CrossRef]
38. Sheldrick, G.M. SHELXL. *Acta Crystallogr.* **2015**, *C71*, 3–8. [CrossRef]
39. McCann, M.; McGinley, J.; Ni, K.; O'Connor, M.; Kavanagh, K.; McKee, V.; Colleran, J.; Devereux, M.; Gathergood, N.; Barron, N.; et al. A new phenanthroline-oxazine ligand: Synthesis, coordination chemistry and atypical DNA binding interaction. *Chem. Commun.* **2013**, *49*, 2341–2343. [CrossRef]
40. Slator, C.; Molphy, Z.; McKee, V.; Kellett, A. Triggering Autophagic Cell Death with a di-Manganese(II) Developmental Therapeutic. *Redox Biol.* **2017**, *12*, 150–161. [CrossRef]
41. Addison, A.W.; Rao, T.N.; Reedijk, J.; van Rijn, J.; Verschoor, G.C. Synthesis, structure, and spectroscopic properties of copper(II) compounds containing nitrogen–sulphur donor ligands; the crystal and molecular structure of aqua[1,7-bis(N-methylbenzimidazol-2′-yl)-2,6-dithiaheptane]copper(II) perchlorate. *J. Chem. Soc. Dalton Trans.* **1984**, 1349–1356. [CrossRef]
42. Groom, C.R.; Bruno, I.J.; Lightfoot, M.P.; Ward, S.C. The Cambridge Structural Database. *Acta Cryst. B.* **2016**, *72*, 171–179. [CrossRef] [PubMed]
43. Roy, M.; Dhar, S.; Maity, B.; Chakravarty, A.R. Dicopper(II) complexes showing DNA hydrolase activity and monomeric adduct formation with bis(4-nitrophenyl)phosphate. *Inorg. Chim. Acta* **2011**, *375*, 173–180. [CrossRef]
44. Ghosh, M.; Biswas, P.; Flörke, U. Structural, spectroscopic and redox properties of transition metal complexes of dipyrido[3,2-f:2′,3′-h]-quinoxaline (dpq). *Polyhedron* **2007**, *26*, 3750–3762. [CrossRef]
45. Reddy, P.A.N.; Santra, B.K.; Nethaji, M.; Chakravarty, A.R. Synthesis, crystal structure and nuclease activity of bis(dipyridoquinoxaline)copper(I) perchlorate. *Indian J. Chem. A.* **2003**, *42A*, 2185–2190.
46. Biswas, P.; Dutta, S.; Ghosh, M. Influence of counter anions on structural, spectroscopic and electrochemical behaviours of copper(II) complexes of dipyrido[3,2-f: 2′,3′-h]-quinoxaline (dpq). *Polyhedron* **2008**, *27*, 2105–2112. [CrossRef]
47. Kellett, A.; Molphy, Z.; Slator, C.; McKee, V.; Farrell, N.P. Molecular methods for assessment of non-covalent metallodrug–DNA interactions. *Chem. Soc. Rev.* **2019**, *48*, 971–988. [CrossRef]

Sample Availability: Please contact the corresponding author as required.

© 2019 by the authors. Licensee MDPI, Basel, Switzerland. This article is an open access article distributed under the terms and conditions of the Creative Commons Attribution (CC BY) license (http://creativecommons.org/licenses/by/4.0/).

Article

Why Does the Type of Halogen Atom Matter for the Radiosensitizing Properties of 5-Halogen Substituted 4-Thio-2′-Deoxyuridines?

Paulina Spisz [1], Magdalena Zdrowowicz [1], Samanta Makurat [1], Witold Kozak [1], Konrad Skotnicki [2], Krzysztof Bobrowski [2] and Janusz Rak [1,*]

1. Laboratory of Biological Sensitizers, Faculty of Chemistry, University of Gdańsk, Wita Stwosza 63, 80-308 Gdańsk, Poland
2. Centre of Radiation Research and Technology, Institute of Nuclear Chemistry and Technology, Dorodna 16, 03-195 Warsaw, Poland
* Correspondence: janusz.rak@ug.edu.pl

Academic Editor: Chryssostomos Chatgilialoglu
Received: 17 July 2019; Accepted: 31 July 2019; Published: 2 August 2019

Abstract: Radiosensitizing properties of substituted uridines are of great importance for radiotherapy. Very recently, we confirmed 5-iodo-4-thio-2′-deoxyuridine (ISdU) as an efficient agent, increasing the extent of tumor cell killing with ionizing radiation. To our surprise, a similar derivative of 4-thio-2′-deoxyuridine, 5-bromo-4-thio-2′-deoxyuridine (BrSdU), does not show radiosensitizing properties at all. In order to explain this remarkable difference, we carried out a radiolytic (stationary and pulse) and quantum chemical studies, which allowed the pathways to all radioproducts to be rationalized. In contrast to ISdU solutions, where radiolysis leads to 4-thio-2′-deoxyuridine and its dimer, no dissociative electron attachment (DEA) products were observed for BrSdU. This observation seems to explain the lack of radiosensitizing properties of BrSdU since the efficient formation of the uridine-5-yl radical, induced by electron attachment to the modified nucleoside, is suggested to be an indispensable attribute of radiosensitizing uridines. A larger activation barrier for DEA in BrSdU, as compared to ISdU, is probably responsible for the closure of DEA channel in the former system. Indeed, besides DEA, the XSdU anions may undergo competitive protonation, which makes the release of X^- kinetically forbidden.

Keywords: radiosensitizers; stationary radiolysis; pulse radiolysis; modified nucleosides; cellular response

1. Introduction

Trojan horse radiotherapy employs a nucleoside radiosensitizer, a "Trojan horse", that is activated only due to DNA exposure to ionizing radiation [1]. Such radiosensitizers are usually electrophilic nucleosides, incorporated into DNA during replication and repair, and undergoing efficient dissociative electron attachment (DEA) that leaves behind a nucleoside radical, which in secondary reactions is able to produce damage to the biopolymer (frequently a strand break) [1]. Although the purine derivatives of nucleosides were proposed as potential radiosensitizers [2–5], most of the reported examples comprise uridines substituted at the C5 position. This is because thymidine kinase accepts a broad set of modified uridines [6], which after phosphorylation may be incorporated into DNA [7]. The modified DNA sensitivity to hydrated electron attachment is especially important for radiotherapy, since cells of solid tumors (80% of cases [8]) are hypoxic, which make them resistant to hydroxyl radicals, a major damaging agent of native DNA produced during radiotherapy [9]. In the normoxic cells, damage produced by the •OH radicals becomes "fixed" due to reaction with oxygen, while under hypoxia, naturally occurring radioprotectors like cysteine or glutathione can restore DNA through hydrogen donation [10]. Moreover, hydrated electrons, the second most abundant product of water radiolysis, are not harmful to the natural DNA [11,12]. The situation becomes quite different when DNA is labeled with nucleosides undergoing efficient DEA. Indeed, Sanche et al. [13] demonstrated efficient formation of single strand breaks in DNA oligonucleotides labeled with BrdU, when an aqueous solutions of these biopolymers were irradiated with X-rays in the presence of a hydroxyl radical scavenger. Similarly, a radiolysis of a solution containing TXT oligonucleotides (where X stands for 5-bromo-2′-deoxyuridine (BrdU), 5-iodo-2′-deoxyuridine (IdU), 5-bromo-2′-deoxycitidine (BrdC), 5-iodo-2′-deoxycitidine (IdC), 8-bromo-2′-deoxyadenosine (BrdA), or 8-bromo-2′-deoxyguanosine (BrdG)) and t-butyl alcohol (t-BuOH) as •OH scavenger led to strand breaks besides other types of DNA damage [11,12].

BrdU and IdU are well-known radiosensitizers, which are phosphorylated in cytoplasm forming the respective 5′-triphosphates and then incorporated into the cellular DNA by human DNA polymerases [7]. Their promising radiosensitizing properties were investigated in numerous in vitro [14–16] and in vivo [17] studies and even in clinical trials [18]. In one of the most extensive clinical studies on brain tumor patients, no positive effects were observed in patients exposed to the specific doses of BrdU besides radiotherapy [18]. To this end, it is worth emphasizing that a swift and efficient metabolism (most radiosensitizers, as other chemotherapeutics, are applied systemically) of a sensitizer, may lead to its lower cellular concentration in vivo than in vitro. This, at least partially, explains a high radiosensitizing activity of BrdU in vitro, and practically, the lack of such activity in the clinical studies.

This situation calls for new radiosensitizers of superior pharmacokinetic and/or better radiosensitizing properties. Recently, we proposed several new C5-pyrimidine derivatives that have not been studied in animal models or in clinic to date [19–23]. In terms of electron-induced degradation yields, they are all more prone to dissociative electron attachment (DEA) than BrdU. These compounds comprise 5-thiocyanato-2′-deoxyuridine (SCNdU), 5-selenocyanato-2′-deoxyuridine (SeCNdU), 5-selenocyanatouracil (SeCNU), and 5-trifluoromethanesulfonyl-2′-deoxyuridine (OTfdU). Other promising candidates of this type of radiosensitizers seem to be derivatives of 4-thio-2′-deoxyuridine. The latter compound, similarly to BrdU and IdU, is incorporated into genomic DNA by the cellular enzymatic machinery [24]. It is worthwhile to note that BrdU works both as a DNA radio- and photosensitizer [1]. On the other hand, the photosensitizing properties of BrSdU and ISdU were proven in the past [25,26]. By the same token, one may conclude that 5-haloderivatives of 4-thio-2′-deoxyuridine could also work as radiosensitizers. Indeed, we recently demonstrated the radiosensitizing properties of ISdU [27]. Under the same conditions, the yield of damage produced by 140 Gy of X-ray was 1.5-fold larger than that assayed in the irradiated BrdU aqueous solutions. Simultaneously, in vitro studies demonstrated a significant increase of the mortality in cells treated with ISdU after irradiation.

In the current paper, studies on BrSdU–similar to those shown in [27] on ISdU–are described. To our surprise, BrSdU does not possess increased radiosensitizing properties. It is decomposed during radiolysis by X-ray, but the stable products resulted only from the reactions between the compound studied and H_2O_2 or radicals forming in the reaction between t-BuOH and the •OH radicals. We did not observe the characteristic pattern of DEA, i.e., the formation of 4-thio-2′-deoxyuridine, in this case. In accordance with this finding, the clonogenic assay does not differentiate the cells that were grown with and without BrSdU. We explain this striking difference between ISdU and BrSdU with the height of activation barrier for DEA, which is almost twice as much as in BrSdU.

2. Results and Discussion

A radiosensitizing nucleoside working under hypoxia must be sensitive to hydrated electrons, which are the second most abundant product of water radiolysis. In order to assess the radiosensitizing potential of a nucleoside, one must expose its aqueous, deoxygenated solution to ionizing radiation. If hydroxyl radicals are scavenged during irradiation, only the reaction between hydrated electrons and potential radiosensitizer may lead to serious damage associated with the formation of radical products. If the radiolysis proceeds in the cells containing DNA labeled with radiosensitizer, those radicals will produce DNA damage, hopefully single/double strand breaks, leading to apoptosis as an ultimate cellular response. This is why the radiolysis of a nucleoside in aqueous solution and the qualitative and quantitative analysis of radiolytic products is an indispensable step for assessing the radiosensitizing potential of the derivative under investigations.

2.1. Stationary Radiolysis

The high-performance liquid chromatography (HPLC) traces of radiolytes, originating from X-ray irradiation of buffered aqueous solution, containing 10^{-4} M of BrSdU in the presence of a hydroxyl radical scavenger (t-BuOH, 0.03 M) with the dose of 140 Gy, are depicted in Figure 1.

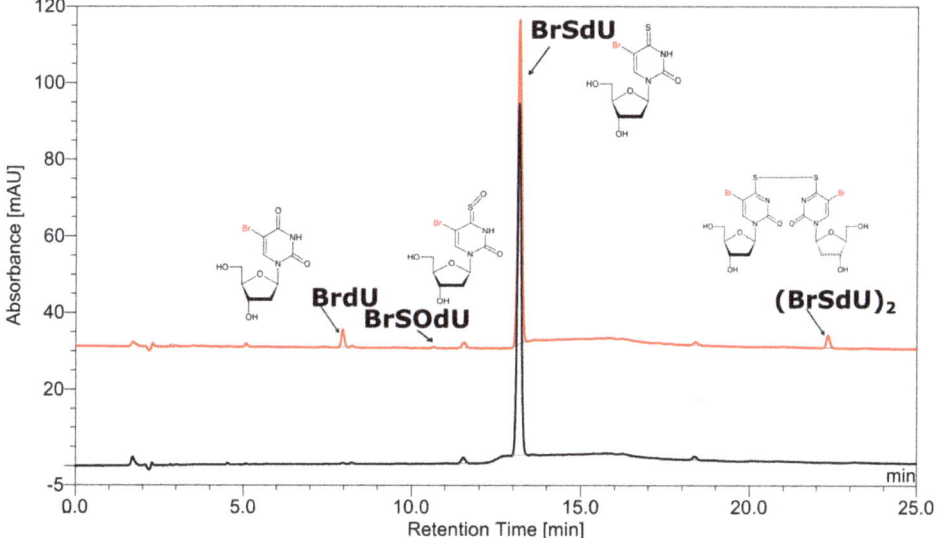

Figure 1. High-performance liquid chromatography (HPLC) traces for a solution of BrSdU before (black) and after irradiation (red). The chemical structures of products, as indicated by the liquid chromatography–mass spectrometry (LC–MS/MS) analysis, are depicted at particular peaks.

As indicated by the comparison of two chromatograms (Figure 1), three main products are formed due to radiolysis. The liquid chromatography–mass spectrometry (LC–MS) analysis enabled the identification of all these species. Figure 1 also depicts the chemical structures of the identified decomposition products, while the MS/MS spectra (shown in Figures S1–S4 in Supplementary Materials) confirm the assignment of particular structures. These radiolysis products are a dimer and two oxidation products, whose probable mechanism of formation was modelled for 5-bromo-1-methyl-4-thiouracil (see the Computational section) at the B3LYP(PCM)/DGDZVP++ level and is shown in Figure 2. The (BrSU)$_2$ dimer is suggested to be the product of two BrSU• radicals recombination (Figure 2A), which are created in the reaction of BrSU with •CH$_2$(CH$_3$)$_2$COH (the •CH$_2$(CH$_3$)$_2$COH radicals are formed in the reaction between t-BuOH, present in the solution as a hydroxyl radical scavenger, and the •OH radicals–a primary product of water radiolysis). The reaction is associated with a kinetic barrier of 76.1 kJ/mol and is favorable thermodynamically (Figure 2A). The second reaction, leading to BrSOU (Figure 2B) due to oxidation of BrSU by H$_2$O$_2$ [28–30] (H$_2$O$_2$ is produced during water radiolysis [31]), was modelled with three explicitly added water molecules, which is the approach that had been suggested in the literature [32]. It is worthwhile to note that under the experimental conditions, BrSOU

(BrSOdU) appears to be the least abundant product, which probably results from the fact that it is also the substrate for the most abundant one, i.e., for BrU (Figure 2C). The latter is formed via a cyclic oxathiirane followed by the sulfur extrusion reaction [33,34]. In our calculations, we were unable to obtain the stable oxathiirane intermediate. During the optimization, the ring opened to give the BrOSU structure shown in Figure 2C. Also the sulfur extrusion does not show any intermediates. After the TS structure is achieved (4.4 kJ/mol barrier), one of the sulfur atoms attaches to the other one and the S-O bond breaks leading to BrU and S_2 (Figure 2).

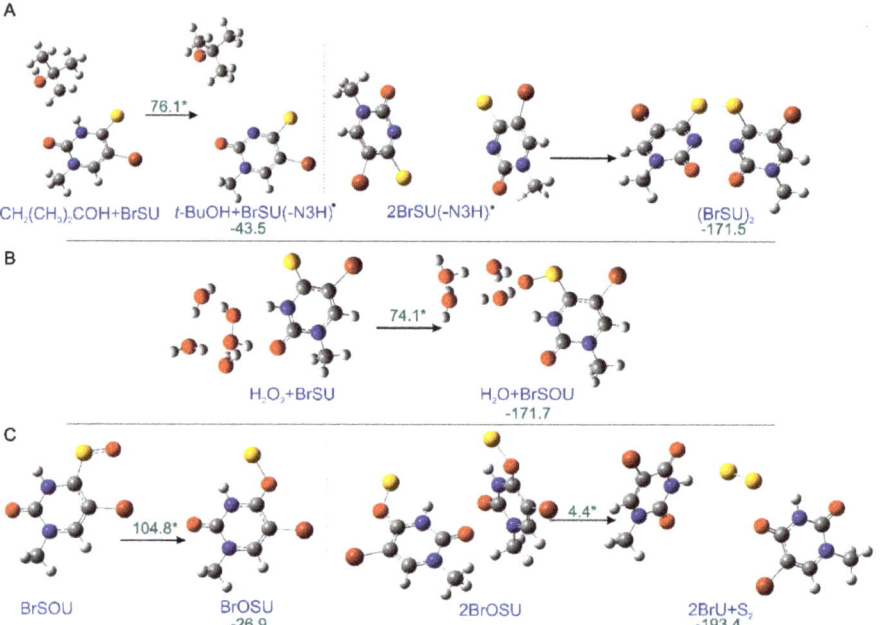

Figure 2. The radiolysis products. (**A**) (BrSU)$_2$, (**B**) BrSOU, and (**C**) BrU formation, as suggested by calculations. The optimized reactants in ball and stick representation are shown along with their kinetic barriers (marked with asterisks) and thermodynamic stimulus (kJ/mol). All the reactions shown correspondence to the most favorable pathways, as obtained by the IRC procedure. The transition states structures can be found in Supplementary Materials (Figure S5).

To our surprise, one of the expected products, 4-thio-2′-deoxyuridine, that should form due to DEA to BrSdU, was not detected in the BrSdU radiolytes (see Figure 1). DEA, occurring in many similar systems including BrdU, IdU, and ISdU, is thought to be the main reason of DNA damage in the irradiated cells, i.e., it is responsible for the radiosensitizing potential of modified nucleosides [12,35–37]. However, neither 4-thio-2′-deoxyuridine nor dimer with the substrate (both were observed in radiolytes of ISdU) were observed among the radiolysis products.

In order to explain why the radiolysis of ISdU leads to SdU, while this reaction channel is actually closed for BrSdU under the same experimental conditions, we calculated the respective DEA profile (Figure 3). We found that the kinetic barrier for the C5-Br bond had breakage as much as 26.0 kJ/mol–more than two times higher than that for C5-I in ISU (12.6 kJ/mol [27]) dissociation calculated at the same level of theory. Similarly, the thermodynamic stimulus for the release of the bromide anion from BrSU$^{•-}$ amounts to only −10.5 kJ/mol as compared to −24.3 kJ/mol for ISU$^{•-}$ [27]. These differences, especially a significant difference in the height of activation barriers in the two compared systems, explains formation of SdU only in the latter derivative. The estimated lifetime of BrSdU$^{•-}$ at the ambient temperature is ca. 200-fold longer than that of ISU$^{•}$, which results from the above-mentioned activation barriers and transition state theory [38]. It is probably sufficiently long to allow the anion to be protonated, which prevents completion of the DEA process [39], and in consequence, SdU is not formed.

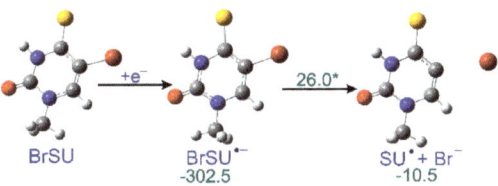

Figure 3. Calculated dissociative electron attachment (DEA) profile for BrSU. After the initial electron attachment to BrSU, the anion radical BrSU$^{\bullet-}$ is formed, and subsequently dissociates via a transition state giving SU$^\bullet$ and Br$^-$. The thermodynamic and kinetic characteristics [kJ/mol], shown in green, were calculated as the difference between the given state and the previous stable one, the transition state barrier marked with asterisk.

2.2. Pulse Radiolysis

The hypothesis, explaining different behavior of BrSdU and ISdU based on the computational results and discussed in the previous section, is confirmed by the results of our pulse radiolysis experiments. Figures 4 and 5 depict the transient spectra, as well as the respective decays and growths in microsecond-time domain for ISdU and BrSdU, respectively.

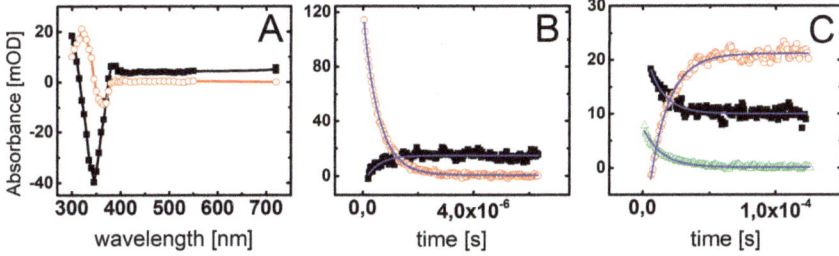

Figure 4. (**A**) Transient absorption spectra recorded in deoxygenated and buffered with phosphate (10 mM, pH = 7.0) ISdU solution (5 · 10^{-5} M), in the presence of 0.5 M t-BuOH, after 2 μs (■) and 120 μs (○) after the electron pulse. (**B**) Short-time profiles representing the growth at λ = 305 nm (■) and the decay at λ = 720 nm (○) of transient absorptions and their least-square fits to the first order formation and decay, respectively. (**C**) Long-time profiles representing the growth at λ = 320 nm (○) and decays at λ = 300 nm (△) and λ = 385 nm (■) of transient absorptions and their least-square fits to the first order formation and decays, respectively.

The resulting transient spectrum, obtained 2 μs after the electron pulse, exhibits a rising absorption toward 300 nm with no defined maximum and negative absorption in the wavelength range 320–380 nm (Figure 4A). This time delay for spectra recording was chosen on purpose in order to get rid of participation of hydrated electrons in the spectrum. Nonetheless, the registration of "pure" spectrum of this transient product was not possible due to the bleaching related to the consumption of ISdU. The decay at λ = 720 nm represents the decay of hydrated electrons in the presence of 5 · 10^{-5} M ISdU with the pseudo-first order rate constant k_{720} = 1.8 · 10^6 s^{-1}. In turn, the growth at λ = 300 nm represents the formation of a transient product with the pseudo-first order rate constant k_{300} = 2.0 · 10^6 s^{-1} (Figure 4B). Since these rate constants are very similar, this species could be a direct product of the hydrated electron attachment to ISdU if the lifetime of ISdU$^{\bullet-}$ was long enough and included the microsecond-time domain. However, as indicated by the B3LYP/DGDZVP++ barrier height, its lifetime should be very short and is rather in the nanosecond-time domain (see the previous section). Therefore, we probably observed the product of DEA to ISdU, i.e., the SdU$^\bullet$ radical formed via Reaction (1).

$$e_{aq}^- + \text{ISdU} \rightarrow \text{ISdU}^{\bullet-} \rightarrow \text{SdU}^\bullet + \text{I}^- \tag{1}$$

Indeed, the calculated UV spectrum for SdU$^\bullet$ is characterized by λ_{max} located at 300 nm. With the time elapsed, the absorption spectrum underwent further changes and 120 μs after the electron pulse is characterized by a transient absorption band with λ_{max} = 320 nm, which can be assigned to a new product (Figure 4A). The growth at λ = 320 nm is mono-exponential and occurs with the pseudo-first order rate constant k_{320} = 7.0 · 10^4 s^{-1}. Interestingly, the decays at λ = 300 nm and 385 nm are also mono-exponential, with the respective pseudo-first order rate constants k_{300} = 8.2 · 10^4 s^{-1} and k_{385} = 7.1 · 10^4 s^{-1}, which are reasonably close to k_{320} (Figure 4C). Since these decays seem to represent the decay of the SdU$^\bullet$ radical, the growth observed at λ = 320 nm can be tentatively assigned to the formation of SdU via reaction of SdU$^\bullet$ with hydrogen atom donor, which is t-BuOH present in the system in a large excess (Reaction 2).

$$\text{SdU}^\bullet + t\text{-BuOH} \rightarrow \text{SdU} + {}^\bullet\text{CH2(CH3)2COH} \tag{2}$$

The calculated (and measured) UV spectrum of 4-thiouracil strongly supports this assignment. Moreover, the absorption at λ = 320 nm is stable in the time window of our experiment (up to 1.5 ms) with no signs of disappearance, which might suggest, however not directly, a high stability of the product.

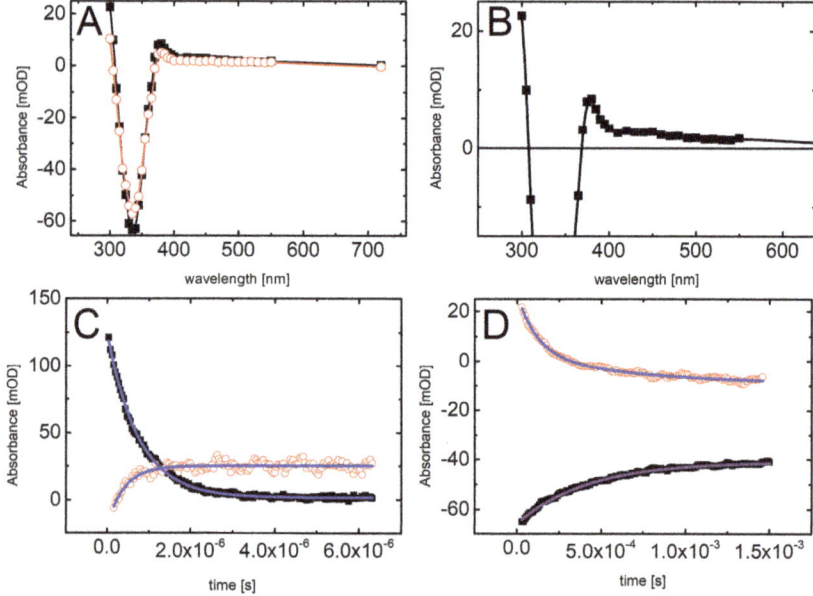

Figure 5. Transient absorption spectra recorded in deoxygenated and buffered with phosphate (10 mM, pH = 7.0) BrSdU solution (5 · 10^{-5} M), in the presence of 0.5 M t-BuOH (**A**) after 12 μs (■) and 120 μs (○) and (**B**) (■) 12 μs after the electron pulse. (**C**) Short-time profiles representing the growth at λ = 300 nm (○) and the decay at λ = 720 nm (■) of transient absorptions and their least-square fits to the first order formation and decay, respectively. (**D**) Long-time profiles representing the growth at λ = 335 nm (■) and decays at λ = 300 nm (○) of transient absorptions and their least-square fits to the second order formation and decays, respectively.

The main difference between spectral features observed during pulse radiolysis of ISdU and BrSdU aqueous solutions is the lack of the transient absorption band with λ_{max} = 320 nm for the latter system (compare Figure 5A with Figure 4A). This finding clearly shows that the formation of SdU does not occur in BrSdU via analogous reaction 2, which requires the presence of SdU$^\bullet$ radical. This fact remains in a good accordance with the results of stationary γ-radiolysis, where we did not observe the product of the bromide anion (Br$^-$) release from BrSdU$^{\bullet-}$. In case of ISdU, we also observed an increase in the integrated XIC signal of the iodide anions (a background signal of iodide anions is observed due to synthesis-related contamination, unavoidable degradation of the sample in an aqueous solution, as well as fragmentation of the studied nucleosides in the MS source [40,41]) due to stationary radiolysis. Indeed, the ratio of the signals after and before irradiation is equal to 2.041 ± 0.016. When it comes to the BrSdU sample, the integrated XIC signal of Br$^-$ anions is the same before and after irradiation, 0.987 ± 0.0791, within the error bar. Using the calibration curve (not shown), we found out that the measured increase in the concentration of I$^-$ due to irradiation means that ca. 27% of ISdU decay occurs in the DEA pathway. Thus, the observed changes in the concentration of halogen anions support that DEA process is operative only in ISdU solutions. This fact can be rationalized by the slower dissociation of BrSdU$^{\bullet-}$ (Reaction 3) in comparison to dissociation of ISdU$^{\bullet-}$, which allows BrSdU$^{\bullet-}$ to be involved in another competitive process, for instance, its protonation by water or phosphate anions (Reaction 4):

$$BrSdU^{\bullet-} \rightarrow SdU^\bullet + Br^- \quad (3)$$

$$BrSdU^{\bullet-} \xrightarrow{H_2O,\ H_2PO_4^-} BrSdUH^\bullet \quad (4)$$

The resulting absorption spectrum, obtained 12 μs after the electron pulse, exhibits a rising absorption toward 300 nm with no defined maximum and negative absorption in the wavelength range 320–380 nm (Figure 5A). This time delay for spectra recording was again chosen on purpose in order to get rid of participation of the most hydrated electrons in the spectrum. Since the absorption spectrum recorded 120 μs after the electron pulse does not exhibit the absorption band with λ_{max} = 320 nm (in contrast to ISdU), the absorption spectrum recorded 2 μs after the electron pulse cannot consequently be assigned to SdU$^\bullet$ radical. Moreover, it is worthwhile to note that the pseudo-first rate constant of the decay of hydrated electrons in the presence of $5 \cdot 10^{-5}$ M BrSdU ($k_{720} = 1.4 \cdot 10^6$ s^{-1}) is nearly two-fold lower than the pseudo-first order rate constant of the formation of the transient measured at λ = 300 nm ($k_{300} = 2.4 \cdot 10^6$ s^{-1}) (Figure 5C). This observation suggests that this transient product cannot be formed in an analogous Reaction (2), as observed for ISdU, but might result from the protonation of BrSdU$^{\bullet-}$ (Reaction 4) (*vide supra*). Therefore, the absorption spectrum recorded after 12 μs can be tentatively assigned to BrSdUH$^\bullet$. The UV-VIS spectrum of BrSdUH$^\bullet$ (calculated by us) possesses the absorption bands with λ_{max} = 280, 350, and 480 nm. Therefore, for better visualization of the experimental spectrum assigned by us to BrSdUH$^\bullet$ radical, the spectrum recorded 12 μs after the pulse was zoomed to see expected spectral features (Figure 5B). The transient absorption depicted in Figure 5B shows the maxima below 300 nm and at 450 nm. The maximum at 350 nm is concealed by the "negative" absorption of BrSdU, but a tail at 380 nm is quite clear.

Interestingly, the decay at λ = 300 nm and the formation at λ = 335 nm within 1.5 ms time domain can be fitted by the second-order kinetics and occur with the respective second-order rate constants $2k_{300} = 1.2 \cdot 10^9$ M^{-1}s^{-1} and $2k_{335} = 1.0 \cdot 10^9$ M^{-1}s^{-1}, which are very similar (Figure 5D). One of the processes, which can be described by the second-order kinetics is disproportionation reaction involving BrSdUH$^\bullet$ radicals, should lead to the partial recovery of the substrate (BrdSU) (Reaction 5).

$$BrSdUH^\bullet + BrSdUH^\bullet \rightarrow BrSdU + BrSdUH_2 \quad (5)$$

Thus, the decay observed at λ = 300 nm represents the decay of BrSdUH$^\bullet$ radicals and the growth observed at λ = 335 nm represents recovery of BrSdU substrate. This mechanism can, at least in part, explain lower consumption of BrSdU compared to ISdU using the same dose delivered by X-rays.

2.3. Biological Assessments

2.3.1. Incorporation of BrSdU and ISdU into Genomic DNA

According to the concept of the Trojan horse therapy, a radiosensitizer should easily incorporate into DNA. For this reason the incorporation of BrSdU and ISdU into genomic DNA was assessed. The MCF-7 cells treated with BrSdU and ISdU at the concentration of 10^{-4} M were incubated for 48 h. Purified DNA was enzymatically digested and analyzed by HPLC (Figure S6) and LC-MS method. The results of LC-MS analysis shows that both derivatives incorporate to DNA (see extracted-ion chromatograms, MS and MS/MS spectra for BrSdU/ISdU in Figures S7–S10 in Supplementary Materials), but the efficiency of this process is very low, which has been already observed by others [24]. To our surprise, BrdU and IdU beside BrSdU and ISdU were also observed in the digested material, which suggests an enzymatic degradation of BrSdU/ISdU to BrdU/IdU in the cell with the rate similar to that of BrSdU/ISdU incorporation into DNA.

2.3.2. Clonogenic Assay

In order to determine the simultaneous effect of BrSdU and ionizing radiation on the survival and proliferation of cancer cells, the clonogenic assay (based on the ability of a single cell to grow into a colony) [42] was performed. The test was carried out on human breast cancer cells (MCF-7 line) treated with BrSdU at the concentration of 0, 10, and 100 µM and/or ionizing radiation (IR) in four doses of 0.5, 1, 2, and 3 Gy. Figure 6 shows that the studied compound does not affect the survival of cancer cells. We only observed reduction of survival fraction caused by IR. For example, in case of dose of 2 Gy, survival was 29.9 ± 4.6%, 32.7 ± 6.3%, and 28.2 ± 2.8% for 0, 10, and 100 µM BrSdU pretreatment, respectively. This colony formation assay demonstrates that the BrSdU does not sensitize the MCF-7 cells to X-ray. In our previous studies, we demonstrated a significant radiosensitizing effect of ISdU [27]. The addition of the latter derivative to cell culture resulted in a significant decrease (about 20%) of their survival after irradiation, even with doses as low as 0.5 Gy.

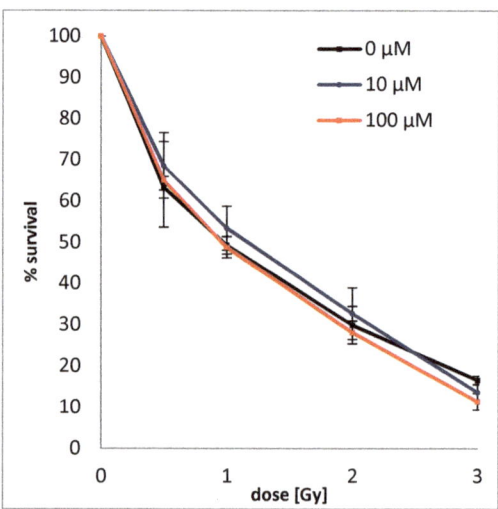

Figure 6. Dose response curves of MCF-7 cells treated with (10 µM or 100 µM solutions of BrSdU) or without BrSdU. The average plating efficiencies for the controls with and without pretreatment are equal to 28.16% (0 µM), 25.63% (10 µM), and 24.78% (100 µM). Experiments were performed at least in two independent experiments in duplicate and the results are expressed as mean ± standard deviation.

2.3.3. Cytotoxicity Assay

One of the properties that good radiosensitizer should possess is low cytotoxicity without IR treatment. To identify cytotoxicity of BrSdU toward MCF-7 (cancer cells) and HDFa (normal cells) line, MTT assay [43] was carried out (Figure 7). BrSdU was tested at six concentrations (0, 10^{-8}, 10^{-7}, 10^{-6}, 10^{-5}, 10^{-4}, $5 \cdot 10^{-4}$ M) and two time variants (24 and 48 h of incubation). Figure 7Ib shows statistically significant reduction of viability up to 93% (for 48 h incubation) for MCF-7 line only in case of the highest tested concentration equal to $5 \cdot 10^{-4}$ M. For lower concentrations, the decrease in vitality was not statistically significant. We also did not observe a meaningful difference between the studied cell lines. These results show that the cytotoxicity of BrSdU is very low both for normal human dermal fibroblasts and human breast cancer cells.

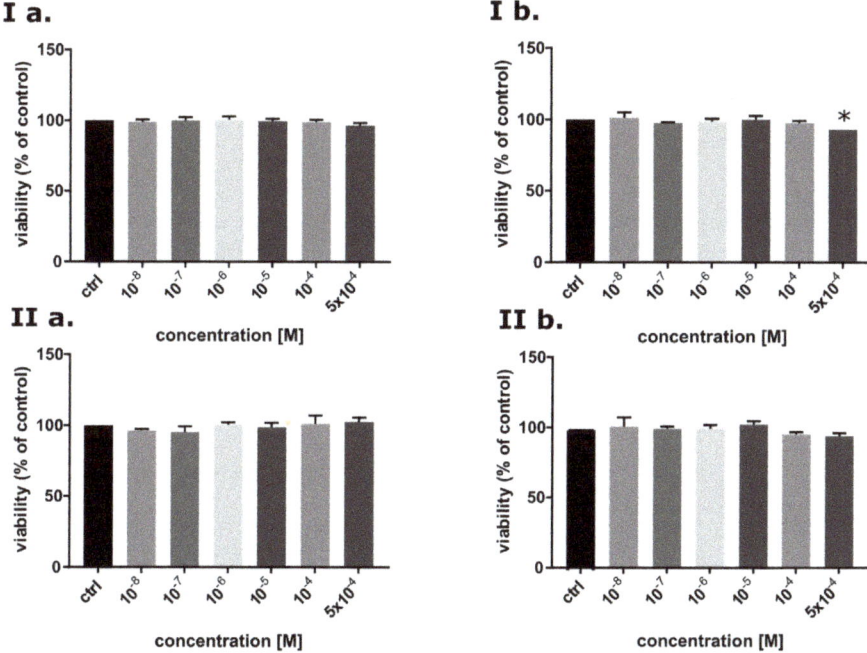

Figure 7. The viability of MCF-7 (**I**) and HDFa (**II**) cells after 24 (**a**) and 48 h treatment (**b**) with BrSdU in a range of concentrations from 0 to $5 \cdot 10^{-4}$ M. Results are shown as mean ± SD of three independent experiments performed in triplicate. *statistically significant difference is present between treated culture compared with control (untreated culture).

2.3.4. Analysis of Histone H2A.X Phosphorylation and Cell Death

One of the most common types of DNA damage related to radiosensitization is double-strand breaks formation. Phosphorylation of histone γH2A.X is the marker of such a damage [44]. The assay was performed for human breast cancer cells treated with BrSdU at concentration of 10^{-4} M and/or irradiated with a dose of 0, 1, or 2 Gy. Analysis of H2A.X phosphorylation was carried out 1 h after irradiation (this time was optimized in previous experiments). The cells were fixed and analyzed by flow cytometry. Our studies show that treatment with BrSdU results in a tiny increase in the population of γH2A.X positive cells after irradiation with the doses of 1 and 2 Gy (Figure 8 and Figure S11). After BrSdU pretreatment and irradiation with the dose equal to 1 Gy, the level of γH2A.X was changed from 26.48% (nontreated cells) to 27.64 ± 0.34%). The exposure of treated cells to the dose of 2 Gy results in a slight enhancement of the γH2A.X fraction from 31.84 ± 3.74% to 36.15 ± 1.25%.

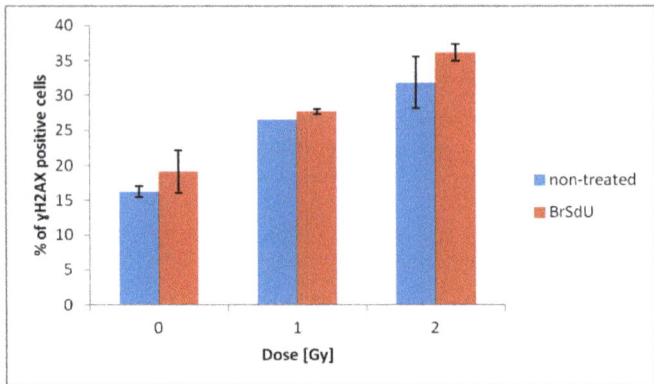

Figure 8. Flow cytometric analysis of H2A.X phosphorylation. γH2A.X was measured 1 h after irradiation. Results are shown as the mean ± standard deviation of at least two independent flow cytometry experiments.

Additionally, we performed the multidimensional test to quantify the number of viable, early apoptotic, late apoptotic, and dead cells. Our results (Figures S12 and S13 in Supplementary Materials) confirm that pretreatment with BrSdU does not affect the sensitivity of MCF-7 cells to ionizing radiation. A significant influence of the studies analog on population of viable, early apoptotic, late apoptotic, and dead cells was not observed. In the case of ISdU, the situation was completely different. The histone H2A.X phosphorylation test showed that ISdU sensitized breast cancer cells to ionizing radiation, at least in part, by formation of DSBs, while the cell death assay confirmed that pretreatment of culture with ISdU led to the IR-induced reduction of cell viability and increase in the population of early apoptotic cells [27].

3. Materials and Methods

3.1. Chemicals

5-bromo-2′-deoxyuridine, acetic anhydride, P_2S_5, 1,4-dioxane, and sodium hydride were commercially available from Sigma–Aldrich (Saint Louis, MO, USA). Nuclear magnetic resonance (NMR) spectrum was recorded on a Bruker AVANCE III (Bruker, Billerica, MA, USA), 500 MHz spectrometer. Chemical shifts are reported in ppm relative to the residual solvent peak (DMSO-d_6 2.49 ppm for ^{1}H and 39.5 ppm for ^{13}C). Column chromatography was performed using silica gel NORMASIL 60 (40–63 mesh, VWR Chemicals, Gdańsk, Poland). Preparative thin-layer chromatography was performed with silica gel plates, 60G, F254 (Sigma–Aldrich).

3.2. Synthesis of 3′,5′-di-O-acetyl-5-bromo-2′-deoxyuridine

A solution of 5-bromo-2′-deoxyuridine (500 mg, 1.63 mmol) in pyridine (6 mL) was stirred at room temperature with acetic anhydride (339 μL, 3.59 mmol) for 24 h. The sirupus residue was co-evaporated with three portions of aqueous ethanol (5 mL) and *n*-heptane to remove the pyridine residue. The raw 3′,5′-di-O-acetyl derivative (580 mg) was obtained in a 91% yield.

3.3. Synthesis of 3′,5′-di-O-acetyl-5-bromo-4-thio-2′-deoxyuridine

3′,5′-di-O-acetyl-5-bromo-2′-deoxyuridine (580 mg, 1.48 mmol) was dissolved in 1,4-dioxane (20 mL) and P_2S_5 (990 mg, 4.45 mmol) was added. The mixture was refluxed until thin-layer chromatography (TLC) analysis ($CHCl_3$:CH_3OH, 30:1) showed complete disappearance of the substrate (2–3 h). Solvent was removed under reduced pressure and the residue was treated several times with $CHCl_3$. The combined chloroform extracts were evaporated, and the residue was separated on silica gel column, which was eluted with CHCl$_3$:CH$_3$OH, 20:1. After evaporation, the desired product was obtained as a yellow solid (567 mg, 94%).

3.4. Synthesis of 5-bromo-4-thio-2'-deoxyuridine

3′,5′-di-O-acetyl-5-bromo-4-thio-2′-deoxyuridine (567 mg, 1.39 mmol) was dissolved in methanol (10 mL) and stirred at 0 °C. A methanolic sodium methoxide solution (111 mg, 2.78 mmol), freshly prepared from NaH and anhydrous methanol, was added in portions. The mixture was stirred at room temperature until TLC analysis showed complete disappearance of the substrate (10 min). The mixture was purified on silica gel column, which was eluted with CHCl$_3$:CH$_3$OH, 20:1. The final product, 5-bromo-4-thio-2′-deoxyuridine, was obtained as a yellow solid (185 mg, 41%).

^{1}H NMR (Bruker AVANCE III, 500 MHz, DMSO), δ: 13.09 (s, 1H), 8.53 (s, 1H), 6.03 (t, 1H), 4.25 (q, 1H), 3.82 (q, 1H), 3.55–3.69 (m, 2H), 2.14–2.26 (m, 2H); ^{13}C NMR (125 MHz, DMSO), δ: 186.8, 147.6, 137.7, 107.1, 88.2, 86.1, 69.9, 60.8, 40.8. HRMS (TripleTOF 5600+, SCIEX), m/z: [M–H]$^-$ calculated for C$_9$H$_{11}$BrN$_2$O$_4$S 323.1636, found 322.9918; UV spectrum (water), λmax: 345 nm.

3.5. Stationary Radiolysis

A mixture of BrSdU (10^{-4} M), 0.03 M t-BuOH (used as a scavenger of the •OH radicals), and phosphate buffer (10 mM, pH = 7.0) was purging with argon for ca. 3 min in order to remove oxygen from the solution. The dose absorbed by all samples during irradiation was 140 Gy (4.14 Gy·min^{-1}, 130.0 kV, 5.0 mA). The studied samples were analyzed in triplicate. Radiolysis was performed in a Cellrad X-ray cabinet (Faxitron X-ray Corporation, Tucscon, AZ, USA).

3.5.1. HPLC Analysis

Irradiated and non-irradiated samples of BrSdU were analyzed with reversed-phase HPLC method. For the separation of analytes a C18 column (Wakopak Handy ODS, 4.6 · 150 mm, 5 μm in particle size and 100 Å in pore size), gradient elution with 80% ACN and 0.1% HCOOH (from 0 to 50% ACN in 30 min), flow rate 1 mL·min^{-1} were used. The HPLC analysis was performed on a DionexUltiMate 3000 System (Dionex Corporation, Sunnyvale, CA, USA) with a Diode Array Detector set at 260 nm.

3.5.2. LC-MS and LC-MS/MS Analysis

The solution of BrSdU containing t-BuOH and phosphate buffer was analyzed by LC-MS and LC-MS/MS methods, before and after irradiation. Conditions of separations: Kinetex column (Phenomenex, 1.7 μm, C18, 100 Å, 2.1 × 150 mm); flow rate 0.3 mL·min^{-1}; a gradient elution with 80% ACN and 0.1% HCOOH (from 0 to 50% acetonitrile); the oven temperature was maintained at 25 °C. The effluent was diverted to waste for 2 min after injection. Conditions for MS and MS/MS analysis: the spray voltage was −4.5 kV, the nebulizer gas (N$_2$) pressure was 25 psi, the flow rate was 11 L·min^{-1}, and the source temperature was 300 °C. Each spectrum was obtained by averaging three scans and the time of each scan was 0.25 s. The LC-MS and LC-MS/MS analysis was performed TripleTOF 5600+ (SCIEX) mass spectrometer (operated in negative mode) coupled with Ultra-High Performance Liquid Chromatography (UHPLC) system Nexera X2 (Shimadzu, Canby, OR, USA).

3.6. Pulse Radiolysis

Pulse radiolysis experiments were performed with the INCT LAE 10 MeV linear electron accelerator with a typical pulse length of 8 ns. A detailed description of the experimental setup can be found in [45], along with the basic details of the equipment and the data collection system. Absorbed doses per pulse were on the order of 20 Gy (1 Gy = 1 J·kg^{-1}). Dosimetry was based on the solutions saturated with nitrous oxide, containing 10^{-2} M KSCN, taking a radiation chemical yield of G = 0.635 μmol·J^{-1} and a molar absorption coefficient of 7580 M^{-1}·cm^{-1} at 472 nm for the (SCN)$_2$•$^-$ radical [45]. Experiments were performed with a continuous flow of sample solutions at room temperature (~23 °C). All solutions were made with triply distilled water provided by a Millipore Direct-Q 3-UV system. The typical concentration of BrSdU and ISdU was $5 \cdot 10^{-5}$ M, and 0.5 M of *t*-BuOH was used as a hydroxyl radical scavenger. Solutions were deoxygenated by purging with high purity argon, and 10 mM of phosphate buffer was added to maintain pH = 7.

3.7. Clonogenic Assay

Adherent cell line MCF-7 (human breast cancer cells obtained from Cell Line Service–CLS, Eppelheim, Germany), which was treated with BrSdU in concentration of 10^{-4} and 10^{-5}, respectively, was plated on 60 mm dishes in a density of 10^6 cells per dish. After 48 h incubation under 37 °C and 5% CO_2, the cells were exposed to 0.5, 1, 2, and 3 Gy, respectively (1.27 Gy $\cdot$ min^{-1}, 130.0 kV, 5.0 mA). After 6 h, the cells were trypsinized and plated on 100 mm dishes in a density of 800 cells per dish. After 16 days, formed colonies were fixed with 6.0% (*v*/*v*) glutaraldehyde and 0.5% crystal violet. Stained colonies were counted manually, and colony size was rated using inverted fluorescence microscope (Olympus, IX73, Tokyo, Japan). The cells were grown in the RPMI medium supplemented with 10% FBS (fetal bovine serum) and with antibiotics (streptomycin and penicillin) at concentration of 100 U $\cdot$ mL^{-1}. Irradiation has been performed in a Cellrad X-ray cabinet (Faxitron X-ray Corporation). Plating efficiencies are shown in Table S1 in Supplementary Materials.

3.8. Cytotoxicity Assay

The MTT assay was used to identify the cytotoxic activity. Adherent MCF-7 and HDFa cell lines were seeded into 96-well plate in a density of $4 \cdot 10^3$ per well and incubated under 37 °C and 5% CO_2, overnight. After that, the medium was replaced to fresh and the cells were treated with BrSdU at concentration of 0 (control), 10^{-4}, 10^{-5}, 10^{-6}, 10^{-7}, 10^{-8} M. Plates with cells were incubated (under the same conditions) with compound at 24 and 48 h. After this time, the aqueous solution of MTT salt at concentration 4 mg $\cdot$ mL^{-1} was added and incubate for 4 h. Then, the medium was removed and dimethylsulfoxide was added to each well in 200 µL volume. The absorbance was measured at 570 nm (and 660 nm, reference wavelength). Absorbance measurement has been performed with use of EnSpire microplate reader (PerkinElmer, Waltham, MA, USA). The liveliness of control was taken as 100%. The results were analyzed with the use of GraphPad Prism software. The statistical evaluation of treated samples and untreated control was calculated using one-way analysis of variance (ANOVA) followed by Dunnett's multiple comparison test. The data was obtained from three independent experiments and each treatment condition assayed in triplicate. The differences were considered significant at $p < 0.05$. The cells were grown in the RPMI (MCF-7)/DMEM (HDFa) medium supplemented with 10% FBS (fetal bovine serum) and with antibiotics (streptomycin and penicillin) at concentration of 100 U $\cdot$ mL^{-1}.

3.9. Incorporation of BrSdU and ISdU into Genomic DNA

The MCF-7 cell line was seeded into plate and incubated under 37 °C and 5% CO_2 overnight. After that, the medium was replaced with fresh one and the cells were treated with BrSdU and ISdU at concentration of 0 (control) and 10^{-4} M. Plates with cells were incubated (under the same conditions) with compound for 48 h. After this time, the cells were pulled from the plates and isolation was carried out according to the protocol provided by the manufacturer (GeneMATRIX Cell Culture DNA Purification Kit, EURX, Gdańsk, Poland). After that, the purified DNA was enzymatically digested by the simultaneous action of DNase I, snake venom phosphodiesterase (SVP) and bacterial alkaline phosphatase (BAP).

3.9.1. HPLC Analysis

The mixture of nucleoside (dC, dA, dG, dT, BrdU, BrSdU, and ISdU), nontreated and treated with BrSdU/ISdU samples, were analyzed with reversed-phase HPLC method. For the separation of analytes, a C18 column (Wakopak Handy ODS, 4.6 × 150 mm, 5 μm in particle size and 100 Å in pore size), gradient elution with 80% ACN, and 0.1% HCOOH (from 0 to 50% ACN in 30 min), flow rate 1 mL·min^{-1} were used. The HPLC analysis was performed on the DionexUltiMate 3000 System with a Diode Array Detector (Dionex Corporation, Sunnyvale, CA, USA) set at 260 nm.

3.9.2. LC-MS Analysis

DNA samples, isolated from the ISdU/BrSdU-treated culture, after enzymatic digestion were analyzed by LC-MS and LC-MS/MS methods. Conditions of separations: Kinetex column (Phenomenex, 1.7 μm, C18, 100 Å, 2.1 × 150 mm); flow rate 0.3 mL·min^{-1}; a gradient elution with 80% ACN and 0.1% HCOOH (from 0 to 50% acetonitrile); the oven temperature was maintained at 25 °C. The effluent was diverted to waste for 2 min after injection. Conditions for MS and MS/MS analysis: the spray voltage was −4.5 kV and the source temperature was 300 °C. The LC-MS and LC-MS/MS analyses were performed with the use of TripleTOF 5600+ (SCIEX) mass spectrometer (operated in negative mode) coupled with Ultra High Performance Liquid Chromatography (UHPLC) system Nexera X2.

3.10. Flow Cytometry Analysis of Histone H2A.X Phosphorylation and CellDdeath

3.10.1. Analysis of Histone H2A.X Phosphorylation

Human cells of breast cancer MCF-7 were grown in RPMI medium supplemented by the 10% FBS and antibiotics (streptomycin and penicillin) at a concentration of 100 U·mL^{-1}. Cells, at a density of $0.2 \cdot 10^6$ per plate, were incubated for 24 h (37 °C, 5% CO$_2$). After this time, the cells were treated with BrSdU at a concentration of 10^{-4} M and incubated for next 48 h under the same conditions. Then plate cultures were irradiated (Cellrad X-ray cabinet, Faxitron X-ray Corporation) with 1 and 2 Gy doses (1.27 Gy·min^{-1}, 130.0 kV, 5.0 mA) and incubated for 1 h. After this time, the MCF-7 cells were dissociated with 1x Accutase solution, fixed, permeabilized, and stained. The last step was cytometric analysis (Guava easyCyte™ 12, Merck, Hayward, CA, USA). Fixation, permeabilization, and staining were carried out according to the manufacturer's protocol (FlowCellect™ Histone H2A.X Phosphorylation Assay Kit, Merck). The experiment was carried out in duplicate. Nontreated cultures were used as controls.

3.10.2. Cell Death

After 24 h incubation, cells were treated with BrSdU at a concentration of 10^{-4} M and again incubated for 48 h (37 °C, 5% CO$_2$). Then, the cells were irradiated (Cellrad X-ray cabinet, Faxitron X-ray Corporation) with a dose of 5 Gy (1.27 Gy·min^{-1}, 130.0 kV, 5.0 mA) and incubated for 24 h under the same conditions. After this time, the cells were dissociated with 1x Accutase solutions and analyzed by flow cytometry (Guava easyCyte™ 12, Merck, Warsaw, Poland) using the manufacturer's protocol (FlowCellect™ MitoDamage Kit, Merck).

3.11. Computational

All calculations were performed with the B3LYP functional and DGDZVP++ basis set using the Gaussian09 package. Water environment was simulated with the polarizable continuum model (PCM) implemented therein, and in one of the reactions, three water molecules were included explicitly. In order to reduce the cost of calculations, 2'-deoxyribose moiety in nucleoside was substituted with the methyl group in the computational model (i.e., we used 5-bromo-1-methyl-4-thiouracil (BrSU), see Figures 2 and 3) The sugar moiety does not take part in any of the considered reactions, hence the conversion of the studied system to BrSU is not expected to affect the computational results. The minima, as well as transition states (TSs), were proven with the vibrational frequency calculations, and the IRC procedure was applied to show that particular TSs are connected to the appropriate minima.

4. Conclusions

Radiotherapy is one of the most common modalities employed against cancer diseases. Unfortunately, its efficacy is seriously impaired due to hypoxia of solid tumors. Therefore, to be efficient, radiotherapy should be combined with radiosensitizers–compounds that are able to sensitize cells to ionizing radiation. The modified uridines belong to radiosensitizers, which produce radiosensitization by incorporation into DNA. Results of numerous studies, mainly on BrdU and IdU, suggest that dissociative electron attachment to the modified uridines incorporated into DNA is responsible for the increased level of damage.

Our studies seem to confirm this view. Using enzymatic digestion of genomic DNA extracted from the cells incubated with BrSdU and ISdU, which was followed by the HPLC and LC-MS analysis of the lysates, we demonstrated that both compounds are incorporated into the DNA of the studied cell lines. Although BrSdU/ISdU are nontoxic, as shown by the results of the MTT test, their radiosensitizing efficacy turned out to be very different. Thus, clonogenic assay, the method of choice to determine cell reproductive death after treatment with ionizing radiation, shows significant activity of ISdU and practical lack of radiosensitization in the case of BrSdU. Similarly, the number of viable, early apoptotic, late apoptotic, and dead cells was not influenced by the pre-incubation with BrSdU, while the incubation with ISdU significantly increased the population of apoptotic and dead cells. Finally, the level of double-stand breaks which are associated with the cell death, assayed with the phosphorylation of histone γH2A.X test, clearly increased only in case of incubation with ISdU.

The significant variation in radiosensitizing activity of such similar derivatives has been explained with our radiolysis and computational studies. The stationary radiolysis of BrSdU demonstrated that although the compound was decomposed, no DEA products were detected. On the other hand, the radiolysis of ISdU solutions led to 4-thio-2'-deoxyuridine and its dimers. A similar conclusion was drawn from pulse radiolysis. Namely, the main dissimilarity between two studied 4-thio-2'-deoxyuridines lies in the fact that the transient absorption of 4SdU$^\bullet$ radical was observed only in the solutions of ISdU.

All of the above-mentioned observations are well-explained by difference in the activation barrier that accompanies the release of the halogen anion from the XSdU$^{\bullet-}$ anion radical. Namely, the barrier for BrSdU$^{\bullet-}$ is more than two-fold larger than that characteristic for DEA in ISdU$^{\bullet-}$. As a consequence, the C–X dissociation process is ca. 200-fold slower in the BrSdU anion. Hence, the lifetime of the anion is sufficiently long for BrdU$^{\bullet-}$ to be protonated forming BrSdUH$^\bullet$, which prevents the DEA process. Thus, our studies confirm the crucial role of DEA leading to uracil-5-yl radical in the radiosensitization mechanism of modified uracils. Moreover, they show that the relatively low barrier of ca. 26 kJ/mol is able to inhibit DEA in the studied class of molecules. Our studies demonstrate that even such a modest barrier makes the potential radiosensitizer completely inactive. This finding is of highly valuable for the computational search of radiosensitizing molecules.

Supplementary Materials: The following are available online, LC-MS/MS identification of radiolysis products (Figures S1–S4); Transition state geometries for the reactions identified during stationary radiolysis (Figure S5); Incorporation yield of BrSdU and ISdU into the genomic DNA determined with its digestion and HPLC and LC-Ms analyses (Figures S6–S10); Flow cytometry analysis of histone H2A.X phosphorylation and cell death (Figures S11–S13).

Author Contributions: Conceptualization, J.R. and K.B.; methodology, M.Z., P.S., K.S., W.K., S.M.; investigation, P.S., M.Z., K.S., W.K., S.M.; writing—original draft preparation, P.S., M.Z., S.M., W.K., K.S., J.R. and K.B.; writing—review and editing, P.S., M.Z., S.M., W.K., K.S., J.R. and K.B.; supervision, J.R.; funding acquisition, J.R.

Funding: This work was supported by the Polish National Science Centre under Grants No. 2014/14/A/ST4/00405 (J.R.) and 2018/28/C/ST4/00479 (K.S.). Some of the calculations were performed in the Wroclaw Center for Networking and Supercomputing, grant No. 209.

Conflicts of Interest: The authors declare no conflict of interest.

References

1. Rak, J.; Chomicz, L.; Wiczk, J.; Westphal, K.; Zdrowowicz, M.; Wityk, P.; Żyndul, M.; Makurat, S.; Golon, Ł. Mechanisms of damage to DNA labeled with electrophilic nucleobases induced by ionizing or UV radiation. *J. Phys. Chem. B* **2016**, *119*, 8227–8238. [CrossRef] [PubMed]
2. Chomicz, L.; Rak, J.; Storoniak, P. Electron-induced elimination of the bromide anion from brominated nucleobases. A computational study. *J. Phys. Chem. B* **2012**, *116*, 5612–5619. [CrossRef] [PubMed]
3. Wieczór, M.; Wityk, P.; Czub, J.; Chomicz, L.; Rak, J. A first-principles study of electron attachment to the fully hydrated bromonucleobases. *Chem. Phys. Lett.* **2014**, *595–596*, 133–137. [CrossRef]
4. Park, Y.; Polska, K.; Rak, J.; Wagner, J.R.; Sanche, L. Fundamental mechanisms of DNA radiosensitization: Damage induced by low-energy electrons in brominated oligonucleotide trimers. *J. Phys. Chem. B* **2012**, *116*, 9676–9682. [CrossRef] [PubMed]
5. Polska, K.; Rak, J.; Bass, A.D.; Cloutier, P.; Sanche, L. Electron stimulated desorption of anions from native and brominated single stranded oligonucleotide trimers. *J. Chem. Phys.* **2012**, *136*, 075101. [CrossRef] [PubMed]
6. Jagiello, K.; Makurat, S.; Pereć, S.; Rak, J.; Puzyn, T. Molecular features of thymidine analogues governing the activity of human thymidine kinase. *Struct. Chem.* **2018**, *29*, 1367–1374. [CrossRef]
7. Goz, B. The Effects of Incorporation of 5-halogenated deoxyuridines into the DNA of eukaryotic cells. *Pharmacol. Rev.* **1977**, *29*, 249–272.
8. Visvader, J.E.; Lindeman, G.J. Cancer stem cells in solid tumors: Accumulating evidence and unresolved questions. *Nat. Rev. Cancer* **2008**, *8*, 755–768. [CrossRef]
9. Rockwell, S.; Dobrucki, I.T.; Kim, E.Y.; Marrison, S.T.; Vu, V.T. Hypoxia and radiation therapy: Past history, ongoing research and future promise. *Curr. Mol. Med.* **2009**, *9*, 442–458. [CrossRef]
10. Oronsky, B.T.; Knox, S.J.; Scicinski, J. Six degrees of separation: The oxygen effect in the development of radiosensitizers. *Trans. Oncol.* **2011**, *4*, 189–198. [CrossRef]
11. Westphal, K.; Skotnicki, K.; Bobrowski, K.; Rak, J. Radiation damage to single stranded oligonucleotide trimers labelled with 5-iodopyrimidines. *Org. Biomol. Chem.* **2016**, *14*, 9331–9337. [CrossRef]
12. Westphal, K.; Wiczk, J.; Miloch, J.; Kciuk, G.; Bobrowski, K.; Rak, J. Irreversible electron attachment—A key to DNA damage by solvated electrons in aqueous solution. *Org. Biomol. Chem.* **2015**, *13*, 10362–10369. [CrossRef] [PubMed]
13. Cecchini, S.; Girouard, S.; Huels, M.A.; Sanche, L.; Hunting, D.J. Interstrand cross-links: A new type of γ-ray damage in bromodeoxyuridine-substituted DNA. *Biochemistry* **2005**, *44*, 1932–1940. [CrossRef]
14. Ling, L.L.; Ward, J.F. Radiosensitization of Chinese hamster V79 cells by bromodeoxyuridine substitution of thymidine: Enhancement of radiation-induced toxicity and DNA strand break production by monofilar andbifilar substitution. *Radiat. Res.* **1990**, *121*, 76–83. [CrossRef]
15. Miller, E.M.; Fowler, J.F.; Kinsella, T.J. Linear-quadratic analysis of radiosen-sitization by halogenated pyrimidines. I.Radiosensitization of humancolon cancer cells by iododeoxyuridine. *Radiat. Res.* **1992**, *131*, 81–89. [CrossRef]
16. Miller, E.M.; Fowler, J.F.; Kinsella, T.J. Linear-quadratic analysis of radiosen-sitization by halogenated pyrimidines. II. Radiosensitization of humancolon cancer cells by bromodeoxyuridine. *Radiat. Res.* **1992**, *131*, 90–97. [CrossRef]
17. McGinn, C.J.; Shewach, D.S.; Lawrence, T.S. Radiosensitizing nucleosides. *J. Natl. Cancer Inst.* **1996**, *88*, 1193–1203. [CrossRef] [PubMed]
18. Phillips, T.L.; Scott, C.B.; Leibel, S.A.; Rotman, M.; Weigensberg, I.J. Results of a randomized comparison of radiotherapy and bromodeoxyuridine with radiotherapy alone for brain metastases: Report of RTOG trial 89-05. *Int. J. Radiat. Oncol. Biol. Phys.* **1995**, *33*, 339–348. [CrossRef]
19. Zdrowowicz, M.; Chomicz, L.; Żyndul, M.; Wityk, P.; Rak, J.; Wiegand, T.J.; Hanson, C.G.; Adhikary, A.; Sevilla, M.D. 5-Thiocyanato-2′-deoxyuridine as a possible radiosensitizer: Electron-induced formation of uracil-C5-thiyl radical and its dimerization. *Phys. Chem. Chem. Phys.* **2015**, *17*, 16907–16916. [CrossRef] [PubMed]
20. Sosnowska, M.; Makurat, S.; Zdrowowicz, M.; Rak, J. 5 selenocyanatouracil: A potential hypoxic radiosensitizer. electron attachment induced formation of selenium centered radical. *J. Phys. Chem. B* **2017**, *121*, 6139–6147. [CrossRef]

21. Makurat, S.; Zdrowowicz, M.; Chomicz-Mańka, L.; Kozak, W.; Serdiuk, I.E.; Wityk, P.; Kawecka, A.; Sosnowska, M.; Rak, J. 5-Selenocyanato and 5-trifluoromethanesulfonyl derivatives of 2′-deoxyuridine: Synthesis, radiation and computational chemistry as well as cytotoxicity. *RSC Adv.* **2018**, *8*, 21378–21388. [CrossRef]
22. Ameixa, J.; Arthur-Baidoo, E.; Meißner, R.; Makurat, S.; Kozak, W.; Butowska, K.; Ferreira da Silva, F.; Rak, J.; Denifl, S. Low–energy electron–induced decomposition of 5–trifluoromethanesulfonyl–uracil: A potential radiosensitizer. *J. Chem. Phys.* **2018**, *149*, 164307. [CrossRef] [PubMed]
23. Meißner, R.; Makurat, S.; Kozak, W.; Limão-Vieira, P.; Rak, J.; Denifl, S. Electron-induced dissociation of the potential radiosensitizer 5-selenocyanato-2′-deoxyuridine. *J. Phys. Chem. B* **2019**, *123*, 1274–1282. [CrossRef] [PubMed]
24. Brem, R.; Zhang, X.; Xu, Y.Z.; Karran, P. UVA photoactivation of DNA containing halogenated thiopyrimidines induces cytotoxic DNA lesions. *J. Photochem. Photobiol. B Biol.* **2015**, *145*, 1–10. [CrossRef] [PubMed]
25. Xu, Y.Z.; Zhang, X.; Wu, H.C.; Massey, A.; Karran, P. 4-Thio-5-bromo-20-deoxyuridine: Chemical synthesis and therapeutic potential of UVA-induced DNA damage. *Bioorg. Med. Chem. Lett.* **2004**, *14*, 995–997. [CrossRef] [PubMed]
26. Brem, R.; Guven, M.; Karran, P. Oxidatively-generated damage to DNA and proteins mediated by photosensitized UVA. *Free Rad. Biol. Med.* **2017**, *107*, 101–109. [CrossRef] [PubMed]
27. Makurat, S.; Spisz, P.; Kozak, W.; Rak, J.; Zdrowowicz, M. 5-iodo-4-thio-2′-deoxyuridine as a sensitizer of X-ray induced cancer cell killing. *Int. J. Mol. Sci.* **2019**, *20*, 1308. [CrossRef]
28. Zeida, A.; Babbush, R.; Gonzalez-Lebrero, M.C.; Trujillo, M.; Radi, R.; Estrin, D.A. Molecular basis of the mechanism of thiol oxidation by hydrogen peroxide in aqueous solution: Challenging the SN2 paradigm. *Chem. Res. Toxicol.* **2012**, *25*, 741–746. [CrossRef]
29. Van Bergen, L.A.; Roos, G.; De Proft, F. From thiol to sulfonic acid: Modeling the oxidation pathway of protein thiols by hydrogen peroxide. *J. Phys. Chem. A* **2014**, *118*, 6078–6084. [CrossRef]
30. Bahrami, K.; Khodaei, M.; Tajik, M. Trimethylsilyl chloride promoted selective desulfurization of thiocarbonyls to carbonyls with hydrogen peroxide. *Synthesis* **2010**, *24*, 4282–4286. [CrossRef]
31. Sonntag, C. *Free-Radical-Induced DNA Damage and Its Repair: A Chemical Perspective*; Springer Science & Business Media: Heidelberg, Germany, 2010.
32. Chu, J.W.; Trout, B.L. On the mechanisms of oxidation of organic sulfides by H_2O_2 in aqueous solutions. *J. Am. Chem. Soc.* **2004**, *126*, 900–908. [CrossRef] [PubMed]
33. McCaw, P.G.; Buckley, N.M.; Collins, S.G.; Maguire, A.R. Generation, reactivity and uses of sulfines in organic synthesis. *Eur. J. Org. Chem.* **2016**, *9*, 1630–1650. [CrossRef]
34. Adam, W.; Bargon, R.M. Synthesis of thiiranes by direct sulfur transfer: The challenge of developing effective sulfur donors and metal catalysts. *Chem. Rev.* **2004**, *104*, 251–262. [CrossRef] [PubMed]
35. Chomicz, L.; Zdrowowicz, M.; Kasprzykowski, F.; Rak, J.; Buonaugurio, A.; Wang, Y.; Bowen, K.H. How to find out whether a 5-substituted uracil could be a potential dna radiosensitizer. *J. Phys. Chem. Lett.* **2013**, *4*, 2853–2857. [CrossRef]
36. Wetmore, S.D.; Boyd, R.J.; Eriksson, L.A. A theoretical study of 5-halouracils: Electron affinities, ionization potentials and dissociation of the related anions. *Chem. Phys. Lett.* **2001**, *343*, 151–158. [CrossRef]
37. Li, X.; Sanche, L.; Sevilla, M.D. Dehalogenation of 5- halouracils after low energy electron attachment: A density functional theory investigation. *J. Phys. Chem. A* **2002**, *106*, 11248–11253. [CrossRef]
38. Eyring, H. The Activated Complex in Chemical Reactions. *J. Chem. Phys.* **1935**, *3*, 107. [CrossRef]
39. McAllister, M.; Smyth, M.; Gu, B.; Tribello, G.A.; Kohanoff, J. Understanding the Interaction between Low-Energy Electrons and DNA Nucleotides in Aqueous Solution. *J. Phys. Chem. Lett.* **2015**, *6*, 3091–3097. [CrossRef] [PubMed]
40. Putschew, A.; Jekel, M. Induced in-source fragmentation for the selective detection of organic bound iodine by liquid chromatography/electrospray mass spectrometry. *Rapid Commun. Mass Spectrom.* **2003**, *17*, 2279–2282. [CrossRef]
41. Hütteroth, A.; Putschew, A.; Jekel, M. Selective detection of unknown organic bromine compounds and quantification potentiality by negative-ion electrospray ionization mass spectrometry with induced in-source fragmentation. *Int. J. Environ. Anal. Chem.* **2007**, *87*, 415–424. [CrossRef]
42. Rafehi, H.; Orlowski, C.; Georgiadis, G.T.; Ververis, K.; El-Osta, A.; Karagiannis, T.C. Clonogenic assay: Adherent cells. *J. Vis. Exp.* **2011**, *49*, 2573–2576. [CrossRef] [PubMed]

43. Mosmann, T. Rapid colorimetric assay for cellular growth and survival: Application to proliferation and cytotoxicity assays. *J. Immunol. Methods.* **1983**, *65*, 55–63. [CrossRef]
44. Taneja, N.; Davis, M.; Choy, J.S.; Beckett, M.A.; Singh, R.; Kron, S.J.; Weichselbaum, R.R. Histone H2AX phosphorylation as a predictor of radiosensitivity and target for radiotherapy. *J. Biol. Chem.* **2004**, *279*, 2273–2280. [CrossRef] [PubMed]
45. Bobrowski, K. Free radicals in chemistry, biology and medicine: contribution of radiation chemistry. *Nukleonika* **2005**, *50*, S67.

Sample Availability: Samples of the compounds are available from the authors.

 © 2019 by the authors. Licensee MDPI, Basel, Switzerland. This article is an open access article distributed under the terms and conditions of the Creative Commons Attribution (CC BY) license (http://creativecommons.org/licenses/by/4.0/).

Review

Replication Stress and Consequential Instability of the Genome and Epigenome

Pawlos S. Tsegay [1], Yanhao Lai [2,3] and Yuan Liu [1,2,3,*]

[1] Biochemistry Ph.D. Program, Florida International University, Miami, FL 33199, USA; ptseg001@fiu.edu
[2] Department of Chemistry and Biochemistry, Florida International University, 11200 SW 8th Street, Miami, FL 33199, USA; yalai@fiu.edu
[3] Biomolecular Sciences Institute, Florida International University, Miami, FL 33199, USA
* Correspondence: yualiu@fiu.edu

Received: 1 October 2019; Accepted: 25 October 2019; Published: 27 October 2019

Abstract: Cells must faithfully duplicate their DNA in the genome to pass their genetic information to the daughter cells. To maintain genomic stability and integrity, double-strand DNA has to be replicated in a strictly regulated manner, ensuring the accuracy of its copy number, integrity and epigenetic modifications. However, DNA is constantly under the attack of DNA damage, among which oxidative DNA damage is the one that most frequently occurs, and can alter the accuracy of DNA replication, integrity and epigenetic features, resulting in DNA replication stress and subsequent genome and epigenome instability. In this review, we summarize DNA damage-induced replication stress, the formation of DNA secondary structures, peculiar epigenetic modifications and cellular responses to the stress and their impact on the instability of the genome and epigenome mainly in eukaryotic cells.

Keywords: oxidative DNA damage; DNA replication stress; replication fork stalling; genomic and epigenomic instability; DNA methylation; histone modifications; miRNAs

1. Introduction

Faithful copying of genetic information is vital for cells to maintain genomic and epigenomic stability. During cell division, DNA replication includes the replication of DNA, the DNA methylation pattern, as well as the duplication of histones and their modifications. These allow the genetic and epigenetic information of a cell to be copied and passed to daughter cells. Also, the integrity of the genome and epigenome in cells is maintained through the coordination between DNA replication and cell cycle, which contains the G1, S, G2 and M phases, respectively [1–3]. Replication of the entire genome and its epigenetic modifications, along with the replication of histones and their modifications have to be completed during the S phase before the cell cycle can enter its M phase, where one single cell is divided into two daughter cells. However, during DNA replication, the opened genomic DNA is also susceptible to attack by varieties of DNA damage. This can lead to replication stress, i.e., replication fork stalling that subsequently results in the accumulation of DNA damage and the formation of secondary DNA structures, triggering DNA damage response (DDR) and repair, as well as corresponding epigenetic changes. All these processes can alter the effectiveness and precision of DNA replication, causing genome and epigenome instability that can ultimately lead to human diseases such as cancer [4,5] (Figure 1).

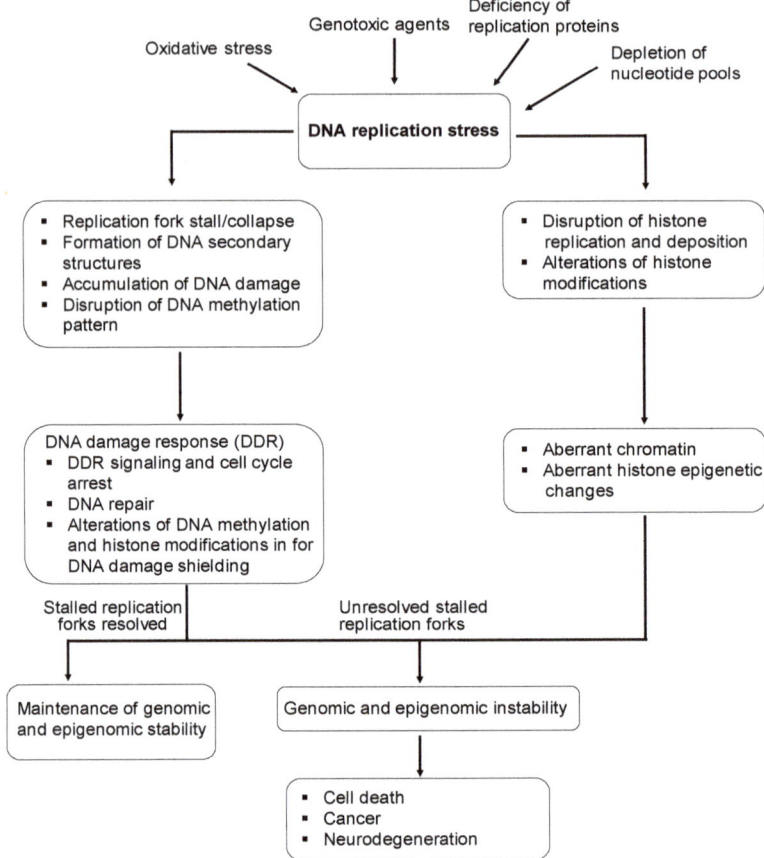

Figure 1. DNA replication stress leads to genomic and epigenomic instability associated with diseases.

2. DNA Replication

The bidirectional DNA replication in eukaryotes starts at multiple replication initiation sites, known as replication origin, that encompasses the DNA sequences recognized and bound by the replication initiator proteins [6]. Subsequently, DNA helicase complex is formed at the replication origin through the assembly of the head to head double hexamer minichromosome maintenance protein (MCM) with the help of cell division cycle 6 (Cdc6), Cdc10-dependent transcript 1 (Cdt1), and origin recognition complex (ORC) [7–9] in the G1 phase of the cell cycle [7,9,10]. The double hexamer MCM helicase complex is then activated in the S phase by cyclin-dependent kinase (CDK) and Dbf4 dependent kinase (DDK) [11–15], forming the functional helicase complex, cell division cycle 45 (CDC45)-MCM-GINS, (CMG complex helicase) with CDC45, and GINS [11,16–18]. Double helical DNA is then unwound upon the recruitment of CMG complex helicase, resulting in the formation of replication forks [12,19–21]. In this process, the ATP-dependent MCM complex serves as the motor of DNA replication by unwinding double-strand DNA.

Unwound single-stranded DNA (ssDNA) is then bound by the ssDNA binding protein, replication protein A (RPA) for protection from the degradation and formation of secondary structures [22,23]. Then the pol α-primase complex is recruited to the replication forks through its interaction with the chromosome transmission fidelity 4 (Ctf4) protein that also interacts with the GINS in the CMG complex [24]. Since DNA is synthesized in a 5' to 3' direction during replication, the DNA synthesis that is carried out by DNA polymerase ε (pol ε) in the leading strand is continuous, whereas the DNA synthesis by DNA polymerase δ (pol δ) on the lagging strand is discontinuous with the synthesis of short Okazaki fragments with 100–250 nucleotides. Both pol ε and pol δ interact with the replication cofactor, proliferating cell nuclear antigen (PCNA), which is loaded on the double-strand DNA by the clamp loader protein complex, replication factor C (RFC). PCNA anchors the polymerases to the template strand, allowing the polymerases to perform the processive DNA synthesis [25,26] and ensuring the high efficiency of DNA replication. Finally, RNA primers in the Okazaki fragments are removed by RNase HI [27] and flap endonuclease 1 (FEN1) [28,29]. Replicative DNA polymerase pol δ also plays a role in removing RNA primers in the Okazaki fragments by coordinating with FEN1 flap cleavage in that pol δ strand displacement synthesis creates a flap containing an RNA primer, which is then cleaved by FEN1 flap cleavage [30,31]. The generated nicked DNA is then sealed by DNA ligase I [32], thereby leading to the completion of DNA replication. While replicative DNA polymerases exhibit a high efficiency of DNA synthesis, they also have a high fidelity of incorporating correct nucleotides. This is because these polymerases bear a catalytic site with a rigid structure, and have their 3' to 5' exonuclease proofreading domain, which can remove mis-paired nucleotides [33–35]. This domain safeguards the accuracy and integrity of the genome and its associated epigenetic modifications. However, the replicative DNA polymerases are susceptible to DNA damage, the distortions of the DNA template, and the secondary structures generated at the replication fork [36–39], leading to polymerase pausing and subsequently replication fork stalling and genome stress. Mutations or the functional deficiency of proteins that are involved in DNA replication and the resolution of stalled replication forks can also cause replication stress and genomic instability, and are associated with diseases [40]. For example, mutations of PRE-RC proteins are associated with the development of Meier Gorlin syndrome [41,42]. This may be because that the deficiency of PRE-RC proteins disrupts the assembly of the PRE-RC complex, thus inhibiting S-phase progression in cells [41]. On the other hand, deficiency of pol ε and GINS is associated with IMAGe (Intrauterine growth restriction, metaphyseal dysplasia, adrenal hypoplasia congenita, and genital anomalies) syndrome and immunodeficiency [43,44]. Also, Mutations in the helicases, including Bloom syndrome protein (BLM), Werner syndrome protein (WRN) and ATP-dependent DNA helicase QL4 (RecQL4) that mediate replication fork remodeling and restart can result in the development of Bloom, Werner and Rothmund-Thomson syndromes, respectively [45]. Bloom and Werner syndrome patients show aging-related symptoms including cancer predisposition, microcephaly, mental retardation, infertility, growth defects and premature aging, atherosclerosis, cataracts, osteoporosis and diabetes [46]. We have also included Table 1 with a list of the diseases that are associated with the deficiency of replication proteins

Table 1. Proteins involved in DNA replication, repair, and replication stress response and associated diseases.

DNA Repair Protein	Function	Human Diseases
CDT1	Facilitates MCM loading on origins	Meier-Gorlin syndrome [40]
Pre-RC (CDT1, ORC1-ORC6, Cdc6, MCM2-7)	Recruitment of DNA polymerase and phosphorylation by both the Cdc7/Dbf4 and CDK2-cyclin A protein kinases	Meier-Gorlin syndrome [40]
Nbs1	ATR/ATM activation	Nijmegen breakage syndrome [40]
Rad50	ATR/ATM activation	Nijmegen breakage syndrome-like disorder [40]
RecQL4	DNA remodeling, replication fork structure resolution	Rothmund-Thomson syndrome [40,47]
RNase H2	Removal of embedded ribonucleotides Resolution of RNA-DNA hybrid	Aicardi-Goutières syndrome [48]

Table 1. Cont.

DNA Repair Protein	Function	Human Diseases
Senataxin	Resolution of RNA-DNA hybrid	Amyotrophic lateral sclerosis [40]
Mre 11	ATM/ATR activation	Ataxia-telangiectasia-like diseases [40]
BLM	DNA remodeling, replication fork stall resolution	Bloom syndrome [49]
FANC family	DNA inter-strand cross-link repair	Fanconi anemia [40,50]
FANCD2	Replication fork protection	Fanconi anemia [40,50]
WRN	DNA remodeling, replication fork structure resolution	Werner syndrome [40]
BRCA1, BRCA2	Checkpoint mediators, DNA repair and recombination	Breast and ovarian carcinoma [51]
MSH2 and MLH1	DNA mismatch repair	Colorectal cancer [51]

3. The Genome Stress Resulting from DNA Replication

There are varieties of sources that can cause genome stress during DNA replication, i.e., replication stress. These include physical impediments of replication fork progression induced by endogenous or/and exogenous DNA damaging agents [52], insufficient synthesis of histone proteins [53], and depletion of dNTPs [49,54,55]. In some occasions, DNA replication and repair enzymes can also create replication stress by inducing DNA lesions, such as abasic sites and ssDNA breaks, as well as by the incorporating damaged nucleotides through repair DNA polymerases [56–59].

Also, repeated DNA sequences in the genome that include microsatellites, minisatellites, isolated repeated motifs comprising homopolymers, elevation transposable elements, pseudogenes and terminal repeats, which constitute 50% of the human genome, can also cause genome stress during DNA replication. Among them, minisatellites and microsatellites are the major sources of causing "dynamic mutations," i.e., repeat deletions and expansions in the genome [60]. These sequences can result in DNA replication fork stalling in the absence of exogenous genome stress [61,62]. They are susceptible to DNA damage and DNA strand breaks and thus known as DNA fragile sites that cause genomic instability [60,62]. Non-canonical or non-B form DNA structures are another source of causing genome stress through DNA replication stalling and DNA damage. The structures include triplex DNA, hairpins, DNA loops, Z-DNA, and G-quadruplexes [63–66]. They form the roadblocks of replicative and repair DNA polymerases to cause polymerase pausing impeding replication fork progression and DNA repair [63–66]

3.1. The DNA Damage that Impedes the Fork Progression

DNA is under constant attack by a variety of endogenous and exogenous DNA damage agents, such as reactive oxygen species (ROS), UV among others, resulting in different types of DNA damage, including oxidized bases, modified sugars, abasic sites, DNA strand breaks, DNA-DNA and DNA-protein crosslinks and thymine dimers, which can result in replication fork stalling [52,56,57,67–70]. It is estimated that 10^4 DNA base lesions are generated in the mammalian genome per day. These lesions can accumulate in the stalled replication fork while they are subject to DNA base excision repair (BER)/single-strand break repair (SSBR) [71,72]. However, repair of the lesions through BER/SSBR results in ssDNA breaks that can terminate the progression of polymerases at a replication fork. Also, unrepaired base lesions and abasic sites can directly block replication polymerases and helicases, leading to disassociation of polymerases from the template as well as helicase uncoupling, causing DNA strand breaks [57,71,73]. Bulky DNA damage, such as DNA-DNA and DNA-protein crosslinks and cyclobutane pyrimidine dimers (CPD) can also directly cause polymerase pausing and terminate replication fork progression [57,74–78]

3.2. Impediment of Replication Fork Progression by Gene Transcription

DNA replication fork stalling can also be induced as a result of gene transcription. In the S phase, genes involved in DNA replication are highly expressed. This may result in a conflict between replication and transcription, i.e., transcription-replication conflicts (TRCs) when both replication and transcription occur simultaneously in the same DNA templates and collide head-on [64,79]. The collision slows down replication fork progression, leading to fork stalling and genome stress and genomic instability [64,79]. Furthermore, gene transcription can impede the replication fork progression through the formation of an R-loop that contains an RNA-DNA hybrid and a single-stranded non-template strand. The structure is involved in the disruption of genomic stability [64,80,81]. The RNA-DNA hybrid in an R-loop can be generated when nascent RNA transcripts reanneal to their template DNA, displacing the non-template strand into ssDNA, and this makes an R-loop become a potent barrier of co-transcription and replication [80,82]. R-loops can be stabilized by a deregulation of DNA replication and transcription proteins and factors [82–84]. The formation of R-loops is also facilitated by trinucleotide repeats including CAG, GAA, CGG repeats that can stabilize the DNA-RNA hybrid in the repeats [85–88]. The persistence of R loops in the GC-rich repeated sequences may facilitate somatic repeat expansion or deletion [89] by causing replication fork stalling, promoting the progression of trinucleotide repeat expansion diseases such as Huntington's Disease (HD) and Friedreich's Ataxia (FRDA) caused by CAG and GAA repeat expansions, respectively [85,86,89,90].

3.3. The Effects of dNTPs and Ribonucleotides on Replication Fork Progression

The progression of the replication fork and fidelity of DNA replication during S phase [49,54,55,91,92] is also regulated by the balance of dNTPs and the size of the nucleotide pool [55,93]. dNTPs are periodically synthesized and degraded at the different phases of the cell cycle [94–96]. A key step for the synthesis of dNTPs is the conversion of ribonucleotides triphosphate (NTPs) to deoxyribonucleotides (dNTPs) by ribonucleotide reductase (RNR), the rate-limiting enzyme for the synthesis of deoxynucleotide [94]. Inhibition of RNR by hydroxyurea (HU) depletes dNTPs, leading to replication fork stalling and genomic instability [55,93]. On the other hand, degradation/hydrolysis of dNTPs also regulates the balance dNTPs and nucleotide pool size to modulate the fidelity of replication and fork progression, impacting genomic stability. For example, knockdown of the dNTP triphosphohydrolase, sterile alpha motif and the HD-domain containing protein 1 (SAMHD1) in the G1 phase, disrupts the dNTP balance, stopping the progression of cell cycle and increasing cellular sensitivity to DNA damage [97,98]. Another important factor is the level of dUTP that can affect the fidelity of DNA replication. This is because replicative DNA polymerases cannot differentiate dUTP from dTTP [99,100]. Thus, the degradation of dUTP to dUMP by dUTP pyrophosphatase (dUTPase) plays a critical role in regulating dUTP to a low level in cells, ensuring the high fidelity of DNA replication. Thus, the rate of DNA replication fork progression and genomic stability is regulated by the balance of dNTPs and nucleotide pool size. Disruption of the balance between purine and pyrimidine can promote nucleotide misincorporations, which generate the source for replication fork stalling, DNA damage and genomic instability [101].

Interestingly, the incorporation of ribonucleotides by DNA polymerases is also associated with genomic instability. It is estimated that about 1 million ribonucleotides are incorporated into the genome during DNA replication by DNA polymerases [102]. Ribonucleotides are removed by RNase H2-mediated ribonucleotide excision repair (RER), which is the primary mechanism to remove ribonucleotides in the genome [103]. Accumulation of ribonucleotides resulting from the deficiency of RNase H2 can lead to replication stress and genomic stability [104]. Interestingly, under the deficiency of RNase H2, incorporated ribonucleotides are removed by DNA topoisomerase I and II [102,105]. However, the removal of incorporated ribonucleotides by topoisomerase can generate ssDNA and dsDNA breaks, deletion at repeated sequence, and genomic instability [106,107].

4. Cellular Responses to the Genome Stress from DNA Replication and Genome Instability

4.1. DNA Damage Response Signaling Induced by Stalled Replication Forks

To combat the unintended adverse consequences from stalled replication forks and the resulted DNA damage and maintain genomic instability and integrity, cells respond to the damage by initiating the DNA damage response signaling pathway that leads to cell cycle arrest [108]. The signaling pathway allows the coordination between DNA damage repair and replication fork processing for preventing stalled replication fork, DNA damage and strand breaks from being passed to the next phase in the cell cycle [57]. The DNA damage-response signaling pathway is activated through the activation of cell cycle checkpoints known as the DNA damage checkpoint (DDC) and DNA replication checkpoint (DRC). DDC is activated by DNA damage recognition, whereas DRC is activated by stalled replication forks [62,109–111]. For the cell cycle checkpoints, G1/S and G2/M [112], the G1/S phase checkpoint plays a major role in preventing the progression of cells carrying replication stress products, such as stalled fork and DNA damage [62,109,110]. Thus, the checkpoint allows DNA damage to be repaired in the S phase, so that DNA replication can proceed to the M phase. Both checkpoints demand that DNA damage generated during the G1 and G2 phases be repaired before the cell cycle can proceed to the next phase [2,51,113].

Activation of DRC is initiated by the slow progression of the replication fork along with the activation of the DNA replication checkpoints [114]. It has been shown that decreased replication fork progression by 5- to 10-fold leads to the activation of the ATR-mediated DNA damage response pathway. Further, it has also been found that a moderate level of replication stress induces the activation of ATR [115]. More severe replication stress induces the activation of both ATR and its downstream target pathways, such as FANC and CHK1 pathways [115–117]. Thus, cell response to replication stress through DRC is dependent on the ATR pathway [114,118,119]. Through the activation of the checkpoints, cell cycles are arrested, and DNA repair machineries are recruited to the damaged sites. Finally, DNA damage is repaired, and the stalled replication forks are resolved, allowing replication and cell division to proceed [120]. Thus, cell cycle checkpoints play vital roles in coordinating DNA damage repair and the resolution of stalled replication forks with cell cycle progression [121], leading to the maintenance of genome stability.

4.2. Resolution of Stalled Replication Forks

Stalled replication forks, if not resolved, will eventually result in replication forks collapse that can cause a series of severe consequences, such as DNA breakage and cell death. To avoid the scenario, stalled replication forks need to be resolved, and DNA replication needs to be restarted for cell survival. One strategy for eukaryotic cells, such as budding yeast to resolve stalled replication forks on the lagging strand, is to create new RNA primers at the downstream of DNA lesions that occur in the forks to restart DNA synthesis, a process named as repriming. It has been found that the repriming mechanism is used in the lagging strand DNA synthesis, as the synthesis of the Okazaki fragments is not affected by DNA damage and fork stalling as long as DNA is unwound continuously [122]. This is because the repriming process is initiated at the downstream of lesions [122]. In this process, a stalled DNA polymerase dissociates from the template strand and rebinds to the newly synthesized primer to synthesize DNA, thereby leading to the restart of stalled forks [123]. It has been found that discontinuous DNA synthesis can occur on both leading and lagging strands after UV damage in budding yeast, suggesting that the repriming mechanism is also used to resolve a stalled replication fork induced by DNA damage in the leading strand [123].

Also, eukaryotic cells can use a backup replication origin, i.e., the licensed replication origin to rescue stalled replication forks [124,125] because the reduced rate of replication fork progression can result in the accumulation of the ssDNAs, causing the uncoupling between DNA polymerase and helicase activities and large ssDNA gaps [126]. In this scenario, pol α-primase can be recruited to the ssDNA gaps and synthesize RNA primers to initiate DNA replication.

Since the recruitment of pol α-primase depends on TopBP1, which also involves in the activation of the ATR/MEC1 pathway [127], this suggests that the reactivation of the replication forks and the signaling pathway are coupled. The licensed origins of replication that are not activated during DNA replication are referred as the dormant origins of replication. They serve as a primary mechanism to restore replication when replication forks are stalled [128]. It has been estimated that about 20–30% of replication origins are activated during DNA replication [129]. Thus, the dormant origins bound by MCM helicases can serve as a backup for initiating the replication, thereby preventing replication stress, chromosome instability and tumorigenesis [130]. Dormant origin firing is regulated by ATR-mediated phosphorylation of FANCI [130]. In response to mild replication stress, unmodified FANCI triggers the firing of adjacent dormant origins to resolve stalled replication fork. In the case of severe replication fork stalling, dormant origin firing is inhibited by the phosphorylation of FANCI [130], and this provides more time for the stalled forks to be resolved, restarting DNA replication. Replication origin firing can also be modulated by claspin protein [131] that recruits Cdc7 kinase to the replication origin, which in turn phosphorylates MCM4, causing unscheduled origin firing in response to replication stress [129,130]. Inhibition of ATR can also result in an unscheduled origin firing, which can be modulated by Cdc7-mediated phosphorylation of MCM4 [129,130]. Figure 2 illustrated the Ataxia telangiectasia and Rad3-related protein/(ataxia-telangiectasia mutated) serine/threonine kinase (ATR/ATM)-activated pathways that are involved in resolution of stalled replication forks (Figure 2)

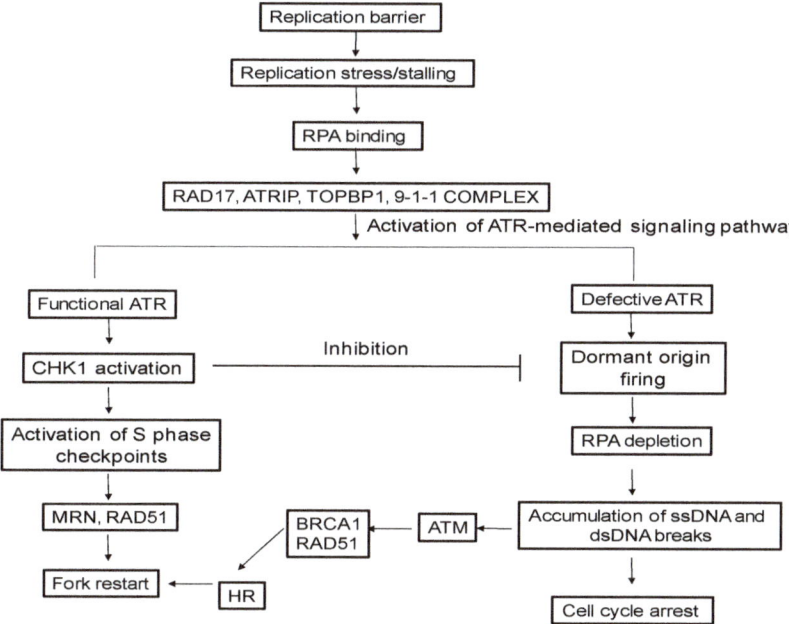

Figure 2. Ataxia telangiectasia and Rad3-related protein/(ataxia-telangiectasia mutated) serine/threonine kinase (ATR/ATM)-activated pathways for resolving stalled replication forks.

Stalled replication forks can also be broken down if not resolved, resulting in genomic instability and carcinogenesis [132,133]. To solve this issue, eukaryotes have evolved the MEC1/ATR pathway to combat the breakdown of the replication forks [134]. In addition, a stalled replication fork is protected by checkpoint and homologous recombination (HR) proteins [135]. Current models propose that the repair protein MRE11 expands the ssDNA gaps at a stalled replication fork behind the replisome, creating the substrate for the post-replicative repair. In contrast, RAD51 is loaded onto the stalled replication fork through BRCA2 to limit the expansion of the ssDNA gaps and protect the stalled forks from being broken [135].

4.3. Bypass of DNA Damage at Stalled Replication Forks

When DNA damage occurs on stalled replication forks, they must be removed by DNA repair, or bypassed by DNA helicases and polymerases, allowing the restart, continuation, and completion of DNA replication. Failure to repair DNA lesions could result in DNA strand breaks, causing chromosomal rearrangement and cell death [136–138]. To ensure cell survival and the completion of the replication and cell cycle, cells may adopt lesion bypass if DNA damage at replication forks fails to be repaired [137]. The lesion bypass mechanisms include template switching, downstream repriming, recombination, lesion bypass through translesion synthesis (TLS) DNA polymerases and FANCJ [125,126,139]. However, the lesion bypass processes are usually error-prone, and can result in a rearrangement of chromosome and genomic instability associated with cancer. Here we discuss the lesion bypass of translesion DNA polymerases and the resulted genomic instability.

Unrepaired DNA lesions can be bypassed by TLS carried out by Y-family DNA polymerases [136,138,140,141] and some of the polymerases from the X- and A-family [142]. Since replicative DNA polymerases pause at DNA lesions, they are dislodged and substituted by TLS DNA polymerases, i.e., polymerase switching. This allows the incorporation of a nucleotide opposite the lesions by TLS polymerases for lesion bypass [40,109,138]. However, lesion bypass by TLS often results in nucleotide misincorporation and mismatches, causing mutations [138]. For example, incorporation of dAMP opposite 8-oxoG by TLS polymerases can induce the T→C transition mutation. DNA base lesions that can be bypassed by TLS polymerases are listed in Table 2. Also, TLS polymerases can incorporate damaged dNTPs and create mismatches to bypass a base lesion. It has been shown that oxidized dGTP can be incorporated opposite to dA on the template strand by TLS polymerases, inducing C→T transition and genomic instability [56–58]. It has been widely accepted that in the leading strand, DNA lesions need to be either repaired or bypassed by TLS polymerases for DNA synthesis to be continued during replication [140,143,144]. However, in the lagging strand, DNA lesions can be bypassed by TLS polymerases and repriming [123,136,138]. Thus, TLS polymerases play a crucial role in bypassing DNA lesions to maintain continuous DNA synthesis in the leading strand [140]. Upon the completion of DNA lesion bypass, TLS polymerases are dislodged by replicative polymerases through polymerase switching, restoring leading strand synthesis [145,146].

Table 2. Translesion DNA polymerases and their bypass of DNA base lesions.

Proteins	DNA Lesions	Nucleotide Preference of Lesion Bypass
Pol η	Thymine dimer	Prefer dA, followed by dG >dT>dC [147]
	8-oxoG	Prefer dC and dA [148]
	Acetyl amino fluorene-dG	Prefer dC followed by dG > dT> dA [147]
	N^6-ethenodeoxyadinosine	Prefer dT followed by dA >dG>dC [149]
	Abasic-site	Prefer A [147]
Pol κ	Thymine dimers	Could not bypass [150]
	N^6-ethenodeoxyadinosine	Prefer dT followed by dA >dC>dG [149]
	Abasic site	Prefer dA followed by dG >dT>dC [150]
Pol ι	Thymine dimer	Prefer T and A followed by dG >dC [151]
	Abasic site	Prefer dA [151]

5. Cellular Responses to Genome Stress and Epigenetic Instability

5.1. Oxidative DNA Damage and Epigenetic Instability

Genome stress, including replication stress induced by oxidative stress and its resulted DNA damage, can also induce epigenetic instability. Typical epigenetic instability includes hypermethylation of the promoter of tumor suppressor genes (TSGs), and hypomethylation of non-promoter CpGs, such as repetitive elements and satellite DNA.

The former causes transcriptional inactivation of TSGs, while the latter induces chromosomal instability and abnormal activation of oncogenes as well as mobile genetic elements. It has been found that a high level of ROS can lead to aberrant DNA hypermethylation in the gene promoter of TSGs, and their silencing suggesting an association between oxidative DNA damage with cancer-associated DNA methylation pattern changes. For example, exposure of hepatocellular carcinoma (HCC) cells to hydrogen peroxide leads to the hypermethylation of the promoter of the E-*Cadherin* gene via Snail-induced recruitment of histone deacetylase 1 (HDAC1) and DNA methyltransferase 1 (DNMT1) [152] to alter the DNA methylation pattern and chromatin structures. Further, oxidative DNA damage can inactivate TSGs through the recruitment of the polycomb repressive complex, which includes DNMT1, histone deacetylase (sirtuin-1) and histone methyltransferase to the CpGs containing 8-oxodGs [153]. It is possible that in responding to oxidative DNA damage, cells may use DNA hypermethylation to create heterochromatin in the genes, such as TSGs that are susceptible to DNA damage. This may shield DNA and protect them from further attack by DNA damaging agents. Interestingly, oxidative DNA damage can also result in DNA demethylation by inhibiting the binding of methyl-CpG binding protein 2 (MBP2) to methyl-CpGs, an epigenetic regulator that recruits DNMTs and histone HDAC to DNA [154]. This is because 8-oxodGs next to the 5mC at the CpGs inhibit the substrate binding of MBP [155,156]. Furthermore, the oxidized 5-methylcytosine, hydroxyl-5-methyl-cytosine can also decrease the binding affinity of MBPs resulting in DNA hypomethylation [157]. Thus, oxidative DNA damage can cause passive DNA demethylation, which in turn results in epigenetic instability leading to cancer and other diseases.

5.2. Histone Modifications at Stalled Replication Forks

Since double-helical DNA is wrapped around histone octamers that consist of H2A, H2B, H3 and H4 histone proteins, respectively [158], histone modifications that govern the structures of chromatin, i.e., opened (euchromatin) and closed (heterochromatin) conformation [159–161] play an important role in shielding DNA during cellular responses to DNA damage. It has been proposed that the formation of heterochromatin induced by genome stress such as replication fork stalling stops DNA replication (Figure 3) and prevents genomic instability. Histone tails are subject to different types of posttranslational modifications for the regulation of chromatin structures upon transcriptional activation or repression or chromatin opening or closing for DNA replication, and DNA damage and repair [162]. Specific histone modifications have also been identified as the response to replication stress [5]. The unscheduled firing of origin, fork stalling and repair of a collapsed fork can result in dramatic changes in chromatin structures. The methylation of newly synthesized histone proteins can be altered as a result of replication stress. This can alter the arrangement of old and newly-synthesized histone proteins, restoration of chromatin and patterning of epigenetic marks. It has been found that when a replication fork is stalled by genome stress, histones along with antisilencing factor 1 (Asf1) fail to be incorporated into chromatin, thereby increasing H3K9me1 [163]. Subsequently, methylation of H3K9 prevents histone acetylation, the active mark. H3K9me1 can also be further methylated into H3K9me3, the suppressive mark. These can then lead to the suppression of replication [5,163–168]. Further, histone methylation can recruit endonucleases to degrade the stalled replication fork. It has been found that methylation of H3K4 triggers MRE11-mediated degradation of replication fork, whereas H3K27me3 recruits MUS81 to cleave stalled forks [169,170]. The results indicate that cells adopt the epigenetic mechanisms to resolve stalled replication forks stalling.

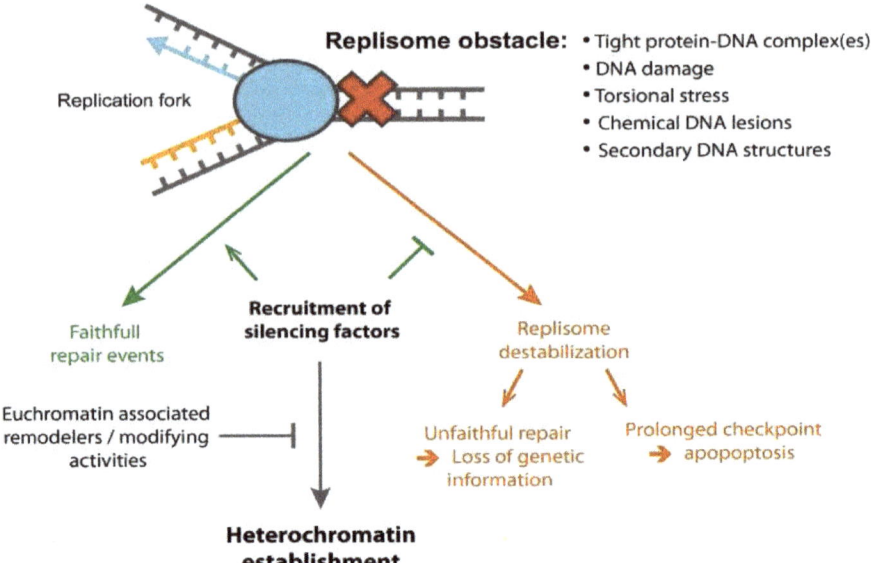

Figure 3. Heterochromatin formation during replication stress to prevent loss of genetic information [171].

Interestingly, the components of replisome, such as pol α, can bind to H2A and H2B [172]. Besides, MCM2 and pol ε can bind to H3 and H4 [173]. The interaction between the replication proteins and histones plays an important role in the redeposition of old histone to the newly-synthesized DNA. The redisposition of old histones helps the maintenance of the epigenetic marks of parental DNA in newly synthesized. Moreover, a challenge of DNA replication at non-B form DNA structures such as G quadruplexes can result in epigenetic instability [174]. Formation of the non-B form DNA structures such as G4 structures during replication can lead to an imbalance of the loading of old and new histones on the leading and lagging strands [175]. It has been shown that old histones are preferentially loaded on the lagging strand, whereas new histones are mainly loaded on the leading strand [175]. This results from the replication polymerases stalling at G4 structures in the leading strand. However, helicases can still keep unwinding the fork, allowing the continuation of the lagging strand synthesis. This subsequently results in the loss of H3K4me3, H3K9ac and H3k14ac marks in the regions at 4.5 kb downstream of the G4 structures in the leading strand [175,176]. The results indicate that cells adopt different epigenetic mechanisms to resolve stalled replication forks stalling.

5.3. DNA Damage and Modulation of miRNA Expression

DNA damage and its resulted replication stress can also alter the expression of microRNAs (miRNAs). MiRNA is short (18–22 nucleotide non-coding RNA molecules that base pair to the 3′ untranslated regions (UTR) of mRNAs) [177]. MiRNAs inhibit protein translation by promoting mRNA degradation or translation repression depending on the degree of their sequences complimentarily with those of their target mRNAs [177,178]. MiRNAs are involved in the regulation of cell proliferation, development, metabolism and gene expression [179,180]. Several classes of miRNAs are associated with the regulation of the genes of replication progression, cell cycle and DNA damage repair. Usually, miRNAs are deregulated by DNA damage [179,181–183]. It has been found that the miRNAs involved in cell cycle control can be upregulated by E2F [184]. MiRNAs that are deregulated by DNA damage include miR-34a, -34b and -34c.

These mRNAs belong to the miR-34 family, and are upregulated in response to DNA damage. They are also the regulators of the expression of the checkpoint genes, such as E2F, CDK4, CDK6 and cyclin E2 [179,185]. In addition, miR-145a and miR-146b that target the tumor suppressor, BRCA1 are also upregulated upon double-strand DNA breaks [179,186]. MiR-155 and miR-21 that target mismatch repair proteins are upregulated during cellular responses to oxidative DNA damage induced by hydrogen peroxide and radiation [179,187]. On the other hand, miR-16 and mir-15 a/b that target the downregulators of checkpoint proteins, Cdc25a and Wip1, are also upregulated upon DNA damage. The Let-7 family miRNAs, *let-7i*, *mir-15b-16-2* and *mir-106b-25*, can also be induced by E2F. The miRNAs in this family are involved in limiting S phase entry as a result of genome stress, thereby preventing mutagenesis [184]. Also, miRNAs can downregulate MCM2-7 in a *Trp53*-dependent manner [183]. The roles of miRNA in mediating cellular response to genome stress warrant further studies in the future.

6. Conclusions

DNA damage and genome stress, including DNA replication stress, can cause genomic and epigenomic instability, which are associated with many diseases such as cancer. The studies summarized here have pointed to a direct link among DNA damage, genome stress, such as replication stress and genomic and epigenomic instability. Since genome stress triggers the alteration of genetic and epigenetic information that can be passed to the next generation, it is important to further explore how genome stress, such as replication stress, can crosstalk with DNA methylation, chromatin structures and miRNA expression in the context of a variety of diseases.

Funding: The work was supported by the National Institutes of Health grant R01ES023569 to Y. Liu.

Conflicts of Interest: The authors declare no conflict of interest.

Abbreviations

ASF	Alternative splicing factor 1
ATM	(ataxia-telangiectasia mutated) serine/threonine kinase
ATR	Ataxia telangiectasia and Rad3-related protein
ATPIP	ATR interacting protein
BRCA1	Breast cancer 1
BLM	Bloom syndrome
BRCA2	Breast cancer type 2 susceptibility protein
Cdc45	Cell division cycle 45
Cdc6	Cell Division Cycle 6
CDK	Cyclin dependent kinase
Cdt1	Cdc10-dependent transcript 1
Chk1	Checkpoint kinase 1
CMG	CDC45-MCM-GINS
CPDs	Pyrimidine dimers
Ctf4	Chromosome transmission fidelity 4
dAMP	Deoxyadenosine monophosphate
Dbf4	Dumbbell former 4
DDC	DNA damage checkpoint
DDK	Dbf4 dependent kinase
DDR	DNA damage response
DDSBs	DNA double-strand breaks
DGCR8	DiGeorge syndrome critical region 8)
dNDPs	Deoxyribonucleotides diphosphate
DNMT1	DNA methyltransferase 1
dNTPs	Deoxynucleotide triphosphates
DRC	DNA replication checkpoint
dUMP	Deoxyuracilmonophosphate
dUTP	Deoxyuraciltriphosphate
dUTPase	dUTP pyrophosphatase

FANCJ	Fanconi Anemia complementation group J Protein
FEN 1	Flap endonuclease 1
FRDA	Friedreich's ataxia
FXN	Frataxin gene
GINS	Go-ichi-ni-san
HCC	Hepatocellular carcinoma
HD	Huntington's disease
HDAC1	Histone deacetylase 1
HR	Homologous recombination
HU	Hydroxyurea
IMAGe	Intrauterine growth restriction, metaphyseal dysplasia, adrenal hypoplasia, and genital anomalies
ICLs	Interstrand DNA crosslinks
MCM	Minichromosome maintenance protein
MEC1	Mitosis entry checkpoint 1
MRE11	Microhomology-mediated end-joining 11
ORC	Origin recognition complex
PCNA	Proliferating cell nuclear antigen
PreRC	Pre-replication complex
PTMs	Post-transcriptional modifications
RecQL4	RecQ like helicase 4
RFC	Replication factor C
RISC	RNA-induced silencing complex
RNR	Ribonucleotide reductase
ROS	Reactive oxygen species
SF2	Splicing factor 2
SRSF1	Serine/arginine-rich splicing factor 1
ssDNA	Single strand DNA
TLS	Translesion polymerases
TopBP1	Topoisomerase II Binding Protein 1
TRCs	Transcription-replication conflicts
TSGs	Tumor suppressor genes
UTR	Untranslated region
SAMHD1	SAM domain and HD domain-containing protein 1
WRN	Werner syndrome

References

1. Kearsey, S.E.; Cotterill, S. Enigmatic variations: Divergent modes of regulating eukaryotic DNA replication. *Mol. Cell* **2003**, *12*, 1067–1075. [CrossRef]
2. Rhind, N.; Russell, P. Signaling pathways that regulate cell division. *Cold Spring Harb. Perspect. Biol.* **2012**, *4*. [CrossRef] [PubMed]
3. Schwob, E. Flexibility and governance in eukaryotic DNA replication. *Curr. Opin. Microbiol.* **2004**, *7*, 680–690. [CrossRef] [PubMed]
4. Gaillard, H.; Garcia-Muse, T.; Aguilera, A. Replication stress and cancer. *Nat. Rev. Cancer* **2015**, *15*, 276–289. [CrossRef] [PubMed]
5. Jasencakova, Z.; Groth, A. Replication stress, a source of epigenetic aberrations in cancer? *Bioessays* **2010**, *32*, 847–855. [CrossRef]
6. Fragkos, M.; Ganier, O.; Coulombe, P.; Mechali, M. DNA replication origin activation in space and time. *Nat. Rev. Mol. Cell Biol.* **2015**, *16*, 360–374. [CrossRef]
7. Coster, G.; Diffley, J.F.X. Bidirectional eukaryotic DNA replication is established by quasi-symmetrical helicase loading. *Science* **2017**, *357*, 314–318. [CrossRef]
8. Ticau, S.; Friedman, L.J.; Ivica, N.A.; Gelles, J.; Bell, S.P. Single-molecule studies of origin licensing reveal mechanisms ensuring bidirectional helicase loading. *Cell* **2015**, *161*, 513–525. [CrossRef]

9. Remus, D.; Beuron, F.; Tolun, G.; Griffith, J.D.; Morris, E.P.; Diffley, J.F. Concerted loading of Mcm2-7 double hexamers around DNA during DNA replication origin licensing. *Cell* **2009**, *139*, 719–730. [CrossRef]
10. Evrin, C.; Clarke, P.; Zech, J.; Lurz, R.; Sun, J.C.; Uhle, S.; Li, H.L.; Stillman, B.; Speck, C. A double-hexameric MCM2-7 complex is loaded onto origin DNA during licensing of eukaryotic DNA replication. *Proc. Natl. Acad. Sci. USA* **2009**, *106*, 20240–20245. [CrossRef]
11. Ilves, I.; Petojevic, T.; Pesavento, J.J.; Botchan, M.R. Activation of the MCM2-7 helicase by association with Cdc45 and GINS proteins. *Mol. Cell* **2010**, *37*, 247–258. [CrossRef] [PubMed]
12. Boos, D.; Frigola, J.; Diffley, J.F. Activation of the replicative DNA helicase: Breaking up is hard to do. *Curr. Opin. Cell Biol.* **2012**, *24*, 423–430. [CrossRef] [PubMed]
13. Sheu, Y.J.; Stillman, B. Cdc7-Dbf4 phosphorylates MCM proteins via a docking site-mediated mechanism to promote S phase progression. *Mol. Cell* **2006**, *24*, 101–113. [CrossRef] [PubMed]
14. Tanaka, S.; Umemori, T.; Hirai, K.; Muramatsu, S.; Kamimura, Y.; Araki, H. CDK-dependent phosphorylation of Sld2 and Sld3 initiates DNA replication in budding yeast. *Nature* **2007**, *445*, 328–332. [CrossRef]
15. Zegerman, P.; Diffley, J.F. Phosphorylation of Sld2 and Sld3 by cyclin-dependent kinases promotes DNA replication in budding yeast. *Nature* **2007**, *445*, 281–285. [CrossRef]
16. Sun, J.; Evrin, C.; Samel, S.A.; Fernandez-Cid, A.; Riera, A.; Kawakami, H.; Stillman, B.; Speck, C.; Li, H. Cryo-EM structure of a helicase loading intermediate containing ORC-Cdc6-Cdt1-MCM2-7 bound to DNA. *Nat. Struct. Mol. Biol.* **2013**, *20*, 944–951. [CrossRef]
17. Yeeles, J.T.; Deegan, T.D.; Janska, A.; Early, A.; Diffley, J.F. Regulated eukaryotic DNA replication origin firing with purified proteins. *Nature* **2015**, *519*, 431–435. [CrossRef]
18. Yeeles, J.T.P.; Janska, A.; Early, A.; Diffley, J.F.X. How the Eukaryotic Replisome Achieves Rapid and Efficient DNA Replication. *Mol. Cell* **2017**, *65*, 105–116. [CrossRef]
19. Costa, A.; Renault, L.; Swuec, P.; Petojevic, T.; Pesavento, J.J.; Ilves, I.; MacLellan-Gibson, K.; Fleck, R.A.; Botchan, M.R.; Berger, J.M. DNA binding polarity, dimerization, and ATPase ring remodeling in the CMG helicase of the eukaryotic replisome. *Elife* **2014**, *3*, e03273. [CrossRef]
20. Soultanas, P. Loading mechanisms of ring helicases at replication origins. *Mol. Microbiol.* **2012**, *84*, 6–16. [CrossRef]
21. McGlynn, P. Helicases at the replication fork. *Adv. Exp. Med. Biol.* **2013**, *767*, 97–121. [CrossRef] [PubMed]
22. Balakrishnan, L.; Bambara, R.A. Eukaryotic lagging strand DNA replication employs a multi-pathway mechanism that protects genome integrity. *J. Biol. Chem.* **2011**, *286*, 6865–6870. [CrossRef] [PubMed]
23. Wold, M.S. Replication protein A: A heterotrimeric, single-stranded DNA-binding protein required for eukaryotic DNA metabolism. *Annu. Rev. Biochem.* **1997**, *66*, 61–92. [CrossRef] [PubMed]
24. Zhang, D.; O'Donnell, M. The Eukaryotic Replication Machine. *Enzymes* **2016**, *39*, 191–229. [CrossRef] [PubMed]
25. Hedglin, M.; Kumar, R.; Benkovic, S.J. Replication clamps and clamp loaders. *Cold Spring Harb. Perspect. Biol.* **2013**, *5*, a010165. [CrossRef] [PubMed]
26. Hedglin, M.; Benkovic, S.J. Regulation of Rad6/Rad18 Activity During DNA Damage Tolerance. *Annu. Rev. Biophys.* **2015**, *44*, 207–228. [CrossRef]
27. Rumbaugh, J.A.; Murante, R.S.; Shi, S.; Bambara, R.A. Creation and removal of embedded ribonucleotides in chromosomal DNA during mammalian Okazaki fragment processing. *J. Biol. Chem.* **1997**, *272*, 22591–22599. [CrossRef]
28. Liu, Y.; Kao, H.I.; Bambara, R.A. Flap endonuclease 1: A central component of DNA metabolism. *Annu Rev. Biochem.* **2004**, *73*, 589–615. [CrossRef]
29. Balakrishnan, L.; Bambara, R.A. Flap endonuclease 1. *Annu. Rev. Biochem.* **2013**, *82*, 119–138. [CrossRef]
30. Garg, P.; Stith, C.M.; Sabouri, N.; Johansson, E.; Burgers, P.M. Idling by DNA polymerase delta maintains a ligatable nick during lagging-strand DNA replication. *Gene Dev.* **2004**, *18*, 2764–2773. [CrossRef]
31. Burgers, P.M.J. Polymerase Dynamics at the Eukaryotic DNA Replication Fork. *J. Biol. Chem.* **2009**, *284*, 4041–4045. [CrossRef] [PubMed]
32. Turchi, J.J.; Huang, L.; Murante, R.S.; Kim, Y.; Bambara, R.A. Enzymatic completion of mammalian lagging-strand DNA replication. *Proc. Natl. Acad. Sci. USA* **1994**, *91*, 9803–9807. [CrossRef] [PubMed]
33. Xia, S.L.; Konigsberg, W.H. RB69 DNA Polymerase Structure, Kinetics, and Fidelity. *Biochemistry* **2014**, *53*, 2752–2767. [CrossRef] [PubMed]

34. Franklin, M.C.; Wang, J.M.; Steitz, T.A. Structure of the replicating complex of a pol alpha family DNA polymerase. *Cell* **2001**, *105*, 657–667. [CrossRef]
35. Doublie, S.; Zahn, K.E. Structural insights into eukaryotic DNA replication. *Front. Microbiol* **2014**, *5*. [CrossRef] [PubMed]
36. Lange, S.S.; Takata, K.; Wood, R.D. DNA polymerases and cancer. *Nat. Rev. Cancer* **2011**, *11*, 96–110. [CrossRef] [PubMed]
37. Sale, J.E.; Lehmann, A.R.; Woodgate, R. Y-family DNA polymerases and their role in tolerance of cellular DNA damage. *Nat. Rev. Mol. Cell Biol.* **2012**, *13*, 141–152. [CrossRef]
38. Washington, M.T.; Carlson, K.D.; Freudenthal, B.D.; Pryor, J.M. Variations on a theme: Eukaryotic Y-family DNA polymerases. *Biochim. Biophys. Acta* **2010**, *1804*, 1113–1123. [CrossRef]
39. Prakash, S.; Johnson, R.E.; Prakash, L. Eukaryotic translesion synthesis DNA polymerases: Specificity of structure and function. *Annu. Rev. Biochem.* **2005**, *74*, 317–353. [CrossRef]
40. Zeman, M.K.; Cimprich, K.A. Causes and consequences of replication stress. *Nat. Cell Biol.* **2014**, *16*, 2–9. [CrossRef]
41. Shen, Z. The origin recognition complex in human diseases. *Biosci. Rep.* **2013**, *33*. [CrossRef] [PubMed]
42. de Munnik, S.A.; Bicknell, L.S.; Aftimos, S.; Al-Aama, J.Y.; van Bever, Y.; Bober, M.B.; Clayton-Smith, J.; Edrees, A.Y.; Feingold, M.; Fryer, A.; et al. Meier-Gorlin syndrome genotype-phenotype studies: 35 individuals with pre-replication complex gene mutations and 10 without molecular diagnosis. *Eur. J. Hum. Genet.* **2012**, *20*, 598–606. [CrossRef] [PubMed]
43. Cottineau, J.; Kottemann, M.C.; Lach, F.P.; Kang, Y.H.; Vely, F.; Deenick, E.K.; Lazarov, T.; Gineau, L.; Wang, Y.; Farina, A.; et al. Inherited GINS1 deficiency underlies growth retardation along with neutropenia and NK cell deficiency. *J. Clin. Invest.* **2017**, *127*, 1991–2006. [CrossRef] [PubMed]
44. Logan, C.V.; Murray, J.E.; Parry, D.A.; Robertson, A.; Bellelli, R.; Tarnauskaite, Z.; Challis, R.; Cleal, L.; Bore, V.; Fluteau, A.; et al. DNA Polymerase Epsilon Deficiency Causes IMAGe Syndrome with Variable Immunodeficiency. *Am. J. Hum. Genet.* **2018**, *103*, 1038–1044. [CrossRef] [PubMed]
45. Munoz, S.; Mendez, J. DNA replication stress: From molecular mechanisms to human disease. *Chromosoma* **2017**, *126*, 1–15. [CrossRef] [PubMed]
46. Bizard, A.H.; Hickson, I.D. The dissolution of double Holliday junctions. *Cold Spring Harb. Perspect. Biol.* **2014**, *6*, a016477. [CrossRef] [PubMed]
47. Bernstein, K.A.; Gangloff, S.; Rothstein, R. The RecQ DNA Helicases in DNA Repair. *Annu. Rev. Genet.* **2010**, *44*, 393–417. [CrossRef]
48. Crow, Y.J.; Leitch, A.; Hayward, B.E.; Garner, A.; Parmar, R.; Griffith, E.; Ali, M.; Semple, C.; Aicardi, J.; Babul-Hirji, R.; et al. Mutations in genes encoding ribonuclease H2 subunits cause Aicardi-Goutieres syndrome and mimic congenital viral brain infection. *Nat. Genet.* **2006**, *38*, 910–916. [CrossRef]
49. Chabosseau, P.; Buhagiar-Labarchede, G.; Onclercq-Delic, R.; Lambert, S.; Debatisse, M.; Brison, O.; Amor-Gueret, M. Pyrimidine pool imbalance induced by BLM helicase deficiency contributes to genetic instability in Bloom syndrome. *Nat. Commun.* **2011**, *2*, 368. [CrossRef]
50. Kim, H.; D'Andrea, A.D. Regulation of DNA cross-link repair by the Fanconi anemia/BRCA pathway. *Genes Dev.* **2012**, *26*, 1393–1408. [CrossRef]
51. Bartek, J.; Lukas, C.; Lukas, J. Checking on DNA damage in S phase. *Nat. Rev. Mol. Cell Biol.* **2004**, *5*, 792–804. [CrossRef] [PubMed]
52. Lopes, M.; Cotta-Ramusino, C.; Pellicioli, A.; Liberi, G.; Plevani, P.; Muzi-Falconi, M.; Newlon, C.S.; Foiani, M. The DNA replication checkpoint response stabilizes stalled replication forks. *Nature* **2001**, *412*, 557–561. [CrossRef] [PubMed]
53. Clemente-Ruiz, M.; Prado, F. Chromatin assembly controls replication fork stability. *EMBO Rep.* **2009**, *10*, 790–796. [CrossRef] [PubMed]
54. Gay, S.; Lachages, A.M.; Millot, G.A.; Courbet, S.; Letessier, A.; Debatisse, M.; Brison, O. Nucleotide supply, not local histone acetylation, sets replication origin usage in transcribed regions. *EMBO Rep.* **2010**, *11*, 698–704. [CrossRef] [PubMed]
55. Bester, A.C.; Roniger, M.; Oren, Y.S.; Im, M.M.; Sarni, D.; Chaoat, M.; Bensimon, A.; Zamir, G.; Shewach, D.S.; Kerem, B. Nucleotide deficiency promotes genomic instability in early stages of cancer development. *Cell* **2011**, *145*, 435–446. [CrossRef] [PubMed]

56. Grollman, A.P.; Moriya, M. Mutagenesis by 8-oxoguanine: An enemy within. *Trends Genet.* **1993**, *9*, 246–249. [CrossRef]
57. Patel, D.R.; Weiss, R.S. A tough row to hoe: When replication forks encounter DNA damage. *Biochem. Soc. Trans.* **2018**, *46*, 1643–1651. [CrossRef]
58. Shibutani, S.; Takeshita, M.; Grollman, A.P. Insertion of specific bases during DNA synthesis past the oxidation-damaged base 8-oxodG. *Nature* **1991**, *349*, 431–434. [CrossRef]
59. Wallace, S.S. Biological consequences of free radical-damaged DNA bases. *Free Radic. Biol. Med.* **2002**, *33*, 1–14. [CrossRef]
60. Sutherland, G.R.; Baker, E.; Richards, R.I. Fragile sites still breaking. *Trends Genet.* **1998**, *14*, 501–506. [CrossRef]
61. Liu, P.F.; Carvalho, C.M.B.; Hastings, P.J.; Lupski, J.R. Mechanisms for recurrent and complex human genomic rearrangements. *Curr. Opin. Genet. Dev.* **2012**, *22*, 211–220. [CrossRef] [PubMed]
62. Techer, H.; Koundrioukoff, S.; Nicolas, A.; Debatisse, M. The impact of replication stress on replication dynamics and DNA damage in vertebrate cells. *Nat. Rev. Genet.* **2017**, *18*, 535–550. [CrossRef] [PubMed]
63. Gordenin, D.A.; Resnick, M.A. Yeast ARMs (DNA at-risk motifs) can reveal sources of genome instability. *Mutat Res.* **1998**, *400*, 45–58. [CrossRef]
64. Mirkin, E.V.; Mirkin, S.M. Replication fork stalling at natural impediments. *Microbiol. Mol. Biol. Rev.* **2007**, *71*, 13–35. [CrossRef] [PubMed]
65. Wang, G.; Vasquez, K.M. Impact of alternative DNA structures on DNA damage, DNA repair, and genetic instability. *DNA Repair (Amst)* **2014**, *19*, 143–151. [CrossRef] [PubMed]
66. Usdin, K.; House, N.C.; Freudenreich, C.H. Repeat instability during DNA repair: Insights from model systems. *Crit Rev. Biochem Mol. Biol.* **2015**, *50*, 142–167. [CrossRef]
67. Branzei, D.; Foiani, M. Maintaining genome stability at the replication fork. *Nat. Rev. Mol. Cell Biol.* **2010**, *11*, 208–219. [CrossRef]
68. Cadet, J.; Delatour, T.; Douki, T.; Gasparutto, D.; Pouget, J.P.; Ravanat, J.L.; Sauvaigo, S. Hydroxyl radicals and DNA base damage. *Mutat Res.* **1999**, *424*, 9–21. [CrossRef]
69. Coluzzi, E.; Leone, S.; Sgura, A. Oxidative Stress Induces Telomere Dysfunction and Senescence by Replication Fork Arrest. *Cells* **2019**, *8*, 19. [CrossRef]
70. Marnett, L.J. Oxy radicals, lipid peroxidation and DNA damage. *Toxicology* **2002**, *181–182*, 219–222. [CrossRef]
71. Dutta, A.; Yang, C.; Sengupta, S.; Mitra, S.; Hegde, M.L. New paradigms in the repair of oxidative damage in human genome: Mechanisms ensuring repair of mutagenic base lesions during replication and involvement of accessory proteins. *Cell Mol. Life Sci.* **2015**, *72*, 1679–1698. [CrossRef] [PubMed]
72. Lindahl, T. Instability and decay of the primary structure of DNA. *Nature* **1993**, *362*, 709–715. [CrossRef] [PubMed]
73. Atkinson, J.; McGlynn, P. Replication fork reversal and the maintenance of genome stability. *Nucleic Acids Res.* **2009**, *37*, 3475–3492. [CrossRef] [PubMed]
74. Lucas-Lledo, J.I.; Lynch, M. Evolution of Mutation Rates: Phylogenomic Analysis of the Photolyase/Cryptochrome Family. *Mol. Biol. Evol.* **2009**, *26*, 1143–1153. [CrossRef]
75. McCready, S.J.; Osman, F.; Yasui, A. Repair of UV damage in the fission yeast Schizosaccharomyces pombe. *Mutat Res.* **2000**, *451*, 197–210. [CrossRef]
76. Zhang, J.; Walter, J.C. Mechanism and regulation of incisions during DNA interstrand cross-link repair. *DNA Repair (Amst)* **2014**, *19*, 135–142. [CrossRef]
77. Stone, J.E.; Kumar, D.; Binz, S.K.; Inase, A.; Iwai, S.; Chabes, A.; Burgers, P.M.; Kunkel, T.A. Lesion bypass by S. cerevisiae Pol zeta alone. *DNA Repair (Amst)* **2011**, *10*, 826–834. [CrossRef]
78. Vare, D.; Groth, P.; Carlsson, R.; Johansson, F.; Erixon, K.; Jenssen, D. DNA interstrand crosslinks induce a potent replication block followed by formation and repair of double strand breaks in intact mammalian cells. *DNA Repair (Amst)* **2012**, *11*, 976–985. [CrossRef]
79. Garcia-Muse, T.; Aguilera, A. Transcription-replication conflicts: How they occur and how they are resolved. *Nat. Rev. Mol. Cell Biol.* **2016**, *17*, 553–563. [CrossRef]
80. Aguilera, A.; Garcia-Muse, T. R loops: From transcription byproducts to threats to genome stability. *Mol. Cell* **2012**, *46*, 115–124. [CrossRef]
81. Santos-Pereira, J.M.; Aguilera, A. R loops: New modulators of genome dynamics and function. *Nat. Rev. Genet.* **2015**, *16*, 583–597. [CrossRef] [PubMed]

82. Huertas, P.; Aguilera, A. Cotranscriptionally formed DNA:RNA hybrids mediate transcription elongation impairment and transcription-associated recombination. *Mol. Cell* **2003**, *12*, 711–721. [CrossRef] [PubMed]
83. Hamperl, S.; Bocek, M.J.; Saldivar, J.C.; Swigut, T.; Cimprich, K.A. Transcription-Replication Conflict Orientation Modulates R-Loop Levels and Activates Distinct DNA Damage Responses. *Cell* **2017**, *170*, 774–786.e19. [CrossRef] [PubMed]
84. Li, X.; Manley, J.L. Inactivation of the SR protein splicing factor ASF/SF2 results in genomic instability. *Cell* **2005**, *122*, 365–378. [CrossRef] [PubMed]
85. Groh, M.; Gromak, N. Out of balance: R-loops in human disease. *PLoS Genet.* **2014**, *10*, e1004630. [CrossRef] [PubMed]
86. Groh, M.; Lufino, M.M.; Wade-Martins, R.; Gromak, N. R-loops associated with triplet repeat expansions promote gene silencing in Friedreich ataxia and fragile X syndrome. *PLoS Genet.* **2014**, *10*, e1004318. [CrossRef] [PubMed]
87. Grabczyk, E.; Mancuso, M.; Sammarco, M.C. A persistent RNA.DNA hybrid formed by transcription of the Friedreich ataxia triplet repeat in live bacteria, and by T7 RNAP in vitro. *Nucleic Acids Res.* **2007**, *35*, 5351–5359. [CrossRef]
88. Reddy, K.; Tam, M.; Bowater, R.P.; Barber, M.; Tomlinson, M.; Edamura, K.N.; Wang, Y.H.; Pearson, C.E. Determinants of R-loop formation at convergent bidirectionally transcribed trinucleotide repeats. *Nucleic Acids Res.* **2011**, *39*, 1749–1762. [CrossRef]
89. Lin, Y.; Dent, S.Y.; Wilson, J.H.; Wells, R.D.; Napierala, M. R loops stimulate genetic instability of CTG.CAG repeats. *Proc. Natl. Acad. Sci. USA* **2010**, *107*, 692–697. [CrossRef]
90. McIvor, E.I.; Polak, U.; Napierala, M. New insights into repeat instability Role of RNA.DNA hybrids. *RNA Biol.* **2010**, *7*, 551–558. [CrossRef]
91. Wilhelm, T.; Ragu, S.; Magdalou, I.; Machon, C.; Dardillac, E.; Techer, H.; Guitton, J.; Debatisse, M.; Lopez, B.S. Slow Replication Fork Velocity of Homologous Recombination-Defective Cells Results from Endogenous Oxidative Stress. *PLoS Genet.* **2016**, *12*, e1006007. [CrossRef] [PubMed]
92. Techer, H.; Koundrioukoff, S.; Carignon, S.; Wilhelm, T.; Millot, G.A.; Lopez, B.S.; Brison, O.; Debatisse, M. Signaling from Mus81-Eme2-Dependent DNA Damage Elicited by Chk1 Deficiency Modulates Replication Fork Speed and Origin Usage. *Cell Rep.* **2016**, *14*, 1114–1127. [CrossRef] [PubMed]
93. Anglana, M.; Apiou, F.; Bensimon, A.; Debatisse, M. Dynamics of DNA replication in mammalian somatic cells: Nucleotide pool modulates origin choice and interorigin spacing. *Cell* **2003**, *114*, 385–394. [CrossRef]
94. Mathews, C.K. Deoxyribonucleotide metabolism, mutagenesis and cancer. *Nat. Rev. Cancer* **2015**, *15*, 528–539. [CrossRef] [PubMed]
95. Nordlund, P.; Reichard, P. Ribonucleotide reductases. *Annu. Rev. Biochem* **2006**, *75*, 681–706. [CrossRef] [PubMed]
96. Pontarin, G.; Fijolek, A.; Pizzo, P.; Ferraro, P.; Rampazzo, C.; Pozzan, T.; Thelander, L.; Reichard, P.A.; Bianchi, V. Ribonucleotide reduction is a cytosolic process in mammalian cells independently of DNA damage. *Proc. Natl. Acad. Sci. USA* **2008**, *105*, 17801–17806. [CrossRef] [PubMed]
97. Clifford, R.; Louis, T.; Robbe, P.; Ackroyd, S.; Burns, A.; Timbs, A.T.; Wright Colopy, G.; Dreau, H.; Sigaux, F.; Judde, J.G.; et al. SAMHD1 is mutated recurrently in chronic lymphocytic leukemia and is involved in response to DNA damage. *Blood* **2014**, *123*, 1021–1031. [CrossRef]
98. Franzolin, E.; Pontarin, G.; Rampazzo, C.; Miazzi, C.; Ferraro, P.; Palumbo, E.; Reichard, P.; Bianchi, V. The deoxynucleotide triphosphohydrolase SAMHD1 is a major regulator of DNA precursor pools in mammalian cells. *Proc. Natl. Acad. Sci. USA* **2013**, *110*, 14272–14277. [CrossRef]
99. Bessman, M.J.; Lehman, I.R.; Adler, J.; Zimmerman, S.B.; Simms, E.S.; Kornberg, A. Enzymatic Synthesis of Deoxyribonucleic Acid. Iii. The Incorporation of Pyrimidine and Purine Analogues into Deoxyribonucleic Acid. *Proc. Natl. Acad. Sci. USA* **1958**, *44*, 633–640. [CrossRef]
100. Chen, C.W.; Tsao, N.; Huang, L.Y.; Yen, Y.; Liu, X.; Lehman, C.; Wang, Y.H.; Tseng, M.C.; Chen, Y.J.; Ho, Y.C.; et al. The Impact of dUTPase on Ribonucleotide Reductase-Induced Genome Instability in Cancer Cells. *Cell Rep.* **2016**, *16*, 1287–1299. [CrossRef]
101. Reijns, M.A.; Rabe, B.; Rigby, R.E.; Mill, P.; Astell, K.R.; Lettice, L.A.; Boyle, S.; Leitch, A.; Keighren, M.; Kilanowski, F.; et al. Enzymatic removal of ribonucleotides from DNA is essential for mammalian genome integrity and development. *Cell* **2012**, *149*, 1008–1022. [CrossRef] [PubMed]

102. Sassa, A.; Yasui, M.; Honma, M. Current perspectives on mechanisms of ribonucleotide incorporation and processing in mammalian DNA. *Genes Environ.* **2019**, *41*, 3. [CrossRef] [PubMed]
103. Sparks, J.L.; Chon, H.; Cerritelli, S.M.; Kunkel, T.A.; Johansson, E.; Crouch, R.J.; Burgers, P.M. RNase H2-initiated ribonucleotide excision repair. *Mol. Cell* **2012**, *47*, 980–986. [CrossRef] [PubMed]
104. Pizzi, S.; Sertic, S.; Orcesi, S.; Cereda, C.; Bianchi, M.; Jackson, A.P.; Lazzaro, F.; Plevani, P.; Muzi-Falconi, M. Reduction of hRNase H2 activity in Aicardi-Goutieres syndrome cells leads to replication stress and genome instability. *Hum. Mol. Genet.* **2015**, *24*, 649–658. [CrossRef]
105. Gao, R.; Schellenberg, M.J.; Huang, S.Y.; Abdelmalak, M.; Marchand, C.; Nitiss, K.C.; Nitiss, J.L.; Williams, R.S.; Pommier, Y. Proteolytic degradation of topoisomerase II (Top2) enables the processing of Top2.DNA and Top2.RNA covalent complexes by tyrosyl-DNA-phosphodiesterase 2 (TDP2). *J. Biol. Chem.* **2014**, *289*, 17960–17969. [CrossRef]
106. Huang, S.N.; Williams, J.S.; Arana, M.E.; Kunkel, T.A.; Pommier, Y. Topoisomerase I-mediated cleavage at unrepaired ribonucleotides generates DNA double-strand breaks. *EMBO J.* **2017**, *36*, 361–373. [CrossRef]
107. Huang, S.Y.N.; Ghosh, S.; Pommier, Y. Topoisomerase I Alone Is Sufficient to Produce Short DNA Deletions and Can Also Reverse Nicks at Ribonucleotide Sites. *J. Biol. Chem.* **2015**, *290*, 14068–14076. [CrossRef]
108. Lanz, M.C.; Dibitetto, D.; Smolka, M.B. DNA damage kinase signaling: Checkpoint and repair at 30 years. *EMBO J.* **2019**, *38*, e101801. [CrossRef]
109. Macheret, M.; Halazonetis, T.D. DNA Replication Stress as a Hallmark of Cancer. *Annu. Rev. Pathol. Mech.* **2015**, *10*, 425–448. [CrossRef]
110. Magdalou, I.; Lopez, B.S.; Pasero, P.; Larnbert, S.A.E. The causes of replication stress and their consequences on genome stability and cell fate. *Semin. Cell Dev. Biol.* **2014**, *30*, 154–164. [CrossRef]
111. Tsai, F.L.; Vijayraghavan, S.; Prinz, J.; MacAlpine, H.K.; MacAlpine, D.M.; Schwacha, A. Mcm2-7 is an active player in the DNA replication checkpoint signaling cascade via proposed modulation of its DNA gate. *Mol. Cell Biol.* **2015**, *35*, 2131–2143. [CrossRef] [PubMed]
112. Kastan, M.B.; Bartek, J. Cell-cycle checkpoints and cancer. *Nature* **2004**, *432*, 316–323. [CrossRef] [PubMed]
113. Rhind, N.; Russell, P. Checkpoints: It takes more than time to heal some wounds. *Curr. Biol.* **2000**, *10*, R908–R911. [CrossRef]
114. Koundrioukoff, S.; Carignon, S.; Techer, H.; Letessier, A.; Brison, O.; Debatisse, M. Stepwise activation of the ATR signaling pathway upon increasing replication stress impacts fragile site integrity. *PLoS Genet.* **2013**, *9*, e1003643. [CrossRef]
115. Dungrawala, H.; Rose, K.L.; Bhat, K.P.; Mohni, K.N.; Glick, G.G.; Couch, F.B.; Cortez, D. The Replication Checkpoint Prevents Two Types of Fork Collapse without Regulating Replisome Stability. *Mol. Cell* **2015**, *59*, 998–1010. [CrossRef]
116. Lossaint, G.; Larroque, M.; Ribeyre, C.; Bec, N.; Larroque, C.; Decaillet, C.; Gari, K.; Constantinou, A. FANCD2 Binds MCM Proteins and Controls Replisome Function upon Activation of S Phase Checkpoint Signaling. *Mol. Cell* **2013**, *51*, 678–690. [CrossRef]
117. Sirbu, B.M.; McDonald, W.H.; Dungrawala, H.; Badu-Nkansah, A.; Kavanaugh, G.M.; Chen, Y.; Tabb, D.L.; Cortez, D. Identification of proteins at active, stalled, and collapsed replication forks using isolation of proteins on nascent DNA (iPOND) coupled with mass spectrometry. *J. Biol. Chem.* **2013**, *288*, 31458–31467. [CrossRef]
118. Dominguez-Kelly, R.; Martin, Y.; Koundrioukoff, S.; Tanenbaum, M.E.; Smits, V.A.; Medema, R.H.; Debatisse, M.; Freire, R. Wee1 controls genomic stability during replication by regulating the Mus81-Eme1 endonuclease. *J. Cell Biol.* **2011**, *194*, 567–579. [CrossRef]
119. Wilhelm, T.; Magdalou, I.; Barascu, A.; Techer, H.; Debatisse, M.; Lopez, B.S. Spontaneous slow replication fork progression elicits mitosis alterations in homologous recombination-deficient mammalian cells. *Proc. Natl. Acad. Sci. USA* **2014**, *111*, 763–768. [CrossRef]
120. Marians, K.J. Lesion Bypass and the Reactivation of Stalled Replication Forks. *Annu Rev. Biochem* **2018**. [CrossRef]
121. Iyer, D.R.; Rhind, N. Replication fork slowing and stalling are distinct, checkpoint-independent consequences of replicating damaged DNA. *PLoS Genet.* **2017**, *13*, e1006958. [CrossRef] [PubMed]
122. Mezzina, M.; Menck, C.F.; Courtin, P.; Sarasin, A. Replication of simian virus 40 DNA after UV irradiation: Evidence of growing fork blockage and single-stranded gaps in daughter strands. *J. Virol.* **1988**, *62*, 4249–4258. [PubMed]

123. Lopes, M.; Foiani, M.; Sogo, J.M. Multiple mechanisms control chromosome integrity after replication fork uncoupling and restart at irreparable UV lesions. *Mol. Cell* **2006**, *21*, 15–27. [CrossRef] [PubMed]
124. Friedberg, E.C. Suffering in silence: The tolerance of DNA damage. *Nat. Rev. Mol. Cell Biol.* **2005**, *6*, 943–953. [CrossRef] [PubMed]
125. Wickramasinghe, C.M.; Arzouk, H.; Frey, A.; Maiter, A.; Sale, J.E. Contributions of the specialised DNA polymerases to replication of structured DNA. *DNA Repair (Amst)* **2015**, *29*, 83–90. [CrossRef] [PubMed]
126. Leon-Ortiz, A.M.; Svendsen, J.; Boulton, S.J. Metabolism of DNA secondary structures at the eukaryotic replication fork. *DNA Repair* **2014**, *19*, 152–162. [CrossRef]
127. Yan, S.; Michael, W.M. TopBP1 and DNA polymerase alpha-mediated recruitment of the 9-1-1 complex to stalled replication forks: Implications for a replication restart-based mechanism for ATR checkpoint activation. *Cell Cycle* **2009**, *8*, 2877–2884. [CrossRef]
128. Blow, J.J.; Ge, X.Q.; Jackson, D.A. How dormant origins promote complete genome replication. *Trends Biochem. Sci.* **2011**, *36*, 405–414. [CrossRef]
129. Courtot, L.; Hoffmann, J.S.; Bergoglio, V. The Protective Role of Dormant Origins in Response to Replicative Stress. *Int J. Mol. Sci.* **2018**, *19*, 3569. [CrossRef]
130. Chen, Y.H.; Jones, M.J.; Yin, Y.; Crist, S.B.; Colnaghi, L.; Sims, R.J., 3rd; Rothenberg, E.; Jallepalli, P.V.; Huang, T.T. ATR-mediated phosphorylation of FANCI regulates dormant origin firing in response to replication stress. *Mol. Cell* **2015**, *58*, 323–338. [CrossRef]
131. Yang, C.C.; Suzuki, M.; Yamakawa, S.; Uno, S.; Ishii, A.; Yamazaki, S.; Fukatsu, R.; Fujisawa, R.; Sakimura, K.; Tsurimoto, T.; et al. Claspin recruits Cdc7 kinase for initiation of DNA replication in human cells. *Nat. Commun.* **2016**, *7*, 12135. [CrossRef] [PubMed]
132. Aguilera, A.; Gomez-Gonzalez, B. Genome instability: A mechanistic view of its causes and consequences. *Nat. Rev. Genet.* **2008**, *9*, 204–217. [CrossRef] [PubMed]
133. Hastings, P.J.; Ira, G.; Lupski, J.R. A microhomology-mediated break-induced replication model for the origin of human copy number variation. *PLoS Genet.* **2009**, *5*, e1000327. [CrossRef] [PubMed]
134. Feng, W.Y. Mec1/ATR, the Program Manager of Nucleic Acids Inc. *Genes-Basel* **2017**, *8*, 10. [CrossRef]
135. Costanzo, V. Brca2, Rad51 and Mre11: Performing balancing acts on replication forks. *DNA Repair (Amst)* **2011**, *10*, 1060–1065. [CrossRef]
136. Lehmann, A.R. Replication of damaged DNA by translesion synthesis in human cells. *FEBS Lett.* **2005**, *579*, 873–876. [CrossRef]
137. Chun, A.C.; Jin, D.Y. Ubiquitin-dependent regulation of translesion polymerases. *Biochem Soc. Trans.* **2010**, *38*, 110–115. [CrossRef]
138. Lehmann, A.R.; Niimi, A.; Ogi, T.; Brown, S.; Sabbioneda, S.; Wing, J.F.; Kannouche, P.L.; Green, C.M. Translesion synthesis: Y-family polymerases and the polymerase switch. *DNA Repair (Amst)* **2007**, *6*, 891–899. [CrossRef]
139. Mendoza, O.; Bourdoncle, A.; Boule, J.B.; Brosh, R.M., Jr.; Mergny, J.L. G-quadruplexes and helicases. *Nucleic Acids Res.* **2016**, *44*, 1989–2006. [CrossRef]
140. Goodman, M.F.; Woodgate, R. Translesion DNA polymerases. *Cold Spring Harb. Perspect. Biol.* **2013**, *5*, a010363. [CrossRef]
141. Vaisman, A.; Woodgate, R. Translesion DNA polymerases in eukaryotes: What makes them tick? *Crit Rev. Biochem. Mol.* **2017**, *52*, 274–303. [CrossRef] [PubMed]
142. Yang, W.; Gao, Y. Translesion and Repair DNA Polymerases: Diverse Structure and Mechanism. *Annu Rev. Biochem.* **2018**, *87*, 239–261. [CrossRef] [PubMed]
143. Courcelle, J.; Crowley, D.J.; Hanawalt, P.C. Recovery of DNA replication in UV-irradiated Escherichia coli requires both excision repair and recF protein function. *J. Bacteriol.* **1999**, *181*, 916–922. [PubMed]
144. Rudolph, C.J.; Upton, A.L.; Lloyd, R.G. Replication fork stalling and cell cycle arrest in UV-irradiated Escherichia coli. *Genes Dev.* **2007**, *21*, 668–681. [CrossRef]
145. Kannouche, P.L.; Wing, J.; Lehmann, A.R. Interaction of human DNA polymerase eta with monoubiquitinated PCNA: A possible mechanism for the polymerase switch in response to DNA damage. *Mol. Cell* **2004**, *14*, 491–500. [CrossRef]
146. Moldovan, G.L.; Pfander, B.; Jentsch, S. PCNA, the maestro of the replication fork. *Cell* **2007**, *129*, 665–679. [CrossRef]

147. Masutani, C.; Kusumoto, R.; Iwai, S.; Hanaoka, F. Mechanisms of accurate translesion synthesis by human DNA polymerase eta. *EMBO J.* **2000**, *19*, 3100–3109. [CrossRef]
148. Haracska, L.; Yu, S.L.; Johnson, R.E.; Prakash, L.; Prakash, S. Efficient and accurate replication in the presence of 7,8-dihydro-8-oxoguanine by DNA polymerase eta. *Nat. Genet.* **2000**, *25*, 458–461. [CrossRef]
149. Washington, M.T.; Johnson, R.E.; Prakash, L.; Prakash, S. Accuracy of lesion bypass by yeast and human DNA polymerase eta. *Proc. Natl. Acad. Sci. USA* **2001**, *98*, 8355–8360. [CrossRef]
150. Haracska, L.; Unk, I.; Johnson, R.E.; Phillips, B.B.; Hurwitz, J.; Prakash, L.; Prakash, S. Stimulation of DNA synthesis activity of human DNA polymerase kappa by PCNA. *Mol. Cell Biol.* **2002**, *22*, 784–791. [CrossRef]
151. Johnson, R.E.; Washington, M.T.; Haracska, L.; Prakash, S.; Prakash, L. Eukaryotic polymerases iota and zeta act sequentially to bypass DNA lesions. *Nature* **2000**, *406*, 1015–1019. [CrossRef] [PubMed]
152. Lim, S.O.; Gu, J.M.; Kim, M.S.; Kim, H.S.; Park, Y.N.; Park, C.K.; Cho, J.W.; Park, Y.M.; Jung, G. Epigenetic changes induced by reactive oxygen species in hepatocellular carcinoma: Methylation of the E-cadherin promoter. *Gastroenterology* **2008**, *135*, 2128–2140.e8. [CrossRef] [PubMed]
153. O'Hagan, H.M.; Wang, W.; Sen, S.; Destefano Shields, C.; Lee, S.S.; Zhang, Y.W.; Clements, E.G.; Cai, Y.; Van Neste, L.; Easwaran, H.; et al. Oxidative damage targets complexes containing DNA methyltransferases, SIRT1, and polycomb members to promoter CpG Islands. *Cancer Cell* **2011**, *20*, 606–619. [CrossRef] [PubMed]
154. Valinluck, V.; Tsai, H.H.; Rogstad, D.K.; Burdzy, A.; Bird, A.; Sowers, L.C. Oxidative damage to methyl-CpG sequences inhibits the binding of the methyl-CpG binding domain (MBD) of methyl-CpG binding protein 2 (MeCP2). *Nucleic Acids Res.* **2004**, *32*, 4100–4108. [CrossRef] [PubMed]
155. Weitzman, S.A.; Turk, P.W.; Milkowski, D.H.; Kozlowski, K. Free radical adducts induce alterations in DNA cytosine methylation. *Proc. Natl. Acad. Sci. USA* **1994**, *91*, 1261–1264. [CrossRef]
156. Turk, P.W.; Laayoun, A.; Smith, S.S.; Weitzman, S.A. DNA adduct 8-hydroxyl-2'-deoxyguanosine (8-hydroxyguanine) affects function of human DNA methyltransferase. *Carcinogenesis* **1995**, *16*, 1253–1255. [CrossRef]
157. Donkena, K.V.; Young, C.Y.; Tindall, D.J. Oxidative stress and DNA methylation in prostate cancer. *Obstet Gynecol. Int.* **2010**, *2010*, 302051. [CrossRef]
158. Kornberg, R.D.; Lorch, Y. Twenty-five years of the nucleosome, fundamental particle of the eukaryote chromosome. *Cell* **1999**, *98*, 285–294. [CrossRef]
159. Li, G.; Reinberg, D. Chromatin higher-order structures and gene regulation. *Curr. Opin. Genet. Dev.* **2011**, *21*, 175–186. [CrossRef]
160. Luger, K.; Hansen, J.C. Nucleosome and chromatin fiber dynamics. *Curr. Opin. Struct. Biol.* **2005**, *15*, 188–196. [CrossRef]
161. Zhou, V.W.; Goren, A.; Bernstein, B.E. Charting histone modifications and the functional organization of mammalian genomes. *Nat. Rev. Genet.* **2011**, *12*, 7–18. [CrossRef] [PubMed]
162. Bannister, A.J.; Kouzarides, T. Regulation of chromatin by histone modifications. *Cell Res.* **2011**, *21*, 381–395. [CrossRef] [PubMed]
163. Jasencakova, Z.; Scharf, A.N.D.; Ask, K.; Corpet, A.; Imhof, A.; Almouzni, G.; Groth, A. Replication Stress Interferes with Histone Recycling and Predeposition Marking of New Histones. *Mol. Cell* **2010**, *37*, 736–743. [CrossRef] [PubMed]
164. Bernstein, B.E.; Kamal, M.; Lindblad-Toh, K.; Bekiranov, S.; Bailey, D.K.; Huebert, D.J.; McMahon, S.; Karlsson, E.K.; Kulbokas, E.J.; Gingeras, T.R.; et al. Genomic maps and comparative analysis of histone modifications in human and mouse. *Cell* **2005**, *120*, 169–181. [CrossRef] [PubMed]
165. Loyola, A.; Bonaldi, T.; Roche, D.; Imhof, A.; Almouzni, G. PTMs on H3 variants before chromatin assembly potentiate their final epigenetic state. *Mol. Cell* **2006**, *24*, 309–316. [CrossRef] [PubMed]
166. Wang, Z.B.; Zang, C.Z.; Rosenfeld, J.A.; Schones, D.E.; Barski, A.; Cuddapah, S.; Cui, K.R.; Roh, T.Y.; Peng, W.Q.; Zhang, M.Q.; et al. Combinatorial patterns of histone acetylations and methylations in the human genome. *Nat. Genet.* **2008**, *40*, 897–903. [CrossRef] [PubMed]
167. Loyola, A.; Tagami, H.; Bonaldi, T.; Roche, D.; Quivy, J.P.; Imhof, A.; Nakatani, Y.; Dent, S.Y.R.; Almouzni, G. The HP1 alpha-CAF1-SetDB1-containing complex provides H3K9me1 for Suv39-mediated K9me3 in pericentric heterochromatin. *EMBO Rep.* **2009**, *10*, 769–775. [CrossRef]
168. Singh, R.K.; Kabbaj, M.H.M.; Paik, J.; Gunjan, A. Histone levels are regulated by phosphorylation and ubiquitylation-dependent proteolysis. *Nat. Cell Biol.* **2009**, *11*, 925–933. [CrossRef]

169. Chaudhuri, A.R.; Callen, E.; Ding, X.; Gogola, E.; Duarte, A.A.; Lee, J.E.; Wong, N.; Lafarga, V.; Calvo, J.A.; Panzarino, N.J.; et al. Replication fork stability confers chemoresistance in BRCA-deficient cells. *Nature* **2016**, *539*, 456. [CrossRef]
170. Rondinelli, B.; Gogola, E.; Yucel, H.; Duarte, A.A.; van de Ven, M.; van der Sluijs, R.; Konstantinopoulos, P.A.; Jonkers, J.; Ceccaldi, R.; Rottenberg, S.; et al. EZH2 promotes degradation of stalled replication forks by recruiting MUS81 through histone H3 trimethylation. *Nat. Cell Biol.* **2017**, *19*, 1371–1378. [CrossRef]
171. Nikolov, I.; Taddei, A. Linking replication stress with heterochromatin formation. *Chromosoma* **2016**, *125*, 523–533. [CrossRef] [PubMed]
172. Evrin, C.; Maman, J.D.; Diamante, A.; Pellegrini, L.; Labib, K. Histone H2A-H2B binding by Pol alpha in the eukaryotic replisome contributes to the maintenance of repressive chromatin. *EMBO J.* **2018**, *37*. [CrossRef] [PubMed]
173. Yu, C.; Gan, H.; Serra-Cardona, A.; Zhang, L.; Gan, S.; Sharma, S.; Johansson, E.; Chabes, A.; Xu, R.M.; Zhang, Z. A mechanism for preventing asymmetric histone segregation onto replicating DNA strands. *Science* **2018**, *361*, 1386–1389. [CrossRef] [PubMed]
174. Lerner, L.K.; Sale, J.E. Replication of G Quadruplex DNA. *Genes (Basel)* **2019**, *10*, 95. [CrossRef] [PubMed]
175. Sarkies, P.; Reams, C.; Simpson, L.J.; Sale, J.E. Epigenetic Instability due to Defective Replication of Structured DNA. *Mol. Cell* **2010**, *40*, 703–713. [CrossRef]
176. Svikovic, S.; Sale, J.E. The Effects of Replication Stress on S Phase Histone Management and Epigenetic Memory. *J. Mol. Biol.* **2017**, *429*, 2011–2029. [CrossRef]
177. Bartel, D.P. MicroRNAs: Genomics, biogenesis, mechanism, and function. *Cell* **2004**, *116*, 281–297. [CrossRef]
178. Perron, M.P.; Provost, P. Protein interactions and complexes in human microRNA biogenesis and function. *Fron. Biosci* **2008**, *13*, 2537–2547. [CrossRef]
179. Wang, Y.; Taniguchi, T. MicroRNAs and DNA damage response: Implications for cancer therapy. *Cell Cycle* **2013**, *12*, 32–42. [CrossRef]
180. Carthew, R.W.; Sontheimer, E.J. Origins and Mechanisms of miRNAs and siRNAs. *Cell* **2009**, *136*, 642–655. [CrossRef]
181. Zhang, X.; Wan, G.; Berger, F.G.; He, X.; Lu, X. The ATM kinase induces microRNA biogenesis in the DNA damage response. *Mol. Cell* **2011**, *41*, 371–383. [CrossRef] [PubMed]
182. He, M.; Zhou, W.; Li, C.; Guo, M. MicroRNAs, DNA Damage Response, and Cancer Treatment. *Int. J. Mol. Sci.* **2016**, *17*, 87. [CrossRef] [PubMed]
183. Bai, G.; Smolka, M.B.; Schimenti, J.C. Chronic DNA Replication Stress Reduces Replicative Lifespan of Cells by TRP53-Dependent, microRNA-Assisted MCM2-7 Downregulation. *PLoS Genet.* **2016**, *12*, e1005787. [CrossRef] [PubMed]
184. Bueno, M.J.; Gomez de Cedron, M.; Laresgoiti, U.; Fernandez-Piqueras, J.; Zubiaga, A.M.; Malumbres, M. Multiple E2F-induced microRNAs prevent replicative stress in response to mitogenic signaling. *Mol. Cell Biol.* **2010**, *30*, 2983–2995. [CrossRef] [PubMed]
185. Huang, Y.; Chuang, A.; Hao, H.; Talbot, C.; Sen, T.; Trink, B.; Sidransky, D.; Ratovitski, E. Phospho-Delta Np63 alpha is a key regulator of the cisplatin-induced microRNAome in cancer cells. *Cell Death Differ.* **2011**, *18*, 1220–1230. [CrossRef]
186. Garcia, A.I.; Buisson, M.; Bertrand, P.; Rimokh, R.; Rouleau, E.; Lopez, B.S.; Lidereau, R.; Mikaelian, I.; Mazoyer, S. Down-regulation of BRCA1 expression by miR-146a and miR-146b-5p in triple negative sporadic breast cancers. *EMBO Mol. Med.* **2011**, *3*, 279–290. [CrossRef]
187. Valeri, N.; Gasparini, P.; Fabbri, M.; Braconi, C.; Veronese, A.; Lovat, F.; Adair, B.; Vannini, I.; Fanini, F.; Bottoni, A.; et al. Modulation of mismatch repair and genomic stability by miR-155. *Proc. Natl. Acad. Sci. USA* **2010**, *107*, 6982–6987. [CrossRef]

© 2019 by the authors. Licensee MDPI, Basel, Switzerland. This article is an open access article distributed under the terms and conditions of the Creative Commons Attribution (CC BY) license (http://creativecommons.org/licenses/by/4.0/).

Review

Protein Chemical Labeling Using Biomimetic Radical Chemistry

Shinichi Sato * and Hiroyuki Nakamura *

Laboratory for Chemistry and Life Science, Institute of Innovative Research, Tokyo Institute of Technology, Yokohama 226-8503, Japan
* Correspondence: shinichi.sato@res.titech.ac.jp (S.S.); hiro@res.titech.ac.jp (H.N.);
  Tel.: +81-45-924-5245 (S.S.); +81-45-924-5244 (H.N.)

Academic Editor: Chryssostomos Chatgilialoglu
Received: 5 October 2019; Accepted: 31 October 2019; Published: 3 November 2019

Abstract: Chemical labeling of proteins with synthetic low-molecular-weight probes is an important technique in chemical biology. To achieve this, it is necessary to use chemical reactions that proceed rapidly under physiological conditions (i.e., aqueous solvent, pH, low concentration, and low temperature) so that protein denaturation does not occur. The radical reaction satisfies such demands of protein labeling, and protein labeling using the biomimetic radical reaction has recently attracted attention. The biomimetic radical reaction enables selective labeling of the C-terminus, tyrosine, and tryptophan, which is difficult to achieve with conventional electrophilic protein labeling. In addition, as the radical reaction proceeds selectively in close proximity to the catalyst, it can be applied to the analysis of protein–protein interactions. In this review, recent trends in protein labeling using biomimetic radical reactions are discussed.

Keywords: biomimetic radical reaction; bioinspired chemical catalysis; protein labeling

1. Introduction

The development of a technique for covalent bond formation between a specific amino acid residue of a protein and a low-molecular-weight compound is an important issue in protein chemical labeling and the design of protein-based biomaterials. It is also indispensable for the development of antibody–drug conjugates (ADCs) that have attracted attention in recent years. In addition, a technique for selectively labeling a specific protein in a complex protein mixture is useful for the target identification of bioactive molecules. In order to achieve protein chemical labeling, it is essential to develop reactions that result in the formation of covalent bonds with natural proteins in water, at near-neutral pH, at temperatures below 37 °C, and within a short reaction time of a few hours. Methods for labeling nucleophilic amino acid residues (lysine and cysteine residues) using compounds with electrophilic properties have been developed and have greatly contributed to the advancement of biochemistry. Additionally, site-selective protein labeling techniques [1] and enzymatic protein labeling techniques have been developed in recent years [2]. On the other hand, the chemical modification of amino acid residues, other than lysine and cysteine residues, has been extensively studied in recent years. The selective modification of tyrosine residue [3–12], tryptophan residue [3,13–18], methionine residue [19,20], peptide chain N-terminus [21,22], and the C-terminus [23] can also be used for protein functionalization. Radical reactions can modify amino acid residues that cannot be modified by conventional electrophilic methods, or modify proteins/peptides with a novel binding mode (e.g., stable C–C bond formation). In this review, we focus on protein labeling reactions using the bioinspired single-electron transfer (SET) reaction.

2. Biomimetic Tyrosine Radical Labeling Using Enzymes

In the biological radical reaction called radiolysis, water breaks down to highly reactive radicals such as hydroxyl radical, superoxide anion radical, and H_2O_2 [24]. Although the disulfide bond forming reaction is widely known as a response to oxidative stress in living systems, a dityrosine structure resulting from an oxidative cross-linking reaction of a tyrosine residue has also been reported as a protein oxidative modification marker [25,26]. Tyrosine readily undergoes SET under oxidative conditions to produce a highly reactive tyrosyl radical. A dityrosine structure is formed by the dimerization of tyrosine residues through the generation of tyrosyl radicals. Tyramide, a labeling agent that mimics tyrosine, forms a covalent bond with a tyrosine residue in a manner similar to dityrosine (Figure 1). Mimicking the biological response of dityrosine formation, metal complexes such as Ni(III) and Ru(III) were also reported to generate tyrosyl radicals and the radical species of tyramide. They were also used for protein cross-linking and protein labeling [27,28]. Several types of metalloenzymes, including peroxidase, tyrosinase [29–31], and laccase [32,33], catalyze the oxidation of tyrosine residues. As tyrosyl radical generation is efficiently catalyzed by peroxidases such as horseradish peroxidase (HRP), peroxidase was utilized as the catalyst in the dityrosine cross-linking reaction (Figure 1) [34–40]. HRP is activated by H_2O_2, and heme in the HRP molecule is transformed into a highly reactive species called compound I ([PPIX]·+Fe(IV)O), which can abstract a single electron from tyrosine or tyramide with ~1.1 V redox potential [41].

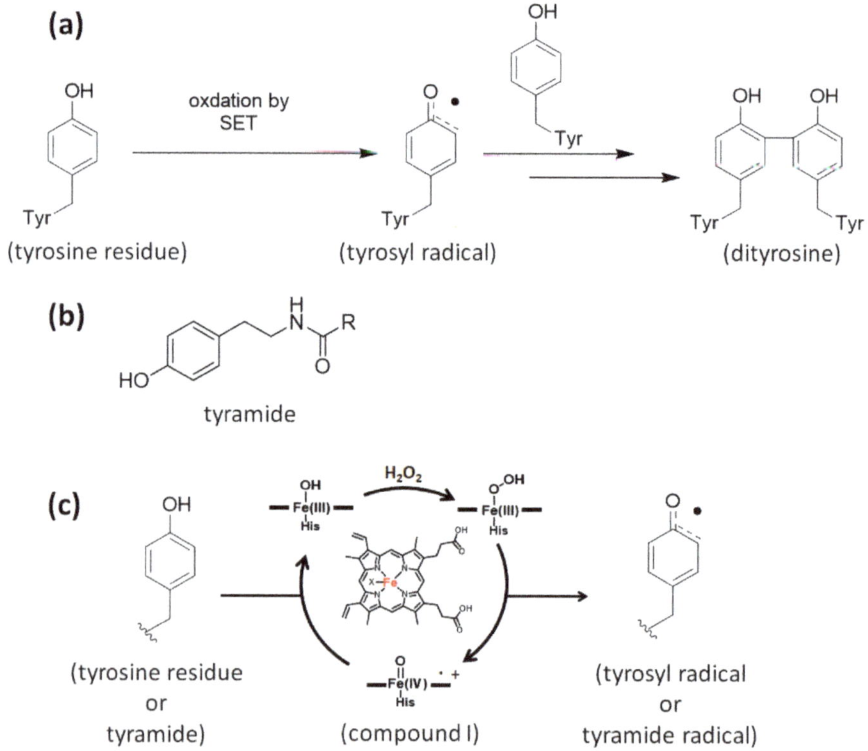

Figure 1. Generation of tyrosyl radical and tyramide radical. (a) Mechanism of dityrosine generation via single-electron transfer (SET). (b) Tyramide, a labeling agent that mimics tyrosine (c) Mechanism of oxidation in the active site of horseradish peroxidase (HRP).

Aside from the tyrosine labeling reactions, other than mimicking dityrosine formation reaction, a tyrosine labeling reaction that uses 4-phenyl-1,2,4-triazoline-3,5-dione (PTAD) as the labeling agent was reported [10,42]. However, PTAD easily decomposes in water to form isocyanate, an active electrophile. Therefore, the resulting isocyanate reacts not only with tyrosine residues but also with electrophilic amino acid residues and the N-terminus. To achieve tyrosine-specific labeling, we developed tyrosine labeling agents based on the structure of luminol and found that tyrosine-specific labeling can be achieved under biomimetic radical oxidation conditions [43,44]. The idea originated from a reactive intermediate of the luminol chemiluminescence reaction, which has a cyclic diazodicarboxamide structure in common with PTAD. However, unlike PTAD, the luminol derivative selectively reacts with tyrosine residues without generating an electrophilic by-product. Various heme proteins and enzymes were tested as catalysts for oxidative tyrosine labeling reactions, and it was found that HRP effectively catalyzes the oxidative activation of luminol derivatives and induces tyrosine-specific modifications (Figure 2). Through the structure–activity relationship studies of luminol derivatives as tyrosine labeling agents, we revealed that N-methylated luminol derivatives labeled tyrosine residues efficiently, instead of showing chemiluminescent properties. The redox potential of activated HRP (~1.1 V) is sufficient to activate SET reactions between compound I (Figure 1) and N-methylated luminol derivatives, resulting in a radical activation labeling agent. Tyrosine residues in proteins and peptides were selectively and efficiently labeled with N-methylated luminol derivatives under HRP-activated conditions.

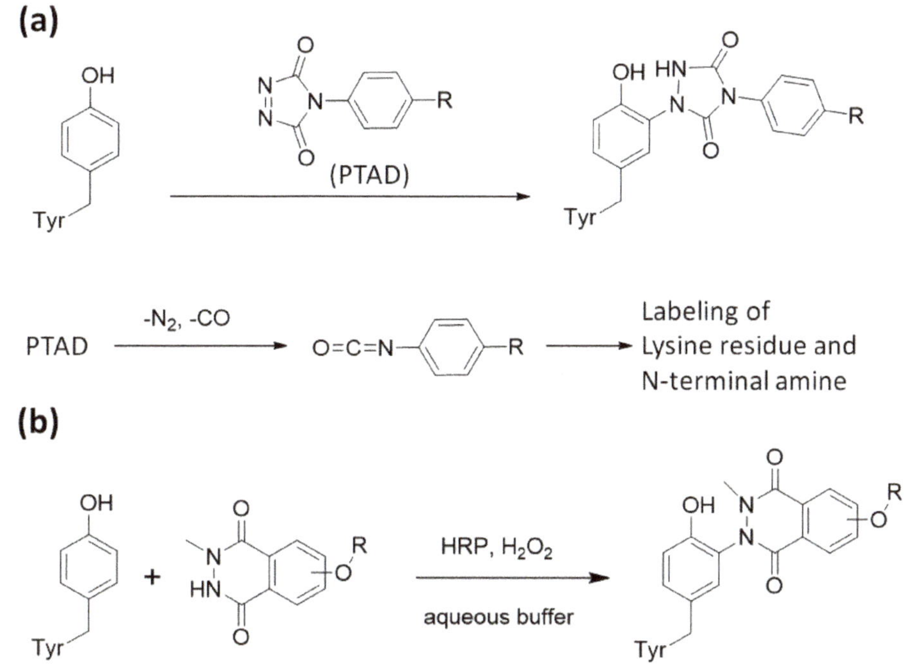

Figure 2. Tyrosine labeling with PTAD and N-methylated luminol derivatives. (**a**) Tyrosine labeling with PTAD and side reaction with amine group via isocyanate generation. (**b**) Tyrosine labeling with N-methylated luminol derivative in the presence of HRP and H_2O_2.

3. Peroxidase-Proximity Protein Labeling

Radical protein labeling using peroxidase has been employed in various applications in biological research. In general, the biomimetic radical reaction proceeds selectively in close proximity to the catalyst because of the short lifetime of the generated radical species. This concept is called proximity-dependent labeling (PDL). PDL catalyzed by HRP bound on the secondary antibody is also used as a signal amplification method (tyramide signal amplification—TSA) for immunostaining in biochemistry [45]. Although several signal amplification methods have been reported [46–50], TSA using HRP and tyramide derivatives is the most widely used. The generated tyramide radical reacts with amino acid residues such as tyrosine, tryptophan, histidine, and cysteine [51,52], in close proximity to HRP [53]. We found the novel signal amplification agent N'-acyl-N-methylphenylenediamine instead of tyramide, and revealed that it could be applied to signal amplification using HRP with comparable efficiency to tyramide (Figure 3) [54].

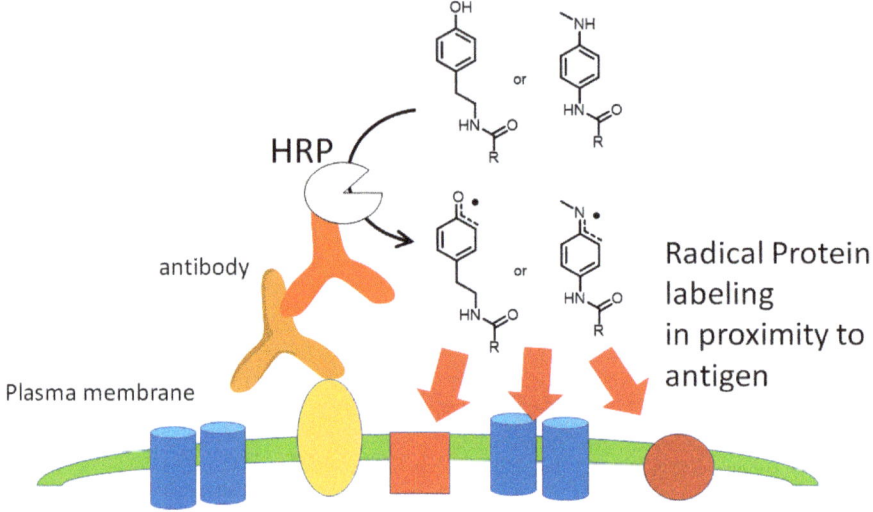

Figure 3. Immunohistochemical signal amplification using HRP-proximity protein labeling. Tyramide and N'-acyl-N-methylphenylenediamine were reported as HRP-proximity protein labeling agents.

PDL has also been applied to the analysis of protein–protein interactions. Methods using HRP have been reported, including selective proteomic proximity labeling assay using tyramide (SPPLAT) [53] and enzyme-mediated activation of radical sources (EMARS) [55]. With SPPLAT, proteins on the cell membrane can be labeled with biotin-tyramide using HRP-conjugated antibodies or HRP-conjugated ligands (e.g., HRP–transferrin). Membrane proteins labeled by proximity labeling can be enriched by streptavidin beads capture. Enriched proteins are identified by MS/MS analysis. Li and co-workers labeled membrane proteins using the SPPLAT method targeting the B cell receptor (BCR) and succeeded in identifying not only known proteins that interact with BCR but also proteins whose interactions were unknown [53]. EMARS is a method that uses biotin-aryl azide as the labeling agent. HRP activates aryl azide to produce short-lived aryl nitrene. Nitrenes are known to react with various amino acid residues, such as tyrosine, tryptophan, lysine, threonine, isoleucine, and proline [56]. Honke and co-workers demonstrated that many kinds of receptor tyrosine kinases (RTKs) formed clusters with beta-integrin by a combination of the EMARS method and antibody array analysis [55].

The labeling radius from HRP by these methods ranges from less than 200 nm to 300 nm [53,55], which is suitable for analyzing protein clusters on cell membranes. However, HRP is inactive when expressed in mammalian cytosol. Considering that disulfide bonds and Ca^{2+} binding sites in the structure of HRP are not formed under intracellular reducing conditions and a Ca^{2+}-scarce environment, Ting and co-workers focused on ascorbate peroxidase that lacks a disulfide bond and a Ca^{2+} binding site, and developed an engineered ascorbate peroxidase (APEX) that functions as peroxidase even in an intracellular environment [57]. In an intracellular environment, APEX catalyzes the generation of tyramide radical. The tyramide radical is short-lived (<1 ms) [58] and has a labeling radius of less than 20 nm [59,60] in the cells (Figure 4).

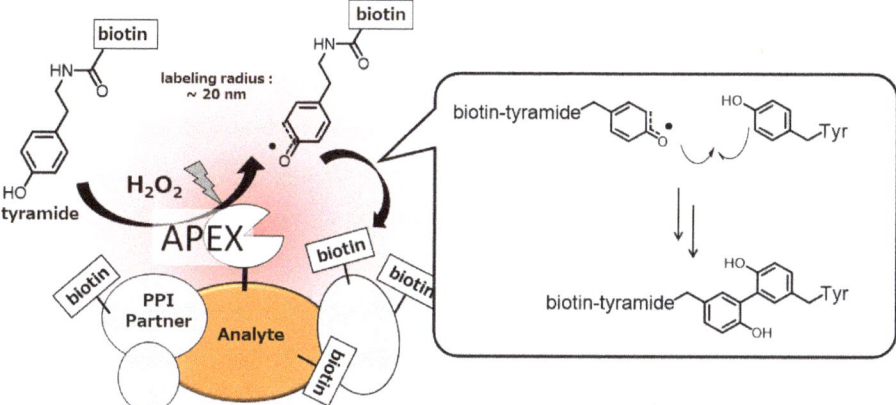

Figure 4. Ascorbate peroxidase (APEX) -proximity labeling of endogenous proteins in living cells.

Ting and co-workers established a method for comprehensively labeling and identifying proteins expressed in specific organelles by fusing APEX to proteins expressed in specific organelles [52]. Furthermore, they developed APEX2, which showed much higher peroxidase activity than APEX among 10^6 APEX mutants by the yeast-display evolution technique [61].

APEX2 has attracted much attention as a powerful tool for protein interaction analysis, and its applications include revealing proteomes in subcellular compartments [51,52,62–65], G-protein-coupled receptor complexes [66,67], subcellular transcriptome mapping [68,69], and APEX2-proximity RNA labeling [70,71].

4. Protein Labeling Using Photocatalyst

Not only radical enzymes but also small photocatalysts are used as protein labeling catalysts. Photocatalysts generate reactive oxygen species (ROS) and catalyze SET reactions in response to light stimulus [72]. Utilizing the SET mechanism, Noël and co-workers reported cysteine labeling using eosin Y and aryldiazonium salt [73], and Molander and co-workers reported a method that uses Ni/ruthenium photocatalyst and arylbromide [74]. MacMillan and co-workers developed a photocatalyst-mediated C-terminal labeling technique [23]. They focused on the redox potential of the carboxylic acid structures contained in the protein structure and hypothesized that the carboxyl group at the C-terminus would be selectively activated. The $E_{1/2}^{red}$ value of the carboxyl group in aspartic acid and glutamic acid residues is ~1.25 V (vs. saturated calomel electrode (SCE)), whereas the $E_{1/2}^{red}$ value of the C-terminal carboxyl group that exists at a single site in the protein sequence is ~0.95 V (vs. SCE). Slightly acidic reaction conditions (pH 3.5) are required in order to achieve efficient conversion, but the selective labeling of the C-terminus proceeded in the presence of aspartic acid and glutamic acid residues. MacMillan and co-workers tuned the reactivity of the Michael acceptor, a labeling agent, so that the labeling reaction with nucleophilic amino acid residues (lysine, serine, threonine, and histidine) would not proceed. The proposed reaction mechanism is shown in Figure 5. Flavin photocatalyst **1** is excited by visible light and undergoes subsequent intersystem crossing (quantum yield $\Phi_{ISC} = 0.38$ for flavin in water at pH 7) and conversion into triplet-excited state **2**. Triplet-excited flavin is a strong single-electron oxidant ($E_{1/2}^{red} = 1.5$ V vs. SCE in water) and should undergo facile SET with C-terminal carboxylate. Subsequent loss of CO_2 from **4** furnishes nitrogen atom stabilized carbon-centered radical **5**. Radical **5** reacts with Michael acceptor **6** to produce carbonyl α-radical **7**.

The photocatalyst in the radical anion state **3** reduces radical **7** to give product **8**, and regenerates ground-state photocatalyst **1**.

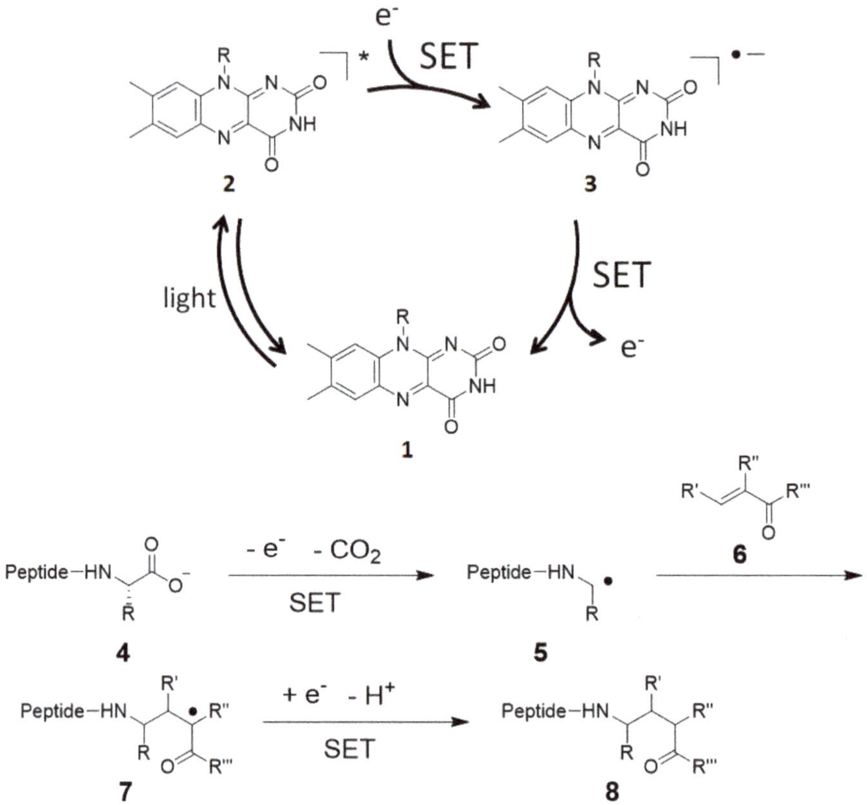

Figure 5. Proposed reaction mechanism for C-terminal labeling with flavin photocatalyst.

Shi and co-workers reported a SET-mediated tryptophan modification at the β-position through C–H activation using Ir[dF(CF$_3$)ppy]$_2$(dtbbpy) complex as the photocatalyst [75]. They proposed a possible mechanism as shown in Figure 6. SET of the indole nitrogen atom generates radical cation **13**. The benzylic proton (β-position) of the tryptophan can be extracted by a base (K$_2$HPO$_4$) to form **14**, and subsequent electron transfer results to form more stable tryptophan radical **15**. The mechanism of generating N radicals by the dehydrogenation of indole NH from **13** can also be considered, but the β-position radical **15** contributes to the reaction. Radical **15** reacts with methyl acrylate **16** to generate another radical, and this is reduced by the iridium catalyst to afford labeled tryptophan product **17**. Although the Michael addition reactions with the amine group of lysine and the imidazole of histidine were also observed as side reactions, the modification proceeded selectively at the β-position of tryptophan and not at the β-position of tyrosine or phenylalanine in the reaction that used a peptide as the substrate.

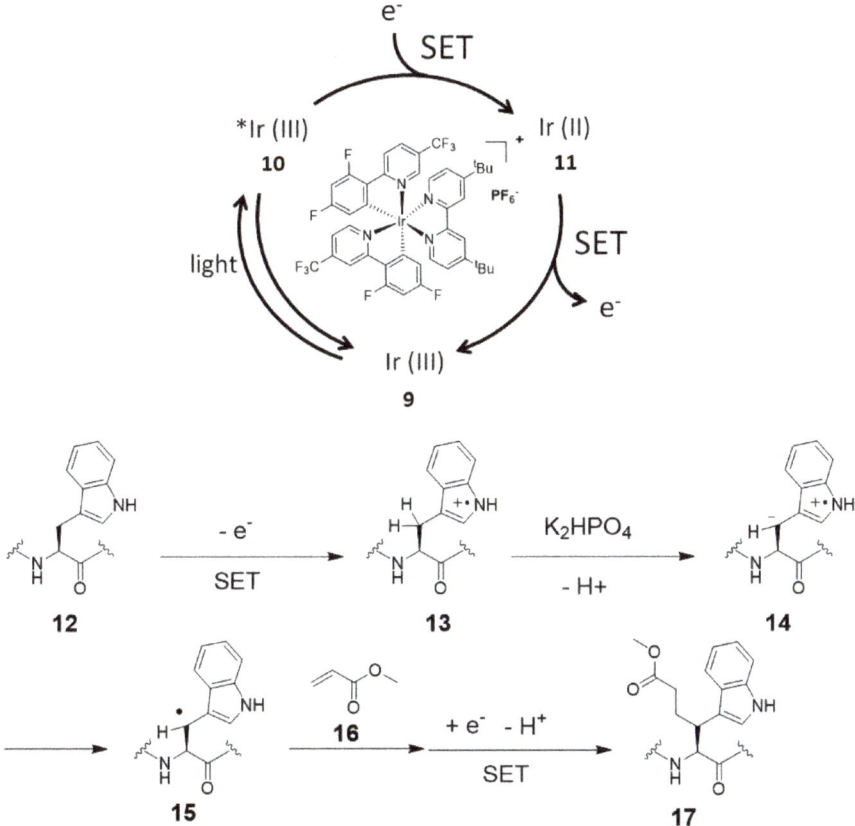

Figure 6. Proposed reaction mechanism for tryptophan β-position labeling with iridium photocatalyst.

We also developed a tyrosine labeling method that uses Ru(bpy)$_3$ complex and N'-acyl-N,N-dimethyl-1,4-phenylenediamine **23** as the photoredox catalyst and the labeling agent, respectively [11]. Under visible light irradiation, a stable carbon–carbon bond is formed between the ortho-carbon atom of the phenolic oxygen of the tyrosine residue and the ortho-carbon atom of the phenylenediamine derivative. Regarding the mechanism, in the absence of a labeling agent, $^1O_2$ is generated by the catalyst that functions as a photosensitizer. $^1O_2$ is involved in the production of Ru(III) active species **20**. Ru(III) active species **20** (1.1 V vs. SCE) can abstract a single electron from the tyrosine residue (~0.7 V vs. SCE) [76] and labeling agent **23** (0.63 V vs. SCE) [54]. Radical species **22** or **24** can react with **23** or **21**, respectively, to give product **25** through subsequent oxidation by SET (Figure 7) [77].

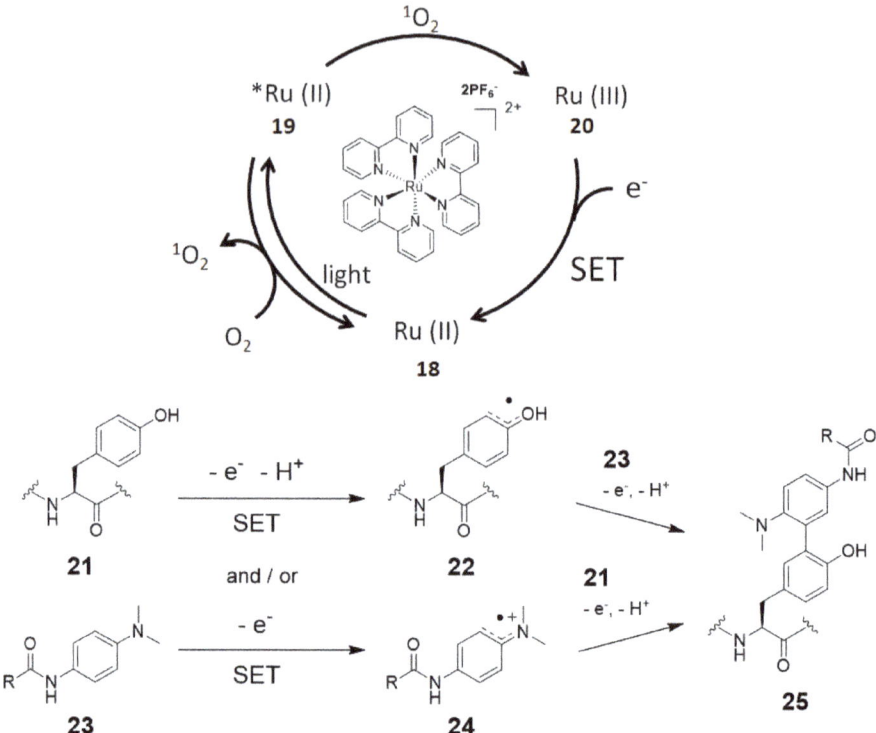

Figure 7. Proposed reaction mechanism for tyrosine labeling with ruthenium photocatalyst.

5. Photocatalyst-Proximity Labeling

As mentioned in Section 3, photocatalyst-catalyzed radical protein labeling proceeds selectively in close proximity to a catalyst. Using this property, we designed a ligand-conjugated catalyst in which a ligand and a ruthenium catalyst were linked, and using this catalyst, we selectively labeled ligand-binding proteins in a protein mixture. As a proof-of-concept model, benzenesulfonamide-conjugated ruthenium complex **26** was synthesized for targeting carbonic anhydrase (CA). Mouse erythrocytes were incubated with **26** and photo-irradiated in the presence of the labeling agent. Despite the presence of various proteins in erythrocytes, CA was selectively labeled [11]. We also synthesized gefitinib-conjugated ruthenium catalyst **27**, which targets the epidermal growth factor receptor (EGFR) expressed in A431 cells, and succeeded in the selective labeling of EGFR in A431 cells [77]. Furthermore, we developed a method for target-selective purification and labeling using ruthenium-catalyst-functionalized affinity beads targeting CA and dihydrofolate reductase (DHFR) (Figure 8) [72].

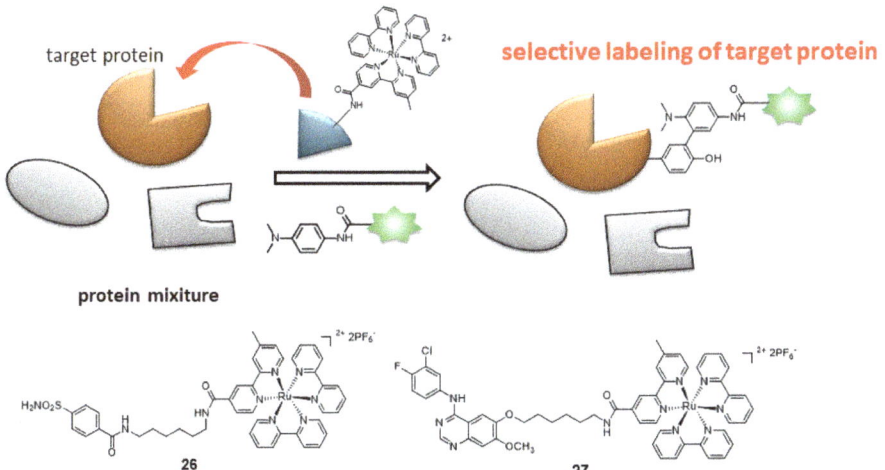

Figure 8. Target selective labeling by proximity labeling using ligand-conjugated photocatalysts **26** and **27**.

In these applications, protein labeling in close proximity to the ruthenium photocatalyst was accomplished using N'-acyl-N,N-dimethyl-1,4-phenylenediamine **23** as the labeling agent. We also found a novel labeling agent that labels efficiently and selectively in nanometer-scale catalyst proximity. Using model substrate **28**, in which a tyrosine residue is linked to a ruthenium photocatalyst, the reaction efficiencies of various labeling agents were evaluated. It was found by LC-MS analysis that **28** was efficiently labeled with 1-methyl-4-aryl-urazole (MAUra, **29**) and converted into **30** and **31** (Figure 9). Furthermore, in order to estimate the labeling radius from the ruthenium complex, a ruthenium complex conjugated to tyrosine was synthesized with a rigid proline linker, in which the distance between ruthenium and tyrosine is several nanometers, as shown in Figure 9. MAUra (**29**) labeled tyrosine when a ruthenium complex and a tyrosine residue were in close proximity, and its distance dependence is not contradicted by the reported SET distance in a physiological environment (~1.4 nm) [78]. Desthiobiotin-conjugated MAUra **32** was used to selectively label CA in a protein mixture. The CA labeled with **32** was also successfully enriched using streptavidin beads (18.5% in two steps of labeling and enrichment). Identification of the labeling site by MS revealed that the tyrosine residue closest to the ligand binding site was selectively labeled, suggesting nanometer-scale proximity dependence of MAUra labeling (Figure 10) [12].

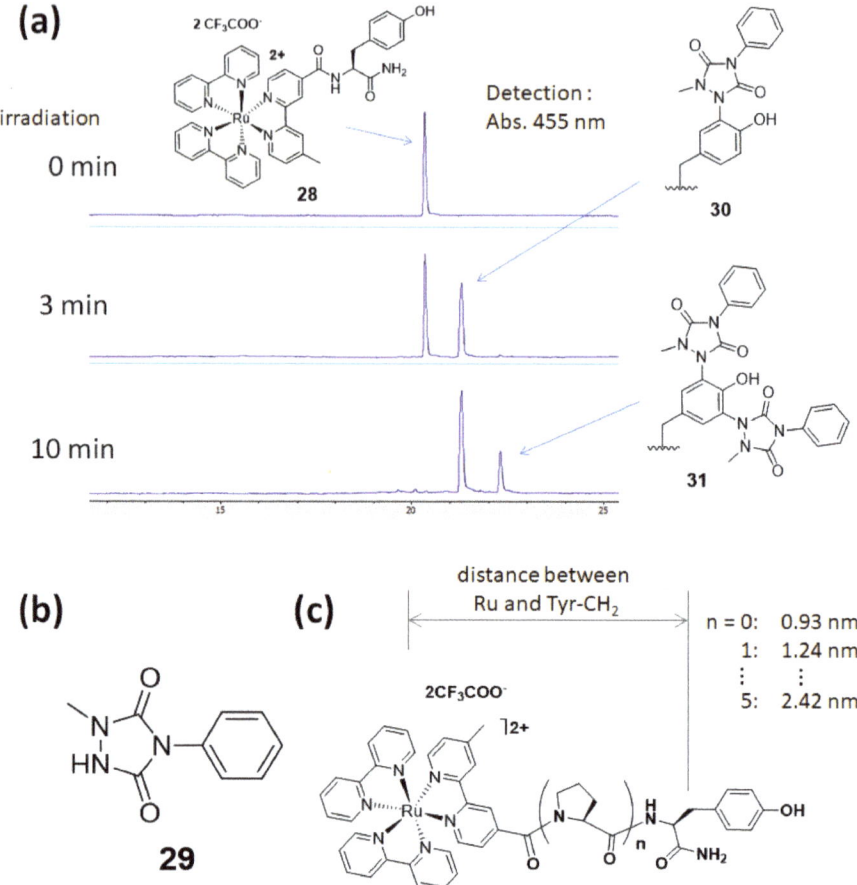

Figure 9. Photocatalyst-proximity tyrosine labeling. (**a**) Model substrate **28** was labeled with **29**. (**b**) Structure of labeling agent MAUra **29**. (**c**) Model substrate with a rigid proline linker with a distance of several nanometers between ruthenium and tyrosine.

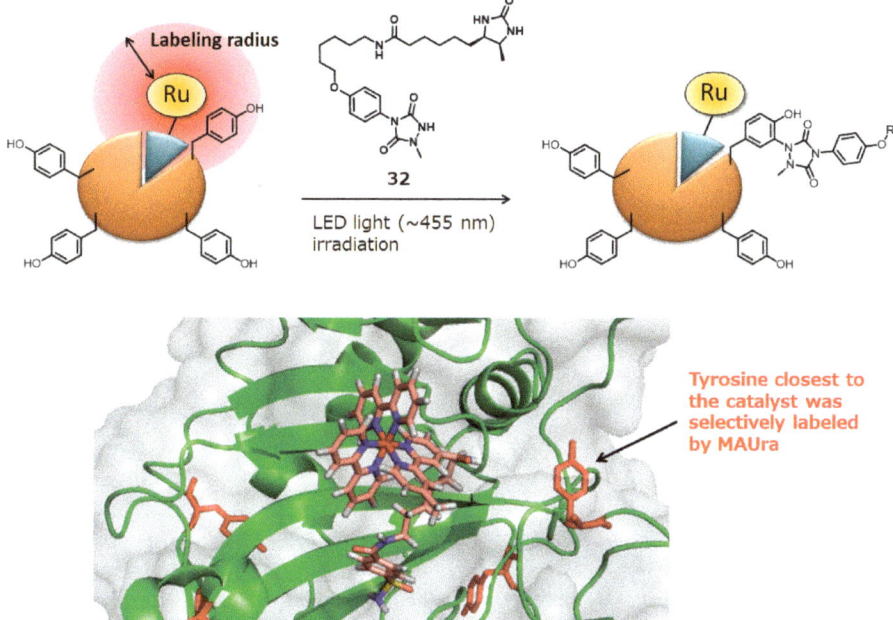

Figure 10. Photocatalyst-proximity labeling with MAUra.

6. Electrochemical Protein Labeling

Protein modification using peroxidase or a photocatalyst is suitable for analyzing protein association and protein–protein interactions. However, it is necessary to develop an appropriate protein labeling agent according to the oxidation potential of each catalyst. Moreover, these methods sometimes require the addition of an oxidant, which is often a cause for concern about the oxidative damage of proteins. In recent years, protein labeling methods using electrochemistry have been reported to overcome this disadvantage. At present, electrochemistry is limited to labeling purified proteins, but in the case of electrochemical organic chemistry, the voltage applied to the reaction system can be adjusted easily and the reaction proceeds efficiently even in an aqueous buffer. It can be used for the functionalization of proteins because of its high amino acid residue selectivity and low oxidative damage.

An electrochemical tyrosine-selective modification reaction (e-Y-Click) was reported by Alvarez-Dorta, Boujtita, Gouin, and co-workers (Figure 11) [79]. In this method, phenylurazole **33** is electrochemically oxidized and PTAD (Figures 2 and 11) is gradually produced in the reaction system. Because the PTAD generated by anode oxidation reacts with tyrosine instantaneously, side reactions with nucleophilic residues and N-terminus via isocyanate formation can be suppressed. As phenylurazole undergoes anodic oxidation at 0.36 V (vs. SCE), peptides and proteins are labeled without severe oxidative damage. Using glucose oxidase (GOx) as the substrate, they confirmed that the enzymatic activity of GOx was not affected by tyrosine labeling through the e-Y-Click reaction. Lei and co-workers also reported tyrosine-selective electrochemical labeling using phenothiazine **34** as the labeling agent. (Figure 11) [80].

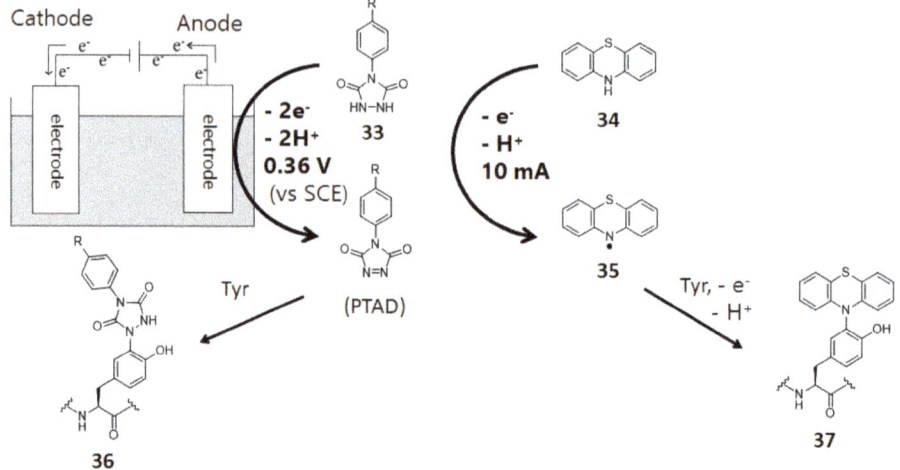

Figure 11. Electrochemical tyrosine labeling. SCE: saturated calomel electrode.

Kanai, Oisaki, and co-workers reported that 9-azabicyclo [3.3.1]nonane-3-one-*N*-oxyl (keto-ABNO, Figure 12) selectively labels tryptophan residues in the presence of 0.1% acetic acid and NaNO$_2$ [18]. Although keto-ABNO is oxidized by NOx in this method, they recently reported a method for activating the reaction by electrochemical oxidation [81]. They added 4-oxo-TEMPO as the electrochemical mediator to suppress both the anodic overoxidation of proteins and the cross reactivity to other amino acid residues (Figure 12).

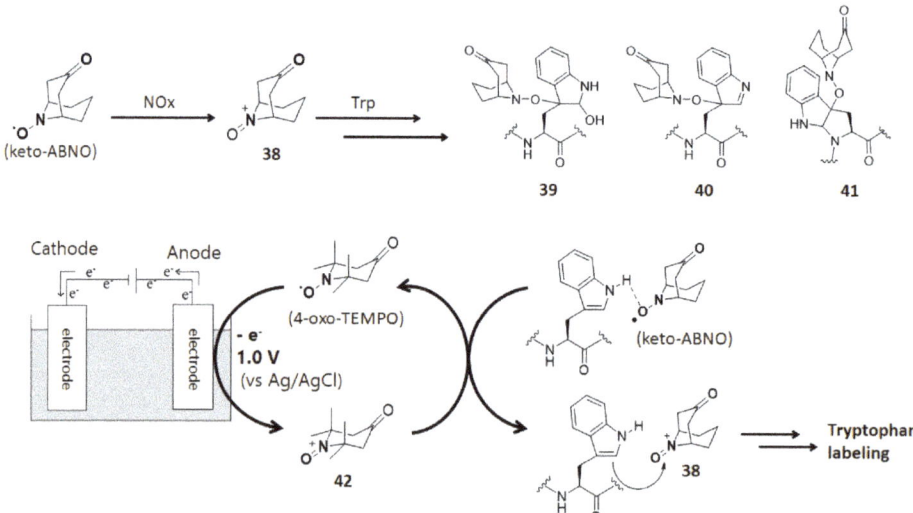

Figure 12. Tryptophan labeling with keto-ABNO and the electrochemical activation of tryptophan labeling.

7. Conclusions

In this review, protein labeling methods using biomimetic radical reactions were reviewed. Protein labeling techniques using electrophilic agents have been extensively employed. However, protein labeling targeting other amino acid residues is a challenging and attractive research topic. In recent years, in order to resolve several related issues, protein labeling using radical reactions has been actively developed targeting tyrosine and tryptophan residues and the C-terminus. Enzymes, particularly peroxidase, have been utilized as the catalyst for radical protein labeling, and peroxidase-proximity labeling has recently been used as an analytical method for protein association, protein–protein interaction, and transcriptome. In addition, protein modification using photocatalysts has been developed for the target identification of bioactive small molecules, and it is expected in the future to be used in not only the selective modification of target proteins in protein-mixed systems but also proximity labeling in cells. Furthermore, labeling with an electrochemical technique for precise voltage control has recently been developed and will be useful for labeling functional proteins. Table 1 summarizes representative protein labeling methods using biomimetic radical reactions. Future developments in radical protein modification will contribute to research on the elucidation of biological phenomena and drug delivery systems, and protein labeling using radical reactions will be a breakthrough technique in the development of these research areas.

Table 1. Overview of protein labeling methods using biomimetic radical reactions.

	Strategy	Labeling Agent	Target	Advantage	Disadvantage
Enzyme	Peroxidase	(tyramide)	Tyr (Trp, His, Cys)	Various biological applications Proximity labeling (see Section 3)	Use of H_2O_2 (1 mM) Low efficiency
	Peroxidase	(N-Me luminol)	Tyr	High conversion Tyr selectivity	Use of H_2O_2 (~5 equiv.)
Photocatalyst	flavin	etc. (Michael acceptor)	C-terminus	Site-selective labeling	Low pH condition
	Ir[dF(CF$_3$)ppy]$_2$(dtbbpy)	etc. (Michael acceptor)	Trp	β-position labeling Stable C-C bond	Cross reaction (with Lys, His)
	Ru(bpy)$_3$	(phenylenediamine)	Tyr	Stable C-C bond Application to signal amplification (see Section 3)	Low membrane permeability of Ru catalyst
	Ru(bpy)$_3$	(MAUra)	Tyr	High efficiency Proximity labeling	Low membrane permeability of Ru catalyst
Electrochemical	0.36 V (vs. SCE)	(phenylurazole)	Tyr	Mild condition	(All electrochemical methods) Not applicable to intracellular condition)
	10 mA	(phenothiazine)	Tyr	Tyrosine selectivity	Need > 50% CH_3CN
	~1.0 V (vs. Ag/AgCl)	(keto-ABNO)	Trp	Tryptophan selectivity	Need high voltage

121

Funding: Financial support in the form of a "Grant-in-Aid for Scientific Research (B) (19H02848 to S. Sato)" and "Chemistry for Multimolecular Crowding Biosystems (18H04542 to H. Nakamura)" from MEXT, Japan is gratefully acknowledged.

Conflicts of Interest: The authors declare no conflict of interest.

References

1. Krall, N.; Da Cruz, F.P.; Boutureira, O.; Bernardes, G.J.L. Site-selective protein-modification chemistry for basic biology and drug development. *Nat. Chem.* **2016**, *8*, 103–113. [CrossRef]
2. Zhang, Y.; Park, K.Y.; Suazo, K.F.; Distefano, M.D. Recent progress in enzymatic protein labelling techniques and their applications. *Chem. Soc. Rev.* **2018**, *47*, 9106–9136. [CrossRef]
3. Seim, K.L.; Obermeyer, A.C.; Francis, M.B. Oxidative Modification of Native Protein Residues Using Cerium(IV) Ammonium Nitrate. *J. Am. Chem. Soc.* **2011**, *133*, 16970–16976. [CrossRef]
4. Hooker, J.M.; Kovacs, E.W.; Francis, M.B. Interior Surface Modification of Bacteriophage MS2. *J. Am. Chem. Soc.* **2004**, *126*, 3718–3719. [CrossRef]
5. Joshi, N.S.; Whitaker, L.R.; Francis, M.B. A three-component Mannich-type reaction for selective tyrosine bioconjugation. *J. Am. Chem. Soc.* **2004**, *126*, 15942–15943. [CrossRef]
6. Tilley, S.D.; Francis, M.B. Tyrosine-selective protein alkylation using π-allylpalladium complexes. *J. Am. Chem. Soc.* **2006**, *128*, 1080–1081. [CrossRef] [PubMed]
7. Schlick, T.L.; Ding, Z.; Kovacs, E.W.; Francis, M.B. Dual-surface modification of the tobacco mosaic virus. *J. Am. Chem. Soc.* **2005**, *127*, 3718–3723. [CrossRef] [PubMed]
8. Struck, A.W.; Bennett, M.R.; Shepherd, S.A.; Law, B.J.C.; Zhuo, Y.; Wong, L.S.; Micklefield, J. An Enzyme Cascade for Selective Modification of Tyrosine Residues in Structurally Diverse Peptides and Proteins. *J. Am. Chem. Soc.* **2016**, *138*, 3038–3045. [CrossRef] [PubMed]
9. Ohata, J.; Miller, M.K.; Mountain, C.M.; Vohidov, F.; Ball, Z.T. A Three-Component Organometallic Tyrosine Bioconjugation. *Angew. Chemie Int. Ed.* **2018**, *57*, 2827–2830. [CrossRef] [PubMed]
10. Ban, H.; Gavrilyuk, J.; Barbas, C.F. Tyrosine Bioconjugation through Aqueous Ene-Type Reactions: A Click-Like Reaction for Tyrosine. *J. Am. Chem. Soc.* **2010**, *132*, 1523–1525. [CrossRef] [PubMed]
11. Sato, S.; Nakamura, H. Ligand-directed selective protein modification based on local single-electron-transfer catalysis. *Angew. Chemie Int. Ed.* **2013**, *52*, 8681–8684. [CrossRef] [PubMed]
12. Sato, S.; Hatano, K.; Tsushima, M.; Nakamura, H. 1-Methyl-4-aryl-urazole (MAUra) labels tyrosine in proximity to ruthenium photocatalysts. *Chem. Commun.* **2018**, *54*, 5871–5874. [CrossRef] [PubMed]
13. Antos, J.M.; Francis, M.B. Selective tryptophan modification with rhodium carbenoids in aqueous solution. *J. Am. Chem. Soc.* **2004**, *126*, 10256–10257. [CrossRef] [PubMed]
14. Ruiz-Rodriguez, J.; Albericio, F.; Lavilla, R. Postsynthetic modification of peptides: Chemoselective C-arylation of tryptophan residues. *Chem. A Eur. J.* **2010**, *16*, 1124–1127. [CrossRef] [PubMed]
15. Popp, B.V.; Ball, Z.T. Proximity-driven metallopeptide catalysis: Remarkable side-chain scope enables modification of the Fos bZip domain. *Chem. Sci.* **2011**, *2*, 690. [CrossRef]
16. Williams, T.J.; Reay, A.J.; Whitwood, A.C.; Fairlamb, I.J.S. A mild and selective Pd-mediated methodology for the synthesis of highly fluorescent 2-arylated tryptophans and tryptophan-containing peptides: A catalytic role for Pd0 nanoparticles? *Chem. Commun.* **2014**, *50*, 3052–3054. [CrossRef] [PubMed]
17. Hansen, M.B.; Hubálek, F.; Skrydstrup, T.; Hoeg-Jensen, T. Chemo- and Regioselective Ethynylation of Tryptophan-Containing Peptides and Proteins. *Chem. A Eur. J.* **2016**, *22*, 1572–1576. [CrossRef]
18. Seki, Y.; Ishiyama, T.; Sasaki, D.; Abe, J.; Sohma, Y.; Oisaki, K.; Kanai, M. Transition Metal-Free Tryptophan-Selective Bioconjugation of Proteins. *J. Am. Chem. Soc.* **2016**, *138*, 10798–10801. [CrossRef]
19. Lin, S.; Yang, X.; Jia, S.; Weeks, A.M.; Hornsby, M.; Lee, P.S.; Nichiporuk, R.V.; Iavarone, A.T.; Wells, J.A.; Toste, F.D.; et al. Redox-based reagents for chemoselective methionine bioconjugation. *Science* **2017**, *355*, 597–602. [CrossRef]
20. Taylor, M.T.; Nelson, J.E.; Suero, M.G.; Gaunt, M.J. A protein functionalization platform based on selective reactions at methionine residues. *Nature* **2018**, *562*, 563–568. [CrossRef]
21. Obermeyer, A.C.; Jarman, J.B.; Francis, M.B. N-Terminal Modification of Proteins with o-Aminophenols. *J. Am. Chem. Soc.* **2014**. [CrossRef] [PubMed]

22. Rosen, C.B.; Francis, M.B. Targeting the N terminus for site-selective protein modification. *Nat. Chem. Biol.* **2017**, *13*, 697–705. [CrossRef] [PubMed]
23. Bloom, S.; Liu, C.; Kölmel, D.K.; Qiao, J.X.; Zhang, Y.; Poss, M.A.; Ewing, W.R.; MacMillan, D.W.C. Decarboxylative alkylation for site-selective bioconjugation of native proteins via oxidation potentials. *Nat. Chem.* **2018**, *10*, 205–211. [CrossRef] [PubMed]
24. Le Caër, S. Water Radiolysis: Influence of Oxide Surfaces on H2 Production under Ionizing Radiation. *Water* **2011**, *3*, 235–253. [CrossRef]
25. DiMarco, T.; Giulivi, C. Current Analytical Methods for the Detection of Dityrosine, a Biomarker of Oxidative Stress, in Biological Samples. *Mass Spectrom Rev* **2007**, *26*, 108–120. [CrossRef] [PubMed]
26. Houée-Lévin, C.; Bobrowski, K.; Horakova, L.; Karademir, B.; Schöneich, C.; Davies, M.J.; Spickett, C.M. Exploring oxidative modifications of tyrosine: An update on mechanisms of formation, advances in analysis and biological consequences. *Free Radic. Res.* **2015**, *49*, 347–373. [CrossRef]
27. Fancy, D.A.; Melcher, K.; Johnston, S.A.; Kodadek, T. New chemistry for the study of multiprotein complexes: The six- histidine tag as a receptor for a protein crosslinking reagent. *Chem. Biol.* **1996**, *3*, 551–559. [CrossRef]
28. Meunier, S.; Strable, E.; Finn, M.G. Crosslinking of and Coupling to Viral Capsid Proteins by Tyrosine Oxidation. *Chem. Biol.* **2004**, *11*, 319–326. [CrossRef]
29. Lewandowski, A.T.; Small, D.A.; Chen, T.; Payne, G.F.; Bentley, W.E. Tyrosine-based "activatable pro-tag": Enzyme-catalyzed protein capture and release. *Biotechnol. Bioeng.* **2006**, *93*, 1207–1215. [CrossRef]
30. Lewandowski, A.T.; Yi, H.; Luo, X.; Payne, G.F.; Ghodssi, R.; Rubloff, G.W.; Bentley, W.E. Protein assembly onto patterned microfabricated devices through enzymatic activation of fusion pro-tag. *Biotechnol. Bioeng.* **2008**, *99*, 499–507. [CrossRef]
31. Wu, H.C.; Shi, X.W.; Tsao, C.Y.; Lewandowski, A.T.; Fernandes, R.; Hung, C.W.; DeShong, P.; Kobatake, E.; Valdes, J.J.; Payne, G.F.; et al. Biofabrication of antibodies and antigens via IgG-binding domain engineered with activatable pentatyrosine pro-tag. *Biotechnol. Bioeng.* **2009**, *103*, 231–240. [CrossRef] [PubMed]
32. Mattinen, M.L.; Kruus, K.; Buchert, J.; Nielsen, J.H.; Andersen, H.J.; Steffensen, C.L. Laccase-catalyzed polymerization of tyrosine-containing peptides. *FEBS J.* **2005**, *272*, 3640–3650. [CrossRef] [PubMed]
33. Mattinen, M.L.; Hellman, M.; Permi, P.; Autio, K.; Kalkkinen, N.; Buchert, J. Effect of protein structure on laccase-catalyzed protein oligomerization. *J. Agric. Food Chem.* **2006**, *54*, 8883–8890. [CrossRef] [PubMed]
34. Gross, A.J.; Sizer, I.W. The oxidation of tyramine, tyrosine, and related compounds by peroxidase. *J. Biol. Chem.* **1959**, *234*, 1611–1614. [PubMed]
35. Matheis, G.; Whitaker, J.R. Peroxidase-catalyzed cross linking of proteins. *J. Protein Chem.* **1984**, *3*, 35–48. [CrossRef]
36. Jacob, J.S.; Cistola, D.P.; Hsu, F.F.; Muzaffar, S.; Mueller, D.M.; Hazen, S.L.; Heinecke, J.W. Human phagocytes employ the myeloperoxidase-hydrogen peroxide system to synthesize dityrosine, trityrosine, pulcherosine, and isodityrosine by a tyrosyl radical-dependent pathway. *J. Biol. Chem.* **1996**, *271*, 19950–19956. [CrossRef]
37. Malencik, D.A.; Anderson, S.R. Dityrosine formation in calmodulin: Cross-linking and polymerization catalyzed by Arthromyces peroxidase. *Biochemistry* **1996**, *35*, 4375–4386. [CrossRef]
38. Michon, T.; Chenu, M.; Kellershon, N.; Desmadril, M.; Guéguen, J. Horseradish peroxidase oxidation of tyrosine-containing peptides and their subsequent polymerization: A kinetic study. *Biochemistry* **1997**, *36*, 8504–8513. [CrossRef]
39. Oudgenoeg, G.; Hilhorst, R.; Piersma, S.R.; Boeriu, C.G.; Gruppen, H.; Hessing, M.; Voragen, A.G.J.; Laane, C. Peroxidase-mediated cross-linking of a tyrosine-containing peptide with ferulic acid. *J. Agric. Food Chem.* **2001**, *49*, 2503–2510. [CrossRef]
40. Minamihata, K.; Goto, M.; Kamiya, N. Site-specific protein cross-linking by peroxidase-catalyzed activation of a tyrosine-containing peptide tag. *Bioconjug. Chem.* **2011**, *22*, 74–81. [CrossRef]
41. Kersten, P.J.; Kalyanaraman, B.; Hammel, K.E.; Reinhammar, B.; Kirk, T.K. Comparison of lignin peroxidase, horseradish peroxidase and laccase in the oxidation of methoxybenzenes. *Biochem. J.* **1990**, *268*, 475–480. [CrossRef] [PubMed]
42. Ban, H.; Nagano, M.; Gavrilyuk, J.; Hakamata, W.; Inokuma, T.; Barbas, C.F. Facile and Stabile Linkages through Tyrosine: Bioconjugation Strategies with the Tyrosine-Click Reaction. *Bioconjug. Chem.* **2013**, *24*, 520–532. [CrossRef] [PubMed]
43. Sato, S.; Nakamura, K.; Nakamura, H. Tyrosine-Specific Chemical Modification with in Situ Hemin-Activated Luminol Derivatives. *ACS Chem. Biol.* **2015**, *10*, 2633–2640. [CrossRef] [PubMed]

44. Sato, S.; Nakamura, K.; Nakamura, H. Horseradish-Peroxidase-Catalyzed Tyrosine Click Reaction. *ChemBioChem* **2017**, *18*, 475–478. [CrossRef] [PubMed]
45. Bobrow, M.N.; Shaughnessy, K.J.; Litt, G.J. Catalyzed reporter deposition, a novel method of signal amplification. II. Application to membrane immunoassays. *J. Immunol. Methods* **1991**, *137*, 103–112. [CrossRef]
46. Hsu, S.-M.; Raine, L.; Fange, H. Use of Avidin-Biotin-Peroxidase Immunoperoxidase Techniques: A Comparison Complex (ABC) in between Unlabeled Antibody (PAP). *J. Histochem. Histochem. Cytochem.* **1981**, *29*, 577–580. [CrossRef]
47. Toda, Y.; Kono, K.; Abiru, H.; Kokuryo, K.; Endo, M.; Yaegashi, H.; Fukumoto, M. Application of tyramide signal amplification system to immunohistochemistry: A potent method to localize antigens that are not detectable by ordinary method. *Pathol. Int.* **1999**, *49*, 479–483. [CrossRef]
48. Pham, X.H.; Hahm, E.; Kim, T.H.; Kim, H.M.; Lee, S.H.; Lee, Y.S.; Jeong, D.H.; Jun, B.H. Enzyme-catalyzed Ag Growth on Au Nanoparticle-assembled Structure for Highly Sensitive Colorimetric Immunoassay. *Sci. Rep.* **2018**, *8*, 1–7. [CrossRef]
49. Polaske, N.W.; Kelly, B.D.; Ashworth-Sharpe, J.; Bieniarz, C. Quinone Methide Signal Amplification: Covalent Reporter Labeling of Cancer Epitopes using Alkaline Phosphatase Substrates. *Bioconjug. Chem.* **2016**, *27*, 660–666. [CrossRef]
50. Lee, J.; Song, E.K.; Bae, Y.; Min, J.; Rhee, H.W.; Park, T.J.; Kim, M.; Kang, S. An enhanced ascorbate peroxidase 2/antibody-binding domain fusion protein (APEX2-ABD) as a recombinant target-specific signal amplifier. *Chem. Commun.* **2015**, *51*, 10945–10948. [CrossRef]
51. Hung, V.; Zou, P.; Rhee, H.W.; Udeshi, N.D.; Cracan, V.; Svinkina, T.; Carr, S.A.; Mootha, V.K.; Ting, A.Y. Proteomic Mapping of the Human Mitochondrial Intermembrane Space in Live Cells via Ratiometric APEX Tagging. *Mol. Cell* **2014**, *55*, 332–341. [CrossRef] [PubMed]
52. Rhee, H.W.; Zou, P.; Udeshi, N.D.; Martell, J.D.; Mootha, V.K.; Carr, S.A.; Ting, A.Y. Proteomic mapping of mitochondria in living cells via spatially restricted enzymatic tagging. *Science* **2013**, *339*, 1328–1331. [CrossRef] [PubMed]
53. Rees, J.S.; Li, X.W.; Perrett, S.; Lilley, K.S.; Jackson, A.P. Selective proteomic proximity labeling assay using tyramide (SPPLAT): A quantitative method for the proteomic analysis of localized membrane-bound protein clusters. *Curr. Protoc. Protein Sci.* **2015**, *2015*, 19.27.1–19.27.18.
54. Sato, S.; Yoshida, M.; Hatano, K.; Matsumura, M.; Nakamura, H. N'-acyl-N-methylphenylenediamine as a novel proximity labeling agent for signal amplification in immunohistochemistry. *Bioorganic Med. Chem.* **2019**, *27*, 1110–1118. [CrossRef]
55. Kotani, N.; Gu, J.; Isaji, T.; Udaka, K.; Taniguchi, N.; Honke, K. Biochemical visualization of cell surface molecular clustering in living cells. *Proc. Natl. Acad. Sci. USA* **2008**, *105*, 7405–7409. [CrossRef]
56. Kotzyba-Hibert, F.; Kapfer, I.; Goeldner, M. Recent Trends in Photoaffinity Labeling. *Angew. Chemie Int. Ed. English* **1995**, *34*, 1296–1312. [CrossRef]
57. Martell, J.D.; Deerinck, T.J.; Sancak, Y.; Poulos, T.L.; Mootha, V.K.; Sosinsky, G.E.; Ellisman, M.H.; Ting, A.Y. Engineered ascorbate peroxidase as a genetically encoded reporter for electron microscopy. *Nat. Biotechnol.* **2012**, *30*, 1143–1148. [CrossRef]
58. Mortensen, A.; Skibsted, L.H. Importance of Carotenoid Structure in Radical-Scavenging Reactions. *J. Agric. Food Chem.* **1997**, *45*, 2970–2977. [CrossRef]
59. Mayer, G.; Bendayan, M. Biotinyl-tyramide: A novel approach for electron microscopic immunocytochemistry. *J. Histochem. Cytochem.* **1997**, *45*, 1449–1454. [CrossRef]
60. Chen, C.L.; Hu, Y.; Udeshi, N.D.; Lau, T.Y.; Wirtz-Peitz, F.; He, L.; Ting, A.Y.; Carr, S.A.; Perrimon, N. Proteomic mapping in live Drosophila tissues using an engineered ascorbate peroxidase. *Proc. Natl. Acad. Sci. USA* **2015**, *112*, 12093–12098. [CrossRef]
61. Lam, S.S.; Martell, J.D.; Kamer, K.J.; Deerinck, T.J.; Ellisman, M.H.; Mootha, V.K.; Ting, A.Y. Directed evolution of APEX2 for electron microscopy and proximity labeling. *Nat. Methods* **2015**, *12*, 51–54. [CrossRef] [PubMed]
62. Han, S.; Udeshi, N.D.; Deerinck, T.J.; Svinkina, T.; Ellisman, M.H.; Carr, S.A.; Ting, A.Y. Proximity Biotinylation as a Method for Mapping Proteins Associated with mtDNA in Living Cells. *Cell Chem. Biol.* **2017**, *24*, 404–414. [CrossRef] [PubMed]
63. Mick, D.U.; Rodrigues, R.B.; Leib, R.D.; Adams, C.M.; Chien, A.S.; Gygi, S.P.; Nachury, M.V. Proteomics of Primary Cilia by Proximity Labeling. *Dev. Cell* **2015**, *35*, 497–512. [CrossRef] [PubMed]

64. Ting, A.Y.; Stawski, P.S.; Draycott, A.S.; Udeshi, N.D.; Lehrman, E.K.; Wilton, D.K.; Svinkina, T.; Deerinck, T.J.; Ellisman, M.H.; Stevens, B.; et al. Proteomic Analysis of Unbounded Cellular Compartments: Synaptic Clefts. *Cell* **2016**, *166*, 1295–1307.
65. Markmiller, S.; Soltanieh, S.; Server, K.L.; Mak, R.; Jin, W.; Fang, M.Y.; Luo, E.C.; Krach, F.; Yang, D.; Sen, A.; et al. Context-Dependent and Disease-Specific Diversity in Protein Interactions within Stress Granules. *Cell* **2018**, *172*, 590–604.e13. [CrossRef]
66. Lobingier, B.T.; Hüttenhain, R.; Eichel, K.; Miller, K.B.; Ting, A.Y.; von Zastrow, M.; Krogan, N.J. An Approach to Spatiotemporally Resolve Protein Interaction Networks in Living Cells. *Cell* **2017**, *169*, 350–360.e12. [CrossRef]
67. Paek, J.; Kalocsay, M.; Staus, D.P.; Wingler, L.; Pascolutti, R.; Paulo, J.A.; Gygi, S.P.; Kruse, A.C. Multidimensional Tracking of GPCR Signaling via Peroxidase-Catalyzed Proximity Labeling. *Cell* **2017**, *169*, 338–349.e11. [CrossRef]
68. Kaewsapsak, P.; Shechner, D.M.; Mallard, W.; Rinn, J.L.; Ting, A.Y. Live-cell mapping of organelle-associated RNAs via proximity biotinylation combined with protein-RNA crosslinking. *Elife* **2017**, *6*, 1–31. [CrossRef]
69. Benhalevy, D.; Anastasakis, D.G.; Hafner, M. Proximity-CLIP provides a snapshot of protein-occupied RNA elements in subcellular compartments. *Nat. Methods* **2018**, *15*, 1074–1082. [CrossRef]
70. Fazal, F.M.; Han, S.; Parker, K.R.; Kaewsapsak, P.; Xu, J.; Boettiger, A.N.; Chang, H.Y.; Ting, A.Y. Atlas of Subcellular RNA Localization Revealed by APEX-Seq. *Cell* **2019**, *178*, 473–490.e26. [CrossRef]
71. Zhou, Y.; Wang, G.; Wang, P.; Li, Z.; Yue, T.; Wang, J.; Zou, P. Expanding APEX2 Substrates for Spatial-specific Labeling of Nucleic Acids and Proteins in Living Cells. *Angew. Chemie Int. Ed.* **2019**.
72. Sato, S.; Tsushima, M.; Nakamura, H. Target-protein-selective inactivation and labelling using an oxidative catalyst. *Org. Biomol. Chem.* **2018**, *16*, 6168–6179. [CrossRef] [PubMed]
73. Bottecchia, C.; Rubens, M.; Gunnoo, S.B.; Hessel, V.; Madder, A.; Noël, T. Visible-Light-Mediated Selective Arylation of Cysteine in Batch and Flow. *Angew. Chemie Int. Ed.* **2017**, *56*, 12702–12707. [CrossRef] [PubMed]
74. Vara, B.A.; Li, X.; Berritt, S.; Walters, C.R.; Petersson, E.J.; Molander, G.A. Scalable thioarylation of unprotected peptides and biomolecules under Ni/photoredox catalysis. *Chem. Sci.* **2018**, *9*, 336–344. [CrossRef] [PubMed]
75. Yu, Y.; Zhang, L.K.; Buevich, A.V.; Li, G.; Tang, H.; Vachal, P.; Colletti, S.L.; Shi, Z.C. Chemoselective Peptide Modification via Photocatalytic Tryptophan β-Position Conjugation. *J. Am. Chem. Soc.* **2018**, *140*, 6797–6800. [CrossRef] [PubMed]
76. Brabec, V.; Mornstein, V. Electrochemical behaviour of proteins at graphite electrodes. II. Electrooxidation of amino acids. *Biophys Chem.* **1980**, *12*, 159–165. [CrossRef]
77. Sato, S.; Morita, K.; Nakamura, H. Regulation of target protein knockdown and labeling using ligand-directed Ru(bpy)3 photocatalyst. *Bioconjug. Chem.* **2015**, *26*, 250–256. [CrossRef]
78. Page, C.C.; Moser, C.C.; Chen, X.; Dutton, P.L. Natural engineering principles of electron tunnelling in biological oxidation-reduction. *Nature* **1999**, *402*, 47–52. [CrossRef]
79. Alvarez-Dorta, D.; Thobie, C.; Croyal, M.; Bouzelha, M.; MEVEL, M.; Deniaud, D.; Boujtita, M.; Gouin, S.G. Electrochemically promoted tyrosine-click-chemistry for protein labelling. *J. Am. Chem. Soc.* **2018**, *140*, 17120–17126. [CrossRef]
80. Song, C.; Liu, K.; Wang, Z.; Ding, B.; Wang, S.; Weng, Y.; Chiang, C.-W.; Lei, A. Electrochemical oxidation induced selective tyrosine bioconjugation for the modification of biomolecules. *Chem. Sci.* **2019**, 7982–7987. [CrossRef]
81. Toyama, E.; Marumaya, K.; Sugai, T.; Kondo, M.; Masaoka, S.; Saitoh, T.; Oisaki, K.; Kanai, M. Electrochemical Tryptophan-Selective Bioconjugation Electrochemical Tryptophan-Selective Bioconjugation. *ChemRxiv* **2019**. [CrossRef]

© 2019 by the authors. Licensee MDPI, Basel, Switzerland. This article is an open access article distributed under the terms and conditions of the Creative Commons Attribution (CC BY) license (http://creativecommons.org/licenses/by/4.0/).

Review

Thiyl Radical Reactions in the Chemical Degradation of Pharmaceutical Proteins

Christian Schöneich

Department of Pharmaceutical Chemistry, University of Kansas, 2093 Constant Avenue, Lawrence, KS 66047, USA; schoneic@ku.edu; Tel.: +1-785-864-4880

Received: 9 October 2019; Accepted: 18 November 2019; Published: 28 November 2019

Abstract: Free radical pathways play a major role in the degradation of protein pharmaceuticals. Inspired by biochemical reactions carried out by thiyl radicals in various enzymatic processes, this review focuses on the role of thiyl radicals in pharmaceutical protein degradation through hydrogen atom transfer, electron transfer, and addition reactions. These processes can lead to the epimerization of amino acids, as well as the formation of various cleavage products and cross-links. Examples are presented for human insulin, human and mouse growth hormone, and monoclonal antibodies.

Keywords: protein stability; therapeutic proteins; thiyl radicals; oxidation; fragmentation; cross-link

1. Introduction

The physical and chemical stability of proteins are critical for the efficacy and safety of protein therapeutics [1,2]. The reactions of free radicals play an important role in the chemical degradation of peptide and protein therapeutics in pharmaceutical formulations. For example, pharmaceutical excipients, such as polysorbate, are prone to generate peroxyl radicals [3]. Peroxyl radicals (and secondary oxidants derived from peroxyl radicals, such as alkoxyl radicals or hydroperoxides) can oxidize proteins via various pathways [4–6], generating a manifold of radical and non-radical intermediates and products. Free radicals can also be generated in pharmaceutical formulations by mechanical shock [7], leading to cavitation, exposure to light [8–11], or ionizing radiation (e.g., during sterilization) [12]. This article focuses on the role of a specific type of radical, the thiyl radical (RS$^\bullet$), in the degradation of therapeutic proteins, which is primarily induced by exposure to light.

Photochemically, protein thiyl radicals can be generated through the direct homolytic cleavage of a disulfide bond [13] or via the one-electron reduction of a disulfide bond [14,15] (here, we do not consider thiyl radical generation through the oxidation of thiols, as pharmaceutical proteins rarely contain free thiols). The direct photochemical cleavage of disulfides may be relevant for specific conditions, e.g., when chromatographic protein separations are monitored by UV detection (e.g., by UV-C light; $\lambda \leq 280$ nm) or if protein preparations are exposed to UV-C light for viral decontamination [16]. However, photo-induced electron transfer, e.g., from Trp to disulfide, occurs during exposure to UV-B light ($\lambda = 280$–315 nm) [9,17], as well as exposure to photostability testing conditions according to the International Conference on Harmonization (ICH), guideline ICHQ1B [10,11]. Importantly, in the presence of high concentrations of salt, even the exposure to visible light results in the photo-ionization of Trp [18].

Thiyl radicals can engage in a great variety of reactions including hydrogen atom transfer (HAT), electron transfer (ET), and addition/elimination reactions. In this way, they can react with literally all of the 20 essential amino acids, though rate constants, e.g., for HAT reactions, may vary with amino acid structure [19]. Important information about the potential reactions of thiyl radicals in proteins can be gleaned from enzymatic processes, rendering at least some thiyl radical-mediated degradation pathways of pharmaceutical proteins "biomimetic" [20].

Thiyl radicals can also indirectly affect protein structure and function, e.g., via the non-enzymatic chemical transformations of mono- and polyunsaturated fatty acids. These transformation reactions can involve oxidation [21,22] and cis/trans-isomerization [22–24]. Specifically, thiyl radicals, which can partition into the lipid environment, induce the cis/trans-isomerization of unsaturated fatty acids in biological membranes, e.g., the HS• radical [25] (derived from H_2S/HS^-) or the $HO\text{-}CH_2CH_2S^{\bullet}$ radical [26] (derived from 2-mercaptoethanol). On the other hand, thiyl radicals from glutathione (GSH) show little efficiency in the cis/trans isomerization processes of biological membranes due to their hydrophilicity [26]. Changes in lipid structure through oxidation can promote the conformational changes of polypeptides and proteins, e.g., of amyloid-beta [27]—the main component of amyloid plaques present in Alzheimer's disease brains [28]. Lipid peroxidation products chemically modify proteins [29]. Moreover, the presence of trans fatty acids in membranes can modulate the intramembrane proteolysis of the amyloid precursor protein (APP) [30], leading to an enhanced generation of amyloid-beta.

In the following, we provide a general overview on the free radical reactions of thiyl radicals that are relevant for the degradation of proteins and, subsequently, summarize recent results on individual pharmaceutical proteins.

2. Thiyl Radicals in Reversible HAT Reactions

Thiyl radicals engage in reversible HAT reactions, either inter- or intramolecularly. Early results by Walling and Rabinovitz on product formation during the reaction of isobutylthiyl radicals [2-mythylpropane-1-thiyl radicals; $(CH_3)_2CH\text{-}CH_2\text{-}S^{\bullet})$] with cumene suggested the reversibility of Reaction (1) [31].

Relative rate constants for the reaction of cyclohexanethiyl and benzenethiyl radicals with a number of substrates, including cumene, were subsequently provided by Pryor et al. [32,33]. In an elegant study, Akhlaq et al. demonstrated that the exposure of 2,5-dimethyltetrahydrofurane to thiyl radicals resulted in cis/trans isomerization (Reactions (2) and (3)) via a chain reaction, a process from which k_2 and k_{-3} were derived as ca. 10^4 $M^{-1}s^{-1}$ [34]. Similar rate constants were later measured by pulse radiolysis for the reactions of various thiyl radicals with aliphatic alcohols and ethers [35,36] and by a kinetic NMR method for the reaction of thiyl radicals with carbohydrates [37].

In synthetic procedures, thiols are employed as so-called "polarity-reversal" catalysts [38,39], due to the propensity of thiyl radicals to react via HAT with a series of organic substrates such as alcohols, ethers and amines.

The reversibility of HAT between thiyl radicals and amino acids is of significance for glycyl radical enzymes (GRE), such as ribonucleotide reductase (RNR), pyruvate formate lyase (PFL), glycerol dehydrogenase, benzylsuccinate synthase, and 4-hydroxyphenylacetate decarboxylase [40]. In these enzymes, active site Cys thiyl radicals (CysS$^\bullet$) are generated by HAT to glycyl radicals (Gly$^\bullet$) (Reaction (4)), and Gly$^\bullet$ can be restored by the reverse reaction (Reaction (-4)).

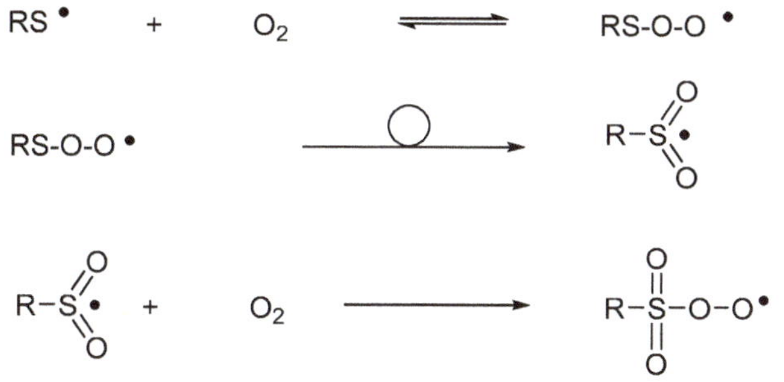

$$RS^\bullet + O_2 \rightleftharpoons RS\text{-}O\text{-}O^\bullet \qquad (5)$$

$$RS\text{-}O\text{-}O^\bullet \longrightarrow R\text{-}S(=O)_2^\bullet \qquad (6)$$

$$R\text{-}S(=O)_2^\bullet + O_2 \longrightarrow R\text{-}S(=O)_2\text{-}O\text{-}O^\bullet \qquad (7)$$

The location of Equilibrium (4) is controlled by conformational properties of Gly$^\bullet$ within the protein environment, illustrated here for the case of PFL. Electron paramagnetic resonance (EPR) spectroscopy has demonstrated that the active form of PFL harbors a Gly$^\bullet$ radical at the Gly^{734} position, where hyperfine coupling constants indicate that Gly$^\bullet$ adopts a planar conformation. This Gly$^\bullet$ radical exchanges its α-hydrogen with the solvent via a HAT reaction with Cys [41]. In the planar conformation, however, Gly$^\bullet$ is more stable than CysS$^\bullet$, based on a ca. 3.4 kcal/mol lower $^\alpha$C–H bond energy of Gly as compared to the S–H bond energy of Cys; as such, Equilibrium (4) is located on the site of Gly$^\bullet$ [42]. In order to afford HAT from Cys to Gly$^\bullet$, generating CysS$^\bullet$, Gly$^\bullet$ has been proposed to adopt a less planar conformation, as supported by its location within the protein framework, rendering Gly$^\bullet$ 4.6 kcal/mol less stable than CysS$^\bullet$ and moving Equilibrium (4) towards the site of CysS$^\bullet$ [42]. Theoretical calculations by Rauk and coworkers suggested that HAT reactions occur between thiyl radicals and the $^\alpha$C–H bonds of amino acids located in random and β-sheet conformations, but these do not occur when amino acids are located in α-helices [43]. Experimentally, the inter- and intramolecular HAT reactions of thiyl radicals have been demonstrated for amino acids, amino acid derivatives and peptides, including glutathione, in solution and in the gas phase [19,44–50]. Importantly, these HAT reactions do not only target $^\alpha$C–H bonds but also C–H bonds of amino acid side chains [51,52]. Together, the experimental data and theoretical calculations on biologically significant HAT reactions in GRE and HAT reactions in amino acids, amino acid derivatives [45,53], and peptides inspired us to consider the possibility of thiyl radical-dependent HAT processes and other reactions in the degradation of protein therapeutics.

3. Thiyl Radical Reactions with Molecular Oxygen

Thiyl radicals reversibly add oxygen to yield thiylperoxyl radicals (RSOO$^\bullet$) (Reaction (5)) [54]. The latter can rearrange to sulfonyl radicals (RS$^\bullet$O$_2$) (Reaction (6)) [54] and further convert into sulfonylperoxyl radicals (RSO$_2$OO$^\bullet$) (Reaction (7)) [55]. In the presence of electron or hydrogen donors, sulfonyl radicals convert into sulfonates, while sulfonylperoxyl radicals ultimately yield sulfonates [56,57].

4. Insulin

4.1. HAT Reactions in Solution

Insulin is a small protein containing two separate chains (A- and B-chain), connected by two interchain disulfide bonds (Cys^{A7}–Cys^{B7} and Cys^{A20}–Cys^{B19}) [58]. A third, intrachain disulfide bond connects Cys^{A6} and Cys^{A11} [58]. The disulfide bonds of insulin are shown in the cartoon in Figure 1.

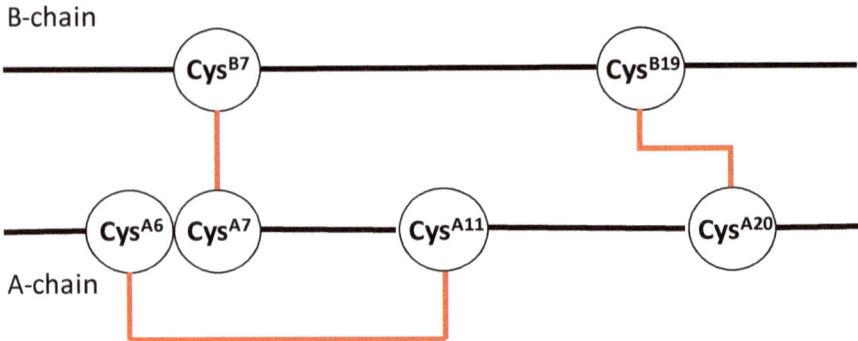

Figure 1. Cartoon displaying the disulfide bonds (in red) of insulin.

It is well known that insulin is sensitive to chemical and physical degradation, such as the photolytic cleavage of disulfide bonds [59,60] and dityrosine formation [60], deamidation [58,61,62], and fibrillation [58,63,64]. The biologically active form of insulin is the monomer, which exists at insulin concentrations <0.1 µM in the absence of Zn^{2+} [58]. At higher concentrations, insulin exists as a dimer [58], where self-association specifically involves the B8, B9, B12, B13, B16, and B23–28 residues [65].

Important for dimer formation is an aromatic triplet, consisting of the residues Phe^{B24}, Phe^{B25}, and Tyr^{B26}, which is part of an antiparallel β-sheet present at the dimer interface [66]. This dimer interface is different from the sequences responsible for fibrillation [63,64], mainly $L^{B11}VEALYL^{B17}$, causing the formation of the cross-β spine motif. The Phe^{B24} residue is equally important for the binding of insulin to the insulin receptor [66]. Interestingly, the substitution of L-Phe^{B24} by D-Phe^{B24} caused a significant increase of insulin affinity to the insulin receptor [67]. Therefore, it was of interest to evaluate whether thiyl radical-mediated intramolecular HAT reactions would proceed in insulin, whether such reactions would be restricted to specific amino acid residues, and whether these would include Phe^{B24}, possibly converting L-Phe^{B24} into D-Phe^{B24}.

The potential for insulin-derived CysS• to engage in intramolecular HAT reactions was monitored by covalent H/D-exchange according to the general Reactions (8)–(10) in Scheme 1, which representatively show the HAT of amino acid $^\alpha$C–H bonds. Solutions of Zn^{2+}-free insulin (50 or 500 µM) in either H_2O or D_2O were exposed to UV photolysis at 253.7 nm, followed by the alkylation of Cys and the HPLC-MS/MS analysis of an endoproteinase GluC-derived peptide map.

The predominant site of the covalent H/D exchange in the A-chain was Cys^{A20}, confirmed by the MS/MS sequencing of the Asn^{A18}–Tyr–Cys–Asn^{A21} peptide after photolysis in D_2O. This Cys residue is located at the end of the α-helix formed between the A13 and A20 residues. On the B-chain, the covalent H/D-exchange was most prevalent between the Leu^{B6} and Ser^{B9} residues and between the Val^{B18} and Gly^{B20} residues. Hence, deuterium incorporation proceeded selectively and did not target Phe^{B24}, suggesting that Phe^{B24} is also not a target for thiyl radical-mediated epimerization. Based on the calculations by Rauk et al. [43], the lack of a covalent H/D exchange at the $^\alpha$C–H bond of Phe^{B24} cannot be rationalized with an effect of the antiparallel β-sheet structure around the aromatic triplet (Phe^{B24}–Phe^{B25}–Tyr^{B26}) in the insulin dimer on the $^\alpha$C–H bond energy of Phe^{B24}. However, it is possible that the protein conformation did not permit the reaction of any of the photolytically generated CysS• radicals with Phe^{B24}, which would have excluded a covalent H/D exchange at both the $^\alpha$C–H and $^\beta$C–H bonds of Phe^{B24}. Importantly, a covalent H/D-exchange occurred either in the vicinity of Cys or on Cys itself. This result is consistent with studies on small Cys-containing model peptides where deuterium incorporation has been found to be most efficient at residues −1 or +1 from Cys. Deuterium incorporation into Cys itself is consistent with the 1,2- and 1,3-HAT reactions of thiyl radicals [45,53], for which rate constants were recently reported [45]. Additional evidence for the 1,2-HAT reactions of CysS• radicals in proteins comes from studies with *Escherichia coli* class III ribonucleotide reductase, where electron spin resonance (ESR) studies revealed the presence of an H/D exchange at the $^\beta$C–H bond of a CysS• radical [68].

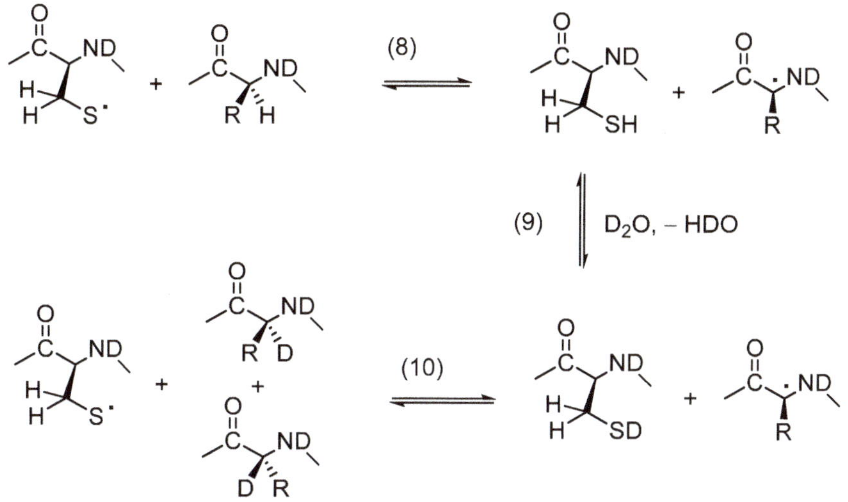

Scheme 1. Covalent H/D-exchange mediated by thiyl radicals, representatively shown for an $^\alpha$C–H bond.

An interesting product detected by MS/MS analysis was a cross-link between Tyr^{A19} and Cys^{B20}. This cross-link can form by the reaction of a CysS• radical with a tyrosyl radical (TyrO•). Under our experimental conditions, a pair of TyrO• and CysS• radicals could be formed via at least two different ways: (i) photo-induced electron transfer from Tyr to cystine, followed by combination of TyrO• and CysS•; and/or (ii) the homolytic cleavage of cystine, followed by electron/hydrogen transfer from Tyr to one CysS• radical and a combination of TyrO• with the second CysS• radical. An alternative pathway for the formation of Cys–Tyr cross-links would be the addition of a CysS• radical to Tyr, followed by the oxidation of this radical adduct. In fact, the potential for an addition of CysS• to aromatic amino acids was experimentally and theoretically demonstrated for the reaction of CysS• with Phe. More recently, the fast reversible additions of various radicals to the aromatic amino acid His have been reported [69,70].

4.2. Additional Reactions of Thiyl Radicals Leading to Cross-Links in Solution

Along with the Cys–Tyr cross-link, the photo-irradiation of insulin in solution generated a dithiohemiacetal cross-link between Cys^{A20} and Cys^{B19} [71]. Such photolytically generated dithiohemiacetal cross-links have also been identified and characterized for various disulfide-containing model peptides and proteins, including human and mouse growth hormone and monoclonal antibodies (see below). Mechanistically, the formation of dithiohemiacetal likely involves the light-induced homolysis of cystine, yielding a CysS• radical pair, which disproportionates to thiol and thioaldehyde, followed by the addition of the thiol to the thioaldehyde (Scheme 2).

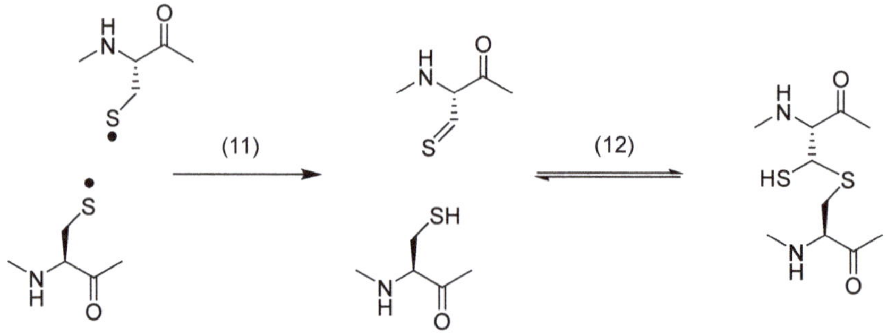

Scheme 2. Formation of dithiohemiacetal subsequent to the disproportionation of a thiyl radical pair.

4.3. Thiyl Radical Reactions in Solids

In order to evaluate the propensity for HAT reactions in solid insulin formulations, we prepared amorphous, crystalline, and microcrystalline human insulin [71]. Photo-irradiation at λ = 253.7 nm yielded a dithiohemiacetal between Cys^{A20} and Cys^{B19}, as well as peptide products with reduced Cys at the Cys^{B7} and Cys^{B19} positions, as characterized by HPLC-MS/MS. The photolysis of an amorphous insulin sample, generated by drying a D_2O solution of insulin, showed no evidence of a covalent H/D exchange, suggesting that the reversible HAT reactions shown in Scheme 1 may not occur to a significant extent in insulin solids. We note, however, that the lack of a covalent H/D exchange at C–H bonds may either be caused by the absence of HAT reactions or by an inefficient H/D exchange of the sulfhydryl group (Scheme 1; Reaction (9)) in solid formulations.

5. Growth Hormone

Human growth hormone (hGH) belongs to the class of four-helix bundle proteins [72] and is used for the treatment of pediatric hypopituitary dwarfism [73], as well as children [73] and adults [74] with hGH deficiencies. HGH is sensitive to deamidation [73,75,76], N-terminal truncation [77], oxidation [4,7,73,75,76,78–81], aggregation [73], and photo-degradation [82–84]. The structures of a trisulfide [76,85–87] and a thioether [88] variant, originating from the biosynthetic pathway, have been characterized by mass spectrometry. HGH contains two disulfides between Cys^{53} and Cys^{165} and between Cys^{182} and Cys^{189} [73]. The Cys^{182}–Cys^{189} disulfide bond defines the small C-terminal loop. A cartoon displaying the disulfide bonds of hGH is shown in Figure 2. Mutants of hGH, in which either Cys^{182} or Cys^{189} or both Cys residues are replaced with Ala, show a significantly reduced binding to the human growth hormone receptor [89].

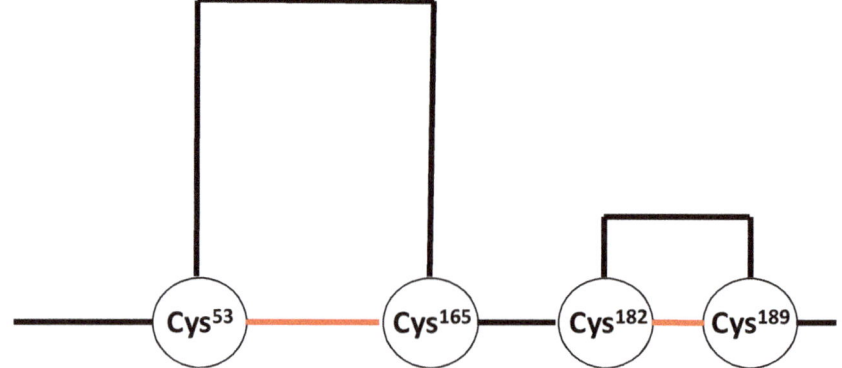

Figure 2. Cartoon displaying the disulfide bonds (in red) of human growth hormone.

The photolysis of hGH with UV light has resulted in a large number of products originating from the disulfide cleavage and subsequent reactions of CysS• radicals [84]. For example, reduced Cys and thioaldehyde were detected for all Cys residues originally present in the disulfide bonds, and a dithiohemiacetal was formed between Cys^{182} and Cys^{189}. In addition, sulfinic and/or sulfonic acid were detected as products of Cys^{165}, Cys^{182}, and Cys^{189}. These oxyacid products are expected from thiyl radicals generated in the presence of oxygen, as described in Section 3. The following discussion shall focus on a few rather unusual degradation products and cross-links, as well as the proposed mechanisms for their formation. In this discussion, the chemical names for structures are given as if they were present in amino acid form (rather than in the protein).

5.1. The Conversion of Cys to Gly

The conversion of Cys to Gly was detected for Cys^{165}, Cys^{182}, and Cys^{189}; at the same time, Cys^{165} and Cys^{189} were also converted to serine semialdehyde (2-amino-3-oxopropanoic acid; 3-oxoalanine; 2-formylglycine) and Ser. The proposed mechanism for the conversion of Cys to Gly (and serine semialdehyde) is shown in Scheme 3, where a 1,2-HAT reaction [45] of a CysS• radical is critical for the formation of a carbon-centered radical at C$_\beta$, followed by the addition of oxygen to yield a peroxyl radical. A series of reactions could transform the peroxyl radical into an alkoxyl radical, such as oxygen transfer reactions or reactions with additional peroxyl radicals. The alkoxyl radical is precursor for a carbon–carbon bond cleavage—yielding a Gly• radical—or a carbon–sulfur bond cleavage—yielding serine semialdehyde (though we note that serine semialdehyde can also be formed by hydrolysis of a Cys thioaldehyde). The proposal of a 1,2-HAT reaction for a protein CysS• radical also suggests that products of a 1,3-HAT reaction might be observed. In fact, for all hGH Cys residues, the formation of dehydroalanine (Dha) was detected. Dha can form via a 1,3-HAT reaction, followed by the elimination of HS• [45] (though care has to be taken during sample preparation for HPLC-MS/MS analysis, as Dha can also form during proteolytic digestion [90]).

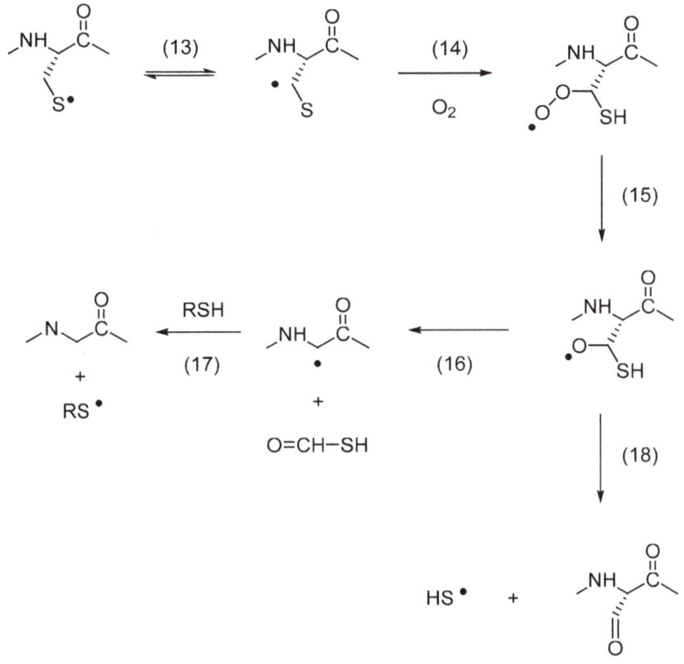

Scheme 3. Thiyl radical mediated conversion of Cys to Gly (glycyl).

5.2. The Formation of Ether and Vinyl ether

The exposure of mouse growth hormone (mGH) to UV light triggered the formation of particles of various sizes [91]. The injection of UV-exposed mGH into Balb/c or nude Balb/c mice caused an immune response, generating antibodies that cross-reacted with unmodified mGH [91]. The mass spectrometry analysis of UV-exposed mGH tentatively revealed the presence of chemical cross-links containing an ether bond between the original Cys^{78} and Ser^{188} residues or a thioether bond between the original Cys^{78} and Cys^{189} residues [91]. Whether these cross-links contribute to the immunogenicity of mGH remains to be shown. However, especially the photochemical conversion of a disulfide into an ether bond involving a neighboring Ser residue, is mechanistically intriguing.

When hGH was exposed to UV light, mass spectrometry analysis revealed the formation of vinyl ether between the original Cys^{189} and Ser^{184} residues and between the original Cys^{160} or Cys^{165} and Tyr^{164} residues [84]. The proposed mechanisms for the formation of these products are displayed in Schemes 4 and 5, respectively. Key to product formation is the homolytic cleavage of a disulfide bond into a pair of CysS•, followed by disproportionation into thiol and thioaldehyde. The latter can react with the hydroxyl group of either Ser or Tyr to yield a thiohemiacetal. Under continuous UV exposure, the C–S bond of the thiohemiacetal is expected to cleave, either homolytically or heterolytically [92–94], and the vinyl ether is generated via subsequent oxidation and deprotonation, respectively [95]. An important difference between the results of both growth hormones is the formation of an ether cross-link in mGH vs. a vinylether cross-link in hGH. This may be rationalized by sequence differences between hGH and mGH [96], as we also observed significant differences in photooxidation between hGH and rat growth hormone (rGH) [97]. We believe that the ether cross-link is ultimately generated via reduction of a vinyl ether. The UV exposure of disulfide-containing peptides leads to the formation of Cys [95,98] and H_2S [95]. Under UV-exposure, the specifically thiolate forms of Cys and H_2S, i.e., CysS$^-$ and HS$^-$, release an electron, which can reduce Dha. The latter was experimentally tested during the photo-irradiation of a disulfide-containing model peptide in the absence and presence of methylene chloride (CH_2Cl_2), a prominent scavenger of hydrated electrons [99]. Hence, we propose that vinyl ether reduction by a hydrated electron, followed by a HAT—likely from a photochemically generated thiol [95,98]—is key to the generation of an ether cross-link. We observed an analogous mechanism for the formation of a thioether cross-link from vinyl thioether, where CH_2Cl_2 inhibited thioether formation during UV-exposure [95]. Interestingly, the UV exposure of hGH also led to both vinyl thioether and thioether cross-links between the original Cys^{182} and Cys^{189} residues [84], i.e., Cys residues which originally form the disulfide bond characterizing the small C-terminal loop of hGH.

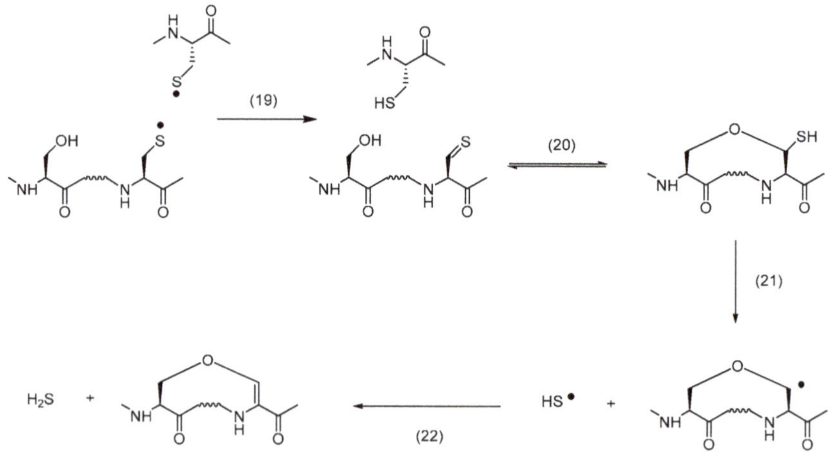

Scheme 4. Formation of vinylether from Ser.

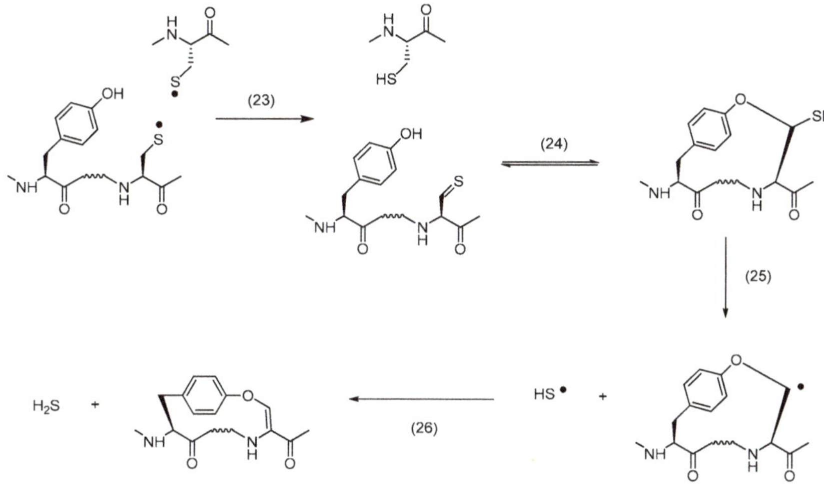

Scheme 5. Formation of vinylether form Tyr.

5.3. Formation of Non-Native Disulfides

The UV exposure of hGH produced significant yields of non-native disulfide bonds, i.e., intramolecularly between Cys^{189} and Cys^{189}, as well as either inter-or intramolecularly between Cys^{53} and Cys^{182}, and Cys^{165} and Cys^{189} [84]. These disulfide bonds can form via homolytic substitution, i.e., an S$_H$2 mechanism that involves the reaction of a thiyl radical with a disulfide bond [100], or via the recombination of two thiyl radicals. In addition, non-native disulfides may form via the reaction of free thiols with disulfide bonds. Free thiols are generated together with thiyl radicals by the one-electron reduction of disulfides. Usually, free thiols are derivatized by alkylation prior to mass spectrometric analysis. However, free thiols can react with disulfide bonds during the time of UV exposure or any other time required for sample preparation; for example, in the case of hGH, sample preparation involved 30 min of reduction of the remaining disulfide bonds at pH 7.5 and 45 °C. The fact that non-native disulfide bonds between Cys^{189} and Cys^{189} and between Cys^{53} and Cys^{182} were not formed in non-photolyzed control solutions of hGH [84] is consistent with a free radical mechanism of disulfide scrambling.

5.4. hGH Cleavage Products

The UV-exposure of hGH results in several backbone cleavage products originating from $^{\alpha}$C$^{\bullet}$ radicals generated at Cys residues or at amino acid residues in the vicinity of Cys residues, e.g., at Cys^{53}, Cys^{165} and Leu^{52} [84]. Mechanistically, these fragmentation products are generated through the well-established diamide or α-amidation pathways. What is of interest here is that these fragmentation products again give testimony to the ability of CysS$^{\bullet}$ radicals to generate $^{\alpha}$C$^{\bullet}$ radicals via intramolecular HAT reactions.

6. Monoclonal Antibodies

While a comprehensive product analysis, such as performed for hGH [84], has not yet been completed for monoclonal antibodies, certain products that are analogous to those generated from insulin or hGH have been detected. For example, the UV exposure of an immunoglobulin 1 (IgG1) resulted in the formation of dithiohemiacetal and thioether cross-links [101]. The light exposure of an IgG1 in an Atlas Suntest CPS+ Xenon test instrument, utilized for photostability studies according to the ICHQ1B guideline, resulted in disulfide scrambling [11]. The reactions leading to the formation of non-native disulfides in the IgG1 molecule are likely analogous to those described for hGH in Section 5.3.

The potential of CysS• radicals to engage in intramolecular HAT reactions in monoclonal antibodies was first demonstrated by a UV light-induced covalent H/D exchange analogous to the reactions presented in Scheme 1 [102]. Subsequent studies on an IgG1 revealed that these HAT reactions have the potential to epimerize amino acids in a protein, e.g., to convert L- into D-amino acids [103] (analogous experiments with model peptides [104] and octreotide [105] had demonstrated the ability of CysS• radicals to epimerize amino acids in peptides). Overall, the exposure of an IgG1 to UV light resulted in the generation of D-Glu and D-Val (and some D-Ala), where the relative yields of D-amino acids depended on the presence of various excipients in these formulations [103]. Experimentally, D-amino acids were recovered from the protein by controlled acid hydrolysis, which converts D-Gln to D-Glu, so that the yields of D-Glu were representative for the combined yields of D-Glu and D-Gln generated by HAT. Through proteolytic digestion, peptide fractionation, and the controlled acid hydrolysis of individual peptides, some locations of D-amino acids were identified such as the heavy chain (HC) sequences HC51–59 and HC287–296. Specifically, the sequence HC51–59 is located in the hypervariable region that is responsible for antigen binding, where conformational changes induced by amino acid epimerization may have biological consequences. We note that the exposure of monoclonal antibodies to light results in aggregation [106,107] and immunogenicity [108]. An important study was able to correlate immunogenicity with the presence of chemical modifications on subvisible particles, while particles not carrying chemical modifications were not found to be immunogenic [109].

7. Conclusions

The preceding sections provide examples for a variety of mechanisms by which thiyl radicals engage in the chemical degradation of pharmaceutical proteins. Noteworthy are the cross-links generated between thiol oxidation products and either Ser or Tyr. The exposure of the small (ca. 22 kDa) protein human growth hormone yielded nearly 60 different products that originated from thiyl radical generation, and a significantly higher number of products may be expected from the light-exposure of a monoclonal antibody. Comprehensive product studies on monoclonal antibodies are ongoing in our laboratory and will be reported in due time.

Funding: Research on pharmaceutical proteins described in this review was funded, in part, by Amgen, Inc. and Genentech, Inc.

Acknowledgments: The author wishes to acknowledge the important contributions of students, postdocs and collaborators to the research described in this review.

Conflicts of Interest: The author declares no conflict of interest.

References

1. Manning, M.C.; Chou, D.K.; Murphy, B.M.; Payne, R.W.; Katayama, D.S. Stability of Protein Pharmaceuticals: An Update. *Pharm. Res.* **2010**, *27*, 544–575. [CrossRef] [PubMed]
2. Grassi, L.; Cabrele, C. Susceptibility of Protein Therapeutics to Spontaneous Chemical Modifications by Oxidation, Cyclization, and Elimination Reactions. *Amino Acids* **2019**, *51*, 1409–1431. [CrossRef] [PubMed]
3. Harmon, P.A.; Kosuda, K.; Nelson, E.; Mowery, M.; Reed, R.A. A Novel Peroxy Radical Based Oxidative Stressing System for Ranking the Oxidizability of Drug Substances. *J. Pharm. Sci.* **2006**, *95*, 2014–2028. [CrossRef] [PubMed]
4. Steinmann, D.; Ji, J.A.; Wang, Y.J.; Schöneich, C. Oxidation of Human Growth Hormone by Oxygen-Centered Radicals: Formation of Leu-101 Hydroperoxide and Tyr-103 Oxidation Products. *Mol. Pharm.* **2012**, *9*, 803–814. [CrossRef]
5. Escobar-Alvarez, E.; Leinisch, F.; Araya, G.; Monasterio, O.; Lorentzen, L.G.; Silva, E.; Davies, M.J.; Lopez-Alarcon, C. The Peroxyl Radical-Induced Oxidation of Escherichia Coli FtsZ and Its Single Tryptophan Mutant (Y222W) Modifies Specific Side-Chains, Generates Protein Cross-Links and Affects Biological Function. *Free Radic. Biol. Med.* **2017**, *112*, 60–68. [CrossRef]

6. Leinisch, F.; Mariotti, M.; Rykaer, M.; Lopez-Alarcon, C.; Hagglund, P.; Davies, M.J. Peroxyl Radical-And Photo-Oxidation of Glucose 6-Phosphate Dehydrogenase Generates Cross-Links and Functional Changes Via Oxidation of Tyrosine and Tryptophan Residues. *Free Radic. Biol. Med.* **2017**, *112*, 240–252. [CrossRef]
7. Randolph, T.W.; Schiltz, E.; Sederstrom, D.; Steinmann, D.; Mozziconacci, O.; Schöneich, C.; Freund, E.; Ricci, M.S.; Carpenter, J.F.; Lengsfeld, C.S. Do Not Drop: Mechanical Shock in Vials Causes Cavitation, Protein Aggregation, and Particle Formation. *J. Pharm. Sci.* **2015**, *104*, 602–611. [CrossRef]
8. Sreedhara, A.; Lau, K.; Li, C.; Hosken, B.; Macchi, F.; Zhan, D.; Shen, A.; Steinmann, D.; Schöneich, C.; Lentz, Y. Role of Surface Exposed Tryptophan as Substrate Generators for the Antibody Catalyzed Water Oxidation Pathway. *Mol. Pharm.* **2013**, *10*, 278–288. [CrossRef]
9. Haywood, J.; Mozziconacci, O.; Allegre, K.M.; Kerwin, B.A.; Schöneich, C. Light-Induced Conversion of Trp to Gly and Gly Hydroperoxide in IgG1. *Mol. Pharm.* **2013**, *10*, 1146–1150. [CrossRef]
10. Bane, J.; Mozziconacci, O.; Yi, L.; Wang, Y.J.; Sreedhara, A.; Schöneich, C. Photo-Oxidation of IgG1 and Model Peptides: Detection and Analysis of Triply Oxidized His and Trp Side Chain Cleavage Products. *Pharm. Res.* **2017**, *34*, 229–242. [CrossRef]
11. Wecksler, A.T.; Yin, J.; Lee Tao, P.; Kabakoff, B.; Sreedhara, A.; Deperalta, G. Photodisruption of the Structurally Conserved Cys-Cys-Trp Triads Leads to Reduction-Resistant Scrambled Intrachain Disulfides in an IgG1 Monoclonal Antibody. *Mol. Pharm.* **2018**, *15*, 1598–1606. [CrossRef] [PubMed]
12. Mohr, D.; Wolff, M.; Kissel, T. Gamma Irradiation for Terminal Sterilization of 17beta-Estradiol Loaded Poly-(D,L-Lactide-Co-Glycolide) Microparticles. *J. Control. Release* **1999**, *61*, 203–217. [CrossRef]
13. Creed, D. The Photophysics and Photochemistry of the Near-Uv Absorbing Amino-Acids. 3. Cystine and Its Simple Derivatives. *Photochem. Photobiol.* **1984**, *39*, 577–583. [CrossRef]
14. Creed, D. The Photophysics and Photochemistry of the Near-Uv Absorbing Amino-Acids. 1. Tryptophan and Its Simple Derivatives. *Photochem. Photobiol.* **1984**, *39*, 537–562. [CrossRef]
15. Dose, K. Photolysis of Free Cystine in Presence of Aromatic Amino Acids. *Photochem. Photobiol.* **1968**, *8*, 331. [CrossRef]
16. Lorenz, C.M.; Wolk, B.M.; Quan, C.P.; Alcala, E.W.; Eng, M.; McDonald, D.J.; Matthews, T.C. The Effect of Low Intensity Ultraviolet-C Light on Monoclonal Antibodies. *Biotechnol. Prog.* **2009**, *25*, 476–482. [CrossRef]
17. Mossoba, M.M.; Makino, K.; Riesz, P. Photo-Ionization of Aromatic-Amino-Acids in Aqueous-Solutions-A Spin-Trapping and Electron-Spin Resonance Study. *J. Phys. Chem.* **1982**, *86*, 3478–3483. [CrossRef]
18. Truong, T.B. Charge-Transfer to a Solvent State. 5. Effect of Solute-Solvent Interaction on the Ionization-Potential of the Solute-Mechanism for Photo-Ionization. *J. Phys. Chem.* **1980**, *84*, 964–970. [CrossRef]
19. Nauser, T.; Schöneich, C. Thiyl Radicals Abstract Hydrogen Atoms from the (alpha)C-H Bonds in Model Peptides: Absolute Rate Constants and Effect of Amino Acid Structure. *J. Am. Chem. Soc.* **2003**, *125*, 2042–2043. [CrossRef]
20. Breslow, R. Biomimetic Chemistry: Biology as An Inspiration. *J. Biol. Chem.* **2009**, *284*, 1337–1342. [CrossRef]
21. Schöneich, C.; Asmus, K.D.; Dillinger, U.; Von Bruchhausen, F. Thiyl Radical Attack on Polyunsaturated Fatty Acids: A Possible Route to Lipid Peroxidation. *Biochem. Biophys. Res. Commun.* **1989**, *161*, 113–120. [CrossRef]
22. Chatgilialoglu, C.; Ferreri, C.; Guerra, M.; Samadi, A.; Bowry, V.W. The Reaction of Thiyl Radical with Methyl Linoleate: Completing the Picture. *J. Am. Chem. Soc.* **2017**, *139*, 4704–4714. [CrossRef] [PubMed]
23. Chatgilialoglu, C.; Ferreri, C.; Melchiorre, M.; Sansone, A.; Torreggiani, A. Lipid Geometrical Isomerism: From Chemistry to Biology and Diagnostics. *Chem. Rev.* **2014**, *114*, 255–284. [CrossRef] [PubMed]
24. Ferreri, C.; Costantino, C.; Landi, L.; Mulazzani, Q.G.; Chatgilialoglu, C. The Thiyl Radical-Mediated Isomerization of Cis-Monounsaturated Fatty Acid Residues in Phospholipids: A Novel Path of Membrane Damage? *Chem. Commun.* **1999**, *5*, 407–408. [CrossRef]
25. Lykakis, I.N.; Ferreri, C.; Chatgilialoglu, C. The Sulfhydryl Radical (HS Center dot/S Center Dot-): A Contender for the Isomerization of Double Bonds in Membrane Lipids. *Angew. Chem. Int. Edit.* **2007**, *46*, 1914–1916. [CrossRef]
26. Ferreri, C.; Kratzsch, S.; Brede, O.; Marciniak, B.; Chatgilialoglu, C. Trans Lipid Formation Induced by Thiols in Human Monocytic Leukemia Cells. *Free Radical. Biol. Med.* **2005**, *38*, 1180–1187. [CrossRef]
27. Koppaka, V.; Axelsen, P.H. Accelerated Accumulation of Amyloid Beta Proteins on Oxidatively Damaged Lipid Membranes. *Biochemistry* **2000**, *39*, 10011–10016. [CrossRef]

28. Musiek, E.S.; Holtzman, D.M. Three Dimensions of the Amyloid Hypothesis: Time, Space and 'Wingmen'. *Nat. Neurosci.* **2015**, *18*, 800–806. [CrossRef]
29. Gesslbauer, B.; Kuerzl, D.; Valpatic, N.; Bochkov, V.N. Unbiased Identification of Proteins Covalently Modified by Complex Mixtures of Peroxidized Lipids Using a Combination of Electrophoretic Mobility Band Shift with Mass Spectrometry. *Antioxidants* **2018**, *7*, 116. [CrossRef]
30. Grimm, M.O.; Rothhaar, T.L.; Grosgen, S.; Burg, V.K.; Hundsdorfer, B.; Haupenthal, V.J.; Friess, P.; Kins, S.; Grimm, H.S.; Hartmann, T. Trans Fatty Acids Enhance Amyloidogenic Processing of the Alzheimer Amyloid Precursor Protein (APP). *J. Nutr. Biochem.* **2012**, *23*, 1214–1223. [CrossRef]
31. Walling, C.; Rabinowitz, R. The Photolysis of Isobutyl Disulfide in Cumene. *J. Am. Chem. Soc.* **1959**, *81*, 1137–1143. [CrossRef]
32. Pryor, W.A.; Gojon, G.; Church, D.F. Relative Rate Constants for Hydrogen-Atom Abstraction by Cyclohexanethiyl and Benzenethiyl Radicals. *J. Org. Chem.* **1978**, *43*, 793–800. [CrossRef]
33. Pryor, W.A.; Gojon, G.; Stanley, J.P. Hydrogen Abstraction by Thiyl Radicals. *J. Am. Chem. Soc.* **1973**, *95*, 945–946. [CrossRef]
34. Akhlaq, M.S.; Schuchmann, H.P.; Von Sonntag, C. The Reverse of the 'Repair' Reaction of Thiols: H-Abstraction at Carbon by Thiyl Radicals. *Int. J. Radiat. Biol. Relat. Stud. Phys. Chem. Med.* **1987**, *51*, 91–102. [CrossRef]
35. Schöneich, C.; Bonifacic, M.; Asmus, K.D. Reversible H-Atom Abstraction from Alcohols by Thiyl Radicals-Determination of Absolute Rate Constants by Pulse-Radiolysis. *Free Radical. Res. Com.* **1989**, *6*, 393–405. [CrossRef] [PubMed]
36. Schöneich, C.; Asmus, K.D.; Bonifacic, M. Determination of Absolute Rate Constants for the Reversible Hydrogen-Atom Transfer Between Thiyl Radicals and Alcohols or Ethers. *J. Chem. Soc. Faraday T.* **1995**, *91*, 1923–1930. [CrossRef]
37. Pogocki, D.; Schöneich, C. Thiyl Radicals Abstract Hydrogen Atoms from Carbohydrates: Reactivity and Selectivity. *Free Radic. Biol. Med.* **2001**, *31*, 98–107. [CrossRef]
38. Dang, H.S.; Roberts, B.P.; Tocher, D.A. Selective Radical-Chain Epimerisation at Electron-Rich Chiral Tertiary C-H Centres Using Thiols as Protic Polarity-Reversal Catalysts. *J. Chem. Soc. Perk. T.* **2001**, *1*, 2452–2461. [CrossRef]
39. Roberts, B.P. Polarity-Reversal Catalysis of Hydrogen-Atom Abstraction Reactions: Concepts and Applications in Organic Chemistry. *Chem. Soc. Rev.* **1999**, *28*, 25–35. [CrossRef]
40. Backman, L.R.F.; Funk, M.A.; Dawson, C.D.; Drennan, C.L. New Tricks for the Glycyl Radical Enzyme Family. *Crit. Rev. Biochem. Mol.* **2017**, *52*, 674–695. [CrossRef]
41. Parast, C.V.; Wong, K.K.; Lewisch, S.A.; Kozarich, J.W.; Peisach, J.; Magliozzo, R.S. Hydrogen-Exchange of the Glycyl Radical of Pyruvate Formate-Lyase Is Catalyzed by Cysteine-419. *Biochemistry* **1995**, *34*, 2393–2399. [CrossRef] [PubMed]
42. Guo, J.D.; Himo, F. Catalytic Mechanism of Pyruvate-Formate Lyase Revisited. *J. Phys. Chem. B* **2004**, *108*, 15347–15354. [CrossRef]
43. Rauk, A.; Yu, D.; Armstrong, D.A. Oxidative Damage to and by Cysteine in Proteins: An ab Initio Study of the Radical Structures, C-H, S-H, and C-C Bond Dissociation Energies, and Transition Structures for H Abstraction by Thiyl Radicals. *J. Am. Chem. Soc.* **1998**, *120*, 8848–8855. [CrossRef]
44. Nauser, T.; Casi, G.; Koppenol, W.H.; Schöneich, C. Reversible Intramolecular Hydrogen Transfer Between Cysteine Thiyl Radicals and Glycine and Alanine in Model Peptides: Absolute Rate Constants Derived from Pulse Radiolysis and Laser Flash Photolysis. *J. Phys. Chem. B* **2008**, *112*, 15034–15044. [CrossRef] [PubMed]
45. Nauser, T.; Koppenol, W.H.; Schöneich, C. Reversible Hydrogen Transfer Reactions in Thiyl Radicals From Cysteine and Related Molecules: Absolute Kinetics and Equilibrium Constants Determined by Pulse Radiolysis. *J. Phys. Chem. B* **2012**, *116*, 5329–5341. [CrossRef]
46. Lesslie, M.; Lau, J.K.C.; Lawler, J.T.; Siu, K.W.M.; Oomens, J.; Berden, G.; Hopkinson, A.C.; Ryzhov, V. Alkali-Metal-Ion-Assisted Hydrogen Atom Transfer in the Homocysteine Radical. *Chem. Eur. J.* **2016**, *22*, 2243–2246. [CrossRef]
47. Lesslie, M.; Lau, J.K.C.; Pacheco, G.; Lawler, J.T.; Siu, K.W.M.; Hopkinson, A.C.; Ryzhov, V. Hydrogen Atom Transfer in Metal Ion Complexes of the Glutathione Thiyl Radical. *Int. J. Mass. Spectrom.* **2018**, *429*, 39–46. [CrossRef]

48. Hofstetter, D.; Thalmann, B.; Nauser, T.; Koppenol, W.H. Hydrogen Exchange Equilibria in Thiols. *Chem. Res. Toxicol.* **2012**, *25*, 1862–1867. [CrossRef]
49. Mozziconacci, O.; Williams, T.D.; Schöneich, C. Intramolecular Hydrogen Transfer Reactions of Thiyl Radicals from Glutathione: Formation of Carbon-Centered Radical at Glu, Cys, and Gly. *Chem. Res. Toxicol.* **2012**, *25*, 1842–1861. [CrossRef]
50. Reid, D.L.; Armstrong, D.A.; Rauk, A.; Von Sonntag, C. H-Atom Abstraction by Thiyl Radicals from Peptides and Cyclic Dipeptides. A Theoretical Study of Reaction Rates. *Phys. Chem. Chem. Phys.* **2003**, *5*, 3994–3999. [CrossRef]
51. Hao, G.; Gross, S.S. Electrospray Tandem Mass Spectrometry Analysis of S-and N-Nitrosopeptides: Facile Loss of NO and Radical-Induced Fragmentation. *J. Am. Soc. Mass. Spectr.* **2006**, *17*, 1725–1730. [CrossRef] [PubMed]
52. Nauser, T.; Pelling, J.; Schoneich, C. Thiyl Radical Reaction with Amino Acid Side Chains: Rate Constants for Hydrogen Transfer and Relevance for Posttranslational Protein Modification. *Chem. Res. Toxicol.* **2004**, *17*, 1323–1328. [CrossRef] [PubMed]
53. Schöneich, C.; Mozziconacci, O.; Koppenol, W.H.; Nauser, T. Intramolecular 1,2-and 1,3-Hydrogen Transfer Reactions of Thiyl Radicals. *Isr. J. Chem.* **2014**, *54*, 265–271. [CrossRef]
54. Zhang, X.; Zhang, N.; Schuchmann, H.P.; Von Sonntag, C. Pulse Radiolysis of 2-Mercaptoethanol in Oxygenated Aqueous Solution. Generation and Reactions of the Thiylperoxyl Radical. *J. Phys. Chem.* **1994**, *98*, 6541–6547. [CrossRef]
55. Tamba, M.; Dajka, K.; Ferreri, C.; Asmus, K.D.; Chatgilialoglu, C. One-Electron Reduction of Methanesulfonyl Chloride. The Fate of $MeSO_2Cl^{\bullet-}$ and $MeSO_2^{\bullet}$ Intermediates in Oxygenated Solutions and Their Role in the Cis-Trans Isomerization of Mono-Unsaturated Fatty Acids. *J. Am. Chem. Soc.* **2007**, *129*, 8716–8723. [CrossRef] [PubMed]
56. Becker, D.; Swarts, S.; Champagne, M.; Sevilla, M.D. An ESR Investigation of the Reactions of Glutathione, Cysteine and Penicillamine Thiyl Radicals: Competitive Formation of RSO. R. RSSR, and RSS. *Int. J. Radiat. Biol. Relat. Stud. Phys. Chem. Med.* **1988**, *53*, 767–786. [CrossRef]
57. Sevilla, M.D.; Yan, M.Y.; Becker, D. Thiol Peroxyl Radical Formation from the Reaction of Cysteine Thiyl Radical with Molecular Oxygen: An ESR Investigation. *Biochem. Biophys. Res. Commun.* **1988**, *155*, 405–410. [CrossRef]
58. Brange, J.; Langkjoer, L. Insulin Structure and Stability. In *Stability and Characterization of Protein and Peptide Drugs*; Wang, Y.J., Pearlman, R., Eds.; Plenum Press: New York, NJ, USA, 1993; pp. 315–350.
59. Mozziconacci, O.; Williams, T.D.; Kerwin, B.A.; Schöneich, C. Reversible Intramolecular Hydrogen Transfer Between Protein Cysteine Thiyl Radicals and Alpha C-H Bonds in Insulin: Control of Selectivity by Secondary Structure. *J. Phys. Chem. B* **2008**, *112*, 15921–15932. [CrossRef]
60. Correia, M.; Neves-Petersen, M.T.; Jeppesen, P.B.; Gregersen, S.; Petersen, S.B. UV-Light Exposure of Insulin: Pharmaceutical Implications Upon Covalent Insulin Dityrosine Dimerization and Disulphide Bond Photolysis. *PLoS ONE* **2012**, *7*, e50733. [CrossRef]
61. Darrington, R.T.; Anderson, B.D. Effects of Insulin Concentration and Self-Association on the Partitioning of its A-21 Cyclic Anhydride Intermediate to Desamido Insulin and Covalent Dimer. *Pharm. Res.* **1995**, *12*, 1077–1084. [CrossRef]
62. Darrington, R.T.; Anderson, B.D. Evidence for a Common Intermediate in Insulin Deamidation and Covalent Dimer Formation: Effects of pH and Aniline Trapping in Dilute Acidic Solutions. *J. Pharm. Sci.* **1995**, *84*, 275–282. [CrossRef] [PubMed]
63. Ivanova, M.I.; Sievers, S.A.; Sawaya, M.R.; Wall, J.S.; Eisenberg, D. Molecular Basis for Insulin Fibril Assembly. *Proc. Natl. Acad. Sci. USA* **2009**, *106*, 18990–18995. [CrossRef] [PubMed]
64. Ivanova, M.I.; Thompson, M.J.; Eisenberg, D. A Systematic Screen of Beta(2)-Microglobulin And Insulin for Amyloid-Like Segments. *Proc. Natl. Acad. Sci. USA* **2006**, *103*, 4079–4082. [CrossRef]
65. Brange, J.; Ribel, U.; Hansen, J.F.; Dodson, G.; Hansen, M.T.; Havelund, S.; Melberg, S.G.; Norris, F.; Norris, K.; Snel, L.; et al. Monomeric Insulins Obtained by Protein Engineering and Their Medical Implications. *Nature* **1988**, *333*, 679–682. [CrossRef]
66. Weiss, M.A.; Lawrence, M.C. A Thing of Beauty: Structure and Function of Insulin's "Aromatic Triplet". *Diabetes Obes. Metab.* **2018**, *20*, 51–63. [CrossRef] [PubMed]

67. Kobayashi, M.; Ohgaku, S.; Iwasaki, M.; Maegawa, H.; Shigeta, Y.; Inouye, K. Supernormal Insulin: [D-PheB24]-Insulin with Increased Affinity for Insulin Receptors. *Biochem. Biophys. Res. Commun.* **1982**, *107*, 329–336. [CrossRef]
68. Wei, Y.; Mathies, G.; Yokoyama, K.; Chen, J.; Griffin, R.G.; Stubbe, J. A Chemically Competent Thiosulfuranyl Radical on the Escherichia Coli Class III Ribonucleotide Reductase. *J. Am. Chem. Soc.* **2014**, *136*, 9001–9013. [CrossRef]
69. Nauser, T.; Carreras, A. Carbon-Centered Radicals Add Reversibly to Histidine-Implications. *Chem. Commun.* **2014**, *50*, 14349–14351. [CrossRef]
70. Santschi, N.; Nauser, T. An Experimental Radical Electrophilicity Index. *Chemphyschem* **2017**, *18*, 2973–2976. [CrossRef]
71. Mozziconacci, O.; Haywood, J.; Gorman, E.M.; Munson, E.; Schöneich, C. Photolysis of Recombinant Human Insulin in the Solid State: Formation of a Dithiohemiacetal Product at the C-Terminal Disulfide Bond. *Pharm. Res.* **2012**, *29*, 121–133. [CrossRef]
72. De Vos, A.M.; Ultsch, M.; Kossiakoff, A.A. Human Growth Hormone and Extracellular Domain of its Receptor: Crystal Structure of the Complex. *Science* **1992**, *255*, 306–312. [CrossRef] [PubMed]
73. Pearlman, R.; Bewley, T.A. Stability and Characterization of Human Growth Hormone. In *Stability and Characterization of Protein and Peptide Drugs*; Wang, Y.J., Pearlman, R., Eds.; Plenum Press: New York, NJ, USA, 1993; pp. 1–58.
74. Williams, B.R.; Cho, J.S. Hormone Replacement: The Fountain of Youth? *Prim. Care* **2017**, *44*, 481–498. [CrossRef] [PubMed]
75. Jiang, H.; Wu, S.L.; Karger, B.L.; Hancock, W.S. Mass Spectrometric Analysis of Innovator, Counterfeit, and Follow-On Recombinant Human Growth Hormone. *Biotechnol. Prog.* **2009**, *25*, 207–218. [CrossRef]
76. Karlsson, G.; Eriksson, K.; Persson, A.; Mansson, H.; Soderholm, S. The Separation of Recombinant Human Growth Hormone Variants by UHPLC. *J. Chromatogr. Sci.* **2013**, *51*, 943–949. [CrossRef] [PubMed]
77. Battersby, J.E.; Hancock, W.S.; Canova-Davis, E.; Oeswein, J.; O'Connor, B. Diketopiperazine Formation and N-Terminal Degradation in Recombinant Human Growth Hormone. *Int. J. Pept. Protein. Res.* **1994**, *44*, 215–222. [CrossRef] [PubMed]
78. Zhao, F.; Ghezzo-Schöneich, E.; Aced, G.I.; Hong, J.; Milby, T.; Schöneich, C. Metal-Catalyzed Oxidation of Histidine in Human Growth Hormone. Mechanism, Isotope Effects, and Inhibition by a Mild Denaturing Alcohol. *J. Biol. Chem.* **1997**, *272*, 9019–9029. [CrossRef] [PubMed]
79. Hovorka, S.W.; Hong, J.; Cleland, J.L.; Schöneich, C. Metal-Catalyzed Oxidation of Human Growth Hormone: Modulation by Solvent-Induced Changes of Protein Conformation. *J. Pharm. Sci.* **2001**, *90*, 58–69. [CrossRef]
80. Mulinacci, F.; Capelle, M.A.; Gurny, R.; Drake, A.F.; Arvinte, T. Stability of Human Growth Hormone: Influence of Methionine Oxidation on Thermal Folding. *J. Pharm. Sci.* **2011**, *100*, 451–463. [CrossRef] [PubMed]
81. Mulinacci, F.; Poirier, E.; Capelle, M.A.; Gurny, R.; Arvinte, T. Influence of Methionine Oxidation on the Aggregation of Recombinant Human Growth Hormone. *Eur. J. Pharm. Biopharm.* **2013**, *85*, 42–52. [CrossRef]
82. Chang, S.H.; Teshima, G.M.; Milby, T.; Gillece-Castro, B.; Canova-Davis, E. Metal-Catalyzed Photooxidation of Histidine in Human Growth Hormone. *Anal. Biochem.* **1997**, *244*, 221–227. [CrossRef]
83. Steinmann, D.; Ji, J.A.; Wang, Y.J.; Schöneich, C. Photodegradation of Human Growth Hormone: A Novel Backbone Cleavage Between Glu-88 and Pro-89. *Mol. Pharm.* **2013**, *10*, 2693–2706. [CrossRef] [PubMed]
84. Steinmann, D.; Mozziconacci, O.; Bommana, R.; Stobaugh, J.F.; Wang, Y.J.; Schöneich, C. Photodegradation Pathways of Protein Disulfides: Human Growth Hormone. *Pharm. Res.* **2017**, *34*, 2756–2778. [CrossRef] [PubMed]
85. Jespersen, A.M.; Christensen, T.; Klausen, N.K.; Nielsen, F.; Sorensen, H.H. Characterisation of a Trisulphide Derivative of Biosynthetic Human Growth Hormone Produced in Escherichia Coli. *Eur. J. Biochem.* **1994**, *219*, 365–373. [CrossRef]
86. Andersson, C.; Edlund, P.O.; Gellerfors, P.; Hansson, Y.; Holmberg, E.; Hult, C.; Johansson, S.; Kordel, J.; Lundin, R.; Mendel-Hartvig, I.B.; et al. Isolation and Characterization of a Trisulfide Variant of Recombinant Human Growth Hormone Formed During Expression in Escherichia Coli. *Int. J. Pept. Protein. Res.* **1996**, *47*, 311–321. [CrossRef] [PubMed]

87. Canova-Davis, E.; Baldonado, I.P.; Chloupek, R.C.; Ling, V.T.; Gehant, R.; Olson, K.; Gillece-Castro, B.L. Confirmation by Mass Spectrometry of a Trisulfide Variant in Methionyl Human Growth Hormone Biosynthesized in Escherichia Coli. *Anal. Chem.* **1996**, *68*, 4044–4051. [CrossRef]
88. Datola, A.; Richert, S.; Bierau, H.; Agugiaro, D.; Izzo, A.; Rossi, M.; Cregut, D.; Diemer, H.; Schaeffer, C.; Van Dorsselaer, A.; et al. Characterisation of a Novel Growth Hormone Variant Comprising a Thioether Link Between Cys182 and Cys189. *ChemMedChem* **2007**, *2*, 1181–1189. [CrossRef]
89. Junnila, R.K.; Wu, Z.; Strasburger, C.J. The Role of Human Growth Hormone's C-Terminal Disulfide Bridge. *Growth Horm. IGF Res.* **2013**, *23*, 62–67. [CrossRef]
90. Wang, Z.; Rejtar, T.; Zhou, Z.S.; Karger, B.L. Desulfurization of Cysteine-Containing Peptides Resulting from Sample Preparation for Protein Characterization by Mass Spectrometry. *Rapid Commun. Mass Spectrom.* **2010**, *24*, 267–275. [CrossRef]
91. Fradkin, A.H.; Mozziconacci, O.; Schöneich, C.; Carpenter, J.F.; Randolph, T.W. UV Photodegradation of Murine Growth Hormone: Chemical Analysis and Immunogenicity Consequences. *Eur. J. Pharm. Biopharm.* **2014**, *87*, 395–402. [CrossRef]
92. Asmus, K.D.; Hug, G.L.; Bobrowski, K.; Mulazzani, Q.G.; Marciniak, B. Transients in the Oxidative and H-Atom-Induced Degradation of 1,3,5-Trithiane. Time-Resolved Studies in Aqueous Solution. *J. Phys. Chem. A* **2006**, *110*, 9292–9300. [CrossRef]
93. Hug, G.L.; Janeba-Bartoszewicz, E.; Filipiak, P.; Pedzinski, T.; Kozubek, H.; Marciniak, B. Evidence for Heterolytic Cleavage of C-S Bonds in the Photolysis of 1,3,5-Trithianes. *Pol. J. Chem.* **2008**, *82*, 883–892.
94. Janeba-Bartoszewicz, E.; Hug, G.L.; Andrzejewska, E.; Marciniak, B. Photochemistry of 1,3,5-Trithianes in Solution-Steady-State and Laser Flash Photolysis Studies. *J. Photoch. Photobio. A Chem.* **2006**, *177*, 17–23. [CrossRef]
95. Mozziconacci, O.; Kerwin, B.A.; Schöneich, C. Photolysis of An Intrachain Peptide Disulfide Bond: Primary and Secondary Processes, Formation of H2S, and Hydrogen Transfer Reactions. *J. Phys. Chem. B* **2010**, *114*, 3668–3688. [CrossRef] [PubMed]
96. Strausberg, R.L.; Feingold, E.A.; Grouse, L.H.; Derge, J.G.; Klausner, R.D.; Collins, F.S.; Wagner, L.; Shenmen, C.M.; Schuler, G.D.; Altschul, S.F.; et al. Generation and Initial Analysis of More Than 15,000 Full-Length Human and Mouse cDNA Sequences. *Proc. Natl. Acad. Sci. USA* **2002**, *99*, 16899–16903. [PubMed]
97. Mozziconacci, O.; Stobaugh, J.T.; Bommana, R.; Woods, J.; Franklin, E.; Jorgenson, J.W.; Forrest, M.L.; Schöneich, C.; Stobaugh, J.F. Profiling the Photochemical-Induced Degradation of Rat Growth Hormone with Extreme Ultra-pressure Chromatography-Mass Spectrometry Utilizing Meter-Long Microcapillary Columns Packed with Sub-2-mum Particles. *Chromatographia* **2017**, *80*, 1299–1318. [CrossRef] [PubMed]
98. Mozziconacci, O.; Sharov, V.; Williams, T.D.; Kerwin, B.A.; Schöneich, C. Peptide Cysteine Thiyl Radicals Abstract Hydrogen Atoms from Surrounding Amino Acids: The Photolysis of a Cystine Containing Model Peptide. *J. Phys. Chem. B.* **2008**, *112*, 9250–9257. [CrossRef] [PubMed]
99. Mozziconacci, O.; Kerwin, B.A.; Schöneich, C. Reversible Hydrogen Transfer Reactions of Cysteine Thiyl Radicals in Peptides: The Conversion of Cysteine into Dehydroalanine and Alanine, and of Alanine Into Dehydroalanine. *J. Phys. Chem. B* **2011**, *115*, 12287–12305. [CrossRef]
100. Bonifacic, M.; Asmus, K.D. Adduct Formation and Absolute Rate Constants in the Displacement Reaction of Thiyl Radicals with Disulfides. *J. Phys. Chem.* **1984**, *88*, 6286–6290. [CrossRef]
101. Mozziconacci, O.; Kerwin, B.A.; Schöneich, C. Exposure of a Monoclonal Antibody, IgG1, to UV-Light Leads to Protein Dithiohemiacetal and Thioether Cross-Links: A Role for Thiyl Radicals? *Chem. Res. Toxicol.* **2010**, *23*, 1310–1312. [CrossRef]
102. Zhou, S.; Mozziconacci, O.; Kerwin, B.A.; Schöneich, C. The Photolysis of Disulfide Bonds in IgG1 and IgG2 Leads to Selective Intramolecular Hydrogen Transfer Reactions of Cysteine Thiyl Radicals, Probed by Covalent H/D Exchange and RPLC-MS/MS Analysis. *Pharm. Res.* **2013**, *30*, 1291–1299. [CrossRef]
103. Bommana, R.; Subelzu, N.; Mozziconacci, O.; Sreedhara, A.; Schöneich, C. Identification of D-Amino Acids in Light Exposed mAb Formulations. *Pharm. Res.* **2018**, *35*, 238. [CrossRef] [PubMed]
104. Mozziconacci, O.; Kerwin, B.A.; Schöneich, C. Reversible Hydrogen Transfer Between Cysteine Thiyl Radical and Glycine and Alanine in Model Peptides: Covalent H/D Exchange, Radical-Radical Reactions, and L-to D-Ala Conversion. *J. Phys. Chem. B* **2010**, *114*, 6751–6762. [CrossRef] [PubMed]

105. Mozziconacci, O.; Schöneich, C. Effect of Conformation on the Photodegradation of Trp-And Cystine-Containing Cyclic Peptides: Octreotide and Somatostatin. *Mol. Pharm.* **2014**, *11*, 3537–3546. [CrossRef] [PubMed]
106. Mason, B.D.; Schöneich, C.; Kerwin, B.A. Effect of pH and Light on Aggregation and Conformation of An IgG1 mAb. *Mol. Pharm.* **2012**, *9*, 774–790. [CrossRef] [PubMed]
107. Shah, D.D.; Zhang, J.; Maity, H.; Mallela, K.M.G. Effect of Photo-Degradation on the Structure, Stability, Aggregation, and Function of an IgG1 Monoclonal Antibody. *Int. J. Pharm.* **2018**, *547*, 438–449. [CrossRef] [PubMed]
108. Bessa, J.; Boeckle, S.; Beck, H.; Buckel, T.; Schlicht, S.; Ebeling, M.; Kiialainen, A.; Koulov, A.; Boll, B.; Weiser, T.; et al. The Immunogenicity of Antibody Aggregates in a Novel Transgenic Mouse model. *Pharm. Res.* **2015**, *32*, 2344–2359. [CrossRef] [PubMed]
109. Boll, B.; Bessa, J.; Folzer, E.; Rios Quiroz, A.; Schmidt, R.; Bulau, P.; Finkler, C.; Mahler, H.C.; Huwyler, J.; Iglesias, A.; et al. Extensive Chemical Modifications in the Primary Protein Structure of IgG1 Subvisible Particles Are Necessary for Breaking Immune Tolerance. *Mol. Pharm.* **2017**, *14*, 1292–1299. [CrossRef]

© 2019 by the author. Licensee MDPI, Basel, Switzerland. This article is an open access article distributed under the terms and conditions of the Creative Commons Attribution (CC BY) license (http://creativecommons.org/licenses/by/4.0/).

Article

Stability and Catalase-Like Activity of a Mononuclear Non-Heme Oxoiron(IV) Complex in Aqueous Solution

Balázs Kripli, Bernadett Sólyom, Gábor Speier and József Kaizer *

Department of Chemistry, University of Pannonia, 8201 Veszprém, Hungary
* Correspondence: kaizer@almos.vein.hu; Tel.: +36-88-62-4720

Academic Editor: Chryssostomos Chatgilialoglu
Received: 27 August 2019; Accepted: 5 September 2019; Published: 5 September 2019

Abstract: Heme-type catalase is a class of oxidoreductase enzymes responsible for the biological defense against oxidative damage of cellular components caused by hydrogen peroxide, where metal-oxo species are proposed as reactive intermediates. To get more insight into the mechanism of this curious reaction a non-heme structural and functional model was carried out by the use of a mononuclear complex [Fe^{II}(N4Py*)(CH$_3$CN)](CF$_3$SO$_3$)$_2$ (N4Py* = N,N-bis(2-pyridylmethyl)-1,2-di(2-pyridyl)ethylamine) as a catalyst, where the possible reactive intermediates, high-valent Fe^{IV}=O and Fe^{III}–OOH are known and spectroscopically well characterized. The kinetics of the dismutation of H$_2$O$_2$ into O$_2$ and H$_2$O was investigated in buffered water, where the reactivity of the catalyst was markedly influenced by the pH, and it revealed Michaelis–Menten behavior with K_M = 1.39 M, k_{cat} = 33 s^{-1} and $k_2(k_{cat}/K_M)$ = 23.9 M^{-1}s^{-1} at pH 9.5. A mononuclear [(N4Py)Fe^{IV}=O]$^{2+}$ as a possible intermediate was also prepared, and the pH dependence of its stability and reactivity in aqueous solution against H$_2$O$_2$ was also investigated. Based on detailed kinetic, and mechanistic studies (pH dependence, solvent isotope effect (SIE) of 6.2 and the saturation kinetics for the initial rates versus the H$_2$O$_2$ concentration with K_M = 18 mM) lead to the conclusion that the rate-determining step in these reactions above involves hydrogen-atom transfer between the iron-bound substrate and the Fe(IV)-oxo species.

Keywords: catalase activity; iron(IV)-oxo; hydrogen peroxide; oxidation; kinetic studies

---

1. Introduction

Superoxide dismutases (SODs), catalase-peroxidases (KatGs) and catalases are specialized oxidoreductase enzymes for the degradation of reactive oxygen species (ROS), e.g., hydrogen peroxide, hydroxyl and superoxide radicals to avoid their accumulation and prevent the oxidative damage of cellular components, that may lead to a number of diseases such as cancer, Alzheimer's diseases and aging [1–4]. For example, the hydroxyl and/or hydroperoxyl radicals may cause lipid peroxidation, membrane damage, DNA oxidation and cell death [5,6]. As a fine coupling of SODs and catalases, the former enzymes catalyze the dismutation of superoxide into dioxygen (1-electon oxidation) and H$_2$O$_2$, whilst the latter enzymes eliminate the H$_2$O$_2$ via its decomposition by disproportionation into O$_2$ (2-electron oxidation) and H$_2$O, resulting in the optimal intracellular concentration of a H$_2$O$_2$ molecule [7–9], which acts as a second messenger in signal-transduction pathways. Otherwise, it is worth to note, that the therapeutic potential of H$_2$O$_2$ makes this molecule also a valuable target in cancer killing via chemo- and radiotherapy, and in stroke therapy [10–12].

Two main classes of catalase enzymes are known, an iron and manganese-containing proteins. Although both types of catalases exhibit high catalytic activities, there are significant differences, including the active sites and the catalytic mechanisms [13]. Monofunctional catalases (EC 1.11.1.6)

are heme-containing enzymes, that catalyze the dismutation of hydrogen peroxide (2H$_2$O$_2$ = 2H$_2$O + O$_2$), where the catalytic mechanism is well-characterized with a high-valent oxoiron(IV) porphyrin π-cation radical, compound I, [(P$^{\bullet+}$)Fe^{IV}=O]$^+$ (P = porphyrinate dianion), being responsible for hydrogen peroxide oxidation [14–16]. Manganese catalases such as *Lactobacillus plantarum* [17,18], *Thermus thermophilus* [19,20], *Thermoleophilium album* [21] and *Pyrobaculum calidifontis VA1* [22] are found in several bacterial organisms, and possess a binuclear manganese center with a cycle between Mn(II)-Mn(II) and Mn(III)-Mn(III) states during turnover.

Synthetic compounds as biomimics of catalase enzymes may have potential biomedical application as therapeutic agents against oxidative stress. Besides the heme-type models, a great number of manganese, copper, ruthenium and non-heme iron complexes have been designed and studied as catalase models [23–35]. However, comparative studies between heme and non-heme models are scarce. The non-heme models are mainly binuclear complexes [27–29], only a small number of mononuclear iron compounds have been studied [12,36,37]. The direct dismutation of H$_2$O$_2$ with terminal and bridging oxo ligands has been described for only a few complexes of Fe, Cr, Mn, V and Ru [38–42]. Mononuclear oxoiron(IV) complexes are of interest from a bioinorganic viewpoint, since similar intermediates are frequently invoked as the active species in the active site of numerous proteins and in biomimetic iron-containing catalytic systems. Most of these results were obtained in organic solvent due to the lack of solubility or activity in aqueous solution. Due to the increasing importance of catalase activity, we have focused on the development of such a non-heme iron-containing system that shows catalase-like activity in aqueous solution. To get more insight into the mechanism of H$_2$O$_2$ dismutation the mononuclear complex [Fe^{II}(N4Py*)(CH$_3$CN)](CF$_3$SO$_3$)$_2$ (1) (N4Py* = *N,N*-bis(2-pyridylmethyl)-1,2-di(2-pyridyl)ethylamine) was chosen as a catalyst, where the possible reactive intermediates high-valent Fe^{IV}=O (2) and Fe^{III}-OOH (3) are known and spectroscopically well characterized (Scheme 1) [43–46].

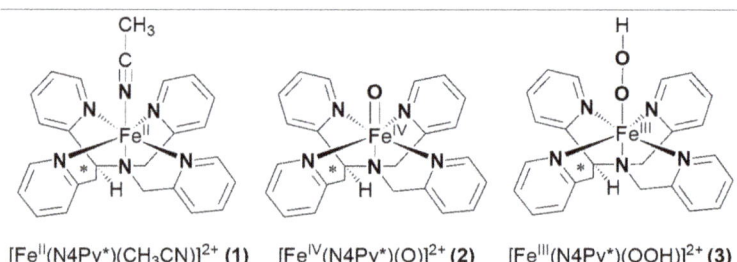

[Fe^{II}(N4Py*)(CH$_3$CN)]$^{2+}$ (1)   [Fe^{IV}(N4Py*)(O)]$^{2+}$ (2)   [Fe^{III}(N4Py*)(OOH)]$^{2+}$ (3)

Scheme 1. Structures of (1), (2) and (3).

2. Results and Discussion

2.1. Catalase-Like Reactivity of [Fe^{II}(N$_4$Py*)(CH$_3$CN)](CF$_3$SO$_3$)$_2$ in Aqueous Solution

The catalase-like activity of the complex [Fe^{II}(N$_4$Py*)(CH$_3$CN)](CF$_3$SO$_3$)$_2$ to disproportionate H$_2$O$_2$ into H$_2$O and O$_2$ was investigated in aqueous solution at 20 °C by gasvolumetric measurements of evolved dioxygen. To gain further information on the mechanism of catalase activity of our iron complex, we first examined pH-dependence of catalase activity. It was reported that the coordination and dissociation of peroxides on metal-porphyrins are pH dependent reactions [47,48]. Moreover, they reported that the coordination is accelerated at a higher pH region and that the subsequent O–O bond cleavage leading to the formation of high-valent oxo-Fe(IV) or oxo-Fe(V) species is pH-independent (only at higher pH region, where the protonation of the distal oxygen in the peroxo-complex can be excluded) irreversible reaction. These results suggest that the coordination of peroxides is a crucial step for the formation of high-valent Fe species, and the mechanism of catalase activity involves the coordination of H$_2$O$_2$, which is considered to be pH-dependent as well. Therefore, we hypothesized that formation of reactive intermediate 2 is accelerated at pH 9.5 and catalase activity is increased as compared at pH 8. As shown in Figure 1, O_2 production of 1 in 50 mM borate buffer (pH 9.5) was significantly higher than that in phosphate buffer (pH 8). V_{in} value under this condition was determined to be $V_{in} = 1.13 \times 10^{-3}$ Ms^{-1}, which is approximately seven times higher than that at pH 8, and 8.5 times higher than that at pH 11. This indicates that the rate-determining step was faster at pH 9.5 than at pH 8, which may be explained by the higher concentration of the more nucleophilic HO_2^-.

The pH dependence of H_2O_2 dismutation was further studied between pH 7 and pH 11. It was found that the initial rate of the disproportionation of H_2O_2 increases with increasing pH and goes through a maximum. The pH profile of 1 exhibits a sharp optimum at pH ~9.5, whereas catalases in general exhibit a broad pH optimum extending from pH 5.6 to 8.5 [48]. In control experiments, in the absence of the complex, the pH of the solution did not change in the presence of H_2O_2, and no significant O_2 volume was evolved. We believe that the activity is influenced by the protonation state of H_2O_2. Assuming that hydrogen peroxide is activated by a direct interaction with the Fe^{IV}=O group of the complex, decomposition is expected to be favored by a high pH because of the larger concentration of the hydroperoxide anion (HOO^- is more nucleophilic than H_2O_2). On the other hand, at higher pH values, the complex may be destroyed by the formation of the mineral forms of iron or catalytically inactive, insoluble μ-oxo-diiron(III) species.

Detailed kinetic studies on the disproportionation of H_2O_2 were performed in aqueous solution (pH 9.5; 0.025 M $Na_2B_4O_7.10H_2O$/0.1 M HCl; I = 0.15 M KNO_3) at 20 °C by volumetric measurements of evolved dioxygen. To determine the dependence of the rates on the substrate concentration, solutions of the complex [$Fe^{II}(N_4Py^*)(CH_3CN)$](CF_3SO_3)$_2$ were treated with increasing amounts of H_2O_2 (1:400–5300). Plots of the amount of dioxygen evolved versus time at [1]$_0$ constant, are shown in Figure 1a. The initial rates values were calculated from the maximum slope of the O_2 versus time curves. Under this experimental condition, saturation kinetics was found for the initial rates ($V_{in} = -d[H_2O_2]/dt$) versus the H_2O_2 concentration (Figure 1b). An analysis of the data based on the Michaelis–Menten model ($V_{in} = k_{cat}[cat][S]_0/(K_M + [S]_0)$), originally developed for enzyme kinetics, was applied. A nonlinear least square fit was applied to calculate the Michaelis–Menten parameters, where k_{cat} is the turnover number, K_M is the Michaelis constant, S is the substrate initial concentration and [cat] is the catalyst concentration. The results were $K_M = 1.39$ M, $k_{cat} = 33$ s^{-1} and $k_2(k_{cat}/K_M) = 23.9$ M^{-1}s^{-1}. The data presented illustrate that the catalyst had a relatively high turnover number (k_{cat}) but appeared to bind peroxide very badly. The K_M value was greater than the values for the natural enzymes from *Thermus thermophilus* ($K_M = 0.083$ M) [19,20], *Tricholoma album* ($K_M = 0.015$ M) [21] and *Lactobacillus plantarum* ($K_M = 0.35$ M) [17,18] indicating a lower affinity to the substrate. The k_{cat} value equaled 33 s^{-1}, however, was 3–4 times magnitudes lower when compared to the natural enzymes *Thermus thermophilus* ($k_{cat} = 2.6 \times 10^5$ s^{-1}), *Tricholoma album* ($k_{cat} = 2.0 \times 10^5$ s^{-1}), *Lactobacillus plantarum* ($k_{cat} = 2.6 \times 10^4$ s^{-1}) and the heme-containing catalases ($k_{cat} = 4 \times 10^7$ s^{-1}). Despite this iron complex presents lower values of catalytic efficiency than other models (Table 1) [49–52], it must be emphasized that this value was obtained in water and in pH close to the natural, representing an advantage of the title complex with respect to most of the published models, whose studies have been conducted in organic solvent due to the lack of solubility or activity in aqueous solution.

Table 1. Kinetic parameters of reported catalase, catalase-peroxidase and their synthetic models.

Entry	Complex/Enzyme	K_M (M)	k_{cat} (s^{-1})	k_{cat}/K_M (s^{-1}M^{-1})	Solvent	Refs.
1	SynKatG [1]	0.0042			H$_2$O, pH 7	[48]
2	BpKatG [2]	0.0059			H$_2$O, pH 7	[48]
3	MtbKatG [3]	0.0025	1.2×10^3	5×10^8	H$_2$O, pH 7	[48]
4	BLC [4]	0.093	4.0×10^7		H$_2$O, pH 7	[53]
5	[Fe^{II}(N4Py*)(CH$_3$CN)](ClO$_4$)$_2$	1.39	33.2	23.9	H$_2$O, pH 9.5	this work.
6	[(N4Py*)Fe^{IV}=O](ClO$_4$)$_2$	0.018	0.014	0.754	CH$_3$CN/H$_2$O, pH 8	this work
7	[Fe$_4$(μ-O) μ-OH)(μ-OAc)$_4$(L$_2$)]$^{3+}$, [5]	1.010	1.41×10^{-4}	1.40×10^{-4}	H$_2$O	[42]
8	[Fe$_4$(μ-O) μ-OH)(μ-OAc)$_4$(L$_2$)]$^{3+}$, [5]	2.882	3.50×10^{-3}	1.21×10^{-3}	H$_2$O, pH 7.2	[42]
9	[Fe$_4$(μ-O) μ-OH)(μ-OAc)$_4$(L$_2$)]$^{3+}$, [5]	0.749	5.37×10^{-2}	7.17×10^{-2}	CH$_3$CN	[42]
10	T. thermophilus	0.083	2.6×10^5	3.13×10^6	H$_2$O	[19,20]
11	T. album	0.015	2.6×10^4	1.73×10^6	H$_2$O	[21]
12	L. plantarum	0.35	2.0×10^5	0.57×10^6	H$_2$O	[17,18]
13	[Mn(indH)Cl$_2$] [6]	0.49	38.9	79.2	H$_2$O, pH 9.5	[30]
14	[Mn(ind)$_2$] [6]	0.019	0.06	3.2	DMF	[51]
15	[Mn(X-salpn)O]$_2$ [7]	10–102	4.2–21.9	305–990	CH$_3$CN	[49,50]

[1] Catalase-peroxidase from *Synechocystis* PCC6803. [2] Catalase-peroxidase from *Burkholderia pseudomallei*. [3] Catalase peroxidase from *Mycobacterium tuberculosis*. [4] Bovine liver catalase. [5] HL = 1,3-bis[2-aminoethyl)amino]-2-propanol. [6] IndH = 1,3-bis(2'-pyridylimino)-isoindoline. [7] H$_2$salpn = N,N'-bis(salicylidene)-1,3-diaminopropane.

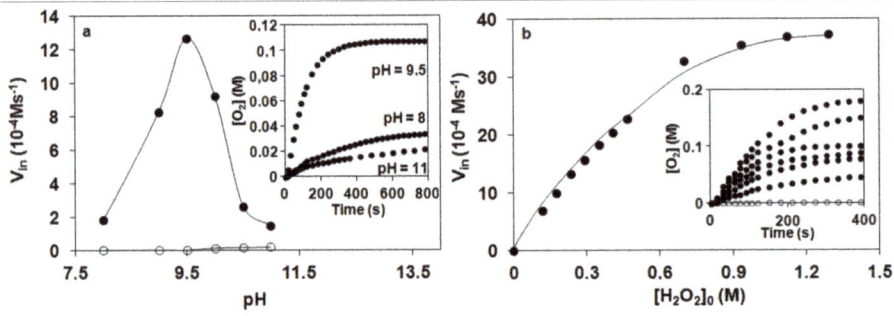

Figure 1. Kinetics of hydrogen peroxide degradation catalyzed by 1 in water: (**a**) pH dependence of hydrogen peroxide degradation determined by volumetrically measuring the evolved dioxygen in the presence (●) and in the absence (○) of 1. The inset shows the time traces for the reaction of 0.275 mM 1 with 0.35 M H$_2$O$_2$ at pH 8, 9.5 and 11 at 20 °C. (**b**) V$_{in}$ versus [H$_2$O$_2$]$_0$ at [1] = 2.75×10^{-4} M, pH 9.5 (borate buffer) and 20 °C. The inset shows the time traces for the reaction of 0.275 mM 1 with H$_2$O$_2$ (0.11–1.29 M).

2.2. Catalase-Like Reactivity Mediated by [(N4Py*)Fe^{IV}=O](ClO$_4$)$_2$ in Aqueous Solution

Rohde and co-workers have shown that the independently prepared [(N4Py)Fe^{IV}=O]$^{2+}$ reacts rapidly with near-stoichiometric H$_2$O$_2$ resulting in dioxygen and [Fe^{II}(N4Py)(CH$_3$CN)]$^{2+}$ in acetonitrile [54]. Later Browne and co-workers have found clear evidence for the reaction of Fe^{III}–OOH with H$_2$O$_2$ in methanol [55]. In their case the oxoiron(IV) intermediate can also be formed by homolytic cleavage of the O–O bond of an Fe^{III}–OOH, but the rate of its formation is much lower than the Fe^{III}–OOH-mediated H$_2$O$_2$ disproportionation observed with high excess H$_2$O$_2$ under catalytic conditions. As a continuity of these studies, we attempted to directly investigate the reactivity of the possible intermediates (Fe^{IV}=O, Fe^{III}–OOH) during the catalase reaction in aqueous solution.

We have shown earlier that complex 1 forms very stable high valent oxoiron(IV) species (2) with PhIO in CH$_3$CN (t$_{1/2}$ = 233 h at R.T., λ$_{max}$ = 705 nm, ε = 400 M^{-1}cm^{-1}) [43]. As a test of our oxoiron(IV) species we firstly investigated its reaction with excess H$_2$O$_2$ (75 equiv.) in acetonitrile at 10 °C, which resulted in the formation of a relatively stable transient purple species with a characteristic absorbance maximum at λ$_{max}$ 535 nm (ε = 1100 M^{-1} cm^{-1}; Figure 2a). It had a half-life of about 3 min even at

25 °C, but its decay can be remarkably enhanced by the addition of H_2O into the Fe^{III}–OOH-containing solution (CH_3CN/H_2O = 1:1) with a k_{obs} value of about 12.3×10^{-3} s^{-1} at 10 °C, resulting in the formation of 2 (Figure 2b). It is worth to note that at higher pH the decay was so fast, that we were not able to follow it. These results might suggest that a high-valent oxoiron(IV) species was one of the possible intermediates that may be responsible for the dismutation of H_2O_2 in aqueous solution.

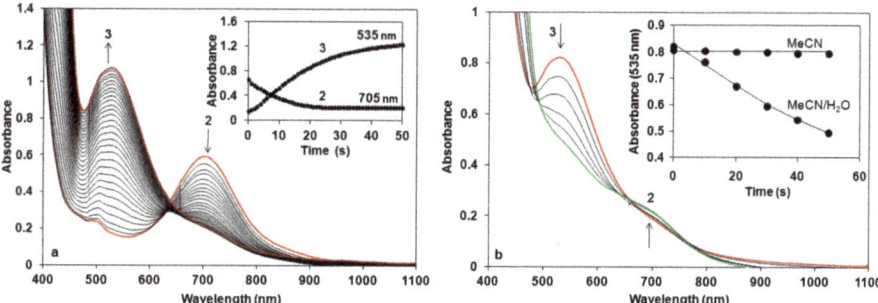

Figure 2. Reaction of 2 with H_2O_2 in acetonitrile: (**a**) UV-Vis spectra of the reaction of 1.5 mM 2 in CH_3CN with 75 equiv of H_2O_2 at 10 °C (path length, 1 cm). Inset: Time course of the reaction monitored at 705 nm (2) and 535 (3). (**b**) UV/Vis spectra of the decay of 3 generated based on (**a**). Inset: Time course of the decay of 3 in CH_3CN and CH_3CN/H_2O (v/v = 1:1) solution at 10 °C.

In the iron-catalyzed oxidation of H_2O_2 with terminal oxidants four processes can be proposed as the rate-controlling step, namely the formation of Fe^{III}–OOH or high-valent oxoiron(IV), or their reaction with the substrate (H_2O_2). To avoid this difficulty, and to get more insight into the mechanism of the H_2O_2 oxidation process we synthesized the oxoiron(IV) complex 2 by an in situ reaction of 1 with PhIO in acetonitrile, and investigated its stability and reactivity with H_2O_2 in a buffered H_2O–CH_3CN mixture (v/v = 1:1). In this way the role of the oxoiron(IV) species could be directly investigated. The UV-vis spectra of 2 in buffered solutions were almost identical to that observed in the acetonitrile. The observed blue shift on the λ_{max} values (from 705 to 697 nm) might be explained by the interaction (H-bridge) of the oxoiron(IV) with the H_2O molecule(s).

The stability of 2 was found to depend significantly on the pH value of reaction solutions, in which 2 was stable at pH 7–8 (k_{sd} = 0.43×10^{-3} s^{-1}, 0.64×10^{-3} s^{-1} with $t_{1/2}$ = 180 and 150 min at pH 7 and 8 at 10 °C, respectively), but decayed at a fast rate with increasing pH at pH 9–11 (k_{sd} = 3.51×10^{-3} s^{-1}, and 7.27×10^{-3} s^{-1}, 23×10^{-3} s^{-1}, 39×10^{-3} s^{-1} and 46×10^{-3} s^{-1} with $t_{1/2}$ = 4, 3, 2, 1.7 and 1 min at pH 9, 9.5, 10, 10.5 and 11 at 10 °C, respectively; Figure 3). This is the second example that the stability of oxoiron(IV) complex is controlled by the pH of reaction solutions [56].

The pH dependence of the reactivity of 2 against H_2O_2 was also examined in the range pH 7–11 in a buffered H_2O–MeCN mixture (v/v = 1:1) at 10 °C (Figure 3). Upon addition of 10 equiv. H_2O_2 to the solution of 2, the characteristic absorption band of 2 (λ_{max} = 697 nm) disappeared rapidly, and no formation of Fe^{III}–OOH was observed. Pseudo-first-order fitting of the kinetic data allowed us to calculate k_{obs} values to be 2.96×10^{-3} s^{-1}, 6.29×10^{-3} s^{-1}, 37.9×10^{-3} s^{-1}, 41.6×10^{-3} s^{-1}, 60.3×10^{-3} s^{-1}, 75.3×10^{-3} s^{-1} and 84×10^{-3} s^{-1} at pH 7, 8, 9, 9.5, 10, 10.5 and 11 at 10 °C, respectively.

The reactivity of 2 was found to depend significantly on the pH value of reaction solutions. The maximum rate of H_2O_2 dismutation, k'_{obs} (k'_{obs} = $k_{obs} - k_{sd}$ from the $-d[2]/dt = k_{obs}[2] = (k_{sd} + k'_{obs})[2]$) could be observed at pH 9, where the self decay process (k_{sd}) could be neglected (Figure 4a). The increase of the k_{obs} at higher pH could be explained by the self decay of 2. Addition of 10 equiv. H_2O_2 at pH 10 resulted in a decrease in absorbance at λ_{max} = 697 nm concomitant with an increase at 490 nm within 40 s at 10 °C, and an isosbestic point obtained at approximately λ_{max} = 620 nm. This spectrum including a weak absorption band at 700 nm with a shoulder around 490 nm corresponded to the spectrum of $[(N4Py^*)Fe^{III}\text{-O-}Fe^{III}(N4Py^*)]^{4+}$ (Figure 4b).

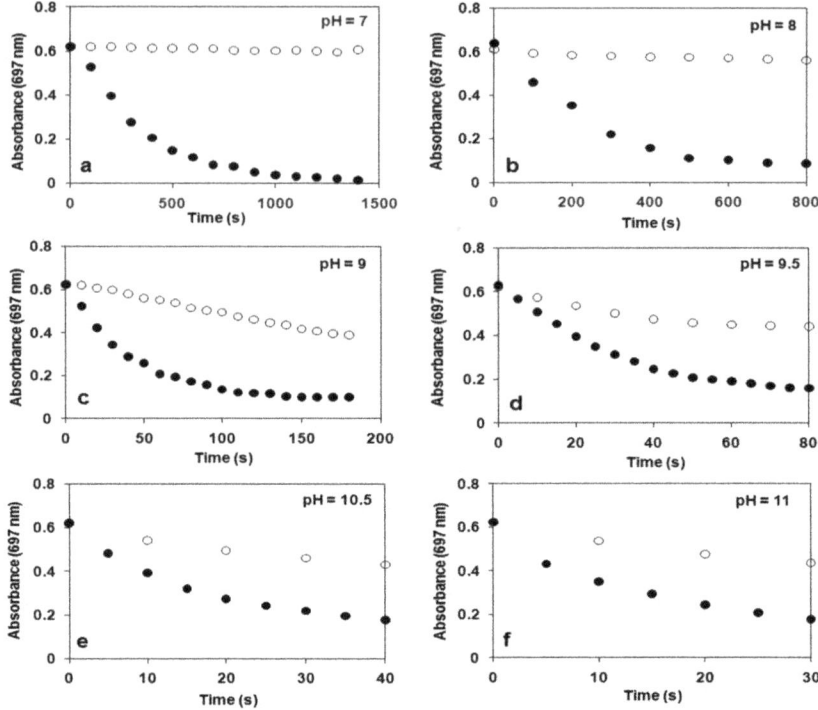

Figure 3. Time course of the decay of 2 monitored at 697 nm at different pH in the presence (●) and in the absence (○) of H_2O_2 at 10 °C. Conditions: [2] = 1.5 mM; $[H_2O_2]_0$ = 15 mM in MeCN/H_2O (2 cm^3, v/v = 1:1, path = 1 cm). (a) pH 7: 0.1 M KH_2PO_4/0.1 M NaOH. (b) pH 8: 0.025 M $Na_2B_4O_7 \cdot 10H_2O$/0.1 M HCl. (c) pH 9: 0.05 M $NaHCO_3$/0.1 M KOH. (d) pH 9.5: 0.05 M $NaHCO_3$/0.1 M KOH. (e) pH 10.5: 0.05 M $NaHCO_3$/0.1 M KOH. (f) pH 11: 0.05 M $NaHCO_3$/0.1 M KOH. I = 0.15 M KNO_3.

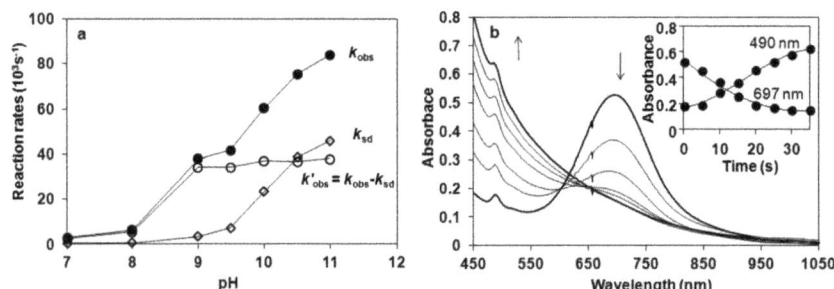

Figure 4. (a) Reaction rates of the decay of 2 monitored at 697 nm at different pH values in the presence (●) and in the absence (○) of H_2O_2 and their normalized values (◊) in buffered CH_3CN/H_2O (v/v = 1:1) solution (pH 7–11) at 10 °C. (b) Reaction of 2 with H_2O_2 in buffered CH_3CN/H_2O: UV-Vis spectra of the reaction of 1.5 mM 2 in buffered CH_3CN/H_2O (pH 10, v/v = 1:1) with 10 equiv of H_2O_2 at 10 °C (path length, 1 cm). Inset: Time course of the reaction monitored at 697 (2) and 490 nm in buffered CH_3CN/H_2O (v/v = 1:1) solution (pH 10) at 10 °C.

Detailed kinetic and mechanistic studies were carried out in buffered water/acetonitril mixture (v/v = 1:1) in pH 8, close to the natural at 10 °C, where the self decay process can be excluded. The reactivity of 2 was monitored by UV-vis spectroscopy and the rate of its rapid decomposition was measured at 697 nm (Figure 5a). Pseudo-first order fitting of the kinetic data allowed us to determine k_{obs} values. These results indicate a direct reaction between 2 and H_2O_2. In order to investigate the possible involvement of a hydrogen atom in the rate-determining step we investigated the reactivity of 2 with H_2O_2 in buffered MeCN/D_2O/H_2O (v/v = 1:0.75:0.25). Solutions of 2 in the presence of D_2O at pH 8 were somewhat less reactive against H_2O_2, yielding a solvent kinetic isotope effect of 6.2. This value was significantly smaller than that was obtained for the H–D isotope effect for [Ru^{IV}O(bpy)$_2$(py)] at pH 2.3 (KIE = 22.1 ± 1.2), but almost identical with that was measured at pH 9.7 (KIE = 8 ± 2.9) at 25 °C [40]. The most straightforward interpretation of the proton dependence was that the pathways involve the acid-base pre-equilibrium of H_2O_2 (H_2O_2 = HO_2^- + H^+) and the concomitant rate-controlling hydrogen-atom-transfer (HAT) between the Fe^{IV}=O species and the OH (or OD) group of H_2O_2 (D_2O_2) [57] forming a peroxyl radical.

To determine the dependence of the rates on the substrate concentration, solutions of the complex [(N_4Py^*)Fe^{IV}=O](CF_3SO_3)$_2$ were treated with increasing amounts of H_2O_2 (1:5–50). Under this experimental condition, saturation kinetics was found for the k_{obs} versus the H_2O_2 concentration (Figure 5b). At low H_2O_2 concentration, a k' value of about 0.47 $M^{-1}s^{-1}$ was obtained at 10 °C ($k' = k_{obs}/[H_2O_2]$ assuming a first order dependence). The reactivity of 2 was lower than that of [(N_4Py)Fe^{IV}=O]$^{2+}$ (N_4Py = N,N'-bis(2-pyridylmethyl)-N-bis(2-pyridyl)methylamine) in CH_3CN (k' value of 8 $M^{-1}s^{-1}$ at 25 °C), but significantly higher than that of [(tmc)(CH_3CN)Fe^{IV}=O]$^{2+}$ (tmc = 1,4,8,11-tetramethyl-1,4,8,11-tetraazacyclotetradecane; k_2 value of 0.035 ± 0.002 $M^{-1}s^{-1}$ at 25 °C) in CH_3CN. Furthermore, a k' value of 12.7 ± 1.3 $M^{-1}s^{-1}$ had been reported for the oxoruthenium(IV) complex [Ru^{IV}O(bpy)$_2$(py)] at 25 °C (H_2O, pH 7.92) [40]. Based on literature data, it can be concluded that [(N_4Py^*)Fe^{IV}=O]$^{2+}$ is more reactive in O–H bond activation (H_2O_2) than in C–H bond activation (hydrocarbons) [46].

Substrates saturation behaviors implied a rapid equilibrium between the unbound substrate and the iron complex as a result of hydrogen bridge bond. Under conditions of high substrate concentration, the primary species in solution was the Fe^{IV}O–H_2O_2 (Fe^{IV}O–HO_2^-) complex. The rate of the reaction was dependent only on the decomposition of the Fe^{IV}O–H_2O_2 (Fe^{IV}O–HO_2^-) complex (r.d.s.) to the product and free precursor complex (Scheme 2) [40,57]. A nonlinear least square fit was applied to calculate the Michaelis–Menten parameters. The results were K_M = 0.018 M, k_{cat} = 0.014 s^{-1} and k_2(k_{cat}/K_M) = 0.754 $M^{-1}s^{-1}$. An apparent K_M value for bovine liver catalase (BLC) was determined to be 0.093 M. By contrast, the K_M values of KatGs (catalase-peroxidase) were much lower (0.0042 M for SynKatG, 0.0025 M for MtbKatG and 0.0059 M for BpKatG, all at pH 7) [48], but was almost identical with the value for the natural enzyme from *Tricholoma album* (K_M = 0.015 M) indicating a high affinity to the substrate, appearing to bind to peroxide very strongly [21].

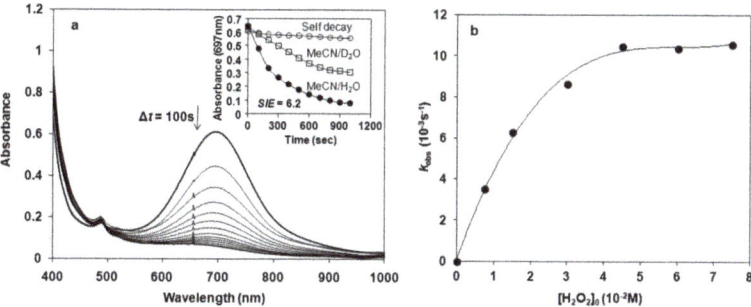

Figure 5. Kinetic studies on the reaction of 2 with H_2O_2 in buffered MeCN/H_2O solution at pH 8 and 10 °C. (**a**) UV-vis spectral change of 1.5 mM 2 upon addition of 10 equiv of H_2O_2. Inset shows time course of the decay of in the absence (○) and in the presence of H_2O_2 in MeCN/D_2O (□) and MeCN/H_2O (●) solution, respectively. (**b**) Plot of k_{obs} versus [H_2O_2]$_0$ at [2] = 1.5 mM, pH 8 and 10 °C.

$$H_2O_2 \rightleftharpoons HO_2^- + H^+ \quad \text{(H/D exchange)}$$

$$[(N4Py^*)Fe^{IV}=O]^{2+} + H_2O_2 \rightleftharpoons [(N4Py^*)Fe^{IV}=O]^{2+},H_2O_2 \quad \text{(Saturation kinetics)}$$

$$[(N4Py^*)Fe^{IV}=O]^{2+},H_2O_2 \xrightarrow{\text{r.d.s.}} [(N4Py^*)Fe^{III}\text{-OH}]^{2+} + {}^\bullet\text{OOH} \quad \text{(HAT, } SIE\text{)}$$

$$([(N4Py^*)Fe^{IV}=O]^{2+},HO_2^-) \xrightarrow{\text{r.d.s.}} [(N4Py^*)Fe^{III}\text{-OH}]^{2+} + {}^\bullet\text{OO}^- \quad \text{(HAT, } SIE\text{)}$$

$$[(N4Py^*)Fe^{IV}=O]^{2+} + {}^\bullet\text{OOH} \longrightarrow [(N4Py^*)Fe^{III}\text{-OH}]^{2+} + O_2$$

$$[(N4Py^*)Fe^{III}\text{-OH}]^{2+} + HO_2^- \longrightarrow [(N4Py^*)Fe^{III}\text{-OOH}]^{2+} + HO^-$$

$$[(N4Py^*)Fe^{III}\text{-OOH}]^{2+} \longrightarrow [(N4Py^*)Fe^{IV}=O]^{2+} + {}^\bullet\text{OH}$$

$$[(N4Py^*)Fe^{III}\text{-OH}]^{2+} \longrightarrow 0.5\,[(N4Py^*)Fe^{III}\text{-O-}Fe^{III}(N_4Py^*)]^{4+} + 0.5\,H_2O$$

Scheme 2. Proposed mechanism for the oxoiron(IV)-mediated H_2O_2 oxidation.

3. Materials and Methods

The N_4Py^* ligand, and its $[Fe^{II}(N_4Py^*)(CH_3CN)](CF_3SO_3)_2$ (1) complex were prepared according to published procedures [31]. UV/Vis spectra were recorded with an Agilent 8453 diode-array spectrophotometer (Agilent Technologies, Hewlett-Packard-Strasse 8, Waldbronn, Germany) with quartz cells.

Catalytic reactions were carried out at 20 °C in a 30 cm³ reactor containing a stirring bar under air. In a typical experiment the appropriate aqueous solution (19 cm³ 0.1 M KH_2PO_4/0.1 M NaOH pH 7, 8; 0.025 M $Na_2B_4O_7 \cdot 10H_2O$/0.1 M HCl pH 9, 9.5, 10; or 0.05 M $NaHCO_3$/0.1 M KOH pH 10.5, 11 buffer and I = 0.15 M KNO_3) was added to the complex dissolved in 1 cm³ DMF, and the flask was closed with a rubber septum. H_2O_2 was injected by syringe through the septum. The reactor was connected to a graduated burette filled with oil, and the evolved dioxygen was measured volumetrically at time intervals of 15 s. Initial rates were expressed as Ms^{-1} by taking the volume of the solution into account, and calculated from the maximum slope of the evolved dioxygen versus time.

Stoichiometric reactions were carried out under thermostated conditions at 10 °C in 1 cm quartz cuvettes. In a typical experiment $[Fe^{II}(N_4Py^*)(CH_3CN)](CF_3SO_3)_2$ (1) (3×10^{-3} M) was dissolved in acetonitrile (1.0 cm³), then iodosobenzene (4.5×10^{-3} M) was added to the solution. The mixture was stirred for 50 min then excess iodosobenzene was removed by filtration. The acetonitril solution was than diluted with the appropriate buffered aqueous solution (1.0 cm³), and the decay of 2 was followed by monitoring the decrease in absorbance at 697 nm ($\varepsilon = 400$ M^{-1} cm^{-1}) in the absence or in the presence of H_2O_2 under a pseudo-first order condition of excess H_2O_2.

4. Conclusions

It was found earlier that non-heme oxoiron(IV) complexes were able to carry out electrophilic transformations including O–H activation of H_2O_2 via homolytic O–H bond cleavage in acetonitrile as a functional catalase model. As a continuity of this study, efforts were made to work out a functional model in aqueous solution, close to the natural, where the postulated oxoiron(IV) intermediate behaved as an electrophilic oxidant. In summary, we reported one of the first examples of catalytic and stoichiometric H_2O_2 dismutation into O_2 and H_2O in aqueous solution mediated by electrophilic oxoiron(IV) intermediate, where the reactivity of 2 was markedly influenced by the pH. Based on detailed mechanistic studies on H_2O_2 oxidation that were investigated with in situ generated oxoiron(IV) species, plausible mechanisms were proposed, in which the H_2O_2 oxidation occurred by the HAT mechanism. To put together the stoichiometric and catalytic results it could be said that the highest catalytic activity of the H_2O_2 dismutation could be observed at pH 9.5, where the concentration of the more nucleophilic hydroperoxide anion (HOO$^-$) was high, and the self-decay of the oxoiron(IV) intermediate could be neglected. These results were in good agreement with the electrophilic reactivity of oxoiron(IV) intermediates proposed for heme-type monoiron catalases, and might help us to understand the mechanism of the detoxification of H_2O_2 in biological systems.

Author Contributions: Individual contribution of authors were as follows: B.K., Organic synthesis; B.S., Reaction kinetics; G.S., Senior supervisor and advisor; and J.K., Project leader, writer of the manuscript.

Funding: This research received no external funding.

Acknowledgments: Financial support of the Hungarian National Research Fund (OTKA K108489), and GINOP-2.3.2-15-2016-00049 are gratefully acknowledged.

Conflicts of Interest: The authors declare no conflict of interest.

References

1. Zamocky, M.; Furtmuller, P.G.; Obinger, C. Evolution of catalases from bacteria to humans. *Antioxid. Redox Signal.* **2008**, *10*, 1527–1548. [CrossRef]
2. Kunsch, C.; Medford, R.M. Oxidative Stress as a Regulator of Gene Expression in the Vasculature. *Circ. Res.* **1999**, *85*, 753–766. [CrossRef]
3. Balaban, R.S.; Nemoto, S.; Finkel, T. Mitochondria, Oxidants, and Aging. *Cell* **2005**, *120*, 483–495. [CrossRef]
4. Giordano, F.J. Oxygen, oxidative stress, hypoxia, and heart failure. *J. Clin. Invest.* **2005**, *115*, 500–508. [CrossRef]
5. Halliwell, B. Free radicals, antioxidants, and human disease: Curiosity, cause, or consequence? *Lancet* **1994**, *344*, 721–724. [CrossRef]
6. Choua, S.; Pacheco, P.; Coquelet, C.; Bienvenüe, E. Catalase-like activity of a water-soluble complex of Ru(II). *J. Inorg. Biochem.* **1997**, *65*, 79–85. [CrossRef]
7. Hempel, N.; Carrico, P.M.; Melendez, J.A. Manganese superoxide dismutase (Sod2) and redoxcontrol of signaling events that drive metastasis. *Anticancer Agents Med. Chem.* **2011**, *11*, 191–201. [CrossRef]
8. Sampson, N.; Koziel, R.; Zenzmaier, C.; Bubendorf, L.; Plas, E.; Jansen-Durr, P.; Berger, P. ROS Signaling by NOX4 Drives Fibroblast-to-Myofibroblast Differentation in the Diseased Prostatic Stroma. *Mol. Endocrinol.* **2011**, *25*, 503–515. [CrossRef]
9. Gao, M.C.; Jia, X.D.; Wu, Q.F.; Cheng, Y.; Chen, F.R.; Zhang, J. Silencing Prx1 and/or Prx5 sensitizes human esophageal cancer cells to ionizing radiation and increases apoptosis via intracellular ROS accumulation. *Acta Pharmacol. Sin.* **2011**, *32*, 528–536. [CrossRef]
10. Zhang, B.; Wang, Y.; Su, Y. Peroxiredoxins, a novel target in cancer radiotherapy. *Cancer Lett.* **2009**, *286*, 154–160. [CrossRef]
11. Holley, A.K.; Miao, L.; St Clair, D.K.; St Clair, W.H. Redox-Modulated Phenomena and Radiation Therapy: The Central Role of Superoxide Dismutases. *Antioxid. Redox Signal.* **2014**, *20*, 1567–1589. [CrossRef]
12. Armogida, M.; Nistico, R.; Mercuri, N.B. Therapeutic potential of targeting hydrogen peroxide metabolism in the treatment of brain ischaemia. *Br. J. Pharmacol.* **2012**, *166*, 1211–1224. [CrossRef]
13. Beyer, W.F.; Fridovich, I. Catalases-with and without heme. *Basic Life Sci.* **1988**, *49*, 651–661.
14. Nicholls, P.; Fita, I.; Loewen, P.C. Enzymology and structure of catalases. *Adv. Inorg. Chem.* **2001**, *51*, 51–106.
15. Ko, T.P.; Day, J.; Malkin, A.J.; McPherson, A. Structure of orthorhombic crystals of beef liver catalase. *Acta Crystallogr.* **1999**, *55*, 1383–1394. [CrossRef]
16. Ivancich, A.; Jouve, H.M.; Sartor, B.; Gaillard, J. EPR investigation of compound I in *Proteus mirabilis* and bovin liver catalases: Formation of porphyrin and tyrosyl radical intermediates. *Biochemistry* **1997**, *36*, 9356–9364. [CrossRef]
17. Kono, Y.; Fridovich, I. Isolation and characterization of the pseudocatalase of *Lactobacillus plantarum*. *J. Biol. Chem.* **1983**, *258*, 6015–6019.
18. Barynin, V.V.; Whittaker, M.M.; Antonyuk, S.V.; Lamzin, V.S.; Harrison, P.M.; Artymiuk, P.J.; Whittaker, J.W. Crystal Structure of Manganese Catalase from Lactobacillus plantarum. *Structure* **2001**, *9*, 725–738. [CrossRef]
19. Antonyuk, S.V.; Melik-Adman, V.R.; Popov, A.N.; Lamzin, V.S.; Hempstead, P.D.; Harrison, P.M.; Artymiuk, P.J.; Barynin, V.V. Three-dimensional structure of the enzyme dimanganese catalase from *Thermus thermophilus* at 1 Å resolution. *Crystallogr. Rep.* **2000**, *45*, 105–113. [CrossRef]
20. Barynin, V.V.; Grebenko, A.I. T-catalase is nonheme catalase of the extremely thermophilic bacterium *Thermus thermophilus* HB8. *Dokl. Akad. Nauk. USSR* **1986**, *286*, 461–464.
21. Allgood, G.S.; Perry, J.J. Characterization of a manganese-containing catalase from the obligate thermophile *Thermoleophilum album*. *J. Bacteriol.* **1986**, *168*, 563–567. [CrossRef]

22. Amo, T.; Atomi, H.; Imanaka, T. Unique Presence of a Manganase Catalase in a Hyperthermophilic Archaeon, *Pyrobaculum calidifontis* VA1. *J. Bacteriol.* **2002**, *184*, 3305–3312. [CrossRef]
23. Gao, J.; Martell, A.E.; Reibenspies, J.H. Novel dicopper(II) catalase-like model complexes: Synthesis, crystal structure, properties and kinetic studies. *Inorg. Chim. Acta* **2003**, *346*, 32–42. [CrossRef]
24. Boelrijk, A.E.M.; Dismukes, G.C. Mechanism of Hydrogen Peroxide Dismutation by a Dimanganese Catalase Mimic: Dominant Role of an Intramolecular Base on Substrate Binding Affinity and Rate Acceleration. *Inorg. Chem.* **2000**, *39*, 3020. [CrossRef]
25. Paschke, J.; Kirsch, M.; Korth, H.G.; Groot, H.; Sustmann, R. Catalase-Like Activity of a Non-Heme Dibenzotetraaza[14]annulene–Fe(III) Complex under Physiological Conditions. *J. Am. Chem. Soc.* **2001**, *123*, 11099–11100. [CrossRef]
26. Okuno, T.; Ito, S.; Ohba, S.; Nishida, Y. µ-Oxo bridged diiron(III) complexes and hydrogen peroxide: Oxygenation and catalase-like activities. *J. Chem. Soc. Dalton. Trans.* **1997**, *24*, 3547–3551. [CrossRef]
27. Mauerer, B.; Crane, J.; Schuler, J.; Wieghardt, K.; Nuber, B. A Hemerythrin Model Complex with Catalase Activity. *Angew. Chem. Int. Ed. Engl.* **1993**, *32*, 289–291. [CrossRef]
28. Ménage, S.; Vincent, J.M.; Lambeaux, C.; Fontecave, M. µ-Oxo-bridged diiron(III) complexes and H_2O_2: Monooxygenase and catalase-like activities. *J. Chem. Soc. Dalton Trans.* **1994**, *21*, 2081–2084. [CrossRef]
29. Sigel, H.; Wiss, K.; Fischer, B.E.; Prijs, B. Metal ions and hydrogen peroxide. Catalase-like activity of copper(2+) ion in aqueous solution and its promotion by the coordination of 2,2′-bipyridyl. *Inorg. Chem.* **1979**, *18*, 1354–1358. [CrossRef]
30. Kaizer, J.; Csonka, R.; Speier, G.; Giorgi, M.; Réglier, M. Synthesis, structure and catalase-like activity of new dicopper(II) complexes with phenylglyoxylate and benzoate ligands. *J. Mol. Catal. A Chem.* **2005**, *236*, 12–17. [CrossRef]
31. Kaizer, J.; Csay, T.; Speier, G.; Réglier, M.; Giorgi, M. Synthesis, structure and catalase-like activity of Cu(N-baa)(2)(phen) (phen=1, 10-phenanthroline, N-baaH = N-benzoylanthranilic acid). *Inorg. Chem. Commun.* **2006**, *9*, 1037–1039. [CrossRef]
32. Pap, J.S.; Horvath, B.; Speier, G.; Kaizer, J. Synthesis and catalase-like activity of dimanganese complexes with phthalazine-based ligands. *Transit. Met. Chem.* **2011**, *36*, 603–609. [CrossRef]
33. Pap, J.S.; Kripli, B.; Bors, I.; Bogáth, D.; Giorgi, M.; Kaizer, J.; Speier, G. Transition metal complexes bearing flexible N-3 or N3O donor ligands: Reactivity toward superoxide radical anion and hydrogen peroxide. *J. Inorg. Biochem.* **2012**, *117*, 60–70. [CrossRef]
34. Kaizer, J.; Csay, T.; Kovari, P.; Speier, G.; Parkanyi, L. Catalase mimics of a manganese(II) complex: The effect of axial ligands and pH. *J. Mol. Catal. A Chem.* **2008**, *280*, 203–209. [CrossRef]
35. Kaizer, J.; Kripli, B.; Speier, G.; Parkanyi, L. Synthesis, structure, and catalase-like activity of a novel manganese(II) complex: Dichloro[1,3-bis(2 ′-benzimidazolylimino) isoindoline] manganese(II). *Polyhedron* **2009**, *28*, 933–936. [CrossRef]
36. Horn, A., Jr.; Parrilha, G.I.; Melo, K.V.; Fernandes, C.; Horner, M.; Visentin, I.C.; Santos, J.A.S.; Santos, M.S.; Eleutherio, E.C.A.; Pereira, M.D. An iron-based cytosolic catalase and superoxide dismutase mimic complex. *Inorg. Chem.* **2010**, *49*, 1274–1276. [CrossRef]
37. Carvalho, N.M.F.; Horn, A., Jr.; Faria, R.B.; Bortoluzzi, A.J.; Drago, V.; Antunes, O.A.C. Synthesis, characterization, X-ray molecular structure and catalase-like activity of a non-heme iron complex: Dichloro[N-propanoate-N,N-bis-(2-pyridylmethyl)amine] iron(III). *Inorg. Chim. Acta* **2006**, *359*, 4250–4258. [CrossRef]
38. Dickman, M.H.; Pope, M.T. Peroxo and Superoxo Complexes of Chromium, Molybdenum, and Tungsten. *Chem. Rev.* **1994**, *94*, 569–584. [CrossRef]
39. Wu, A.J.; Penner-Hahn, J.E.; Pecoraro, V.L. Structural, Spectroscopic, and Reactivity Models for the Manganese Catalases. *Chem. Rev.* **2004**, *104*, 903–938. [CrossRef]
40. Gilbert, J.; Roecker, L.; Meyer, T.J. Hydrogen Atom Transfer in the Oxidation of Hydrogen Peroxide by [(bpy)$_2$(py)Ru^{IV}=O]$^{2+}$ and by [(bpy)$_2$(py)Ru^{III}-OH]$^{2+}$. *Inorg. Chem.* **1987**, *26*, 1126. [CrossRef]
41. Crans, D.C.; Smee, J.J.; Gaidamauskas, E.; Yang, L. The Chemistry and Biochemistry of Vanadium and the Biological Activities Exerted by Vanadium Compounds. *Chem. Rev.* **2004**, *104*, 849–902. [CrossRef]

42. Pires, B.M.; Silva, D.M.; Visentin, L.C.; Drago, V.; Carvalho, N.M.F.; Faria, R.B.; Antunes, O.A.C. Synthesis, characterization and catalase-like activity of the tetranuclear iron(III) complex involving a (μ-oxo)(μ-hydroxo)bis(μ-alkoxo)tetra(μ-carboxylato)tetrairon core. *Inorg. Chim. Acta* **2013**, *407*, 69–81. [CrossRef]
43. Lakk-Bogáth, D.; Csonka, R.; Speier, G.; Reglier, M.; Simaan, A.J.; Naubron, J.V.; Giorgi, M.; Lazar, K.; Kaizer, J. Formation, Characterization, and Reactivity of Nonheme Iron(IV)-Oxo Complex Derived from the Chiral Pentadentate Ligand asN4Py. *Inorg. Chem.* **2016**, *55*, 10090. [CrossRef]
44. Turcas, R.; Lakk-Bogáth, D.; Speier, G.; Kaizer, J. Steric Control and Mechanism of Benzaldehyde Oxidation by Polypyridyl Oxoiron(IV) Complexes: Aromatic versus Benzylic Hydroxylation of Aromatic Aldehydes. *Dalton. Trans.* **2018**, *47*, 3248. [CrossRef]
45. Lakk-Bogáth, D.; Kripli, B.; Meena, B.I.; Speier, G.; Kaizer, J. Catalytic and stoichiometric oxidation of N,N-dimethylanilines mediated by nonheme oxoiron(IV) complex with tetrapyridyl ligand. *Polyhedron* **2019**, *169*, 169–175. [CrossRef]
46. Lakk-Bogáth, D.; Kripli, B.; Meena, B.I.; Speier, G.; Kaizer, J. Catalytic and stoichiometric C-H oxidation of benzylalcohols and hydrocarbons mediated by nonheme oxoiron(IV) complex with chiral tetrapyridyl ligand. *Inorg. Chem. Commun.* **2019**, *104*, 165–170. [CrossRef]
47. Kubota, R.; Imamura, S.; Shimizu, T.; Asayama, S.; Kawakami, H. Synthesis of Water-Soluble Dinuclear Mn-Porphyrin with Multiple Antioxidative Activities. *ACS Med. Chem. Lett.* **2014**, *5*, 639–643. [CrossRef]
48. Jakopitsch, C.; Vlasits, J.; Wiseman, B.; Loewen, P.C.; Obinger, C. Redox Intermediates in the Catalase Cycle of Catalase-Peroxidases from *Synechocystis* PCC 6803, *Burkholderia pseudomallei*, and *Mycobacterium tuberculosis*. *Biochemistry* **2007**, *46*, 1183–1193. [CrossRef]
49. Gelasco, A.; Bensiek, S.; Pecoraro, V.L. The [Mn$_2$(2-OHsalpn)$_2$]$^{2-,1-,0}$ System: An Efficient Functional Model for the Reactivity and Inactivation of the Manganese Catalases. *Inorg. Chem.* **1998**, *37*, 3301–3309. [CrossRef]
50. Larson, E.J.; Pecoraro, V.L. [Mn(III)(2-OHsalpn)]$_2$ is an efficient functional model for the manganese catalases. *J. Am. Chem. Soc.* **1993**, *115*, 7928–7929.
51. Kaizer, J.; Baráth, G.; Speier, G.; Réglier, M.; Giorgi, M. Synthesis, structure and catalase mimics of novel homoleptic manganese(II) complexes of 1,3-bis(2'-pyridylimino) isoindoline, Mn(4R-ind)$_2$ (R= H., Me). *Inorg. Chem. Commun.* **2007**, *10*, 292–294. [CrossRef]
52. Signorella, S.; Palopoli, C.; Ledesma, G. Rationally designed mimics of antioxidant manganoenzymes: Role of structural features in the quest for catalysts with catalaseand superoxide dismutase activity. *Coord. Chem. Rev.* **2018**, *365*, 75–102. [CrossRef]
53. Chance, B.; Greenstein, D.S.; Roughton, F.J. The mechanism of catalase action. I. Steady-state analysis. *Arch. Biochem. Biophys.* **1952**, *37*, 301–321. [CrossRef]
54. Braymer, J.J.; O'Neill, K.P.; Rohde, J.U.; Lim, M.H. The Reaction of a High-Valent Nonheme Oxoiron(IV) Intermediate with Hydrogen Peroxide. *Angew. Chem. Int. Ed.* **2012**, *51*, 1–6. [CrossRef]
55. Chen, J.; Draksharapu, A.; Angelone, D.; Unjaroen, D.; Padamati, S.K.; Hage, R.; Swart, M.; Duboc, C.; Browne, W.R. H$_2$O$_2$ Oxidation by Fe^{III}-OOH Intermediates and Its Effect on Catalytic Efficiency. *ACS Catal.* **2018**, *8*, 9665–9674. [CrossRef]
56. Sastri, C.V.; Seo, M.S.; Park, M.J.; Kim, K.M.; Nam, W. Formation, stability, and reactivity of a mononuclear nonheme oxoiron(IV) complex in aqueous solution. *Chem. Commun.* **2005**, 1405–1407. [CrossRef]
57. Gilbert, J.A.; Gersten, S.W.; Meyer, T.J. H-D Kinetic Isotope Effects of 16 and 22 in the Oxidation of H$_2$O$_2$. *J. Am. Chem. Soc.* **1982**, *104*, 6872–6873. [CrossRef]

© 2019 by the authors. Licensee MDPI, Basel, Switzerland. This article is an open access article distributed under the terms and conditions of the Creative Commons Attribution (CC BY) license (http://creativecommons.org/licenses/by/4.0/).

Review

The Role of Phosphatidylethanolamine Adducts in Modification of the Activity of Membrane Proteins under Oxidative Stress

Elena E. Pohl * and **Olga Jovanovic ***

Institute of Physiology, Pathophysiology and Biophysics, Department of Biomedical Sciences, University of Veterinary Medicine, Vienna A-1210, Austria
* Correspondence: elena.pohl@vetmeduni.ac.at (E.E.P.); olga.jovanovic@vetmeduni.ac.at (O.J.)

Received: 9 November 2019; Accepted: 10 December 2019; Published: 12 December 2019

Abstract: Reactive oxygen species (ROS) and their derivatives, reactive aldehydes (RAs), have been implicated in the pathogenesis of many diseases, including metabolic, cardiovascular, and inflammatory disease. Understanding how RAs can modify the function of membrane proteins is critical for the design of therapeutic approaches in the above-mentioned pathologies. Over the last few decades, direct interactions of RA with proteins have been extensively studied. Yet, few studies have been performed on the modifications of membrane lipids arising from the interaction of RAs with the lipid amino group that leads to the formation of adducts. It is even less well understood how various multiple adducts affect the properties of the lipid membrane and those of embedded membrane proteins. In this short review, we discuss a crucial role of phosphatidylethanolamine (PE) and PE-derived adducts as mediators of RA effects on membrane proteins. We propose potential PE-mediated mechanisms that explain the modulation of membrane properties and the functions of membrane transporters, channels, receptors, and enzymes. We aim to highlight this new area of research and to encourage a more nuanced investigation of the complex nature of the new lipid-mediated mechanism in the modification of membrane protein function under oxidative stress.

Keywords: reactive aldehydes; hydroxynonenal; oxononenal; free fatty acids; mitochondrial uncoupling protein; lipid bilayer membranes

---

1. Reactive Oxygen Species and Their Derivatives, Reactive Aldehydes

With regard to reactive oxygen species (ROS), these include unstable short-lived molecules that contain oxygen ($O_2{}^{\cdot}$, H_2O_2, and OH^-) and are highly reactive in cells. The term "ROS" is often substituted by the phrase "free radicals"; however, strictly speaking, only $O_2{}^{\cdot}$ and $OH^{\cdot}$ are considered free radicals. Besides oxygen, reactive species (RS) may contain nitrogen, carbon, sulfur, and halogens.

Over 90% of ROS in eukaryotic cells are produced by mitochondria [1]. Mitochondria produce a substantial amount of superoxide anion at Complex I and via autoxidation of a ubisemiquinone anion radical at Complex III, where $O_2{}^{\cdot}$ is released on both sides of the membrane (for recent reviews see [2–5]). Superoxide anion may give rise to a variety of reactive carbonyl species (RCS, reactive aldehydes (RAs)), which are three to nine carbons in length (Figure 1). Of these, α,β-unsaturated aldehydes (4-hydroxy-trans-2-nonenal (HNE) and acrolein), di-aldehydes (malondialdehyde (MDA) and glyoxal), and keto-aldehydes (4-oxo-trans-2-nonenal (ONE) and isoketals (IsoK)) are the most toxic RAs (Figure 2).

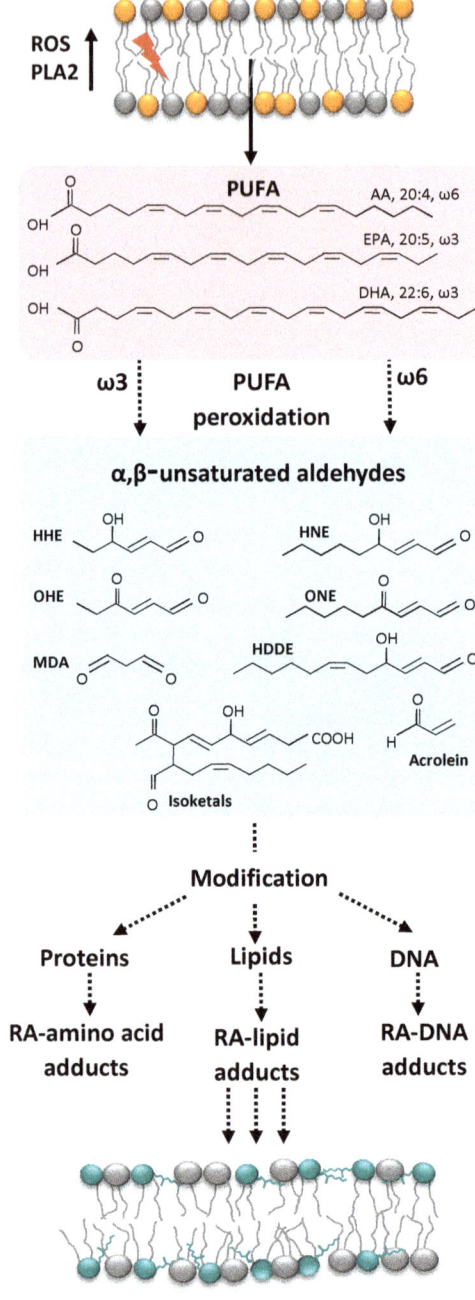

Figure 1. Elevated levels of ROS induce PUFA peroxidation in the cell membrane and the formation of different α, β-unsaturated RAs which can react with proteins, lipids, and DNA. Abbreviations: AA, arachidonic acid; DHA, docosahexaenoic acid; EPA, eicosapentaenoic acid; HDDE, 4-hydroxydodeca-(2E,6Z)-dienal; PLA2, phospholipase A2; PUFA, polyunsaturated fatty acid.

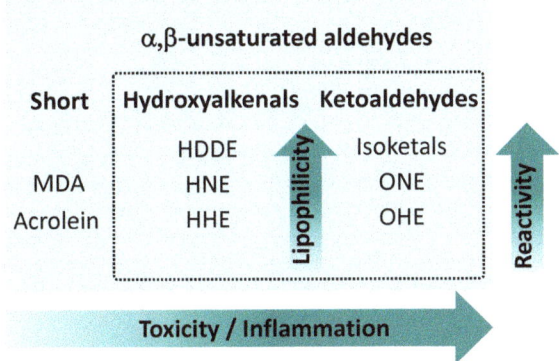

Figure 2. Main types of α,β-unsaturated aldehydes and relations between their lipophilicity, reactivity, and toxicity. Abbreviations as used in Figure 1.

RAs can be more destructive than ROS because (1) they have a much longer half-life (i.e., minutes to hours instead of microseconds to nanoseconds for most free radicals), and (2) the non-charged structure of aldehydes allows them to migrate long distances from the production site through hydrophobic membranes [6–8]. Reactivity between aldehydes differs both qualitatively and quantitatively. The most extensively studied aldehyde, HNE is generated by the oxidation of lipids containing polyunsaturated omega-6 acyl groups, such as arachidonic or linoleic groups, and of the corresponding fatty acids (FAs). This aldehyde elicits deleterious effects primarily by oxidizing intracellular components, including DNA, lipids, and proteins [9]. Another aldehyde, 4-oxo-2-hexenal (OHE), is generated by the oxidation of ω-3 polyunsaturated FAs, which are commonly found in dietary fish oil and soybean oil [10]. OHE is thought to possess DNA-damaging potential similar to that of HNE. In contrast, ONE was shown to be a more reactive protein modifier and cross-linking agent than HNE [11]. It has been proposed that the greater neurotoxicity of ONE may indicate different reactivity characteristics than those of HNE. In comparison, MDA possesses a stronger mutagenic and carcinogenic potential in mammalian cells than HNE [12].

The estimation of exact RA concentrations in cells is difficult since concentrations of several μM, the maximum reported for the cell cytoplasm, are averaged values. However, it is plausible that RA concentration levels may be much higher locally for a short period of time [13]. Whereas cellular concentrations of HNE under physiological conditions can reach 0.3 mM, HNE accumulates at concentrations up to 5 mM in cellular membranes under conditions of oxidative stress [14]. Also, Esterbauer et al. suggested that HNE and other aldehydes are unlikely to reach physiological concentrations of approximately 100 μM [7]. However, much higher levels may be transiently achieved in the vicinity of peroxidizing membranes because of the high lipophilicity of RA. As an example, within the lipid bilayer of isolated peroxidizing microsomes, the concentration HNE is approximately 4.5 mM [13,15]. The concentration of an RA in the membrane is strongly dependent on its lipophilicity and, for example, is higher for ONE and HNE than for 4-hydroxy-2-hexenal (HHE) [16].

2. ROS Detoxification Systems

The cell's own systems of defense from oxidative stress include different ROS detoxification systems, such as superoxide dismutase, catalase, and glutathione peroxidase (for detailed reviews see [12,17]). Several lipophilic or water-soluble, membrane-permeable molecules (tocopherol, carotenoids, anthocyanins, polyphenols, and uric and ascorbic acid [18]) can work as endogenous or nutritional antioxidants (for review see [19]).

The production of ROS in mitochondria is also sensitive to the proton motive force and may be decreased by artificial uncouplers (e.g., dinitrophenol, carbonylcyanid-*m*-chlorphenylhydrazon

(CCCP), carbonilcyanide *p*-triflouromethoxyphenylhydrazone (FCCP)) or proteins transporting protons from the intermembrane space to the matrix, a process termed "mild uncoupling" (Skulachev, 1998).

Several members of a mitochondrial membrane protein superfamily, known as the solute carrier family (SLC25), such as adenine nucleotide transporter (ANT) [20,21], dicarboxylate carrier [22], phosphate, aspartate glutamate carrier [23], and members of the uncoupling protein subfamily (UCP1–4) were proposed to be involved in proton transport.

It is well accepted that the best investigated member of the UCP family—UCP1—uncouples substrate oxidation from mitochondrial ATP synthesis by transporting protons from the intermembrane space to the matrix [24–29]. The protonophoric ability of other uncoupling proteins is a longstanding issue in controversial debates [30]. The proton-transporting capacity of UCP2 and UCP3 was shown to be comparable with the capacity of UCP1 in lipid bilayer membranes reconstituted with recombinant proteins (2–14/s, for a review see [31]). Several studies in multiscale and biomimetic systems indicate that the function of UCP2/UCP3 is associated with the transport of different metabolic substrates [32–34], possibly alongside proton transport [26,35–37]. A recent comparison of wild type and UCP2$^{-/-}$ or UCP3$^{-/-}$ knockout mice [21] implied that both proteins are not involved in H$^+$ transport. Unfortunately, the tissue(s) chosen in this study were previously shown to have no presence or very low abundance of UCP2/UCP3 under physiological conditions [38,39].

Nègre-Salvayre et al. was the first to suggest UCP2 involvement in ROS regulation, combining the largely accepted view that a high mitochondrial membrane potential leads to increased production of ROS (particularly superoxide anion) with the idea that UCP2 diminishes the membrane potential by transporting proteins from intermembrane space to the mitochondrial matrix [40]. This suggestion was extended by Brand and colleagues who proposed that UCP activity was regulated by superoxide [41] or RAs (HNE) [42]. A coherent theory for a feedback mechanism was formulated whereby increased ROS production would activate uncoupling to decrease ROS formation; oxidative damage would thereby be reduced (for review see [43]). In contrast, Cannon's group, using brown-fat mitochondria from UCP1$^{-/-}$ and superoxide dismutase (SOD$^{-/-}$) knock-out mice, showed that HNE could neither (re)activate purine nucleotide-inhibited UCP1, nor induce the additional activation of innately active UCP1 [44]. No support for the tenet that superoxide directly or indirectly regulates UCP1–UCP3 activity could be found [45,46]. In subsequent studies, a high membrane potential was suggested as a requirement for the activation of UCP-mediated uncoupling by HNE [47]. Experiments performed in a well-defined system of bilayer membranes reconstituted with recombinant UCPs [48] revealed that HNE did not directly activate either UCP1 or UCP2. However, HNE strongly potentiated the membrane proton conductance increase mediated by different long-chain FAs in UCP-containing and UCP-free membranes. These results contributed to an understanding of the controversial results observed by different groups in multiscale systems and allowed to investigate the molecular mechanism of HNE–UCP interactions (see Section 4.4).

3. Mechanisms of RA Action

A consensus exists that RAs play a dual role in cellular processes: They are known to modify proteins [49–51], DNA [52,53], and lipids [16,54], but are also involved in important signaling pathways [50]. However, the molecular mechanisms of their action are still far from being well understood. It is becoming increasingly clear that in both cases, RAs form adducts with the nucleophilic groups of proteins, DNA and lipids [55]. Meanwhile, it is evident that RAs not only target a large variety of molecules, but also that the mechanisms of such interactions differ. The latter are still poorly understood and seem to depend on the chemical structure of RAs, interaction molecules, lipid environment, and the distance between the RA source and the target molecule [56].

An increasing number of studies have shown that RAs bind to proteins and impair their function by modification of amino acid residues and protein crosslinking to an extent that depends on their reactivity (for review see [14,57]). Both ONE and HNE covalently bind to cysteine, histidine, and lysine, while ONE also binds to arginines. The reactivity of HNE (k_{HNE}) toward amino acids was reported to be:

cysteine (1.21 M^{-1} s^{-1}) >> histidine (2.14 × 10^{-3} M^{-1} s^{-1}) > lysine (1.33 × 10^{-3} M^{-1} s^{-1}) [58]. This means that the reactivity of thiol group-containing cysteines is higher than that of amino group-containing lysines and histidines. It further implies that HNE and ONE would primarily attack the thiol groups of proteins disabling disulfide bridges formation and affecting thereby protein function(s) [59]. ONE was reported to be more reactive than HNE, given k_{ONE}/k_{HNE}: cysteine 153 >> histidine 10.3 > lysine 5.61. In contrast, extremely reactive IsoK rapidly reacts with positively charged lysine residues rather than with thiols [60]. The modification and crosslinking of amino acid residues, proteins, and peptides are perceived as major toxic effects of RA. The selective and reversible oxidation of key residues in proteins that presumably leads to conformational changes and the alteration of protein activity and function [61,62] is a physiological mechanism well-studied in cytosolic (hydrophilic) proteins.

An important role for the membrane lipid, phosphatidylethanolamine (PE), in the functions of cell membranes and transmembrane proteins (discussed in Section 4.2) implies that its modification affects different processes in the cell. Whereas rate constants for the reaction of HNE with amino acids have been intensively studied, no binding kinetic data exists concerning the reaction rate of HNE with amino groups of lipids (PE, phosphatidylserine (PS), and sphingomyelin (SML)). However, the interaction with amino groups of lipids seems to be highly relevant for membrane proteins, especially in membranes with a low protein/lipid ratio (e.g., oligodendrocytes). Previously it was shown that the function of membrane uncoupling proteins is altered only in the presence of PE [16], although western blot analysis revealed that HNE was also bound to cysteines [48].

4. Phosphatidylethanolamine as a Crucial Target for Reactive Aldehydes

4.1. Phosphatidylethanolamine and Its Physiological Functions

Phosphatidylethanolamine (PE) is the second most abundant phospholipid, after phosphatidylcholine (PC), in the membranes of all mammalian cells. On average, it makes up 25% of the total phospholipid mass [63]. The highest amount of PE, up 45% of all phospholipids, is found in the membranes of tissues of the neuronal system, such as white matter of the brain, nerves, and spinal cord [64]. PE is a non-bilayer lipid, more abundant in the inner than in the outer leaflet of the cell membranes [65]. Due to its conical shape, PE modulates membrane curvature and lateral pressure [66,67] and thus supports membrane fusion [68–71] and function of several membrane proteins [67,71,72].

PE is a fundamental component of biological membranes, needed for many cellular functions. Besides being a precursor for other lipids [73], PE is involved in a multitude of physiological functions. Among others, PE (1) supports chaperoning membrane proteins to their folded state [74], (2) activates oxidative phosphorylation [75,76], (3) is involved in apoptotic [77] and ferroptotic [78] cell death pathways, (4) mediates the modification of prions from a nontoxic to toxic conformation [79], and (5) is crucial for the synthesis of glycosylphosphatidylinositol-anchored proteins essential for cell viability [80]. The importance of PE for cell function is evident in the existence of four separate PE biosynthetic pathways [81], one of which takes place in the inner mitochondrial membrane [76].

Disorders in PE metabolism have been implicated in many chronic diseases, such as Alzheimer's disease, Parkinson's disease, and nonalcoholic liver disease [82], as well as metabolic disorders such as atherosclerosis, insulin resistance, and obesity [63]. Increased levels of PE have been described in cancer cells leading to PE being regarded as a target in the development of anticancer therapies [83].

4.2. PE Adducts

To date, only a few groups have studied the ability of reactive aldehydes (RAs) to modify the headgroup of amino-phospholipids (amino-PLs), predominantly PE, and characterized formed adducts. Reactions of α,β-unsaturated aldehydes (HHE, HNE, and ONE) with amino-PLs lead to the formation of different adducts, such as Michael adducts (MAs) and Schiff base adducts (SBs) (Figure 3). Depending on experimental conditions (for example, incubation) more complex types of adducts, such as double-MAs, double-SBs, and pyrrole adducts, may be formed [84–89]. Initially, studies of modifications of amino-PLs by α,β-unsaturated aldehydes and hydroxyalkenals, and subsequently ketoaldehydes (IsoLGs) and short- and long-chain aldehydes, were performed. Recently, it was demonstrated that primary amines can react with glucose [90] and amide linkages [91], and modify the head group of amino-PLs in a similar manner (s. below).

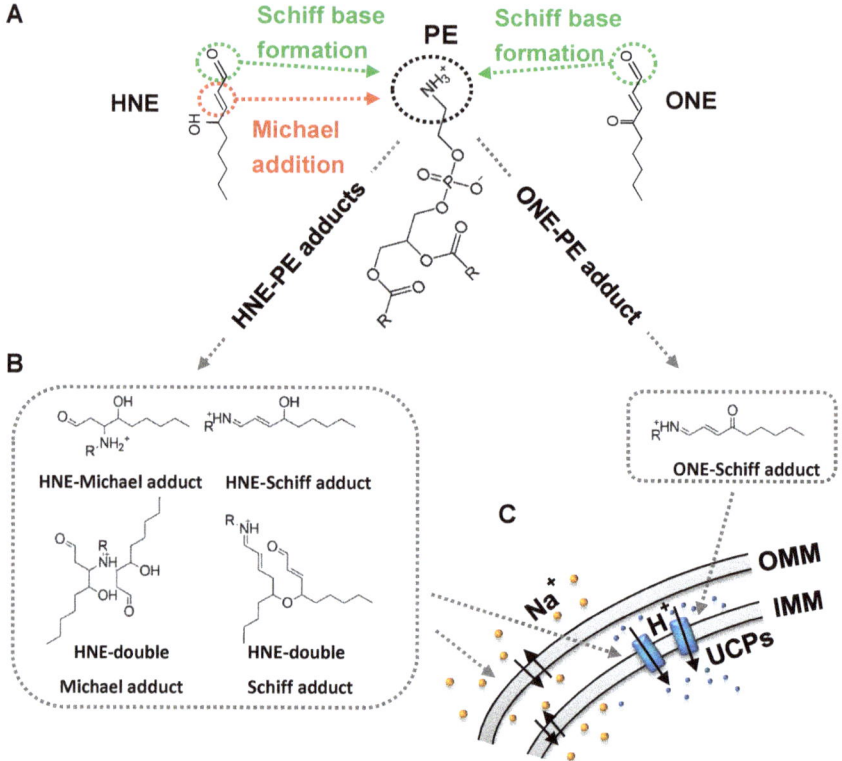

Figure 3. Mechanisms of modification of the phosphatidylethanolamine (PE) by RAs and their impact on function of mitochondrial and other cell membranes. HNE and ONE covalently bind to the PE primary amine group (**A**), forming different RA-PE adducts (Michael or Schiff base type) (**B**). HNE-PE adducts in lipid bilayer membrane decrease free energy barrier ΔG and increase permeability for cations. Localization of ONE-PE and HNE-PE adducts in the cellular membrane change bending properties and lateral pressure profile of the membrane that results in increased proton translocation mediated by uncoupling protein (UCPs) (**C**). Abbreviations: OMM, outer mitochondrial membrane; IMM, inner mitochondrial membrane.

The first described covalent modifications of the lipid head group by an RA were the reactions of the HNE with PE and PS. As the main products, PE-MAs and PS-MAs were identified. Imine and pyrrole adducts were detected only in PE, but to a much lesser extent [84]. Other authors reported covalent modification of the PE headgroup by long chain saturated alkenals (e.g., pentadecanal, heptadecenal), and α-hydroxyalkenals (α-hydroxyhexadecanal, α-hydroxyoctadecanal), produced during oxidation of plasmalogen, resulting in PE-SB adducts, also known as N-alkyl-PEs [92–94].

Evaluation of the role of the acyl chain length of α,β unsaturated hydroxyalkenals (4-HHE, 4-HNE and 4-HDDE) on their ability to covalently modify different types of PEs revealed (1) a correlation between their reactivity to PE with their increasing hydrophobicity in the order HHDE > HNE >

HHE (Figure 2), and (2) their selectivity towards different PEs: all three hydroxyalkenals favored modification of plasmalogen-PE over other PEs [85].

Comparison of the covalent modification of PE due to a reaction with HNE and the more toxic ketoaldehyde, ONE, which has the same length but with a carbonyl instead of a hydroxyl group on C4, revealed that this difference led to the formation of different adducts (Figure 3, B). While HNE formed four types of PE-adducts (MAs, SBs, double-MAs, and double-SBs), only one ONE-PE adduct (SB type) was detected [16]. These results highlight how the toxicity of ONE can be explained by the formation of only one type of ONE-PE adduct compared to a joint effect of several types of HNE-PE adducts.

With increased lipophilicity and complexity, the reactivity of ketoaldehydes also increases (Figure 2). Reactive γ-ketoaldehydes (γKA), also named IsoK or isolevuglandins, are peroxidation products of arachidonic acid formed via the isoprostane pathway [60]. In vitro experiments have revealed that IsoK covalently modified the PE headgroup at a higher rate than the well-characterized HNE, forming IsoK-PE SBs and IsoK-PE pyrrole adducts [95]. Further, it was shown that the reaction rate of IsoK with PE is significantly higher than those with protein or DNA [96]. Recent evidence indicates that IsoK-PE adducts act as inflammatory mediators in the cell [97,98]. However, the molecular mechanisms are largely unknown. One can speculate that many of the effects previously attributed to protein modification due to IsoKs [99,100] could, in fact, be due to their ability to modify PE. It has been shown that even more simple products of arachidonate oxidation, such as diverse carboxyacyls, chemically react with the PE amine group, making a family of so-called amide-linked PEs and forming predominantly SBs and the pyrrole type of adducts. These adducts have been implicated in the inflammation of endothelial cells [91,94].

In addition, the "smallest" aldehydes, such as MDA and acrolein, are capable of modifying the PE headgroup through the initial formation of SB, ending in more complex products. For example, the predominant product of the incubation MDA and PE was identified as dihydropyridine-PE (DHP-PE) [91], while two acroleins in reactions with PE formed a compound termed (3-formyl-4-hydroxy)-piperidine-PE (FDP-PE) [101]. The involvement of MDA-PE and acrolein-PE adducts in the inflammatory process is moderate compared to HNE- or IsoK-PE adducts [91]. Due to their higher hydrophilicity, such RAs are thought to easily leave the lipid membrane and react with cytosolic proteins to a greater extent than with membrane lipids.

It should be mentioned that the PE amine headgroup can be covalently modified by glucose and several fungal products. Glucose has an aldehyde group that can react with the primary amine of aminophospholipids via Maillard reactions to form Amadori adducts, (e.g., glucose phosphatidylethanolamine (gPE) and glucose phosphatidylserine (gPS)) [90,102,103]. Under conditions of oxidative stress, Amadori adducts undergo degradation to form advanced glycation products (goxPE) [104]. Several authors suggest gPEs and goxPEs are involved in diabetic and related neurodegenerative diseases [105,106]. Ophiobolin A (OPA) is a compound found in a fungus that is toxic to plant cells. OPA reacts with the primary amine of PE and forms pyrrole-containing OPA-PE adducts that show cytotoxic effects on some cancer cells [83].

4.3. Modification of Membrane Properties by PE and PE Adducts

Because of its conical shape, PE is essential for the processes of membrane budding, fission, and fusion [64,107,108]. In the lipid bilayer membrane, PE affects a lateral pressure profile and modulates membrane curvature; together with other lipids, PE provides an environment for optimal conformation and function of transmembrane proteins [67].

Although different studies have reported that RAs form adducts with aminophospholipids, their impact on the lipid bilayer membrane has been poorly studied. Recently, Jovanovic et al. [16] showed that PE adducts, formed after incubation of PC/PE lipid membranes with α,β-unsaturated aldehydes, significantly increased negative membrane ζ-potential in the order HHE < HNE << ONE. Notably, RAs did not influence the ζ-potential in PC lipid bilayers. An evaluation of the influence of PE adducts on the order parameter, S, revealed that only modification of PE by ONE leads to an increase in the bilayers' fluidity, caused by alterations in the spatial arrangement of aliphatic chains in the lipid membrane. In contrast, HHE-PE adducts and HNE-PE adducts did not change the order parameter. Covalent modification of PE by HNE increased sodium permeability across the phospholipid bilayer by four orders of magnitude, while in the absence of PE the effect was not observed [89]. A calculation of the Nernst potential in the presence of a proton gradient revealed that the HNE-mediated total membrane conductance, G_m, in PE-containing lipid membranes was mainly caused by cations (2/3 G_m) rather than by protons (1/3 G_m). Surprisingly, this effect was not recorded for the more toxic ONE. Molecular dynamic (MD) simulations of a lipid bilayer membrane composed of PC and either HNE or ONE adducts suggested that all types of HNE-PE adducts (especially the double adducts, D-SB-HNE and D-MA-HNE) became anchored deeper in the hydrophobic region, while ONE-PE adducts were entirely localized in the headgroup region of the lipid membrane. Study of the structural properties of the lipid bilayer revealed that double HNE adducts caused an increase in the area per lipid and a decrease in hydrophobic core thickness. The decrease of lipid dipoles per unit surface area diminishes membrane dipole potential [109]. As a consequence, the free energy barrier (ΔG) for cations should decrease [110]. In turn, the permeability for sodium ions was increased [89].

Guo et al. [94] measured T_H-shifts in the bilayer to a hexagonal phase transition temperature of DiPoPE (T_H) incubated with 4-oxo-pentanal (OPA), γKA and glutaryl (glt) using differential scanning calorimetry to prove whether the formation of such adducts altered the curvature of the lipid membrane. γKA-PE and OPA-PE adducts showed similar behavior, and increased negative membrane curvature, while an N-glt-PE adduct showed the opposite effect to promote a positive membrane curvature. The observed change in the membrane curvature was consistent with the suggested localization of PE adducts. While γKA-PE and OPA-PE adducts are supposedly localized in the hydrophobic region, the N-glt-PE adduct is localized in the headgroup region. These results confirmed the assumption that modification of the PE headgroup alters lipid bilayer membrane properties, such as membrane curvature and, consequently, lateral pressure profile.

Modification of the lipid shape due to the formation of ONE-PE adducts was reported to affect membrane curvature, which then altered the elastic properties of the lipid bilayer and the lateral pressure profile [111]. In general, due to the difference in RA-PE adduct distribution between the two leaflets, asymmetric changes of spontaneous membrane curvature may arise. In turn, the stability of membrane domains (lipid rafts) may be altered [112].

4.4. Modification of Membrane Transporter Function in the Presence of PE Adducts

Both proteins and lipids were identified as targets of RA activity. However, while modification of cytosolic proteins by the activity of RAs has been extensively studied, and is directly related to protein dysfunction [49], an investigation of the impact of RAs on transmembrane proteins has only been made to a very modest extent, mostly due to their hydrophobicity. The observed alteration of membrane protein function was interpreted in the same way, assuming a direct connection between the modification of certain amino acid residues and protein function.

The investigation of RA–protein interactions using artificial lipid membranes reconstituted with several transporters (mitochondrial transporter UCP1, potassium transporter valinomycin and uncoupler CCCP) surprisingly demonstrated that RA altered the transport activity of these molecules only when in the presence of the PE [16]. The greatest effect was elicited by ONE, which was more toxic in cell experiments, followed by HNE. HHE showed a much weaker effect, probably due to its lower hydrophobicity. Experiments further revealed that covalent modification of the PE headgroup causes changes in the electrical and mechanical properties of the lipid membrane, such as the boundary potential, order parameter and membrane bending rigidity [111,113]. According to MD simulations, the position of the RA-PE adduct in the lipid bilayer was responsible for the observed changes. Similar to the dipole potential modifier, phloretin [114], ONE- and HNE-PE adducts altered the boundary potential in the lipid membrane, and decreased the positive membrane energy barrier [115]. This resulted in increased valinomycin-mediated potassium transport and decreased proton transport mediated by CCCP. The same molecular mechanism could not explain the RA-PE action on UCP1 since UCP1-mediated proton conductance was not affected in the presence of dipole potential modifiers. However, MD simulations suggested that formation of RA-PE adducts change the form of PE from an originally negative intrinsic curvature to the opposite one, which was confirmed by observed changes of membrane bending rigidity [116]. Notably, a decrease of membrane bending rigidity in the presence of RA-PE adducts was in the order ONE > HNE > HHE, consistent with their effect on UCP1. The change in membrane curvature by the formation of RA-PEs, and related changes in the membrane lateral pressure profile, were made responsible for the modification of UCP1 transport function. In contrast to that previously shown for cytosolic proteins, modification of UCP1 and UCP2 by RAs [16,48] cannot activate the proteins directly, but rather by a described novel PE-mediated mechanism.

Interestingly, glycated and glycoxidized PEs also alter the transport function of valinomycin in the same direction as RA-PEs, but to a more moderate extent [90]. PE glycation led to a similar change in negative membrane surface potential, as shown for RA-PEs. It indicates that glycated and glycoxidized PEs may decrease the positive energy membrane barrier in the lipid bilayer for cations comparable to the membrane dipole modifier, phloretin [114] and RA-PEs. An observed change in melting temperature upon PE glycation indicates a change in the membrane curvature [90], which allows us to hypothesize that such glucose-derived modifications on PEs could also affect the function of transmembrane proteins.

Unfortunately, we didn't find examples demonstrating the impact of PE adducts on other transmembrane proteins than UCPs [16,48,90]. Although few groups demonstrated the modifications of the PE in cells and tissues [91,94,95,99], the possible impact of PE-adducts on the function of the membrane proteins was neither studied nor discussed. Guided by the hypothesis that the formation of RA-PE adducts could be involved in the pathogenesis of diseases associated with oxidative damage, authors focused on their involvement in the signaling and inflammatory processes. Considering the emerging role of lipid shape and membrane curvature on the function of transmembrane proteins, as well as their distribution in the membrane [117,118], the modification of PEs and their impact on the other membrane proteins have to be seriously studied.

5. Conclusions and Outlook

The question of how the functions of membrane transporters are modified under oxidative stress is a central issue that remains unexplained at the molecular level. ROS and their derivatives, RAs, are implicated in many diseases and, furthermore, in many signaling pathways. Current research has mainly focused on the aldehyde-mediated modification of protein amino acids, such as cysteine, lysine, and histidine, which supposedly affects the conformation of proteins. Recently, it was hypothesized that this mechanism may be more relevant for cytosolic proteins [16]. In contrast, the mechanism by which RAs modify the functions of membrane proteins may fundamentally differ from that of hydrophilic proteins. We have recently demonstrated that the initial binding of aldehydes to PE is a crucial step for alteration of the RA-mediated activity of different membrane transporters, such as mitochondrial inner membrane UCP1, the ionophore valinomycin, and the protonophore CCCP [16]. A lipid-mediated mechanism seems to be even more relevant for membranes abundant in PE, PS, or SML (e.g., mitochondria, bacteria) and for membranes with a low protein/lipid ratio, such as the membranes of oligodendrocytes.

Whereas one can argue that short- and middle-chain aldehydes have approximately equal affinity in binding to the primary amine of an amino phospholipid or amino acid, very reactive long chain IsoK (products of the AA, 20:4, ω-6) bind to PE at a significantly higher rate than to proteins or DNA due to their strong hydrophobicity, as already experimentally shown [96]. Moreover, IsoKs have been detected in brain and nervous tissue as a consequence of oxidative damage. The cells of these tissues meet two conditions for preferential IsoK-PE adduct formation: they are rich in AA acyl chains, which are a source for isoketal formation, and in PE [97,99,119]. This makes an investigation of the mechanisms by which PE adducts influence the function of membrane proteins very important.

Author Contributions: Conceptualization, E.E.P. and O.J.; writing, E.E.P. and O.J.; funding acquisition, E.E.P.

Funding: This research was funded by the Austrian Research Fund (FWF), grant number P25123.

Acknowledgments: We accept Open Access Funding by the Austrian Science Fund (FWF)

Conflicts of Interest: The authors declare no conflicts of interest.

References

1. Skulachev, V.P. Mitochondria-Targeted Antioxidants as Promising Drugs for Treatment of Age-Related Brain Diseases. *J. Alzheimers Dis.* **2012**, *28*, 283–289. [CrossRef] [PubMed]
2. Wong, H.S.; Dighe, P.A.; Mezera, V.; Monternier, P.A.; Brand, M.D. Production of Superoxide and Hydrogen Peroxide from Specific Mitochondrial Sites under Different Bioenergetic Conditions. *J. Biol. Chem.* **2017**, *292*, 16804–16809. [CrossRef] [PubMed]
3. Brand, M.D. Mitochondrial Generation of Superoxide and Hydrogen Peroxide as the Source of Mitochondrial Redox Signaling. *Free Radic. Biol. Med.* **2016**, *100*, 14–31. [CrossRef] [PubMed]
4. Murphy, M. Mitochondrial Superoxide Production in Health and Disease. *Biochim. Biophys. Acta* **2016**, *1857*, E6–E7. [CrossRef]
5. Vinogradov, A.D.; Grivennikova, V.G. Oxidation of Nadh And Ros Production By Respiratory Complex, I. *Biochim. Biophys. Acta* **2016**, *1857*, 863–871. [CrossRef]
6. Naudi, A.; Jove, M.; Ayala, V.; Cabre, R.; Portero-Otin, M.; Pamplona, R. Non-Enzymatic Modification of Aminophospholipids by Carbonyl-Amine Reactions. *Int. J. Mol. Sci.* **2013**, *14*, 3285–3313. [CrossRef]
7. Esterbauer, H.; Schaur, R.J.; Zollner, H. Chemistry and Biochemistry Of 4-Hydroxynonenal, Malonaldehyde and Related Aldehydes. *Free Radic. Biol. Med.* **1991**, *11*, 81–128. [CrossRef]
8. Roede, J.R.; Jones, D.P. Reactive Species and Mitochondrial Dysfunction: Mechanistic Significance of 4-Hydroxynonenal. *Environ Mol Mutagen* **2010**, *51*, 380–390. [CrossRef]
9. Zarkovic, N. 4-Hydroxynonenal as a Bioactive Marker of Pathophysiological Processes. *Mol. Asp. Med.* **2003**, *24*, 281–291. [CrossRef]
10. Kasai, H.; Maekawa, M.; Kawai, K.; Hachisuka, K.; Takahashi, Y.; Nakamura, H.; Sawa, R.; Matsui, S.; Matsuda, T. 4-Oxo-2-Hexenal, A Mutagen Formed By Omega-3 Fat Peroxidation, Causes Dna Adduct Formation In Mouse Organs. *Ind Health* **2005**, *43*, 699–701. [CrossRef]
11. Lin, D.; Lee, H.G.; Liu, Q.; Perry, G.; Smith, M.A.; Sayre, L.M. 4-Oxo-2-Nonenal Is Both More Neurotoxic And More Protein Reactive Than 4-Hydroxy-2-Nonenal. *Chem. Res. Toxicol.* **2005**, *18*, 1219–1231. [CrossRef]
12. Valko, M.; Leibfritz, D.; Moncol, J.; Cronin, M.T.; Mazur, M.; Telser, J. Free Radicals And Antioxidants In Normal Physiological Functions And Human Disease. *Int. J. Biochem. Cell Biol.* **2007**, *39*, 44–84. [CrossRef]
13. Benedetti, A.; Comporti, M.; Fulceri, R.; Esterbauer, H. Cytotoxic Aldehydes Originating from the Peroxidation OF Liver Microsomal Lipids. Identification Of 4,5-Dihydroxydecenal. *Biochim. Biophys. Acta* **1984**, *792*, 172–181. [CrossRef]
14. Uchida, K. 4-Hydroxy-2-Nonenal: A Product and Mediator of Oxidative Stress. *Prog. Lipid Res.* **2003**, *42*, 318–343. [CrossRef]
15. Koster, J.F.; Slee, R.G.; Montfoort, A.; Lang, J.; Esterbauer, H. Comparison of The Inactivation of Microsomal Glucose-6-Phosphatase By In Situ Lipid Peroxidation-Derived 4-Hydroxynonenal And Exogenous 4-Hydroxynonenal. *Free Radic. Res. Commun.* **1986**, *1*, 273–287. [CrossRef]
16. Jovanovic, O.; Pashkovskaya, A.A.; Annibal, A.; Vazdar, M.; Burchardt, N.; Sansone, A.; Gille, L.; Fedorova, M.; Ferreri, C.; Pohl, E.E. The Molecular Mechanism behind Reactive Aldehyde Action on Transmembrane Translocations of Proton and Potassium Ions. *Free Radic. Biol. Med.* **2015**, *89*, 1067–1076. [CrossRef]
17. Droge, W. Free Radicals in the Physiological Control of Cell Function. *Physiol. Rev.* **2002**, *82*, 47–95. [CrossRef]
18. Hannesschlaeger, C.; Pohl, P. Membrane Permeabilities of Ascorbic Acid and Ascorbate. *Biomolecules* **2018**, *8*, 73. [CrossRef]
19. Lushchak, V.I. Free Radicals, Reactive Oxygen Species, Oxidative Stress and Its Classification. *Chem. Biol. Interact.* **2014**, *224*, 164–175. [CrossRef]
20. Tikhonova, I.M.; Andreyev, A.Y.; Kaulen, A.D.; Komrakov, A.Y.; Skulachev, V.P. Ion Permeability Induced In Artificial Membranes By The Atp/Adp Antiporter. *FEBS Lett.* **1994**, *337*, 231–234. [CrossRef]

21. Bertholet, A.M.; Chouchani, E.T.; Kazak, L.; Angelin, A.; Fedorenko, A.; Long, J.Z.; Vidoni, S.; Garrity, R.; Cho, J.; Terada, N.; et al. H(+) Transport Is An Integral Function of The Mitochondrial Adp/Atp Carrier. *Nature* **2019**, *571*, 515–520. [CrossRef]
22. Wieckowski, M.R.; Wojtczak, L. Involvement of the Dicarboxylate Carrier in the Protonophoric Action of Long-Chain Fatty Acids In Mitochondria. *Biochem. Biophys. Res. Commun.* **1997**, *232*, 414–417.
23. Samartsev, V.N.; Marchik, E.I.; Shamagulova, L.V. Free Fatty Acids As Inducers And Regulators Of Uncoupling Of Oxidative Phosphorylation In Liver Mitochondria With Participation Of Adp/Atp- And Aspartate/Glutamate-Antiporter. *Biochemistry* **2011**, *76*, 217–224. [CrossRef]
24. Shabalina, I.G.; Jacobsson, A.; Cannon, B.; Nedergaard, J. Native Ucp1 Displays Simple Competitive Kinetics between the Regulators Purine Nucleotides and Fatty Acids. *J. Biol. Chem.* **2004**, *279*, 38236–38248. [CrossRef]
25. Urbankova, E.; Voltchenko, A.; Pohl, P.; Jezek, P.; Pohl, E.E. Transport Kinetics Of Uncoupling Proteins. Analysis of Ucp1 Reconstituted In Planar Lipid Bilayers. *J. Biol. Chem.* **2003**, *278*, 32497–32500. [CrossRef]
26. Macher, G.; Koehler, M.; Rupprecht, A.; Kreiter, J.; Hinterdorfer, P.; Pohl, E.E. Inhibition Of Mitochondrial Ucp1 And Ucp3 By Purine Nucleotides And Phosphate. *Biochim. Biophys. Acta Biomembr.* **2018**, *1860*, 664–672. [CrossRef]
27. Jezek, P.; Orosz, D.E.; Garlid, K.D. Reconstitution of the Uncoupling Protein of Brown Adipose Tissue Mitochondria. Demonstration of Gdp-Sensitive Halide Anion Uniport. *J. Biol. Chem.* **1990**, *265*, 19296–19302.
28. Bouillaud, F.; Ricquier, D.; Gulik-Krzywicki, T.; Gary-Bobo, C.M. The Possible Proton Translocating Activity of the Mitochondrial Uncoupling Protein of Brown Adipose Tissue. Reconstitution Studies in Liposomes. *FEBS Lett.* **1983**, *164*, 272–276. [CrossRef]
29. Bienengraeber, M.; Echtay, K.S.; Klingenberg, M. H^+ Transport by Uncoupling Protein (Ucp-1) Is Dependent On A Histidine Pair, Absent In Ucp-2 and Ucp-3. *Biochemistry* **1998**, *37*, 3–8. [CrossRef]
30. Cannon, B.; Shabalina, I.G.; Kramarova, T.V.; Petrovic, N.; Nedergaard, J. Uncoupling Proteins: A Role In Protection Against Reactive Oxygen Species-Or Not? *Biochim. Biophys. Acta* **2006**, *1757*, 449–458. [CrossRef]
31. Pohl, E.E.; Rupprecht, A.; Macher, G.; Hilse, K.E. Important Trends In Ucp3 Investigation. *Front. Physiol.* **2019**, *10*, 470. [CrossRef]
32. Vozza, A.; Parisi, G.; De Leonardis, F.; Lasorsa, F.M.; Castegna, A.; Amorese, D.; Marmo, R.; Calcagnile, V.M.; Palmieri, L.; Ricquier, D.; et al. Ucp2 Transports C4 Metabolites Out Of Mitochondria, Regulating Glucose And Glutamine Oxidation. *Proc. Natl. Acad. Sci. USA* **2014**, *111*, 960–965. [CrossRef]
33. Rupprecht, A.; Moldzio, R.; Modl, B.; Pohl, E.E. Glutamine Regulates Mitochondrial Uncoupling Protein 2 To Promote Glutaminolysis In Neuroblastoma Cells. *Biochim. Biophys. Acta Bioenerg.* **2019**, *1860*, 391–401. [CrossRef]
34. Hilse, K.E.; Rupprecht, A.; Egerbacher, M.; Bardakji, S.; Zimmermann, L.; Wulczyn, A.; Pohl, E.E. The Expression Of Uncoupling Protein 3 Coincides With The Fatty Acid Oxidation Type Of Metabolism In Adult Murine Heart. *Front. Physiol.* **2018**, *9*, 747. [CrossRef]
35. Beck, V.; Jaburek, M.; Demina, T.; Rupprecht, A.; Porter, R.K.; Jezek, P.; Pohl, E.E. Polyunsaturated Fatty Acids Activate Human Uncoupling Proteins 1 And 2 In Planar Lipid Bilayers. *FASEB J.* **2007**, *21*, 1137–1144. [CrossRef]
36. Zackova, M.; Jezek, P. Reconstitution of Novel Mitochondrial Uncoupling Proteins Ucp2 and Ucp3. *Biosci. Rep.* **2002**, *22*, 33–46. [CrossRef]
37. Zackova, M.; Skobisova, E.; Urbankova, E.; Jezek, P. Activating Omega-6 Polyunsaturated Fatty Acids And Inhibitory Purine Nucleotides Are High Affinity Ligands For Novel Mitochondrial Uncoupling Proteins Ucp2 And Ucp3. *J. Biol. Chem.* **2003**, *278*, 20761–20769. [CrossRef]
38. Rupprecht, A.; Sittner, D.; Smorodchenko, A.; Hilse, K.E.; Goyn, J.; Moldzio, R.; Seiler, A.E.; Brauer, A.U.; Pohl, E.E. Uncoupling Protein 2 And 4 Expression Pattern During Stem Cell Differentiation Provides New Insight Into Their Putative Function. *PLoS ONE* **2014**, *9*, E88474. [CrossRef]
39. Hilse, K.E.; Kalinovich, A.V.; Rupprecht, A.; Smorodchenko, A.; Zeitz, U.; Staniek, K.; Erben, R.G.; Pohl, E.E. The Expression Of Ucp3 Directly Correlates To Ucp1 Abundance In Brown Adipose Tissue. *Biochim. Biophys. Acta* **2016**, *1857*, 72–78. [CrossRef]
40. Negre-Salvayre, A.; Hirtz, C.; Carrera, G.; Cazenave, R.; Troly, M.; Salvayre, R.; Penicaud, L.; Casteilla, L. A Role for Uncoupling Protein-2 as a Regulator of Mitochondrial Hydrogen Peroxide Generation. *Faseb. J.* **1997**, *11*, 809–815. [CrossRef]

41. Echtay, K.S.; Roussel, D.; St-Pierre, J.; Jekabsons, M.B.; Cadenas, S.; Stuart, J.A.; Harper, J.A.; Roebuck, S.J.; Morrison, A.; Pickering, S.; et al. Superoxide Activates Mitochondrial Uncoupling Proteins. *Nature* **2002**, *415*, 96–99. [CrossRef]
42. Echtay, K.S.; Esteves, T.C.; Pakay, J.L.; Jekabsons, M.B.; Lambert, A.J.; Portero-Otin, M.; Pamplona, R.; Vidal-Puig, A.J.; Wang, S.; Roebuck, S.J.; et al. A Signalling Role for 4-Hydroxy-2-Nonenal in Regulation of Mitochondrial Uncoupling. *EMBO J.* **2003**, *22*, 4103–4110. [CrossRef]
43. Krauss, S.; Zhang, C.Y.; Lowell, B.B. The Mitochondrial Uncoupling-Protein Homologues. *Nat. Rev. Mol. Cell Biol.* **2005**, *6*, 248–261. [CrossRef]
44. Shabalina, I.G.; Petrovic, N.; Kramarova, T.V.; Hoeks, J.; Cannon, B.; Nedergaard, J. Ucp1 And Defense Against Oxidative Stress. 4-Hydroxy-2-Nonenal Effects On Brown Fat Mitochondria Are Uncoupling Protein 1-Independent. *J. Biol. Chem.* **2006**, *281*, 13882–13893. [CrossRef]
45. Couplan, E.; Mar Gonzalez-Barroso, M.; Alves-Guerra, M.C.; Ricquier, D.; Goubern, M.; Bouillaud, F. No Evidence for A Basal, Retinoic, or Superoxide-Induced Uncoupling Activity of the Uncoupling Protein 2 Present In Spleen or Lung Mitochondria. *J. Biol.Chem.* **2002**, *277*, 26268–26275. [CrossRef]
46. Lombardi, A.; Grasso, P.; Moreno, M.; De Lange, P.; Silvestri, E.; Lanni, A.; Goglia, F. Interrelated Influence Of Superoxides And Free Fatty Acids Over Mitochondrial Uncoupling In Skeletal Muscle. *Biochim. Biophys. Acta* **2008**, *1777*, 826–833. [CrossRef]
47. Parker, N.; Vidal-Puig, A.; Brand, M.D. Stimulation of Mitochondrial Proton Conductance by Hydroxynonenal Requires A High Membrane Potential. *Biosci. Rep.* **2008**, *28*, 83–88. [CrossRef]
48. Malingriaux, E.A.; Rupprecht, A.; Gille, L.; Jovanovic, O.; Jezek, P.; Jaburek, M.; Pohl, E.E. Fatty Acids Are Key In 4-Hydroxy-2-Nonenal-Mediated Activation Of Uncoupling Proteins 1 And 2. *PLoS ONE* **2013**, *8*, E77786. [CrossRef]
49. Zarkovic, N.; Cipak, A.; Jaganjac, M.; Borovic, S.; Zarkovic, K. Pathophysiological Relevance of Aldehydic Protein Modifications. *J. Proteomics.* **2013**, *92*, 239–247. [CrossRef]
50. Fritz, K.S.; Petersen, D.R. An Overview of the Chemistry and Biology Of Reactive Aldehydes. *Free Radic. Biol. Med.* **2013**, *59*, 85–91. [CrossRef]
51. Castro, J.P.; Jung, T.; Grune, T.; Siems, W. 4-Hydroxynonenal (Hne) Modified Proteins in Metabolic Diseases. *Free Radic. Biol. Med.* **2017**, *111*, 309–315. [CrossRef] [PubMed]
52. Voulgaridou, G.P.; Anestopoulos, I.; Franco, R.; Panayiotidis, M.I.; Pappa, A. Dna Damage Induced By Endogenous Aldehydes: Current State Of Knowledge. *Mutat. Res.* **2011**, *711*, 13–27. [CrossRef] [PubMed]
53. Gentile, F.; Arcaro, A.; Pizzimenti, S.; Daga, M.; Cetrangolo, G.P.; Dianzani, C.; Lepore, A.; Graf, M.; Ames, P.R.J.; Barrera, G. Dna Damage By Lipid Peroxidation Products: Implications In Cancer, Inflammation And Autoimmunity. *AIMS Genet.* **2017**, *4*, 103–137. [CrossRef] [PubMed]
54. Guo, L.; Davies, S.S. Bioactive Aldehyde-Modified Phosphatidylethanolamines. *Biochimie* **2013**, *95*, 74–78. [CrossRef]
55. Sousa, B.C.; Pitt, A.R.; Spickett, C.M. Chemistry and Analysis of Hne and Other Prominent Carbonyl-Containing Lipid Oxidation Compounds. *Free Radic. Biol. Med.* **2017**, *111*, 294–308. [CrossRef] [PubMed]
56. Sokolov, V.S.; Batishchev, O.V.; Akimov, S.A.; Galimzyanov, T.R.; Konstantinova, A.N.; Malingriaux, E.; Gorbunova, Y.G.; Knyazev, D.G.; Pohl, P. Residence Time Of Singlet Oxygen In Membranes. *Sci. Rep.* **2018**, *8*, 14000. [CrossRef]
57. Gueraud, F.; Atalay, M.; Bresgen, N.; Cipak, A.; Eckl, P.M.; Huc, L.; Jouanin, I.; Siems, W.; Uchida, K. Chemistry And Biochemistry Of Lipid Peroxidation Products. *Free Radic. Res.* **2010**, *44*, 1098–1124. [CrossRef]
58. Doorn, J.A.; Petersen, D.R. Covalent Modification of Amino Acid Nucleophiles by the Lipid Peroxidation Products 4-Hydroxy-2-Nonenal and 4-Oxo-2-Nonenal. *Chem. Res. Toxicol.* **2002**, *15*, 1445–1450. [CrossRef]
59. Carbone, D.L.; Doorn, J.A.; Kiebler, Z.; Petersen, D.R. Cysteine Modification by Lipid Peroxidation Products Inhibits Protein Disulfide Isomerase. *Chem. Res. Toxicol.* **2005**, *18*, 1324–1331. [CrossRef]
60. Brame, C.J.; Salomon, R.G.; Morrow, J.D.; Roberts, L.J., 2nd. Identification Of Extremely Reactive Gamma-Ketoaldehydes (Isolevuglandins) As Products Of The Isoprostane Pathway And Characterization of Their Lysyl Protein Adducts. *J. Biol. Chem.* **1999**, *274*, 13139–13146. [CrossRef]
61. Bleier, L.; Wittig, I.; Heide, H.; Steger, M.; Brandt, U.; Drose, S. Generator-Specific Targets Of Mitochondrial Reactive Oxygen Species. *Free Radic. Biol. Med.* **2015**, *78*, 1–10. [CrossRef] [PubMed]

62. Bogeski, I.; Kappl, R.; Kummerow, C.; Gulaboski, R.; Hoth, M.; Niemeyer, B.A. Redox Regulation of Calcium Ion Channels: Chemical And Physiological Aspects. *Cell Calcium* **2011**, *50*, 407–423. [CrossRef] [PubMed]
63. Van Der Veen, J.N.; Kennelly, J.P.; Wan, S.; Vance, J.E.; Vance, D.E.; Jacobs, R.L. The Critical Role of Phosphatidylcholine and Phosphatidylethanolamine Metabolism in Health and Disease. *Biochim. Biophys. Acta Biomembr.* **2017**, *1859*, 1558–1572. [CrossRef] [PubMed]
64. Vance, J.E.; Tasseva, G. Formation and Function of Phosphatidylserine and Phosphatidylethanolamine in Mammalian Cells. *Biochim. Biophys. Acta* **2013**, *1831*, 543–554. [CrossRef] [PubMed]
65. Devaux, P.F.; Morris, R. Transmembrane Asymmetry and Lateral Domains in Biological Membranes. *Traffic* **2004**, *5*, 241–246. [CrossRef]
66. Epand, R.M.; Fuller, N.; Rand, R.P. Role of the Position of Unsaturation on The Phase Behavior And Intrinsic Curvature Of Phosphatidylethanolamines. *Biophys. J.* **1996**, *71*, 1806–1810. [CrossRef]
67. Van Den Brink-Van Der Laan, E.; Killian, J.A.; De Kruijff, B. Nonbilayer Lipids Affect Peripheral and Integral Membrane Proteins via Changes in the Lateral Pressure Profile. *Biochim. Biophys. Acta* **2004**, *1666*, 275–288. [CrossRef]
68. Cullis, P.R.; De Kruijff, B. Lipid Polymorphism and the Functional Roles of Lipids in Biological Membranes. *Biochim. Biophys. Acta* **1979**, *559*, 399–420. [CrossRef]
69. Verkleij, A.J.; Leunissen-Bijvelt, J.; De Kruijff, B.; Hope, M.; Cullis, P.R. Non-Bilayer Structures in Membrane Fusion. *Ciba Found. Symp.* **1984**, *103*, 45–59.
70. Siegel, D.P.; Epand, R.M. The Mechanism of Lamellar-To-Inverted Hexagonal Phase Transitions in Phosphatidylethanolamine: Implications for Membrane Fusion Mechanisms. *Biophys. J.* **1997**, *73*, 3089–3111. [CrossRef]
71. Martens, C.; Shekhar, M.; Borysik, A.J.; Lau, A.M.; Reading, E.; Tajkhorshid, E.; Booth, P.J.; Politis, A. Direct Protein-Lipid Interactions Shape The Conformational Landscape Of Secondary Transporters. *Nat. Commun.* **2018**, *9*, 4151. [CrossRef] [PubMed]
72. Van Den Brink-Van Der Laan, E.; Chupin, V.; Killian, J.A.; De Kruijff, B. Stability of Kcsa Tetramer Depends On Membrane Lateral Pressure. *Biochemistry* **2004**, *43*, 4240–4250. [CrossRef] [PubMed]
73. Vance, J.E. Phospholipid Synthesis and Transport in Mammalian Cells. *Traffic* **2015**, *16*, 1–18. [CrossRef] [PubMed]
74. Bogdanov, M.; Dowhan, W. Lipid-Assisted Protein Folding. *J. Biol. Chem.* **1999**, *274*, 36827–36830. [CrossRef]
75. Shinzawa-Itoh, K.; Aoyama, H.; Muramoto, K.; Terada, H.; Kurauchi, T.; Tadehara, Y.; Yamasaki, A.; Sugimura, T.; Kurono, S.; Tsujimoto, K.; et al. Structures And Physiological Roles Of 13 Integral Lipids Of Bovine Heart Cytochrome C Oxidase. *EMBO J.* **2007**, *26*, 1713–1725. [CrossRef]
76. Calzada, E.; Avery, E.; Sam, P.N.; Modak, A.; Wang, C.; Mccaffery, J.M.; Han, X.; Alder, N.N.; Claypool, S.M. Phosphatidylethanolamine Made In The Inner Mitochondrial Membrane Is Essential For Yeast Cytochrome Bc1 Complex Function. *Nat. Commun.* **2019**, *10*, 1432. [CrossRef]
77. Ichimura, Y.; Kirisako, T.; Takao, T.; Satomi, Y.; Shimonishi, Y.; Ishihara, N.; Mizushima, N.; Tanida, I.; Kominami, E.; Ohsumi, M.; et al. A Ubiquitin-Like System Mediates Protein Lipidation. *Nature* **2000**, *408*, 488–492. [CrossRef]
78. Kagan, V.E.; Mao, G.; Qu, F.; Angeli, J.P.; Doll, S.; Croix, C.S.; Dar, H.H.; Liu, B.; Tyurin, V.A.; Ritov, V.B.; et al. Oxidized Arachidonic And Adrenic Pes Navigate Cells To Ferroptosis. *Nat. Chem. Biol.* **2017**, *13*, 81–90. [CrossRef]
79. Deleault, N.R.; Piro, J.R.; Walsh, D.J.; Wang, F.; Ma, J.; Geoghegan, J.C.; Supattapone, S. Isolation Of Phosphatidylethanolamine As A Solitary Cofactor For Prion Formation In The Absence Of Nucleic Acids. *Proc. Natl. Acad. Sci. USA* **2012**, *109*, 8546–8551. [CrossRef]
80. Menon, A.K.; Eppinger, M.; Mayor, S.; Schwarz, R.T. Phosphatidylethanolamine Is the Donor of the Terminal Phosphoethanolamine Group in Trypanosome Glycosylphosphatidylinositols. *EMBO J.* **1993**, *12*, 1907–1914. [CrossRef]
81. Vance, J.E. Historical Perspective: Phosphatidylserine and Phosphatidylethanolamine from the 1800s to the Present. *J. Lipid Res.* **2018**, *59*, 923–944. [CrossRef] [PubMed]
82. Calzada, E.; Onguka, O.; Claypool, S.M. Phosphatidylethanolamine Metabolism in Health And Disease. *Int. Rev. Cell Mol. Biol.* **2016**, *321*, 29–88. [PubMed]

83. Chidley, C.; Trauger, S.A.; Birsoy, K.; O'shea, E.K. The Anticancer Natural Product Ophiobolin A Induces Cytotoxicity By Covalent Modification of Phosphatidylethanolamine. *Elife* **2016**, *5*, E14601. [CrossRef] [PubMed]
84. Guichardant, M.; Taibi-Tronche, P.; Fay, L.B.; Lagarde, M. Covalent Modifications of Aminophospholipids by 4-Hydroxynonenal. *Free Radic. Biol. Med.* **1998**, *25*, 1049–1056. [CrossRef]
85. Bacot, S.; Bernoud-Hubac, N.; Baddas, N.; Chantegrel, B.; Deshayes, C.; Doutheau, A.; Lagarde, M.; Guichardant, M. Covalent Binding Of Hydroxy-Alkenals 4-Hdde, 4-Hhe, And 4-Hne To Ethanolamine Phospholipid Subclasses. *J. Lipid Res.* **2003**, *44*, 917–926. [CrossRef]
86. Guo, L.; Amarnath, V.; Davies, S.S. A Liquid Chromatography-Tandem Mass Spectrometry Method for Measurement of N-Modified Phosphatidylethanolamines. *Anal. Biochem.* **2010**, *405*, 236–245. [CrossRef]
87. Annibal, A.; Schubert, K.; Wagner, U.; Hoffmann, R.; Schiller, J.; Fedorova, M. New Covalent Modifications Of Phosphatidylethanolamine By Alkanals: Mass Spectrometry Based Structural Characterization And Biological Effects. *J. Mass Spectrom.* **2014**, *49*, 557–569. [CrossRef]
88. Vazdar, K.; Vojta, D.; Margetic, D.; Vazdar, M. Reaction Mechanism of Covalent Modification of Phosphatidylethanolamine Lipids by Reactive Aldehydes 4-Hydroxy-2-Nonenal And 4-Oxo-2-Nonenal. *Chem. Res. Toxicol.* **2017**, *30*, 840–850. [CrossRef]
89. Jovanovic, O.; Skulj, S.; Pohl, E.E.; Vazdar, M. Covalent Modification of Phosphatidylethanolamine by 4-Hydroxy-2-Nonenal Increases Sodium Permeability across Phospholipid Bilayer Membranes. *Free Radic. Biol. Med.* **2019**, *143*, 433–440. [CrossRef]
90. Annibal, A.; Riemer, T.; Jovanovic, O.; Westphal, D.; Griesser, E.; Pohl, E.E.; Schiller, J.; Hoffmann, R.; Fedorova, M. Structural, Biological And Biophysical Properties of Glycated and Glycoxidized Phosphatidylethanolamines. *Free Radic. Biol. Med.* **2016**, *95*, 293–307. [CrossRef]
91. Guo, L.; Chen, Z.; Amarnath, V.; Davies, S.S. Identification Of Novel Bioactive Aldehyde-Modified Phosphatidylethanolamines Formed By Lipid Peroxidation. *Free Radic. Biol.Med.* **2012**, *53*, 1226–1238. [CrossRef] [PubMed]
92. Stadelmann-Ingrand, S.; Favreliere, S.; Fauconneau, B.; Mauco, G.; Tallineau, C. Plasmalogen Degradation by Oxidative Stress: Production and Disappearance of Specific Fatty Aldehydes and Fatty Alpha-Hydroxyaldehydes. *Free Radic. Biol. Med.* **2001**, *31*, 1263–1271. [CrossRef]
93. Stadelmann-Ingrand, S.; Pontcharraud, R.; Fauconneau, B. Evidence For The Reactivity Of Fatty Aldehydes Released From Oxidized Plasmalogens With Phosphatidylethanolamine To Form Schiff Base Adducts In Rat Brain Homogenates. *Chem. Phys. Lipids* **2004**, *131*, 93–105. [CrossRef] [PubMed]
94. Guo, L.; Chen, Z.; Cox, B.E.; Amarnath, V.; Epand, R.F.; Epand, R.M.; Davies, S.S. Phosphatidylethanolamines Modified By Gamma-Ketoaldehyde (Gammaka) Induce Endoplasmic Reticulum Stress And Endothelial Activation. *J. Biol. Chem.* **2011**, *286*, 18170–18180. [CrossRef]
95. Bernoud-Hubac, N.; Fay, L.B.; Armarnath, V.; Guichardant, M.; Bacot, S.; Davies, S.S.; Roberts II, L.J.; Lagarde, M. Covalent Binding Of Isoketals to Ethanolamine Phospholipids. *Free Radic. Biol. Med.* **2004**, *37*, 1604–1611. [CrossRef]
96. Sullivan, C.B.; Matafonova, E.; Roberts II, L.J.; Amarnath, V.; Davies, S.S. Isoketals Form Cytotoxic Phosphatidylethanolamine Adducts In Cells. *J. Lipid Res.* **2010**, *51*, 999–1009. [CrossRef]
97. Guo, L.; Chen, Z.; Amarnath, V.; Yancey, P.G.; Van Lenten, B.J.; Savage, J.R.; Fazio, S.; Linton, M.F.; Davies, S.S. Isolevuglandin-Type Lipid Aldehydes Induce The Inflammatory Response Of Macrophages By Modifying Phosphatidylethanolamines And Activating The Receptor For Advanced Glycation Endproducts. *Antioxid. Redox Signal.* **2015**, *22*, 1633–1645. [CrossRef]
98. Davies, S.D.; May-Zhang, L.S.; Boutaud, O.; Amarnath, V.; Kirabo, A.; Harrison, D.G. Isolevuglandins as Mediators of Disease and the Development of Dicarbonyl Scavengers as Pharmaceutical Interventions. *Pharmacol. Ther* **2019**, 107418. [CrossRef]
99. Salomon, R.G.; Bi, W. Isolevuglandin Adducts In Disease. *Antioxid. Redox Signal.* **2015**, *22*, 1703–1718. [CrossRef]
100. May-Zhang, L.S.; Yermalitsky, V.; Huang, J.; Pleasent, T.; Borja, M.S.; Oda, M.N.; Jerome, W.G.; Yancey, P.G.; Linton, M.F.; Davies, S.S. Modification By Isolevuglandins, Highly Reactive Gamma-Ketoaldehydes, Deleteriously Alters High-Density Lipoprotein Structure And Function. *J. Biol. Chem.* **2018**, *293*, 9176–9187. [CrossRef]
101. Zemski Berry, K.A.; Murphy, R.C. Characterization of Acrolein-Glycerophosphoethanolamine Lipid Adducts Using Electrospray Mass Spectrometry. *Chem. Res. Toxicol.* **2007**, *20*, 1342–1351. [CrossRef] [PubMed]

102. Ravandi, A.; Kuksis, A.; Marai, L.; Myher, J.J. Preparation and Characterization of Glucosylated Aminoglycerophospholipids. *Lipids* **1995**, *30*, 885–891. [CrossRef] [PubMed]
103. Fountain, W.C.; Requena, J.R.; Jenkins, A.J.; Lyons, T.J.; Smyth, B.; Baynes, J.W.; Thorpe, S.R. Quantification of N-(Glucitol)Ethanolamine and N-(Carboxymethyl)Serine: Two Products of Nonenzymatic Modification of Aminophospholipids Formed in Vivo. *Anal. Biochem.* **1999**, *272*, 48–55. [CrossRef] [PubMed]
104. Requena, J.R.; Ahmed, M.U.; Fountain, C.W.; Degenhardt, T.P.; Reddy, S.; Perez, C.; Lyons, T.J.; Jenkins, A.J.; Baynes, J.W.; Thorpe, S.R. Carboxymethylethanolamine, A Biomarker Of Phospholipid Modification During The Maillard Reaction In Vivo. *J. Biol. Chem.* **1997**, *272*, 17473–17479. [CrossRef]
105. Lapolla, A.; Fedele, D.; Traldi, P. Glyco-Oxidation in Diabetes and Related Diseases. *Clin. Chim. Acta* **2005**, *357*, 236–250. [CrossRef]
106. Simoes, C.; Silva, A.C.; Domingues, P.; Laranjeira, P.; Paiva, A.; Domingues, M.R. Modified Phosphatidylethanolamines Induce Different Levels Of Cytokine Expression In Monocytes And Dendritic Cells. *Chem. Phys. Lipids* **2013**, *175*, 57–64. [CrossRef]
107. Van Meer, G.; Voelker, D.R.; Feigenson, G.W. Membrane Lipids: Where they are and How They Behave. *Nat. Rev. Mol. Cell Biol.* **2008**, *9*, 112–124. [CrossRef]
108. Patel, D.; Witt, S.N. Ethanolamine and Phosphatidylethanolamine: Partners in Health and Disease. *Oxid. Med. Cell Longev.* **2017**, *2017*, 4829180. [CrossRef]
109. Peterson, U.; Mannock, D.A.; Lewis, R.N.; Pohl, P.; Mcelhaney, R.N.; Pohl, E.E. Origin of Membrane Dipole Potential: Contribution of The Phospholipid Fatty Acid Chains. *Chem. Phys. Lipids* **2002**, *117*, 19–27. [CrossRef]
110. Hannesschlaeger, C.; Horner, A.; Pohl, P. Intrinsic Membrane Permeability to Small Molecules. *Chem. Rev.* **2019**, *119*, 5922–5953. [CrossRef]
111. Chekashkina, K.; Jovanovic, O.; Kuzmin, P.; Pohl, E.; Pavel, B. The Changes of Physical Parameters of Lipid Membrane Caused By Lipid Peroxidation-Derived Aldehydes. *Biophys. J.* **2017**, *112*, 520a. [CrossRef]
112. Galimzyanov, T.R.; Kuzmin, P.I.; Pohl, P.; Akimov, S.A. Undulations Drive Domain Registration From The Two Membrane Leaflets. *Biophys. J.* **2017**, *112*, 339–345. [CrossRef] [PubMed]
113. Jovanovic, O.; Chekashkina, K.; Bashkirov, P.; Vazdar, M.; Pohl, E.E. Uncoupling Proteins Are Highly Sensitive to the Membrane Lipid Composition. *Eur. Biophys. J. Biophys. Lett.* **2017**, *46*, S287.
114. Pohl, P.; Rokitskaya, T.I.; Pohl, E.E.; Saparov, S.M. Permeation Of Phloretin Across Bilayer Lipid Membranes Monitored By Dipole Potential And Microelectrode Measurements. *Biochim. Biophys. Acta* **1997**, *1323*, 163–172. [CrossRef]
115. Pohl, E.E.; Krylov, A.V.; Block, M.; Pohl, P. Changes of the Membrane Potential Profile Induced By Verapamil and Propranolol. *Biochim. Biophys. Acta* **1998**, *1373*, 170–178. [CrossRef]
116. Jovanovic, O.; Chekashkina, K.; Bashkirov, P.; Skuljc, S.; Vazdar, M.; Pohl, E.E. Lipid Curvature Modulates Function of Mitochondrial Membrane Proteins. *Eur. Biophys. J. Biophys. Lett.* **2019**, *48*, S51.
117. Aimon, S.; Callan-Jones, A.; Berthaud, A.; Pinot, M.; Toombes, G.E.; Bassereau, P. Membrane Shape Modulates Transmembrane Protein Distribution. *Dev. Cell* **2014**, *28*, 212–218. [CrossRef]
118. Phillips, R.; Ursell, T.; Wiggins, P.; Sens, P. Emerging Roles for Lipids in Shaping Membrane-Protein Function. *Nature* **2009**, *459*, 379–385. [CrossRef]
119. Farooqui, A.A.; Horrocks, L.A.; Farooqui, T. Modulation of Inflammation in Brain: A Matter of Fat. *J. Neurochem.* **2007**, *101*, 577–599. [CrossRef]

© 2019 by the authors. Licensee MDPI, Basel, Switzerland. This article is an open access article distributed under the terms and conditions of the Creative Commons Attribution (CC BY) license (http://creativecommons.org/licenses/by/4.0/).

Review

The Combination of Whole Cell Lipidomics Analysis and Single Cell Confocal Imaging of Fluidity and Micropolarity Provides Insight into Stress-Induced Lipid Turnover in Subcellular Organelles of Pancreatic Beta Cells

Giuseppe Maulucci [1,2], Ofir Cohen [3], Bareket Daniel [3], Carla Ferreri [4] and Shlomo Sasson [3,*]

1. Fondazione Policlinico Universitario A. Gemelli IRCSS, 00136 Rome, Italy; Giuseppe.Maulucci@unicatt.it
2. Istituto di Fisica, Università Cattolica del Sacro Cuore, 00168 Rome, Italy
3. Institute for Drug Research, Faculty of Medicine, The Hebrew University, 911210 Jerusalem, Israel; ofirco55@gmail.com (O.C.); bareketk@gmail.com (B.D.)
4. ISOF, Consiglio Nazionale delle Ricerche, 40129 Bologna, Italy; carla.ferreri@isof.cnr.it
* Correspondence: shlomo.sasson@mail.huji.ac.il; Tel.: +972-2-6758798

Received: 5 September 2019; Accepted: 14 October 2019; Published: 17 October 2019

Abstract: Modern omics techniques reveal molecular structures and cellular networks of tissues and cells in unprecedented detail. Recent advances in single cell analysis have further revolutionized all disciplines in cellular and molecular biology. These methods have also been employed in current investigations on the structure and function of insulin secreting beta cells under normal and pathological conditions that lead to an impaired glucose tolerance and type 2 diabetes. Proteomic and transcriptomic analyses have pointed to significant alterations in protein expression and function in beta cells exposed to diabetes like conditions (e.g., high glucose and/or saturated fatty acids levels). These nutritional overload stressful conditions are often defined as glucolipotoxic due to the progressive damage they cause to the cells. Our recent studies on the rat insulinoma-derived INS-1E beta cell line point to differential effects of such conditions in the phospholipid bilayers in beta cells. This review focuses on confocal microscopy-based detection of these profound alterations in the plasma membrane and membranes of insulin granules and lipid droplets in single beta cells under such nutritional load conditions.

Keywords: beta cells; diabetes; confocal microscopy; lipidomics; membrane fluidity maps; cell micropolarity maps

---

1. Introduction

The composition of phospholipids in biological membranes determines their cell barrier and cellular communication functions as well as subcellular organelles structure and functions. These properties are determined by the nature of the various phospholipid species and the availability of free fatty acids (FFA) from cellular metabolism and the diet. Yet, the composition of phospholipids in membranes of different subcellular compartments in any given cell may differ greatly. For instance, MacDonald et al. [1] found significant changes in the distribution of phosphatidylserine (PS), phosphoinositol (PI), phosphatidylethanolamine (PE), phosphatidylcholine (PC), sphingomyelin (SM) amongst insulin granules, mitochondria and the whole insulin secreting beta cell (INS-1-832/13 cell line). Moreover, the abundance of different saturated (SFA), mono- (MUFA) and polyunsaturated fatty acids (PUFA) in these phospholipids also varied among the different compartments. Similarly interesting is the observation that glucose stimulation of beta cells induced reversible changes in the composition of fatty acid moieties of phospholipids in insulin granules. Such modifications and remodeled specific signatures of phospholipids in the membranes of insulin granules may alter their biophysical properties. This in turn may modify granules interactions with soluble proteins (e.g., SNAP receptors, SNAREs), with target membrane proteins within the granules (e.g., Vesicle associated membrane protein, VAMP) or with plasma membrane docking proteins (e.g., syntaxins) and affect insulin secretion. Indeed, Pearson et al. [2] observed changes in the turnover of arachidonic-containing phospholipids and diacylglycerols in glucose-stimulated beta cells. These findings have led to the theory that higher abundance of shorter length fatty acids and of unsaturated fatty acids in phospholipids may enhance the fusion and docking of insulin granules membrane bilayers to the plasma membrane upon glucose stimulation due to reversible changes in membrane fluidly and curvature. Moreover, glucolipotoxic conditions may also increase the oxidative burden and lead to endogenous oxidation of free and phospholipid-bound PUFA, as well as transforming *cis* configuration of double bonds to the *trans* configuration in their hydrocarbon backbone. This may lead to modified cellular functions, including insulin granule trafficking [3,4].

The basis for these theories was laid by earlier lipidomic investigations of beta cells, such as by Fex and Lernmark [5] or Cortizo et al. [6] who followed phospholipid turnover in resting and stimulated beta cells. Best et al. reviewed in 1984 [7] pioneering studies on the role of arachidonic acid metabolites in the regulation of beta cell function and insulin secretion. Metz suggested in 1986 [8] a key role for arachidonic acid metabolites in potentiating stimulus-secretion coupling in beta cells. Intensive research over the last 35 years have established significant roles of various enzymatic metabolites of arachidonic acid (e.g., prostaglandins, eicosanoids) and non-enzymatic products (e.g., 4-hydroxyalkenals) in the regulation of insulin secretion [9–16].

In addition to the inherent composition of phospholipids and their turnover in subcellular organelles in beta cells, it is equally important to emphasize the critical role of increased availability of dietary (essential and non-essential) FFA and their incorporation into phospholipids. This is of paramount consequence upon exposure of beta cells to high levels of SFA (e.g., palmitic acid) that ensues alone, or in combination with high glucose levels, an array of (gluco)lipotoxic effects that often contribute to the decline in the mass and function of beta cells in islets of Langerhans [17–20].

Our recent studies on the effect of high glucose and high palmitic acid levels on the phospholipid lipidome of rat insulinoma-derived INS-1E beta cells revealed profound changes in the abundance and distribution of various fatty acids in phospholipids. These studies reveal organelle-specific channeling of polyunsaturated fatty acids (PUFA), arachidonic acid in particular, to nonenzymatic peroxidation and the generation of 4hydroxyalkenals, which affect the cells in several ways [11,13]. Furthermore, advanced confocal microscopy imaging of the plasma membrane of the cells under such conditions detected minimal alterations in their biophysical properties. In contrast, membranes of insulin granules underwent significant remodeling that changed their fluidity. These methods also depicted neogenesis of lipid droplets in live cells upon exposure to excessive levels of palmitic acid [21–23]. This study aims at integrating these findings with standard lipidomics analyses to follow lipid turnover single beta cells and in their subcellular organelles and compartments.

2. Phospholipid Turnover in Cells

The fatty acid composition in membrane phospholipids is constantly remodeled by the influence of free fatty acid availability, enzymatic activity of phospholipases, stressful condition (e.g., nutritional deficiencies or overload conditions) or metabolic diseases. The remodeling is a dynamic and fast process that changes the equilibrium between fatty acid hydrolysis from phospholipids by phospholipase A2 (PLA_2), on one hand, and their acylation to the phospholipid backbone by lysophospholipid acyl transferase (LPAT), on the other [24]. Once PUFA are hydrolyzed from the phospholipid backbone they serve as substrates for enzymatic conversions to plethora of metabolites. Hitherto, hundreds metabolites of arachidonic acid and other PUFA have been identified, many of which constitute distinct groups of ligands to known receptors and transcription factors [12,25–28]. Different mammalian cells express enzymatic pathways that convert arachidonic acid and other PUFA to discrete cell-specific repertoire of bioactive metabolites in a cell-specific manner. These metabolites subsequently regulate various cellular functions in autocrine and/or paracrine fashions. It has been shown that endogenous PUFA metabolites, such as 20-hydroperoxyeicosatetraenoic acid (20-HETE), prostaglandin E1, E3, J2 and I2, or endocannabinoids regulate beta cell functions [14,16,29–39]. Some of these mediators are also generated in beta cells by direct enzymatic transformation of exogenously available unsaturated fatty acids; it has been shown that certain metabolites improved insulin secretion and ameliorate obesity- and cytokine-induced beta cell damage [16,40,41]. Equally important are the findings that assigned key regulatory roles for activated fatty acid receptors, such as GPR41, in modulating insulin secretion upon binding of fatty acid ligands [42–48]. In addition, it has been shown that intracellular n-3 PUFA transformation by elongases (e.g., docosahexaenoic acid formation) may protect against glucolipotoxicity-induced apoptosis in rodent and human islets [49]. Nonetheless, Johnston et al. [50] have recently pointed to an association between long-term increase in circulating non-esterified fatty acids and lower beta cell function.

The enzymatic conversions of PUFA occur in cells along with non-enzymatic transformations. The potency of these non-enzymatic pathways is determined foremost by the levels of oxygen free radicals, which initiate the peroxidation of PUFA and lead to the generation of a group of chemically and biologically reactive aldehydes, of which 4-hydroxyalkenals are prominent. There are two contrasting effects of 4-hydroxyalkenals in cells: numerous studies have shown that these reactive electrophiles form adducts with macromolecules, alter their function and contribute to the etiology and progression of pathological processes [51]. However, when present at physiological and non-toxic levels they interact with receptors and ligand-activated transcription factors in a specific manner and modulate cell functions in autocrine or paracrine manners. Indeed, Poganik el al. [52] have recently compiled evidence to propose that native reactive electrophiles (e.g., 4-hydroxyalkenals) are signaling molecules. Moreover, some of these intracellular interactions were found to evoke hormetic responses that induced or augmented cellular defense mechanisms that ultimately enhance the elimination of the same electrophiles [12,23,51,53–57].

These studies were mostly based upon whole cell lipidomic analyses that usually do not detect subcellular membrane-specific phospholipid turnover. Apparently, disparate remodeling of phospholipids in subcellular compartments or organelles may affect cellular functions in various ways. Monitoring and understanding such variable subcellular remodeling may reveal for instance: (i) whether plasma membranes of cells are inherently protected against major remodeling of phospholipids and thus preserve their barrier and communication capabilities with the surrounding environment; (ii) to what extent the remodeling of mitochondria membranes may affect their permeability, membrane potential and oxidative phosphorylation capacity; (iii) how the remodeling of phospholipids in the ER modulates protein sorting and chaperoning or alters their capacity to generate mono-layered lipid droplets within this compartment; (iv) could lysosomal function be influenced upon phospholipid remodeling or (v) to what level neurotransmitter or hormone secretion from vesicles or granules in neurons and endocrine cells, respectively, is disrupted of enhanced due to alterations in their tethering, docking and fusion with the plasma membrane and subsequent internalization. Tedious fractionation, separation and isolation techniques of subcellular organelle fractions were practiced in the past to answer such questions. Often, the amount and cross-contamination of the isolated fractions resulted in erroneous analyses. We showed that what was considered a standard and efficient purification method of the plasma membrane fraction of skeletal muscle cells carried in fact a substantial cross-contamination of intracellular microsomal membrane that could greatly obscure the experimental results [58].

3. Lipidomic Analyses of Beta Cells

We have employed non-targeted lipidomics analysis to study the impact of high glucose and high palmitic acid levels on the turnover of fatty acids in phospholipids of INS-1E cells. These studies discovered significant changes in the content of SFA, MUFA and PUFA [13]. Figure 1 shows the changes in their abundance after exposure of the cells to 11 and 25 mM glucose in comparison to cells that were maintained at 5 mM glucose. The abundance of PUFA was significantly decreased, MUFA increased and SFA levels remained constant under the high glucose incubations. Noteworthy, the total fatty acid content remained unaltered under these experimental conditions. This study also showed that that the released PUFA (i.e., arachidonic and linoleic acids) were avidly peroxidized to 4-hydroxynonenal (4-HNE). The latter in turn activated peroxisome proliferator-activated receptor-δ (PPARδ) that further augmented glucose-stimulated insulin secretion (GSIS). Thus, increasing glucose concentrations have not been previously considered to have specific stressful effects on membranes; in fact, 5 or 11 mM glucose were indifferently reported for beta cell culture conditions without affecting cell viability. In our study we showed for the first time that cellular membranes were not just spectators but were the source for lipid precursors of signaling molecules such as PUFA.

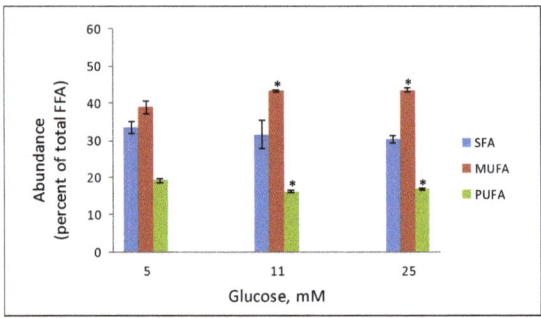

Figure 1. Glucose-induced remodeling of phospholipid in INS-1E cells. INS-1E cells were incubated in serum-free medium supplemented with the indicated glucose concentration for 16 h, and processed for lipidomics analysis as described [13]. The abundance of Saturated- (SFA), Monounsaturated- (MUFA) and polyunsaturated fatty acids (PUFA) is given as percent of total fatty acid content. Mean ± SEM, $n = 4$. * $p < 0.05$ significantly different from the corresponding abundance at the 5 mM glucose incubation (adapted from [13]).

Concomitant exposure of the cells to increasing levels of palmitic acid and glucose further modified the abundance of fatty acids in phospholipids. Figure 2 shows the expected increase of the abundance of SFA (i.e., palmitic acid) that was accompanied with nearly 50% depletion in the amount of PUFA in phospholipids. Increasing glucose levels in the incubation intensified these palmitic acid-induced phospholipid remodeling effects. This study also showed that at these ranges the peroxidation of the released arachidonic and linoleic acids to 4-HNE also activated PPARδ and evoked augmented GSIS [11]. It is important to note that the upper limit of non-toxic concentrations of palmitic acid that did not compromise cell viability upon prolonged incubations were 300, 150, and 100 μM at 5, 11 and 25 mM glucose, respectively.

As mentioned above, these whole cell lipidomic analyses could not detect changes in fatty acid composition in membranes of subcellular compartments in the cells. This limitation also applies to other whole-cell analyses, such as shotgun lipidomics [59]. This method detects and reports the cellular content of most commonly known lipids in cells such as, free fatty acids and their metabolites (e.g., eicosanoids), glycerophospholipids (e.g., PC, PE, PS, PG, PI, PA), glycerolipids (e.g., TAG, DAG, MAG), diphosphatidylglycerol lipids (cardiolipins), sphingolipids (e.g., sphingomyelin, sphingosines, ceramides, cerebrosides, gangliosides) or sterol lipids (e.g., steroids, sterols) [60]. Several studies used this technique in diabetes research and found alterations in myocardial cardiolipin content and composition at the early stages of the disease [61]. Others correlated alterations in the plasma lipidome of diabetic patients or in tissues of diabetic mice (ob/ob) to the progression of impaired glucose tolerance [62,63]. Nevertheless, such shotgun lipidomic analyses have not yet revealed alterations in subcellular compartments of pancreatic beta cells under normal or stressful stimuli.

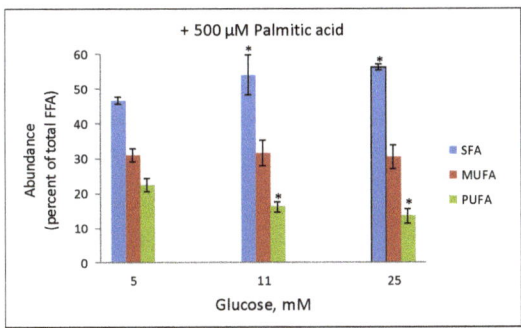

Figure 2. Impact of palmitic acid on fatty acid abundance in phospholipids of INS-1E cells. INS-1E cells were incubated in serum-free medium supplemented with the indicated glucose concentration for 32 h. Palmitic acid (500 µM) was added during the last 16 h of incubation. The cells were then harvested and processed for lipidomics analysis as described [11]. The abundance of SFA, MUFA and PUFA is given as percent of total fatty acid content. Mean ± SEM, $n = 4$. * $p < 0.05$ significantly different from the corresponding values at the 5 mM glucose incubation (adapted from [11]).

Other mass spectrometry methods have also been used to study beta cells. Recent temporal analysis of palmitic acid-treated INS-1 beta cells was based on isobaric labeling-based mass spectrometry and bioinformatics [64]. It highlighted altered cholesterol and fatty acid metabolism as early toxic events associated with ER stress. This quantitative strategy provided insight into general molecular events (lipid metabolism) and pathway adaptation in the cells. Other groups employed non-targeted mass-spectrometric lipidomics to study beta cells [65]. For instance, electrospray ionization mass spectrometric analysis that was employed to analyze phospholipids in INS-1 beta cells [66] discovered changes in total PUFA and MUFA contents, which were quite similar to the results of the lipidomics analysis we have performed, as described above. Interestingly, treatment of the cells with palmitic acid in this study had little effect on the content of both classes of fatty acids in the cells. Recent advances in mass spectrometric methods, such as the matrix-assisted laser desorption/ionization (MALDI) imaging mass spectroscopy (IMS), have been used to obtain molecular profiling of mouse pancreatic tissues [67]. Immunofluorescent images that were acquired from serial pancreatic sections were co-registered with the MS images of the sections and enabled molecular identification of specific phospholipid and glycolipid isoforms. The region selective molecular specificity afforded by this method revealed profound differences between endocrine and exocrine cells. Yet, the capacity of the method to detect clearly changes in the distribution of lipids within insulin granules or other subcellular organelles remains limited. Others have employed nanospray desorption electrospray ionization mass spectrometry imaging (nano-DESI-MSI) to identify different lipid classes in individual islets of Langerhans and the surrounding exocrine cells in sections of mouse pancreatic tissues [68]. The study found some disparate distribution of certain lipid species (including PUFA rich phospholipids, such as PC 34:2, PC 36:2, PC 36:4, PC 38:4) between the two types of cells. Using this method for the analysis of pancreatic tissues from normal and diabetic animals may enable estimation of the content of the phospholipids and other lipids. Both the MALDI-IMS and nano-DESI-MSI techniques analyze islets in pancreatic tissue sections without discriminating amongst the different types of endocrine cells (alpha, beta and delta cells) and without providing clear intracellular maps of the distribution of the different lipid species in subcellular organelles. Recent advances in enhancing the power of resolution of such single-cell analysis by different mass spectrometric platforms may contribute to comprehensive analysis of lipid turnover in beta cells [69].

4. Confocal Imaging-Based Fluidity and Micropolarity Maps of Single Beta Cells

Our interests in ascertaining the impact of phospholipid remodeling in the beta cell lipidome under nutritional overload conditions led to two independent confocal imaging strategies of INS-1E beta cells. The first employed spectral analysis of the fluorescent probe Laurdan to provide fluidity maps of single cell membranes [22]. The second exploited the micropolarity-sensitive emission profile of the dye Nile red to give intracellular maps of neutral and polar lipids in subcellular organelles [21].

In the first case the advantages of fluorescence temporal imaging-based detection methods were exploited to obtain high resolution imaging of subcellular organelles in beta cells. For this purpose, we expressed in INS-1E cells IAPP-mCherry protein that is targeted to insulin granules [70] (probe courtesy of Dr. Patrick E. MacDonald, University of Alberta, Edmonton, Canada). mCherry-expressing cells were incubated at 5, 11 and 25 mM and loaded with the fluorescent dye Laurdan, which integrates into lipid phases in membranes. Its excited-state relaxation is highly sensitive to the presence and mobility of water molecules within the membrane bilayer, while being insensitive to the head-group type in phospholipids [71,72]. By using two-photon infrared excitation techniques and dual-wavelength ratio measurements, we detected Laurdan emission spectrum of coexisting lipid domains in the cells and obtained information on membrane fluidity by following the shift from ordered (gel) phases (yellow-orange emission) to disordered (liquid-crystalline) phases (violet-purple emission). Membrane fluidity in the confocal images was then reported in terms of ratio of emission intensities for each pixel by using Generalized Polarization (GP) value. This is defined as $GP = (I_G - I_R)/(I_G + I_R)$; I_G, emission in the range of 400–460 nm; I_R, emission at the range of 470–530 nm. The GP value ranges from −1 (fluid, liquid disordered state) to 1 (gel-like, solid ordered state). Figure 3 shows such Laurdan spectral analysis of mCherry expressing INS-1E cells that were incubated at 5, 11 and 25 mm glucose for 32 h without or with 500 µM palmitic acid during the last 16 h of incubation.

The fluidity maps of plasma membranes revealed that the GP values remained unaltered (GP = 0.38) in cells that were incubated with increasing glucose concentrations. This indicates that the release of PUFA from phospholipids in the cell sunder high glucose conditions (reported in [13]) did not involve significant remodeling of the plasma membrane. Thus, the barrier and communication properties of the plasma membrane, as well as the capacity to interact with secretory insulin granules upon glucose stimulation were preserved. Of interest are the lower GP values (0.26) of the insulin granules (mCherry positive organelles) that are indicative of a more fluid state than that of the plasma membranes, at all glucose concentrations. This seems to result from a higher abundance of PUFA and the corresponding liquid disordered phase. The latter results from the non-linear geometrical configuration of double bonds in *cis* positions [73]. The GP value increased slightly following the incubation with 11 and 25 mM glucose due to the hydrolysis of PUFA from phospholipids, as we observed in the abovementioned lipidomics analysis. The incubation with 500 µM palmitic acid had significant effects on the GP value in plasma membranes and insulin granules, which steadily increased in both compartments in a glucose-dependent manner. These dramatic changes in membrane fluidity of insulin granules reflect well the capacity of the incorporated saturated palmitic acid to induce transition to a solid-ordered gel-like phase of phospholipids in membrane bilayers. Furthermore, depletion of PUFA from insulin granules' phospholipids may also contribute to the higher levels of the observed GP values. Indeed, this scenario correlates well with the results on the lipidomics analysis in similarly-treated cells that showed marked loss of PUFA from phospholipids (Figure 2). The impact of this phenomenon on the recruitment of insulin granules for secretion upon glucose stimulation is complex. We have recently shown [73] that non-toxic palmitic acid levels augmented GSIS, enabling the organism to facilitate insulin-mediated glucose and fatty acid disposal and thereby reduce the risk of developing peripheral diabetes complications. However, once reaching the toxic range of palmitic acid concentrations, which is reciprocally related to increasing glucose concentrations, the insulin secretory capacity is impaired, partly due to the rigidification of the insulin granule membranes. This study shows that Laurdan-based fluidity maps of cells complements with whole cell lipidomic analysis and provides insight to localized remodeling of phospholipids (i.e., plasma membranes and insulin granules).

In the second case the solvatochromic and lipophilic properties of the fluorescent probe Nile Red were exploited to obtain high-resolution micropolarity maps of individual INS-1E cells [21]. The probe exhibits an emission shift from yellow to red when the degree of polarity of the lipid environment increases [74]. This property enables the detection of the degree of polarity of lipids in cells by evaluating the quantitative ratio of red and yellow emissions. In this study INS-1E cells were incubated with 1 μM of Nile Red for 30 min in the dark and then placed on the inverted confocal microscope equipped with a live chamber and 32 channel spectral images were obtained using a 60X objective under 488 nm excitation for the probe. Internal photon multiplier tubes collected images in 16-bit, unsigned images at 0.25 ms dwell time. This procedure, which enables the assessment of differences of the polarity of the various compartments in the cell, allows to refine the investigation of the subcellular distribution of neutral and polar lipids [75,76]. Figure 4 depicts the principles of phasor analysis that was employed to analyze images of Nile red-treated cells. The method is explained in details in our recent study [21] and elsewhere [77,78]. The cells that were incubated with 300 μM palmitic acid handled the increased influx of palmitic acid by incorporating it into triglycerides that were then sequestered in newly formed lipid droplets.

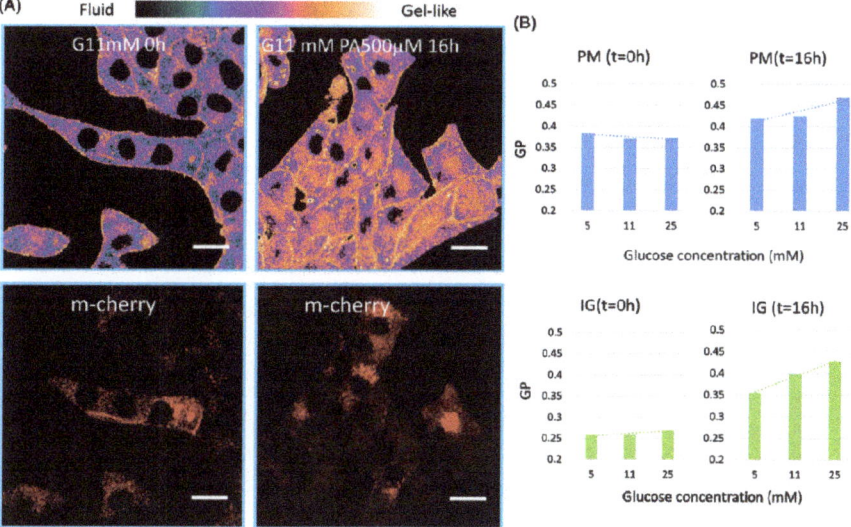

Figure 3. Generalized Polarization (GP) values of the plasma membranes (PM) and insulin granules (IG, mCherry positive organelles) in glucose treated INS-1E cells. (**A**), Representative high-resolution fluorescence images of Laurdan emission for fluidity investigation along with mCherry emission images in INS-1E cells exposed to 11 mM glucose for 32 h and 500 μM palmitic acid (PA) during the last 16 h of incubation. mCherry labeled insulin granules, which have spherical shapes of about 0.5–1 μm diameter. Scale bar is 10 μm. (**B**) Summary of GP values of plasma membrane and insulin granules in cells exposed to different glucose levels without (left) and with palmitic acid (right). Copied with permission from [22].

In the magnification (Figure 4) it is shown that the lipid droplets are composed of a core of non-polar (NP) lipids (blue spots; triglycerides) and a surrounding monolayer of polar (P) lipids (green coating), which is typical of lipid droplets [79]. This analysis also revealed that P lipids are localized in the plasma membrane. Interestingly, hyperpolar (HP) lipids are associated with nuclear membranes, or unevenly compartmentalized in internal membranes throughout the cytoplasm. We propose that these compartments may also serve a major target for phospholipid turnover in nutritionally-challenged beta cells and could therefore be the source for the PUFA required for non-enzymatic peroxidation and generation of 4-HNE. This method, which is very useful for lipid droplet research, can also be applied to Nile red spectral analysis of phospholipid remodeling in subcellular compartments. This may be used for instance for simultaneous fluorescent labeling of insulin granules and other organelles. While mCherry labeling may pose limitation due to overlapping emission spectrum with Nile red, other probes, such as phogrin-fluorescents proteins [80] or neuropeptide Y-pHluorin [81] may be useful for labelling the granules.

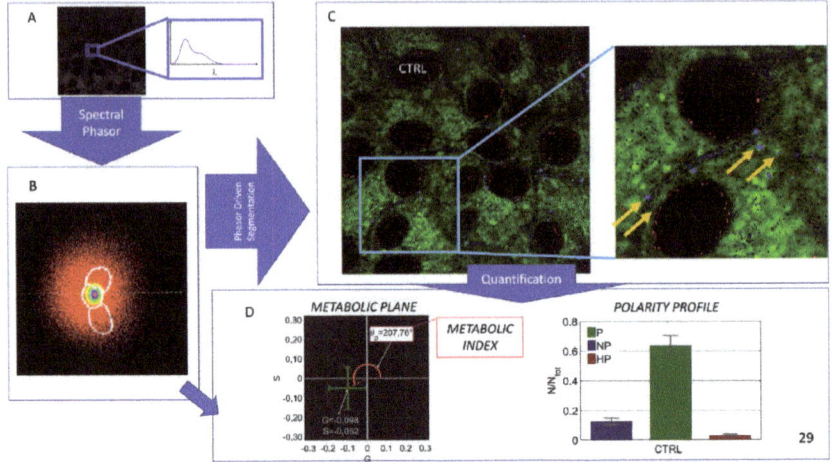

Figure 4. Workflow of the method based on the phasor driven segmentation of Nile Red spectral images. INS-1E cells were maintained at 11 mM glucose received 300 μM palmitic acid for 16 h. (**A**), Nile Red spectral images of live INS-1E cells. Each pixel of the spectral image is associated with the Nile Red emission spectrum. (**B**), Phasor plot generation: the cloud appears as broad and elliptical, because the co-existence of the three classes of lipids (NP, neutral lipids; P, polar lipids and HP, highly polar lipids), which are simultaneously present in the cells. By selecting on the phasor plane the domains corresponding to the three classes (white lines), it becomes possible to remap them to the original fluorescence image. (**C**), Segmentation of the three lipid classes: NP are reported in Blue, P in Green, HP in red. In the magnification, lipid droplets are visualized as spherical particles. P lipids are also localized in the plasma membrane, of which phospholipids is the main component. (**D**), Extraction of the angle θP formed by the center of mass of the cloud in the phasor plot with the g-axis provides information about the average polarity. From the segmented channels, the fractional contribution of the different lipid classes, in terms of the relative fraction of pixels belonging to a particular class was retrieved. Copied with permission from [21].

5. Conclusions

Current lipidomic analyses of beta cells show that hyperglycemic- and hyperlipidemic-like conditions induce fast remodeling of phospholipids. These modifications may reflect substantial structural and functional changes in the cells. These methods have not yet progressed to allow for subcellular lipidomic analysis in fixed or lived cells. Confocal imaging that provide high resolution maps of subcellular membrane fluidity and lipid micropolarity maps of live cells may complement the lipidomic analyses by depicting membrane remodeling upon various stressful stimuli. Our recent results in lipidomics and lipid imaging in beta cells highlight this potential. This has also been demonstrated in a study that employed mass-spectrometry based oxidative lipidomics and lipid imaging in traumatic brain injury models [82]. Finally, recent reports on beta cells heterogeneity [83] attest to the limitations of the whole cell lipidomic analysis and the clear advantages of individual cell analysis by confocal imaging used in this study, which may distinguish among different beta cell populations in islets of Langerhans.

Author Contributions: This Review articles was conceptualized and written by G.M., C.F. and S.S. S.S. and C.F. supervised and carried the experiments described in Figures 1 and 2. G.M. and S.S. supervised and carried the experiments described in Figures 1 and 2 with the graduate students O.C. and B.D.

Funding: This research was funded by grants from Legacy Heritage Biomedical Science Partnership of the Israel Science Foundation of the Israel Academy of Sciences and Humanities (1429/13), the Vigevani Foundation (HUJI-2017) and the Brettler Center for Research of Molecular Pharmacology and Therapeutics in the Hebrew University (BC-2016) (SS). GM would like to acknowledge Università Cattolica del Sacro Cuore, [Linea D1 2018]. CF would like to acknowledge the Di Bella Foundation for the support given to research (2019).

Acknowledgments: All authors gratefully acknowledge the support and networking opportunities they received and enjoyed from COST (European Cooperation in Science and Technology) Action CM1201 on "Biomimetic Radical Chemistry". This COST Action also provided Short-Term Scientific Mission (STSM) fellowships to BD for training in confocal microscopy at GM Laboratory in Rome and for G.M. training in beta cell research at the laboratory in Jerusalem.

Conflicts of Interest: The authors declare no conflicts of interest.

References

1. Macdonald, M.J.; Ade, L.; Ntambi, J.M.; Ansari, I.-U.H.; Stoker, S.W. Characterization of Phospholipids in Insulin Secretory Granules and Mitochondria in Pancreatic Beta Cells and Their Changes with Glucose Stimulation. *J. Boil. Chem.* **2015**, *290*, 11075–11092. [CrossRef] [PubMed]
2. Pearson, G.L.; Mellett, N.; Chu, K.Y.; Boslem, E.; Meikle, P.J.; Biden, T.J. A comprehensive lipidomic screen of pancreatic beta-cells using mass spectroscopy defines novel features of glucose-stimulated turnover of neutral lipids, sphingolipids and plasmalogens. *Mol. Metab.* **2016**, *5*, 404–414. [CrossRef] [PubMed]
3. Ferreri, C.; Panagiotaki, M.; Chatgilialoglu, C. Trans fatty acids in membranes: The free radical path. *Mol. Biotechnol.* **2007**, *37*, 19–25. [CrossRef] [PubMed]
4. Fruhwirth, G.O.; Loidl, A.; Hermetter, A. Oxidized phospholipids: From molecular properties to disease. *Biochim. Biophys. Acta* **2007**, *1772*, 718–736. [CrossRef]
5. Fex, G.; Lernmark, A. Effects of insulin secretagogues on phospholipid metabolism in pancreatic beta-cells. *Biochim. Biophys. Acta* **1975**, *388*, 1–4. [CrossRef]
6. Cortizo, A.; Paladini, A.; Diaz, G.; Garcia, M.; Gagliardino, J. Changes induced by glucose in the plasma membrane properties of pancreatic islets. *Mol. Cell. Endocrinol.* **1990**, *71*, 49–54. [CrossRef]
7. Best, L.; Dunlop, M.; Malaisse, W.J. Phospholipid metabolism in pancreatic islets. *Cell. Mol. Life Sci.* **1984**, *40*, 1085–1091. [CrossRef]
8. Metz, S.A. Putative roles for lysophospholipids as mediators and lipoxygenase-mediated metabolites of arachidonic acid as potentiators of stimulus-secretion coupling: Dual mechanisms of p-hydroxymercuribenzoic acid-induced insulin release. *J. Pharmacol. Exp. Ther.* **1986**, *238*, 819–832.
9. Neuman, J.C.; Fenske, R.J.; Kimple, M.E. Dietary polyunsaturated fatty acids and their metabolites: Implications for diabetes pathophysiology, prevention, and treatment. *Nutr. Heal. Aging* **2017**, *4*, 127–140. [CrossRef]
10. Carboneau, B.A.; Breyer, R.M.; Gannon, M. Regulation of pancreatic beta-cell function and mass dynamics by prostaglandin signaling. *J. Cell Commun.* **2017**, *11*, 105–116.
11. Cohen, G.; Shamni, O.; Avrahami, Y.; Cohen, O.; Broner, E.C.; Filippov-Levy, N.; Chatgilialoglu, C.; Ferreri, C.; Kaiser, N.; Sasson, S. Beta cell response to nutrient overload involves phospholipid remodelling and lipid peroxidation. *Diabetologia* **2015**, *58*, 1333–1343. [CrossRef] [PubMed]
12. Cohen, G.; Riahi, Y.; Sunda, V.; Deplano, S.; Chatgilialoglu, C.; Ferreri, C.; Kaiser, N.; Sasson, S. Signaling properties of 4-hydroxyalkenals formed by lipid peroxidation in diabetes. *Free. Radic. Boil. Med.* **2013**, *65*, 978–987. [CrossRef] [PubMed]
13. Cohen, G.; Riahi, Y.; Shamni, O.; Guichardant, M.; Chatgilialoglu, C.; Ferreri, C.; Kaiser, N.; Sasson, S. Role of lipid peroxidation and PPAR-delta in amplifying glucose-stimulated insulin secretion. *Diabetes* **2011**, *60*, 2830–2842. [CrossRef]
14. Luo, P.; Wang, M.H. Eicosanoids, beta-cell function, and diabetes. *Prostaglandins Other Lipid Mediat.* **2011**, *95*, 1–10. [CrossRef]

15. Ma, K.; Nunemaker, C.S.; Wu, R.; Chakrabarti, S.K.; Taylor-Fishwick, D.A.; Nadler, J.L. 12-Lipoxygenase Products Reduce Insulin Secretion and {beta}-Cell Viability in Human Islets. *J. Clin. Endocrinol. Metab.* **2010**, *95*, 887–893. [CrossRef]
16. Keane, D.; Newsholme, P. Saturated and unsaturated (including arachidonic acid) non-esterified fatty acid modulation of insulin secretion from pancreatic beta-cells. *Biochem Soc. Trans.* **2008**, *36*, 955–958. [CrossRef]
17. Imai, Y.; Cousins, R.S.; Liu, S.; Phelps, B.M.; Promes, J.A. Connecting pancreatic islet lipid metabolism with insulin secretion and the development of type 2 diabetes. *Ann. N. Y. Acad. Sci.* **2019**. [CrossRef]
18. Ye, R.; Onodera, T.; Scherer, P.E. Lipotoxicity and beta Cell Maintenance in Obesity and Type 2 Diabetes. *J. Endocr. Soc.* **2019**, *3*, 617–631. [CrossRef]
19. Oh, Y.S.; Bae, G.D.; Baek, D.J.; Park, E.-Y.; Jun, H.-S. Fatty Acid-Induced Lipotoxicity in Pancreatic Beta-Cells During Development of Type 2 Diabetes. *Front. Endocrinol.* **2018**, *9*, 384. [CrossRef]
20. Janikiewicz, J.; Hanzelka, K.; Kozinski, K.; Kolczynska, K.; Dobrzyn, A. Islet beta-cell failure in type 2 diabetes—Within the network of toxic lipids. *Biochem. Biophys. Res. Commun.* **2015**, *460*, 491–496. [CrossRef]
21. Maulucci, G.; Di Giacinto, F.; De Angelis, C.; Cohen, O.; Daniel, B.; Ferreri, C.; De Spirito, M.; Sasson, S.; Cohen, O. Real time quantitative analysis of lipid storage and lipolysis pathways by confocal spectral imaging of intracellular micropolarity. *Biochim. Biophys. Acta* **2018**, *1863*, 783–793. [CrossRef] [PubMed]
22. Maulucci, G.; Cohen, O.; Daniel, B.; Sansone, A.; Petropoulou, P.I.; Filou, S.; Spyridonidis, A.; Pani, G.; De Spirito, M.; Chatgilialoglu, C.; et al. Fatty acid-related modulations of membrane fluidity in cells: Detection and implications. *Free. Radic. Res.* **2016**, *50*, 40–50. [CrossRef] [PubMed]
23. Maulucci, G.; Daniel, B.; Cohen, O.; Avrahami, Y.; Sasson, S. Hormetic and regulatory effects of lipid peroxidation mediators in pancreatic beta cells. *Mol. Asp. Med.* **2016**, *49*, 49–77. [CrossRef] [PubMed]
24. Wang, B.; Tontonoz, P. Phospholipid Remodeling in Physiology and Disease. *Annu. Rev. Physiol.* **2019**, *81*, 165–188. [CrossRef] [PubMed]
25. Das, U.N. Arachidonic acid in health and disease with focus on hypertension and diabetes mellitus: A review. *J. Adv. Res.* **2018**, *11*, 43–55. [CrossRef] [PubMed]
26. Hanna, V.S.; Hafez, E.A.A. Synopsis of arachidonic acid metabolism: A review. *J. Adv. Res.* **2018**, *11*, 23–32. [CrossRef]
27. Powell, W.S.; Rokach, J. Biosynthesis, biological effects, and receptors of hydroxyeicosatetraenoic acids (HETEs) and oxoeicosatetraenoic acids (oxo-ETEs) derived from arachidonic acid. *Biochim. Biophys. Acta* **2015**, *1851*, 340–355. [CrossRef]
28. Brash, A.R. Arachidonic acid as a bioactive molecule. *J. Clin. Investig.* **2001**, *107*, 1339–1345. [CrossRef]
29. Hauke, S.; Keutler, K.; Phapale, P.; Yushchenko, D.A.; Schultz, C. Endogenous Fatty Acids Are Essential Signaling Factors of Pancreatic beta-Cells and Insulin Secretion. *Diabetes* **2018**, *67*, 1986–1998. [CrossRef]
30. Tunaru, S.; Bonnavion, R.; Brandenburger, I.; Preussner, J.; Thomas, D.; Scholich, K.; Offermanns, S. 20-HETE promotes glucose-stimulated insulin secretion in an autocrine manner through FFAR1. *Nat. Commun.* **2018**, *9*, 177. [CrossRef]
31. Zhao, Y.; Fang, Q.; Straub, S.G.; Lindau, M.; Sharp, G.W. Prostaglandin E1 inhibits endocytosis in the beta-cell endocytosis. *J. Endocrinol.* **2016**, *229*, 287–294. [CrossRef] [PubMed]
32. Jourdan, T.; Godlewski, G.; Kunos, G. Endocannabinoid regulation of beta-cell functions: Implications for glycaemic control and diabetes. *Diabetes Obes. Metab.* **2016**, *18*, 549–557. [CrossRef] [PubMed]
33. Batchu, S.N.; Majumder, S.; Bowskill, B.B.; White, K.E.; Advani, S.L.; Liu, Y.; Thai, K.; Lee, W.L.; Advani, A. Prostaglandin I2 Receptor Agonism Preserves beta-Cell Function and Attenuates Albuminuria Through Nephrin-Dependent Mechanisms. *Diabetes* **2016**, *65*, 1398–1409. [CrossRef] [PubMed]
34. Gurgul-Convey, E.; Hanzelka, K.; Lenzen, S. Mechanism of Prostacyclin-Induced Potentiation of Glucose-Induced Insulin Secretion. *Endocrinology* **2012**, *153*, 2612–2622. [CrossRef] [PubMed]
35. Persaud, S.J.; Muller, D.; Belin, V.D.; Papadimitriou, A.; Huang, G.C.; Amiel, S.A.; Jones, P.M. Expression and function of cyclooxygenase and lipoxygenase enzymes in human islets of Langerhans. *Arch. Physiol. Biochem.* **2007**, *113*, 104–109. [CrossRef]
36. Chambers, K.T.; Weber, S.M.; Corbett, J.A. PGJ2-stimulated beta-cell apoptosis is associated with prolonged UPR activation. *Am. J. Physiol. Endocrinol. Metab.* **2007**, *292*, 1052–1061. [CrossRef]
37. Meng, Z.X.; Sun, J.X.; Ling, J.J.; Lv, J.H.; Zhu, D.Y.; Chen, Q.; Sun, Y.J.; Han, X. Prostaglandin E2 regulates Foxo activity via the Akt pathway: Implications for pancreatic islet beta cell dysfunction. *Diabetology* **2006**, *49*, 2959–2968. [CrossRef]

38. Franca, T.; Pinto, E.; Nogueira, F.; Velho, H.V. Malignant localized fibrous tumor of the pleura. *Acta Med Port* **2001**, *14*, 435–440.
39. Jones, P.M.; Persaud, S.J. Arachidonic acid as a second messenger in glucose-induced insulin secretion from pancreatic beta-cells. *J. Endocrinol.* **1993**, *137*, 7–14. [CrossRef]
40. Oliveira, V.; Marinho, R.; Vitorino, D.; Santos, G.; Moraes, J.; Dragano, N.; Sartori-Cintra, A.; Pereira, L.; Catharino, R.; Da Silva, A.; et al. Diets containing alpha-linolenic (omega 3) or oleic (omega 9) fatty acids rescues obese mice from insulin resistance. *Endocrinology* **2015**, *156*, 4033–4046. [CrossRef]
41. Bhaswant, M.; Poudyal, H.; Brown, L. Mechanisms of enhanced insulin secretion and sensitivity with n-3 unsaturated fatty acids. *J. Nutr. Biochem.* **2015**, *26*, 571–584. [CrossRef] [PubMed]
42. Jezek, P.; Jaburek, M.; Holendova, B.; Plecita-Hlavata, L. Fatty Acid-Stimulated Insulin Secretion vs. Lipotoxicity. *Molecules* **2018**, *23*, 1483. [CrossRef] [PubMed]
43. Veprik, A.; Laufer, D.; Weiss, S.; Rubins, N.; Walker, M.D. GPR41 modulates insulin secretion and gene expression in pancreatic beta-cells and modifies metabolic homeostasis in fed and fasting states. *FASEB J.* **2016**, *30*, 3860–3869. [CrossRef] [PubMed]
44. Oh, Y.S. Mechanistic insights into pancreatic beta-cell mass regulation by glucose and free fatty acids. *Anat. Cell Boil.* **2015**, *48*, 16–24. [CrossRef]
45. Prentice, K.J.; Wheeler, M.B. FFAR out new targets for diabetes. *Cell Metab.* **2015**, *21*, 353–354. [CrossRef]
46. Wang, X.; Chan, C.B. n-3 polyunsaturated fatty acids and insulin secretion. *J. Endocrinol.* **2015**, *224*, 97–106. [CrossRef]
47. Hara, T.; Ichimura, A.; Hirasawa, A. Therapeutic Role and Ligands of Medium- to Long-Chain Fatty Acid Receptors. *Front. Endocrinol.* **2014**, *5*, 83. [CrossRef]
48. Nolan, C.J.; Madiraju, M.S.; Delghingaro-Augusto, V.; Peyot, M.L.; Prentki, M. Fatty acid signaling in the beta-cell and insulin secretion. *Diabetes* **2006**, *55*, 16–23. [CrossRef]
49. Bellini, L.; Campana, M.; Rouch, C.; Chacinska, M.; Bugliani, M.; Meneyrol, K.; Hainault, I.; Lenoir, V.; Denom, J.; Veret, J.; et al. Protective role of the ELOVL2/docosahexaenoic acid axis in glucolipotoxicity-induced apoptosis in rodent beta cells and human islets. *Diabetology* **2018**, *61*, 1780–1793. [CrossRef]
50. Johnston, L.W.; Harris, S.B.; Retnakaran, R.; Giacca, A.; Liu, Z.; Bazinet, R.P.; Hanley, A.J. Association of NEFA composition with insulin sensitivity and beta cell function in the Prospective Metabolism and Islet Cell Evaluation (PROMISE) cohort. *Diabetologia* **2018**, *61*, 821–830. [CrossRef]
51. Negre-Salvayre, A.; Auge, N.; Ayala, V.; Basaga, H.; Boada, J.; Brenke, R.; Chapple, S.; Cohen, G.; Fehér, J.; Grune, T.; et al. Pathological aspects of lipid peroxidation. *Free. Radic. Res.* **2010**, *44*, 1125–1171. [CrossRef] [PubMed]
52. Poganik, J.R.; Long, M.J.C.; Aye, Y. Getting the Message? Native Reactive Electrophiles Pass Two Out of Three Thresholds to be Bona Fide Signaling Mediators. *Bioessays* **2018**, *40*, e1700240. [CrossRef] [PubMed]
53. Sasson, S. 4-Hydroxyalkenal-activated PPARdelta mediates hormetic interactions in diabetes. *Biochimie* **2017**, *136*, 85–89. [CrossRef] [PubMed]
54. Sasson, S. Nutrient overload, lipid peroxidation and pancreatic beta cell function. *Free. Radic. Boil. Med.* **2017**, *111*, 102–109. [CrossRef]
55. Davies, K.J. Adaptive homeostasis. *Mol. Aspects Med.* **2016**, *49*, 1–7. [CrossRef]
56. Jaganjac, M.; Tirosh, O.; Cohen, G.; Sasson, S.; Zarkovic, N. Reactive aldehydes—Second messengers of free radicals in diabetes mellitus. *Free. Radic. Res.* **2013**, *47*, 39–48. [CrossRef]
57. Riahi, Y.; Cohen, G.; Shamni, O.; Sasson, S. Signaling and cytotoxic functions of 4-hydroxyalkenals. *Am. J. Physiol. Metab.* **2010**, *299*, 879–886. [CrossRef]
58. Greco-Perotto, R.; Wertheimer, E.; Jeanrenaud, B.; Cerasi, E.; Sasson, S. Glucose regulates its transport in L8myocytes by modulating cellular trafficking of the transporter GLUT-1. *Biochem. J.* **1992**, *286*, 157–163. [CrossRef]
59. Hsu, F.-F. Mass spectrometry-based shotgun lipidomics—A critical review from the technical point of view. *Anal. Bioanal. Chem.* **2018**, *410*, 6387–6409. [CrossRef]
60. Jeucken, A.; Brouwers, J.F. High-Throughput Screening of Lipidomic Adaptations in Cultured Cells. *Biomolecules* **2019**, *9*, 42. [CrossRef]
61. Han, X.; Yang, J.; Yang, K.; Zhongdan, Z.; Abendschein, D.R.; Gross, R.W.; Zhao, Z. Alterations in myocardial cardiolipin content and composition occur at the very earliest stages of diabetes: A shotgun lipidomics study. *Biochemistry* **2007**, *46*, 6417–6428. [CrossRef] [PubMed]

62. Hayakawa, J.; Wang, M.; Wang, C.; Han, R.H.; Jiang, Z.Y.; Han, X. Lipidomic analysis reveals significant lipogenesis and accumulation of lipotoxic components in ob/ob mouse organs. *Prostaglandins Leukot. Essent. Fatty Acids* **2018**, *136*, 161–169. [CrossRef]
63. Kopprasch, S.; Dheban, S.; Schuhmann, K.; Xu, A.; Schulte, K.-M.; Simeonovic, C.J.; Schwarz, P.E.H.; Bornstein, S.R.; Shevchenko, A.; Graessler, J. Detection of Independent Associations of Plasma Lipidomic Parameters with Insulin Sensitivity Indices Using Data Mining Methodology. *PLoS ONE* **2016**, *11*, e0164173. [CrossRef] [PubMed]
64. Li, Z.; Liu, H.; Niu, Z.; Zhong, W.; Xue, M.; Wang, J.; Yang, F.; Zhou, Y.; Zhou, Y.; Xu, T.; et al. Temporal Proteomic Analysis of Pancreatic beta-Cells in Response to Lipotoxicity and Glucolipotoxicity. *Mol. Cell. Proteomics* **2018**, *17*, 2119–2131. [CrossRef] [PubMed]
65. Lee, H.-C.; Yokomizo, T. Applications of mass spectrometry-based targeted and non-targeted lipidomics. *Biochem. Biophys. Res. Commun.* **2018**, *504*, 576–581. [CrossRef] [PubMed]
66. Ramanadham, S.; Hsu, F.-F.; Zhang, S.; Bohrer, A.; Ma, Z.; Turk, J. Electrospray ionization mass spectrometric analyses of phospholipids from INS-1 insulinoma cells: Comparison to pancreatic islets and effects of fatty acid supplementation on phospholipid composition and insulin secretion. *Biochim. Biophys. Acta* **2000**, *1484*, 251–266. [CrossRef]
67. Prentice, B.M.; Hart, N.J.; Phillips, N.; Haliyur, R.; Judd, A.; Armandala, R.; Spraggins, J.M.; Lowe, C.L.; Boyd, K.L.; Stein, R.W.; et al. Imaging mass spectrometry enables molecular profiling of mouse and human pancreatic tissue. *Diabetology* **2019**, *62*, 1036–1047. [CrossRef]
68. Yin, R.; Kyle, J.; Burnum-Johnson, K.; Bloodsworth, K.J.; Sussel, L.; Ansong, C.; Laskin, J. High Spatial Resolution Imaging of Mouse Pancreatic Islets Using Nanospray Desorption Electrospray Ionization Mass Spectrometry. *Anal. Chem.* **2018**, *90*, 6548–6555. [CrossRef]
69. Yin, L.; Zhang, Z.; Liu, Y.; Gao, Y.; Gu, J. Recent advances in single-cell analysis by mass spectrometry. *Analyst* **2019**, *144*, 824–845. [CrossRef]
70. Pigeau, G.M.; Kolic, J.; Ball, B.J.; Hoppa, M.B.; Wang, Y.W.; Rückle, T.; Woo, M.; Manning Fox, J.E.; MacDonald, P.E. Insulin granule recruitment and exocytosis is dependent on p110gamma in insulinoma and human beta-cells. *Diabetes* **2009**, *58*, 2084–2092. [CrossRef]
71. Bagatolli, L.A.; Parasassi, T.; Fidelio, G.D.; Gratton, E. A Model for the Interaction of 6-Lauroyl-2-(N,N-dimethylamino)naphthalene with Lipid Environments: Implications for Spectral Properties. *Photochem. Photobiol.* **1999**, *70*, 557–564. [CrossRef]
72. Parasassi, T.; De Stasio, G.; Ravagnan, G.; Rusch, R.; Gratton, E. Quantitation of lipid phases in phospholipid vesicles by the generalized polarization of Laurdan fluorescence. *Biophys. J.* **1991**, *60*, 179–189. [CrossRef]
73. Barelli, H.; Antonny, B. Lipid unsaturation and organelle dynamics. *Curr. Opin. Cell Boil.* **2016**, *41*, 25–32. [CrossRef] [PubMed]
74. Greenspan, P.; Mayer, E.P.; Fowler, S.D. Nile red: A selective fluorescent stain for intracellular lipid droplets. *J. Cell Boil.* **1985**, *100*, 965–973. [CrossRef]
75. Bongiovanni, M.N.; Godet, J.; Horrocks, M.H.; Tosatto, L.; Carr, A.R.; Wirthensohn, D.C.; Ranasinghe, R.T.; Lee, J.-E.; Ponjavic, A.; Fritz, J.V.; et al. Multi-dimensional super-resolution imaging enables surface hydrophobicity mapping. *Nat. Commun.* **2016**, *7*, 13544. [CrossRef]
76. Diaz, G.; Melis, M.; Batetta, B.; Angius, F.; Falchi, A.M. Hydrophobic characterization of intracellular lipids in situ by Nile Red red/yellow emission ratio. *Micron* **2008**, *39*, 819–824. [CrossRef]
77. Cutrale, F.; Trivedi, V.; Trinh, L.A.; Chiu, C.-L.; Choi, J.M.; Artiga, M.S.; Fraser, S.E. Hyperspectral phasor analysis enables multiplexed 5D in vivo imaging. *Nat. Methods* **2017**, *14*, 149–152. [CrossRef]
78. Stringari, C.; Nourse, J.L.; Flanagan, L.A.; Gratton, E. Phasor Fluorescence Lifetime Microscopy of Free and Protein-Bound NADH Reveals Neural Stem Cell Differentiation Potential. *PLoS ONE* **2012**, *7*, e48014. [CrossRef]
79. Gao, M.; Huang, X.; Song, B.-L.; Yang, H. The biogenesis of lipid droplets: Lipids take center stage. *Prog. Lipid Res.* **2019**, *75*, 100989. [CrossRef]
80. Ferri, G.; Digiacomo, L.; Lavagnino, Z.; Occhipinti, M.; Bugliani, M.; Cappello, V.; Caracciolo, G.; Marchetti, P.; Piston, D.W.; Cardarelli, F. Insulin secretory granules labelled with phogrin-fluorescent proteins show alterations in size, mobility and responsiveness to glucose stimulation in living beta-cells. *Sci Rep.* **2019**, *9*, 2890. [CrossRef]

81. Makhmutova, M.; Liang, T.; Gaisano, H.; Caicedo, A.; Almaça, J. Confocal Imaging of Neuropeptide Y-pHluorin: A Technique to Visualize Insulin Granule Exocytosis in Intact Murine and Human Islets. *J. Vis. Exp.* **2017**. [CrossRef] [PubMed]
82. Sparvero, L.J.; Amoscato, A.A.; Kochanek, P.M.; Pitt, B.R.; Kagan, V.E.; Bayir, H.; Bayır, H. Mass-spectrometry based oxidative lipidomics and lipid imaging: Applications in traumatic brain injury. *J. Neurochem.* **2010**, *115*, 1322–1336. [CrossRef] [PubMed]
83. Avrahami, D.; Klochendler, A.; Dor, Y.; Glaser, B. Beta cell heterogeneity: An evolving concept. *Diabetology* **2017**, *60*, 1363–1369. [CrossRef] [PubMed]

Sample Availability: Cell lines and probes are commercially available.

© 2019 by the authors. Licensee MDPI, Basel, Switzerland. This article is an open access article distributed under the terms and conditions of the Creative Commons Attribution (CC BY) license (http://creativecommons.org/licenses/by/4.0/).

Article

The Entrapment of Somatostatin in a Lipid Formulation: Retarded Release and Free Radical Reactivity

Anna Vita Larocca [1,†], Gianluca Toniolo [2,†], Silvia Tortorella [3,‡], Marios G. Krokidis [2], Georgia Menounou [3], Giuseppe Di Bella [4], Chryssostomos Chatgilialoglu [1,3] and Carla Ferreri [1,3,*]

1. R&D Laboratory, Lipinutragen srl, Via Piero Gobetti 101, 40129 Bologna, Italy
2. Institute of Nanoscience and Nanotechnology, N.C.S.R. "Demokritos", 15310 Agia Paraskevi Attikis, Greece
3. ISOF, Consiglio Nazionale delle Ricerche, Via P. Gobetti 101, 40129 Bologna, Italy
4. Di Bella Foundation, Via G. Marconi 51, 40122 Bologna, Italy
* Correspondence: carla.ferreri@isof.cnr.it; Tel.: +39-051-639-8289
† These authors contributed equally to the work.
‡ Present address: Università di Bologna, Dipartimento di Chimica Industriale, Viale del Risorgimento 4, 40136 Bologna, Italy.

Academic Editor: Chryssostomos Chatgilialoglu
Received: 10 August 2019; Accepted: 22 August 2019; Published: 25 August 2019

Abstract: The natural peptide somatostatin has hormonal and cytostatic effects exerted by the binding to specific receptors in various tissues. Therapeutic uses are strongly prevented by its very short biological half-life of 1–2 min due to enzymatic hydrolysis, therefore encapsulation methodologies are explored to overcome the need for continuous infusion regimes. Multilamellar liposomes made of natural phosphatidylcholine were used for the incorporation of a mixture of somatostatin and sorbitol dissolved in citrate buffer at pH = 5. Lyophilization and reconstitution of the suspension were carried out, showing the flexibility of this preparation. Full characterization of this suspension was obtained as particle size, encapsulation efficiency and retarded release properties in aqueous medium and human plasma. Liposomal somatostatin incubated at 37 °C in the presence of Fe(II) and (III) salts were used as a biomimetic model of drug-cell membrane interaction, evidencing the free radical processes of peroxidation and isomerization that transform the unsaturated fatty acid moieties of the lipid vesicles. This study offers new insights into a liposomal delivery system and highlights molecular reactivity of sulfur-containing drugs with its carrier or biological membranes for pharmacological applications.

Keywords: liposomal somatostatin; retarded delivery; free radicals; isomerization; trans lipid; peroxidation

1. Introduction

Somatostatin (also known as somatotropin release-inhibiting factor or growth hormone release-inhibiting factor, SST) is a cyclic peptide of 14 amino acids first isolated from ovine hypothalamic and known to inhibit the secretion of multiple hormones (e.g., growth hormone, insulin, glucagon, gastrin), gastric acid and pancreatic enzymes. In the central nervous system, this peptide acts as a neurotransmitter and affects locomotor activity and cognitive functions [1]. SST exerts its activity by binding to at least five different subtypes of specific receptors (SSTR 1-5) located on the target cells with a wide range of biological effects that can be exploited for the treatment of a variety of human diseases [2–4]. Signaling pathways activated by the SST-receptor interaction (such as, mitogen-activated protein kinase pathway, inhibition of adenylyl cyclase, activation of phosphotyrosine phosphatase, changes in plasma membrane calcium and potassium channel activity) are evoked for the antineoplastic and anti-proliferative activities [5,6]. The therapeutic potential of SST is strongly limited by its very short half-life of less than 1–2 min in plasma [7], as expected for several neuropeptide hormones that must be rapidly inactivated after their release and interaction with their receptors. In fact, in neuronal cell cultures the degradation of SST was measured, and membrane-bound proteases or proteases released into the incubation medium were found to be responsible for this inactivation [8].

Solutions to the short lifetime have been proposed based on two main approaches: (a) preparation of SST analogues attaching one or more groups to the peptide molecule, such as the N-methyl group, able to act as a shield for the in vivo hydrolysis [9]; (b) encapsulation of SST or its analogues in polymeric materials, such as poly(alkyl cyanoacrylate) nanocapsules [10], or in natural phospholipid vesicles, eventually coated with agents ensuring circulation in blood, such as polyethylene glycol (PEG), that protects liposomes from recognition and rapid removal from circulation by the phagocyte system [11]. SST, its analogue octreotide or other synthetic analogues are also studied in liposomal formulations combined with antitumoral drugs like daunorubicin or with radiopharmaceuticals, due to their ability to target SSTR-rich tumoral cells [12–15]. The biocompatibility and biodegradability of liposomes have an overwhelming importance thus motivating research to deepen the use of natural phospholipids as drug delivery components. It is worth noting that, as far as natural lipid formulations are concerned, the choice of the fatty acid quality is important for liposome behavior, since it influences fluidity and permeability properties connected with drug release [16]. In the frame of our interest in liposomes as tools for investigations in biomimetic chemistry, we recently proposed trans-double bond-containing liposomes for drug delivery systems with differentiated behavior respect to those made of natural unsaturated lipids, which display cis-geometry [17]. In fact, trans double bonds are well known to change the molecular properties of the lipid assembly, compared to the cis double bonds, shifting toward less fluid and permeable double lipid layers [18]. Moreover, the cis-trans double bond isomerization can occur as an endogenous process in cells, due to the formation of sulfur-centered radicals during oxidative stress and their reaction with unsaturated lipids [19]. Recently, lipid isomerization and peroxidation were observed in liposomes with iron-binding antitumoral drugs such as bleomycin in the presence of thiols [20]. We were intrigued by the SST structure, which contains the disulfide bond between Cys-3 and Cys14 (Figure 1) that under free radical conditions can produce thiyl radicals and react with unsaturated lipids. Such reactivity for SST has not yet been reported.

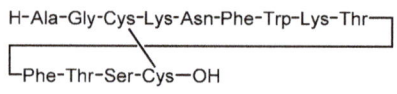

Somatostatin

Figure 1. Amino acid sequence of the peptide somatostatin (SST).

Here we report the entrapment of SST in a liposome formulation with full characterization of size and properties including encapsulation efficiency, retarded release and the possibility of lyophilization-reconstitution of the emulsion. Liposomes containing SST and unsaturated fatty acid moieties were used also as biomimetic model of iron-induced free radical stress, evidencing the occurrence of lipid isomerization and peroxidation processes. This study contributes to new knowledge in the interdisciplinary field of liposomes and drug mechanism.

2. Results and Discussion

Lipid formulations used in this study were made of natural L-α-phosphatidyl choline (PC) from soybean lecithin, of general formula shown in Figure 2A, which has choline as hydrophilic head whereas the hydrophobic tails are composed by different fatty acids. In Figure 2B the fatty acid moieties of soybean lecithin are shown as relative percentages (%rel). In order to study the chemical behavior of SST, we used also the synthetic phospholipid 1-palmitoyl-2-oleoyl phosphatidyl choline (POPC), which contains the saturated fatty acid (SFA) palmitic acid (C16:0) and the monounsaturated fatty acid (MUFA) oleic acid (9*cis*-C18:1) (Figure 2C).

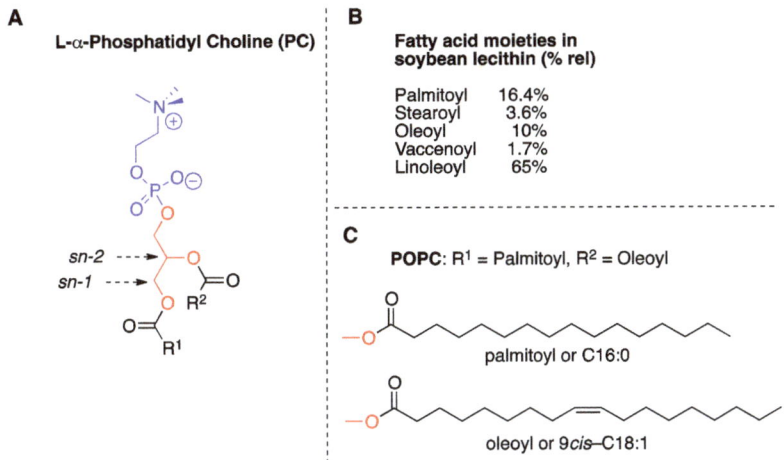

Figure 2. Lipid structures: (**A**) L-α-phosphatidyl choline (PC) general structure; (**B**) list of fatty acids present in soybean lecithin and the relative percentages obtained from the gas chromatographic analysis; (**C**) the structures of palmitic acid and oleic acid present in 1-palmitoyl-2-oleoyl phosphatidyl choline (POPC).

Experiments of SST release and resistance to degradation were carried out with the soybean lecithin liposome preparations, exploring the behavior of the liposomes after lyophilization-reconstitution. POPC liposomes and soybean lecithin liposomes containing SST were then used as biomimetic models of metal-induced free radical stress to highlight the reactivity between peptide and unsaturated lipids.

Soybean lecithin liposomes were prepared for the release experiments of this study. L-α-phosphatidylcholine from soybean was dissolved in ethanol to reach 5 mM concentration, and this ethanolic solution was evaporated under vacuum in order to obtain a lipid film without traces of solvent. In tridistilled water, 0.1 M citrate buffer at pH 5 was prepared and added with 2 mM sorbitol. The peptide SST (0.5 mg/mL) was dissolved in this buffer. At this pH, closer to its isoelectric point (pH = 5.6) [21], the peptide was shown to be stable as established by LC analyses at different time points (for methods see Experimental part). The peptide solution was added to the lipid film, calculating a lipid:peptide ratio of 20:1 in all preparations. After 5 min at the vortex, the suspension was characterized by DLS (Dynamic Light Scattering) for evaluating particle size and polydispersity index (PDI) prior to be lyophilized. The mean diameter was found to be 159.5 ± 12 nm (PDI 0.16 ± 0.01). The measurements of the nanoemulsion (NE) mean diameter were run in an interval time of 5 h, showing that the size does not undergo variations. It is worth noting that after lyophilization and reconstitution the mean diameter was found to be similar to the original preparation. The encapsulation efficiency (EE) in soybean lecithin liposomes was evaluated, resulting to be 66.9 ± 2.9%. Using liposomes formed by soybean lecithin two in vitro experiments were designed for the release of SST: in aqueous citrate buffer (pH 5) and, in order to test the resistance to peptidases, in human plasma.

2.1. Release Experiments in Aqueous Citrate Buffer

For each set of experiments 5 mg of SST were used, and after lyophilization the suspension was reconstituted using 10 mL of tridistilled water.

The release experiment was designed to analyze not only the directly released SST (in the aqueous phase of the nanoemulsion (NE)—direct analysis) but also the amount of non-released SST, still in the lipid droplets (indirect analysis). Since SST can have stability problems, the determination of the peptide remained in the lipid phase was meant to collect important additional information for the characterization of the release profile of the NE. The direct analysis consisted of measuring the concentration of SST in the aqueous fraction of the emulsion after centrifugation, using LC/MS analysis following a published protocol [22]. In the indirect analysis the peptide was extracted from the lipid droplets, separated by centrifugation, and analyzed again by LC/MS.

Moreover, the release experiments were run using the NE emulsions reconstituted from the lyophilized samples, following two different protocols. In Protocol 1 (data reported in Figure 3), the freeze-dried NEs containing SST (0.5 mg/mL) were reconstituted in water to recreate the same conditions of the original preparation and the samples were left for the indicated times; the quantities of SST at each time point were then calculated as shown in Figure 3A,B: (a) the first one from the analyses of each sample treated by centrifugation and then considering both SST quantities in the aqueous and lipid phases. In Figure 3A these two quantities give ca 93% of released SST which is constant along the time; (b) the second quantification considering the % SST released in the aqueous phase. In Figure 3B it is clearly shown that the peptide is diminishing along the time when considering as starting quantity the encapsulated SST (0.5 mg/mL).

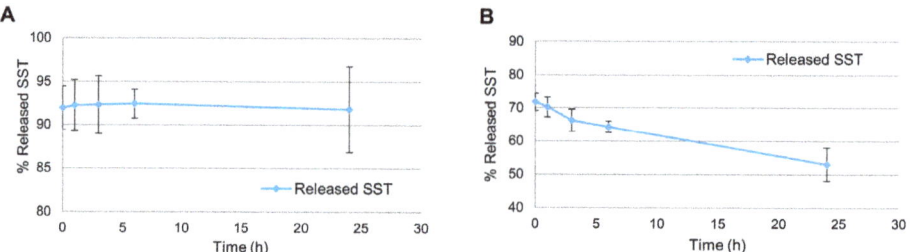

Figure 3. Released somatostatin (SST) from the nanoemulsions (NEs) following protocol 1 in 0.1 M buffer citrate pH 5.0. (**A**) % released SST considering the sum of the SST fraction found in aqueous phase (direct analysis) and the SST fraction in the lipid droplets (indirect analysis) at each time point. (**B**) % released SST in the aqueous phase considering as 100% the initial amount of SST used in the preparation procedure (0.5 mg/mL).

We also wanted to determine the SST released from the NEs with the removal of the aqueous phase replaced at each time point by fresh buffer. This led to the second protocol (Protocol 2—data in Figure 4) which involved the centrifugation step to remove the aqueous phase of the NE; the resulting SST-containing lipid pellet was then suspended in 0.1 M citrate buffer pH 5.0 and the release experiment was carried out at the indicated times. Again, in Figure 4A,B the SST quantities are reported in a complementary way: in Figure 4A it is calculated from the direct and indirect analyses, and in Figure 4B it is reported the release in buffer from the direct analysis.

In all cases LC/MS using SST as calibration and standard reference as described in the Materials and Methods were carried out at the time points (0, 1 h, 3 h, 6 h, 24 h) when samples are collected and analyzed. The % of released SST were calculated considering as 100% the starting amount of SST used for the preparation of the NEs (0.5 mg/mL).

The two protocols gave clearer insight into the behavior of SST and release dynamics of the NE. In particular, the first protocol gave information on the behavior of the NE and the variations in the partition of SST between the two phases along the time; the second one informed on the diffusion of SST from the lipid droplets to the less concentrated aqueous phase. Also, since the experiments were run for 24 h, the stabilizing effect of NE on SST could be evaluated.

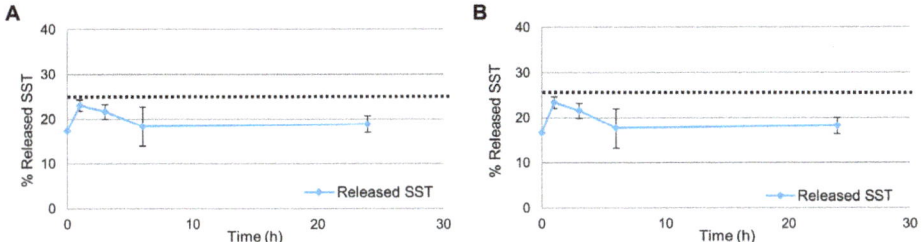

Figure 4. Released SST from the NEs following protocol 2. (**A**) Considering as 100% the sum of the fraction released, the fraction still encapsulated and the non-encapsulated one at each time point. (**B**) Considering as 100% the amount of SST expected in the samples from the centrifugation step. The dotted line represents 25%, the amount of SST available after the centrifugation to remove the non-encapsulated fraction of SST.

In Figure 3A the SST quantity was found ca. 92% at time 0 and its concentration remained steady up to 24 h. Figure 3B shows that the SST released in the aqueous phase at time 0 was ca. 72% and decreased to 53% after 24 h. Interestingly, at time 0 the starting amount of SST was different in the two graphs: part of the SST was already unavailable at time 0, corresponding to roughly 22%, as it can be calculated considering a) the value at time 0 of ca. 92% given by the sum of the fraction released and the fraction in the lipid droplets, as shown in Figure 3A, and b) the value at time 0 of ca. 28% SST not found in aqueous phase from the initial SST concentration, as shown in Figure 3B. The results in Figure 3B also proved that some degradation/aggregation processes can occur to SST from its original preparation during lyophilization and manipulation steps, since there is a decrease in the peptide concentration found after reconstitution. Nevertheless, when SST was removed from the aqueous phase, part of the SST associated to the lipid phase was released and went to the aqueous phase, which is shown in Figure 3A, where the released/total ratio remained constant up to 24 h. By comparing the two graphs it is clear that, while some SST is degraded in the aqueous phase, some SST in the lipid droplets was released in order to maintain a constant presence of peptide in the aqueous fraction. Since SST is constantly released into the aqueous phase of the NE for 24 h, the system can be considered capable of releasing the peptide in a prolonged manner.

In protocol 2 the freeze-dried NEs were first reconstituted using the aqueous phase and then centrifuged to remove the non-encapsulated fraction of SST, to understand the quantity of the drug immediately released by the droplets upon reconstitution. The pellet obtained after the centrifugation was suspended in buffer at pH 5 to start the release follow up. Again, the results of protocol 2 (Figure 4) were expressed in two ways. In Figure 4A 100% is represented by the sum of SST found with the direct analysis and the indirect analysis and in Figure 4B 100% was considered to be amount of SST used for the preparation of the NEs (0.5 mg/mL).

The aqueous phase removed by centrifugation contained 75% of the peptide in feeding, therefore only 25% was calculated for the release experiments (marked with the dotted line in Figure 4). There is no remarkable difference between Figure 4A,B: in both cases at initial time 17% of the total amount of SST was in the aqueous phase of the NE. After 1 h, the peak of concentration was reached in both conditions (23% of total SST, corresponding to 90% of the available amount). Afterward the concentration started to decrease, being 19% after 24 h (85% of the available drug). The similarity between Figure 4A,B suggests that none or very little degradation occurred during this experiment. Probably this is due to the overall lower concentration of the peptide, and consequently the stabilizing effect of the NE is more evident. Also, it is clear that upon reconstitution SST distributes between the two phases immediately and, apart from the peak at 1 h, the concentration of SST remains constant in both compartments for the whole experiment.

2.2. Release Experiments in Human Plasma

Only direct analysis was used to evaluate the concentration of SST in the experiments of the NE release using human plasma, since the indirect analysis proved to be unsuitable for the LC/MS system when working with plasma. Therefore, 100% of released SST corresponds to amount of SST used for the preparation of the NEs (0.5 mg/mL).

The experiments were run following two protocols similar to the ones described above. The first procedure consisted of reconstituting the freeze-dried NE directly in plasma to start the release experiment (Figure 5A). In the second one, the NE was first reconstituted in water, then underwent centrifugation to remove the non-encapsulated fraction of SST, and the resulting pellet was suspended in human plasma to start the experiment (Figure 5B).

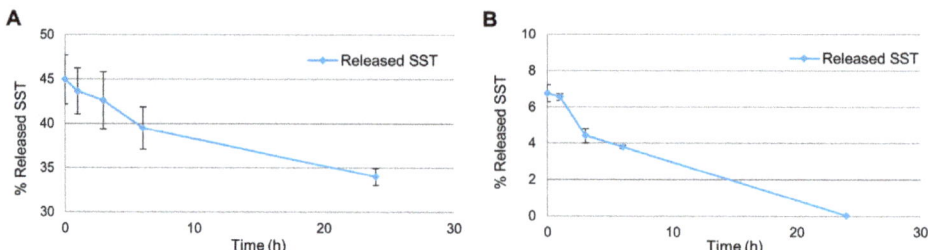

Figure 5. (**A**) Measurement of SST concentration in plasma after reconstituting the NEs directly in plasma (protocol 1). (**B**) Measurement of SST concentration in plasma after removing the non-encapsulated SST fraction (protocol 2). Percentages are calculated from the LC/MS analyses related to the starting concentration of 0.5 mg/mL.

In Figure 5A, the amount of SST measured in the aqueous phase of the NE was 45% of the starting concentration (0.5 mg/mL) at time 0 and decreased to 34% after 24 h. When the NE release profile was studied following protocol 2, roughly 7% of the SST was found in plasma at time 0, and after 24 h only traces of SST were found (0.03%) (Figure 5B). These results imply that there was a decrease in concentration due to degradation of the peptide. Nevertheless, SST is known to have very short half-life in plasma [7] and the NE proved to be efficient in improving the peptide half-life, maintaining its measurable concentration at least for the first 7–8 h.

2.3. Reactivity of SST in Liposome Vesicles

We were interested in the study of the reactivity of the disulfide bridge present in the structure of SST, between the cysteine residues in positions 3 and 14 (Figure 1), since a variety of conditions can influence the reactivity of this functional group, such as the presence of redox-active metal ions. In fact, the chelation ability of metals toward disulfides is known, and it can act to generate catalytic amount of thiyl radicals together as known for Fe(III) and its reduction to Fe(II) [23]. Since we worked under aerobic conditions and it is known that complexes with Fe(II) can be oxidized to Fe(III) under air [24], we used Fe(II) and Fe(III) salts in our experiments in order to follow the liposome reactivity in both cases. On the other hand, as mentioned in the introduction, investigations by some of us demonstrated that iron binding in case of the antitumoral drug bleomycin was able to induce the formation of thiyl radicals, which react with liposome vesicles in terms of cis-trans isomerization and peroxidation of the unsaturated fatty acid moieties of vesicle lipids [20]. We also used the reduced form of SST (red-SST), with the free thiol groups in positions 3 and 14, which is commercially available, to compare the results of the two free cysteine residues with the cystine moiety of the natural peptide in the formation of iron-complex, including the role of thiyl radicals in the consequent isomerization reaction. As matter of facts, the reduction of SST is considered to occur in vivo and is associated to fibril formation relevant to the storage and secretion [25] but was never connected with chemical reactivity.

We used the biomimetic model made of large unilamellar vesicle by the extrusion technique with 200 nm of diameter (LUVET) [26], in the presence of SST or red-SST for investigating membrane phospholipid behavior. Two different compositions of liposomes have been employed: first, the synthetic phospholipid, 1-palmitoyl-2-oleoyl phosphatidyl choline (POPC) was used to study the reactivity of the MUFA moiety of oleic acid (9cis-C18:1, Figure 2C) which is expected to be directly transformed by thiyl radical catalyzed cis-trans isomerization into its corresponding geometrical trans isomer, elaidic acid (9trans-C18:1) by the addition-elimination mechanism represented for the double bond in Figure 6. The second material used for liposome formation was soybean lecithin, containing various percentages of saturated, mono- and poly-unsaturated (MUFA and PUFA) fatty acids (Figure 2B). The PUFA residues of soybean lecithin liposomes can be transformed both by oxidative and isomerization pathways. In particular, linoleic acid (9cis,12cis-C18:2), the most representative PUFA in lecithin (65% of the total fatty acid composition), can give geometrical trans isomers that are separated, recognized and quantified by GC analysis [18,19,26,27]. Lipid peroxidation can be indirectly estimated by the amount of the PUFA linoleic acid in each experiment, evidencing its diminution by using the GC calibration curves vs. the saturated fatty residue of palmitic acid (C16:0), as previously described [20,28]. GC analysis is the gold standard for the fatty acid quantification.

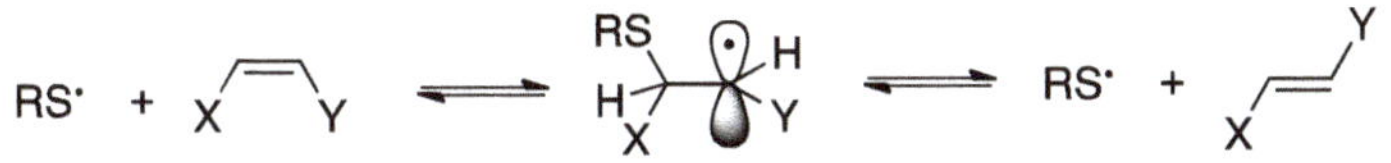

Figure 6. Reaction mechanism for the cis–trans isomerization of a double bond catalyzed by thiyl radicals.

The iron salts used in the experiments were: ferrous (II) ammonium sulfate $Fe(NH_4)(SO_4)_2$ and ferric (III) sulfate $Fe_2(SO_4)_3$. The salts at concentration of 100 µM were added to the liposome suspension (2 mM) and then the addition of SST or red-SST (100 µM) to the vesicle aqueous suspension was carried out by syringe pump, keeping the vials at 37 °C in an orbital shaker for the indicated reaction time. The experiments were carried out in aerobic and anaerobic conditions and the results are summarized in Table 1. Blank experiments included liposome reaction without the peptide and in presence/absence of iron, in order to better estimate the drug contribution to lipid reactivity. The blank experiments without SST with POPC, and with or without Fe salts, gave a not detectable quantity of trans isomers (results not shown).

Table 1. Formation of trans isomers of monounsaturated fatty acid (MUFA) and consumption of polyunsaturated fatty acids (PUFA) residues[1] in liposome aqueous suspensions (2 mM) treated with 100 µM SST/red-SST and 100 µM Fe salts at 37 °C (n = 3 of each experiment).

Entry	Liposome	O_2	Peptide	Fe(II)	Fe(III)	trans-18:1 (%)	trans-18:2 (%)	PUFA Consumption (%)
1	POPC [2]	no	SST		x	tr		
2	POPC [2]	no	SST	x		0.9 ± 0.1		
3	POPC [2]	no	red-SST	x		0.7 ± 0.2		
4	POPC [2]	no	red-SST		x	29.9 ± 0.2		
5	POPC [2]	yes	red-SST		x	5 ± 0.3		
6	soybean lecithin [3]	yes	SST			nd	nd	2.2 ± 1.4
7	soybean lecithin [3]	yes	SST	x		tr	tr	95.9 ± 0.3
8	soybean lecithin [3]	yes	SST	x[4]		0.3 ± 0.0	0.2 ± 0.1	75.5 ± 0.4
9	soybean lecithin [3]	yes	red-SST	x[4]		0.4 ± 0.1	0.5 ± 0.1	84.1 ± 2.6
10	soybean lecithin [3]	yes	SST		x[4]	tr	tr	35.9± 0.2
11	soybean lecithin [3]	yes	red-SST		x[4]	tr	tr	92.3± 0.4

Nd = not detectable; tr = traces (<0.1%) [1] The fatty acid residues were obtained from the liposome suspension after lipid extraction and transesterification, as described in the Materials and Methods. [2] 24 h incubation. [3] 8 h incubation. [4] incubation with 30 µM Fe(II) salt.

With SST-Fe(III) salt and SST-Fe(II) salt under anaerobic conditions (entries 1 and 2), MUFA trans isomer was detected at <1% with Fe(II) salt under anaerobic condition. With the reduced form of the peptide and Fe(II) salt still the reactivity for isomerization was low (entry 3). Instead, red-SST with Fe(III) salt under anaerobic conditions gave the most effective isomerization reaction, reaching 30% after 24 h incubation (entry 4). In aerobic conditions the same experiment gave isomerization, although at low extents (5%) (entry 5).

When soybean lecithin liposomes are used, the reactivity of polyunsaturated fatty acids (PUFA) was followed-up, and the scenario changed in particular evidencing the consumption of PUFA due to peroxidation processes under aerobic conditions, as depicted first for bleomycin-Fe complexes [20]. The role of iron salt is important for the reactivity of the system, since SST under aerobic conditions but without Fe salts induced a slight peroxidation process and no isomerization after 24 h (entry 6, Table 1). The reactivity of PUFA-containing vesicles in the presence of both SST and Fe(II) salt increased toward the peroxidation of polyunsaturated residues, shortening the reaction time to 8 h instead of 24 h to detect an almost total PUFA consumption (96%, see entry 7, Table 1). At this time, the strong PUFA consumption did not allow for evidencing any isomerization process. Lowering Fe(II) concentration to 30 µM the peroxidation process decreased to ca. 75% (entry 8). When the low concentration of Fe(II) salt is used with red-SST, peroxidation slightly increased compared to the cyclic form of the peptide (entry 9). Finally, with 30 µM Fe(III) salt the peroxidation process occurred both with SST and red-SST, however the latter was found to be more efficient (cfr., entry 10 and entry 11).

Further work is needed for the full understanding of the complex reactivity involving metal-sulfur adducts in liposomes, taking into account that the different isomerization yields can be also attributed to the diffusion of reactive radical species within the lipid bilayer. It is worth mentioning that the interaction of SST with its membrane receptors [2,5,6] brings this peptide in close contact with the lipid bilayer, therefore its reactivity with lipids cannot be ruled out. The present results obtained with the biomimetic model of liposomes using SST or its reduced form highlight the role of sulfur-containing biomolecules as trigger of oxidative radical-based processes which can transform unsaturated fatty acids, which are important constituents of cell membranes. This is a very interesting reactivity, which should be taken into account since lipid-peptide bioconjugation is used in liposome technology [29]. Lipid peroxidation and isomerization have been already evidenced for antitumoral drugs [20], and nowadays the oxidative aspect is considered not only as a side effect induced by anticancer therapy, but also as a condition to create oxidant-antioxidant unbalance in cancer cells [30]. This is connected with the metal redox state of the cellular environment that can be an important trigger of radical reactions, as shown for the reactivity of copper complexes [31,32]. Thiol compounds are reactive species in this context as in the new model here proposed for SST, expanding the actual knowledge of the molecular properties of this peptide for pharmacological applications.

3. Conclusions

In this article we described the entrapment of SST in a lipid formulation that increased of 10-folds the stability of the peptide in human plasma. These results are promising for application as a drug delivery system by extension to appropriate experiments in biological context, using cells or murine models, which are ongoing. Furthermore, the inclusion of SST or its acyclic derivative in liposome vesicles has been used as biomimetic model in the presence of iron salts, for evidencing the reactivity between the peptide and its lipid carrier with thiyl radical formation and induction of lipid isomerization/peroxidation processes. New molecular insights are provided that can be developed further for synergic pharmacological strategies.

4. Materials and Methods

Somatostatin (SST) and its acyclic form were received from Bachem, Germany and used with no further purification. POPC and L-α-phosphatidylcholine from soybean were obtained from Avanti Polar Lipids, USA and used without further purification. Formic acid and ammonium formate were purchased from Sigma-Aldrich, Milan and used as received. Acetonitrile, sodium citrate, citric acid and absolute ethanol were bought from Sigma-Aldrich, Milan and used without further purification.

4.1. Preparation of Nanoemulsions (NE)

To prepare 10 mL of NE, 0.2 g of L-α-phosphatidylcholine (2% w/v) were dissolved in absolute ethanol, placed in a round-bottom flask and then the solvent was removed under reduced pressure to yield a thin and uniform lipid film on the wall of the flask. Five milligrams of SST (0.5 mg/mL) and 72.8 mg of sorbitol (2 mM) were dissolved in 10 mL of 0.1 M citrate buffer. The SST solution was then added to the lipid film and the flask content was maintained under agitation with vortex for 5 min. The so-obtained NE was freeze-dried and, when needed for further experiments, reconstituted with 10 mL of mq water.

4.2. LC/MS Analysis

Agilent 1260 Infinity automated LC/MS purification system was used, and the analyses were run with a reverse phase C-18 column (ZORBAX SB-C18 Rapid Resolution HT 2.1 × 50 mm 1.8 micron 600 Bar). A Phenomenex HPLC guard cartridge was used as well. The mobile phases were (A) H_2O + 0.1% formic acid and (B) Acetonitrile + 0.1% formic acid. The samples were eluted with a linear gradient of B from 5% to 90% in 15 min, B was then decreased to 5% in 5 min (20 min) and kept so for 5 min (25 min).

4.3. In Vitro Release Experiments in Buffer

The release experiment was designed to analyze not only the directly released SST (detected in the medium) but also the amount of non-released SST, still entrapped in the lipid droplets (indirect analysis). Two different protocols were followed: (1) the freeze-dried NE was reconstituted in citrate buffer pH 5.0 0.1 M to start the release experiment and after each time point the analyses of the SST present in the resulting aqueous and lipid phases were carried out as described below; (2) after reconstitution in water, the NE was immediately centrifuged, the lipid fraction was isolated and resuspended in fresh citrate buffer pH 5.0 0.1 M to start the experiment. SST-containing NE were kept in a horizontal shaker at 37 °C (stirring rate 100 rpm). Reached the time point (0 min, 1 h, 3 h, 6 h, 24 h), three samples for each protocol of 0.2 mL were collected, centrifuged (15,000 rpm × 5 min × 4 °C) and 50 µL of supernatant were directly analyzed via LC/MS (direct analysis). The remaining lipid pellets were then dissolved in a mixture of hexane/formate buffer pH 3.0 (0.2 mL each, total volume 0.4 mL), vortexed for 5 min and after 10 min 50 µL of aqueous phase were withdrawn and analyzed with LC/MS (indirect analysis). The experiment was carried out three times for statistical analysis.

4.4. In Vitro Release Profile in Human Plasma

Two different protocols were followed according also to literature [33]: (1) the freeze-dried NE was reconstituted directly in human plasma to start the release experiment; (2) after reconstitution in tridistilled water, the NE was immediately centrifuged, the lipid fraction was isolated and suspended in human plasma to start the experiment. SST-containing NE were kept in a horizontal shaker at 37 °C (stirring rate 100 rpm). Reached the time point (0 min, 1 h, 3 h, 6 h, 24 h), three samples of 0.2 mL for each protocol were collected and centrifuged (15,000 rpm × 5 min × 4 °C). To 100 µL of the so-obtained supernatant were added 100 µL of TCA 6% (final concentration of TCA 3%) and the product was placed in ice bath for 10 min and centrifuged again (15,000 rpm × 5 min × 4 °C). 100 µL of the resulting supernatant were collected, quenched with a solution of NaOH 1 M and analyzed via LC/MS [1]. The indirect analysis was not performed due to incompatibility with the HPLC column.

4.5. Liposome Experiments

Large unilamellar vesicles were prepared with known methodologies [26]. Briefly, POPC or soybean lecithin were dissolved in chloroform and then evaporated to a thin film in a test tube under argon stream. In the next step the tube remained under vacuum for 30 min. Degassed water was added and multilamellar vesicles were formed by vortexing under argon atmosphere for 7 min. In order to obtain the LUVET vesicles of 200 nm diameter the emulsions were extruded by passage for 19 times through two polycarbonate membranes with the specific pore dimension. In a 4 mL vial 2 mM vesicle suspension was added together with the iron complex (0.1 mM), and the vial was kept under incubation at 37 °C adding somatostatin solution (0.1 mM) dropwise by a syringe pump (0.5 mm/min). The reaction mixture was worked up after the reported time (8 or 24 h) by adding 2:1 chloroform/methanol (1 mL) to extract lipids. The organic phase was dried over anhydrous sodium sulfate and evaporated under vacuum at room temperature to dryness. The resulting phospholipid extract was then treated with 0.5 M KOH/MeOH for 10 min at ambient temperature, converting them to the corresponding fatty acid methyl esters (FAME). The reaction was quenched with the addition of tridistilled water and an extraction with n-hexane followed. The organic layer containing the corresponding FAME was analyzed by GC under known conditions to examine cis and trans fatty acid isomers [25,27]. For the experiments under anaerobic conditions, all the solutions were degassed with argon for 15 min and the addition of all reagents took place under an argon stream. Anaerobic conditions were maintained during the incubation period by creating pressure of argon inside the reaction vial.

Author Contributions: Conceptualization, C.F.; Methodology, C.F., C.C., M.G.K., G.T., S.T., A.V.L., G.M.; Data Analysis, C.C., G.D.B. and C.F.; Writing—Original Draft Preparation, C.F. and C.C.; Writing—Review and Editing, C.F., C.C., G.T., G.D.B.; Supervision, C.F. All authors drafted, read and approved the final version of the manuscript.

Funding: This project has received funding from the European Union's Horizon 2020 research and innovation program under grant agreement No. 642023 (ClickGene) in terms of the salaries for G.T. and G.M. The support of the Di Bella Foundation for the grants given to S.T. and A.V.L. is gratefully acknowledged.

Acknowledgments: We thank Lipinutragen srl for the use of its R&D Laboratory and facilities to realize part of the experimental work herein described.

Conflicts of Interest: The authors declare no conflict of interest. The funders had no role in the design of the study; in the collection, analyses, or interpretation of data; in the writing of the manuscript, or in the decision to publish the results.

References

1. Weckbecker, G.; Raulf, F.; Stolz, B.; Bruns, C. Somatostatin analogs for diagnosis and treatment of cancer. *Pharmacol. Ther.* **1994**, *60*, 245–264. [CrossRef]
2. Raulf, F.; Pérez, J.; Hoyer, D.; Bruns, C. Differential expression of five somatostatin receptor subtypes, SSTR1-5, in the CNS and peripheral tissue. *Digestion* **1994**, *55* (Suppl. 3), 46–53. [CrossRef]
3. Lamberts, S.W.; Krenning, E.P.; Reubi, J.C. The role of somatostatin and its analogs in the diagnosis and treatment of tumors. *Endocr. Rev.* **1991**, *12*, 450–482. [CrossRef] [PubMed]
4. Scarpignato, C.; Pelosini, L. Somatostatin analogs for cancer treatment and diagnosis: An overview. *Chemotherap* **2001**, *47* (Suppl. 2), 1–29. [CrossRef]
5. Florio, T.; Morini, M.; Villa, V.; Arena, S.; Corsaro, A.; Thellung, S.; Culler, M.D.; Pfeffer, U.; Noonan, D.M.; Schettini, G.; et al. Somatostatin inhibits tumor angiogenesis and growth via somatostatin receptor-3-mediated regulation of endothelial nitric oxide synthase and mitogen-activated protein kinase activities. *Endocrinology* **2003**, *144*, 1574–1584. [CrossRef] [PubMed]
6. Lahlou, H.; Guillermet, J.; Hortala, M.; Vernejoul, F.; Pyronnet, S.; Bousquet, C.; Susini, C. Molecular signaling of somatostatin receptors. *Ann. N. Y. Acad. Sci.* **2004**, *1014*, 121–131. [CrossRef] [PubMed]
7. Rai, U.; Thrimawithana, T.R.; Valery, C.; Young, S.A. Therapeutic uses of somatostatin and its analogues: Current view and potential applications. *Pharmacol. Ther.* **2015**, *152*, 98–110. [CrossRef] [PubMed]
8. Lucius, R.; Mentlein, R. Degradation of the neuropeptide somatostatin by cultivated neuronal and glial cells. *J. Biol. Chem.* **1991**, *266*, 18907–18913. [PubMed]

9. Biron, E.; Chatteriee, J.; Ovadia, O.; Langenegger, D.; Brueggen, J.; Hoyer, D.; Schmid, H.A.; Jelinek, R.; Gilon, C.; Hoffman, A.; et al. Improving oral bioavailability of peptides by multiple N-methylation: Somatostatin analogues. *Angew. Chem. Int. Ed.* **2008**, *27*, 2505–2599. [CrossRef]
10. Demgé, C.; Vonderscher, J.; Marbach, P.; Pinget, M. Poly(alkyl cyanoacrylate) nanocapsules as delivery system in the rat for octreotide, a long-acting somatostatin analogue. *J. Pharm. Pharm.* **1997**, *49*, 949–954. [CrossRef]
11. Sanarova, E.; Lantsova, A.; Oborotova, N.; Polozkova, A.; Dmitrieva, M.; Orlova, O.; Nikolaeva, L.; Borisova, L.; Shprakh, Z. Development of a liposomal dosage form for a new somatostatin analogue. *Indian J. Pharm. Sci.* **2019**, *81*, 146–149. [CrossRef]
12. Helbok, A.; Rangger, C.; von Guggenberg, E.; Saba-Lepek, M.; Radolf, T.; Thurner, G.; Andreae, F.; Prassi, R.; Decristoforo, C. Targeting properties of peptide-modified radiolabeled liposomal nanoparticles. *Nanomed. Nanotechnol. Biol. Med.* **2012**, *8*, 112–118. [CrossRef] [PubMed]
13. Perche, F.; Torchilin, V.P. Recent trends in multifunctional liposomal nanocarriers for enhanced tumor targeting. *J. Drug Deliv.* **2013**, *2013*, 705265. [CrossRef] [PubMed]
14. Ju, R.J.; Cheng, L.; Peng, X.M.; Wang, T.; Li, C.Q.; Song, X.L.; Liu, S.; Chao, J.P.; Li, X.T. Octreotide-modified liposomes containing daunorubicin and dihydroartemisinin for treatment of invasive breast cancer. *Artif. Cells Nanomed. Biotechnol.* **2018**, *46* (Suppl. 1), 616–628. [CrossRef]
15. Chatzisideri, T.; Leonidis, G.; Sarli, V. Cancer-targeted delivery systems based on peptides. *Future Med. Chem.* **2018**, *10*, 2201–2226. [CrossRef] [PubMed]
16. Arouri, A.; Mouritsen, O.G. Membrane-perturbing effect of fatty acids and lysolipids. *Prog. Lipid Res.* **2013**, *52*, 130–140. [CrossRef]
17. Giacometti, G.; Marini, M.; Papadopoulos, K.; Ferreri, C.; Chatgilialoglu, C. Trans-double bond-containing liposomes as potential carriers for drug delivery. *Molecules* **2017**, *22*, 2082. [CrossRef]
18. Ferreri, C.; Pierotti, S.; Barbieri, A.; Zambonin, L.; Landi, L.; Rasl, S.; Luisi, P.L.; Barigelletti, F.; Chatgilialoglu, C. Comparison of phosphatidylcholine vesicle properties related to geometrical isomerism. *Photochem. Photobiol.* **2006**, *82*, 274–280. [CrossRef]
19. Chatgilialoglu, C.; Ferreri, C.; Melchiorre, M.; Sansone, A.; Torreggiani, A. Lipid geometrical isomerism: From chemistry to biology and diagnostics. *Chem. Rev.* **2014**, *114*, 255–284. [CrossRef]
20. Cort, A.; Ozben, T.; Sansone, A.; Barata-Vallejo, S.; Chatgilialoglu, C.; Ferreri, C. Bleomycin-induced trans lipid formation in cell membranes and in liposome models. *Org. Biomol. Chem.* **2015**, *13*, 1100–1105. [CrossRef]
21. Isoelectric Point. Available online: https://www.mybiosource.com/sst-recombinant-protein/somatostatin-sst/2011722 (accessed on 24 August 2019).
22. Reithmeier, H.; Herrmann, J.; Göpferich, A. Development and characterization of lipid microparticles as a drug carrier for somatostatin. *Int. J. Pharm.* **2001**, *218*, 133–143. [CrossRef]
23. Sakurai, H.; Yokoyama, A.; Tanaka, H. Studies on the sulfur-containing chelating agents. XXXI. Catalytic effect of copper (II) ion to formation of mixed disulfide. *Chem. Pharm. Bull.* **1971**, *19*, 1416–1423. [CrossRef]
24. Antholine, W.E.; Petering, D.H. On the reaction of iron bleomycin with thiols and oxygen. *Biochem. Biophys. Res. Commun.* **1979**, *90*, 384–389. [CrossRef]
25. Anoop, A.; Ranganathan, S.; Das Dhaked, B.; Nath Jha, N.; Pratihal, S.; Ghosh, S.; Sahav, S.; Kumar, S.; Das, S.; Kombrabail, M.; et al. Elucidating the role of disulfide bond on amyloid formation and fibril reversibility of somatostatin-14. *J. Biol. Chem.* **2014**, *289*, 16884–16903. [CrossRef] [PubMed]
26. Ferreri, C.; Sassatelli, F.; Samadi, A.; Landi, L.; Chatgilialoglu, C. Regioselective cis–trans isomerization of arachidonic double bonds by thiyl radicals: the influence of phospholipid supramolecular organization. *J. Am. Chem. Soc.* **2004**, *126*, 1063–1072. [CrossRef] [PubMed]
27. Ferreri, C.; Chatgilialoglu, C. Trans lipids: The free radical path. *Acc. Chem. Res.* **2005**, *36*, 441–448.
28. Chatgilialoglu, C.; Ferreri, C.; Guerra, M.; Samadi, A.; Bowry, V. The reaction of thiyl radical with methyl linoleate: Completing the picture. *J. Am. Chem. Soc.* **2017**, *139*, 4704–4714. [CrossRef] [PubMed]
29. De la Fuente-Herreruela, D.; Monnappa, A.K.; Muñoz-Úbeda, M.; Morallón-Piñ, A.; Enciso, E.; Sánchez, E.; Giusti, F.; Natale, P.; López-Montero, I. Lipid–peptide bioconjugation through pyridyl disulfide reaction chemistry and its application in cell targeting and drug delivery. *J. Nanobiotechnol.* **2019**, *17*, 77. [CrossRef] [PubMed]

30. Yokohama, C.; Sueyoshi, Y.; Ema, M.; Takaishi, K.; Hisatomi, H. Induction of oxidative stress by anticancer drugs in the presence and absence of cells. *Oncol. Lett.* **2017**, *14*, 6066–6070.
31. Toniolo, G.; Louka, M.; Menounou, G.; Fantoni, N.Z.; Mitrikas, G.; Efthimiadou, E.K.; Masi, A.; Bortolotti, M.; Polito, L.; Bolognesi, A.; et al. [Cu(TPMA)(Phen)](ClO$_4$)$_2$: Metallodrug nanocontainer delivery and membrane lipidomics of a neuroblastoma cell line coupled with a liposome biomimetic model focusing on fatty acid reactivity. *ACS Omega* **2018**, *3*, 15952–15965. [CrossRef] [PubMed]
32. Jomova, K.; Valko, M. Advances in metal-induced oxidative stress and human disease. *Toxicology* **2011**, *283*, 65–87. [CrossRef] [PubMed]
33. Böttger, R.; Hoffmann, R.; Knappe, D. Differential stability of therapeutic peptides with different proteolytic cleavage sites in blood, plasma and serum. *PLoS ONE* **2017**, *12*, e0178943. [CrossRef] [PubMed]

Sample Availability: Samples of the compounds "nanoemulsions of SST" are available from the authors.

© 2019 by the authors. Licensee MDPI, Basel, Switzerland. This article is an open access article distributed under the terms and conditions of the Creative Commons Attribution (CC BY) license (http://creativecommons.org/licenses/by/4.0/).

Article

Formation and Stabilization of Gold Nanoparticles in Bovine Serum Albumin Solution

Iulia Matei, Cristina Maria Buta, Ioana Maria Turcu, Daniela Culita, Cornel Munteanu and Gabriela Ionita *

"Ilie Murgulescu" Institute of Physical Chemistry of the Romanian Academy, Splaiul Independentei 202, 060021 Bucharest, Romania; iulia.matei@yahoo.com (I.M.); butamariacristina@gmail.com (C.M.B.); oana_turcu@yahoo.com (I.M.T.); danaculita@yahoo.co.uk (D.C.); cornel_munteanuro@yahoo.com (C.M.)
* Correspondence: ige@icf.ro

Academic Editor: Chryssostomos Chatgilialoglu
Received: 22 August 2019; Accepted: 17 September 2019; Published: 18 September 2019

Abstract: The formation and growth of gold nanoparticles (AuNPs) were investigated in pH 7 buffer solution of bovine serum albumin (BSA) at room temperature. The processes were monitored by UV-Vis, circular dichroism, Raman and electron paramagnetic resonance (EPR) spectroscopies. TEM microscopy and dynamic light scattering (DLS) measurements were used to evidence changes in particle size during nanoparticle formation and growth. The formation of AuNPs at pH 7 in the absence of BSA was not observed, which proves that the albumin is involved in the first step of Au(III) reduction. Changes in the EPR spectral features of two spin probes, CAT16 and DIS3, with affinity for BSA and AuNPs, respectively, allowed us to monitor the particle growth and to demonstrate the protective role of BSA for AuNPs. The size of AuNPs formed in BSA solution increases slowly with time, resulting in nanoparticles of different morphologies, as revealed by TEM. Raman spectra of BSA indicate the interaction of albumin with AuNPs through sulfur-containing amino acid residues. This study shows that albumins act as both reducing agents and protective corona of AuNPs.

Keywords: gold nanoparticles; albumin; EPR spectroscopy; Raman spectroscopy; circular dichroism

1. Introduction

The research focused on the synthesis and properties of gold nanoparticles (AuNPs) has expanded into the advanced fields of nanomedicine and nanotechnologies [1–4]. The interactions of AuNPs with biomolecules found in biofluids are relevant for medicinal applications like imaging, sensing, drug and gene delivery, and in photothermal therapy. AuNPs are usually stabilized in different solutions by layers of small organic molecules. Sulfur, amino or hydroxyl groups bind to the surface of AuNPs through chemical or physical interactions [5]. Citrate, a nontoxic and relevant species for biological systems, represents both a reducing agent and a stabilization agent for AuNPs in water [6]. Albumin can bind spontaneously to citrate-protected AuNPs [7–10] and the resulting nanosystems are of interest for multicomponent drug system applications.

In recent years, a significant number of studies have reported on the synthesis and properties of AuNPs in the presence of proteins [4,11–14]. Assemblies of AuNPs protected by lysozyme were tested for loading capacity of hydrophilic and hydrophobic molecules like doxorubicin and pyrene, respectively. These nanocarriers have been then coated with albumin in order to facilitate their uptake by cancer cells [11]. Bakshi et al. [12] studied the biochemical properties of BSA-conjugated nanoparticles synthetized in the presence of ionic surfactants using thermally denatured BSA as reducing agent of Au(III). They found that BSA-protected AuNPs do not show any hemolytic response.

In another study, Murawala et al. [4] reported on the ability of BSA to reduce Au^{3+} at low pH, underlying that reduction did not occur in BSA-free Au^{3+} solution, irrespective of the pH. Interestingly, when both Ag$^+$ and Au^{3+} ions are added, Ag^0 acts as reducing agent for Au^{3+} ions (galvanic exchange reaction) and AuNPs are formed. The authors also infer that Trp residues of BSA and not Tyr residues are responsible for Ag$^+$ reduction. The secondary structure of BSA is not altered to a significant extent following its reductive action on Ag$^+$ and AgNPs coating. Similarly, in another study, Xie et al. [13] have shown that the reduction of HAuCl$_4$ with BSA at physiological temperature leads to the formation of triangular and hexagonal gold nanoplates within two days. The reaction rate and particle morphology depend on the temperature, pH and presence of trace amounts of Ag$^+$. The authors concluded that BSA at room temperature has no apparent reduction capability, as no particles were formed within two days.

Basu et al. [14] were the first to report that two proteins, antibodies Rituximab and Cetuximab, can function as reducing and structure-directing agents to produce AuNPs from Au salts. However, they only synthesized AuNPs of 100–1000 nm size and having mostly triangular morphology at acidic pH value. The authors mentioned that experiments at pH 7 and 10 did not lead to the formation of AuNPs. In their analysis, they showed that the type of protecting agent influences the nanoparticle shape. For instance, in the presence of BSA, Au nanotriangles are formed. All NPs were characterized after four weeks of BSA/HAuCl$_4$ incubation at room temperature. In the presence of ascorbic acid as an additional, weak, reducing agent, smaller AuNPs (20–100 nm) of flower-like morphology are obtained at pH 3 in a matter of hours.

In this work, we investigated, through a series of physico-chemical methods, the formation of AuNPs using BSA as reducing and protecting agent, at room temperature. The formation and growth of gold nanoclusters have been monitored by UV-Vis spectroscopy, which allowed us to evidence the appearance of the surface plasmon band and the changes in its position. The process also involves protein denaturation, therefore circular dichroism (CD) spectroscopy was used as a tool to prove the changes in albumin conformation. Dynamic light scattering (DLS) and transmission electron microscopy (TEM) gave information on the particle size in solution and in solid state, respectively. Based on the analyses of the Raman spectra of BSA before and after the formation of AuNPs, we evidenced the interaction of sulfur groups from the albumin chain with AuNPs.

Additionally, electron paramagnetic resonance (EPR) spectroscopy was used to investigate the slow process of Au(III) reduction in albumin solution by following the changes in the parameters of the two spin probes presented in Figure 1: 4-N,N-dimethyl hexadecyl ammonium-2,2,6,6-tetramethylpiperidine-1-oxyl iodide (CAT16) that can report on interactions in various colloidal systems and has affinity for albumin [15–17], and a biradical with a disulfide structure (DIS3) that has been used to study the dynamics of ligands at the AuNPs surface [18,19].

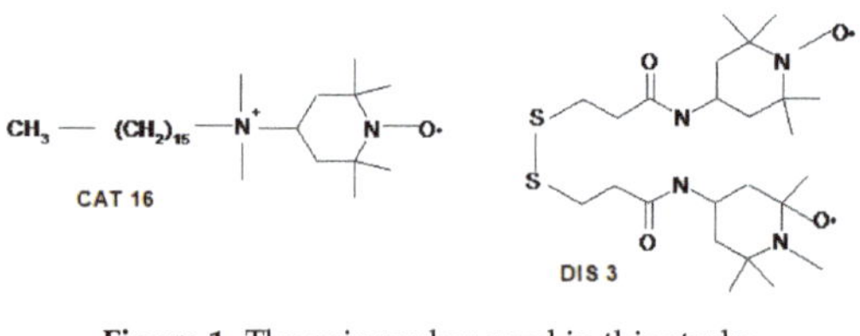

Figure 1. The spin probes used in this study.

Due to its ionic character, CAT16 binds to the albumin binding sites located at the water interface. This spin probe has been previously used to evidence BSA denaturation in the presence of SDS micelles and renaturation following the addition of cyclodextrin [15], to study the interaction of BSA with Pluronic micelles [16] and the thermal denaturation of BSA [17]. We reason that the interaction between albumin and AuNPs can induce changes in the EPR spectrum of CAT16. On the other hand, DIS3 is a biradical with affinity for AuNPs and can thus provide information on the adsorption kinetics on the nanoparticle surface [19].

The investigation of the reduction of Au(III) only in the presence of BSA using a palette of physico-chemical methods evidenced different aspects related to the mechanism of gold nanoparticles formation and their reactivity properties.

2. Results and Discussion

2.1. UV-Vis Measurements

Reduction of Au(III) was not observed in phosphate solution of $HAuCl_4$ (10^{-3} M), but occurred after addition of BSA in a final concentration of 2 mg/mL. The solution underwent a color change from yellow to colorless, corresponding to the reduction process, and then to purple (Figure 2). The reduction of Au(III) in the system containing BSA is due to the presence of specific amino acids in the protein chain, namely 35 threonine and 32 serine units that bear hydroxyl groups [20,21]. The change from colorless to purple color indicates the formation of gold nanoclusters in which the collective oscillation of the metallic surface electrons gives rise to the plasmon resonance band.

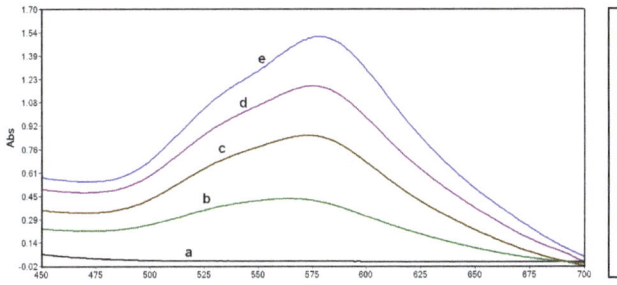 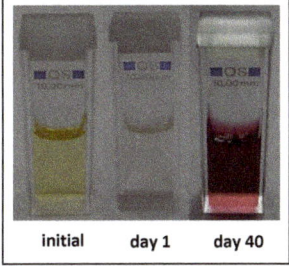

Figure 2. Evolution of the UV-Vis absorption spectrum of the bovine serum albumin (BSA)/$HAuCl_4$ system at different stages of incubation: (a) initial, (b) 3 days, (c) 1 week, (d) 2 weeks and (e) 3 weeks. Inset: Color change accompanying the formation of gold nanoparticles.

The evolution of the reduction process was followed by UV-Vis spectroscopy. It can be observed from Figure 2 that the intensity of the plasmon resonance band of AuNPs increases in time. The optical properties of AuNPs are dependent on the nanoparticle size [22]. The UV-Vis band shown in Figure 2 can be deconvoluted in two components, and this can be an indication that AuNPs have non-uniform size and shape distribution. Assuming a spherical shape of these particles, a particle diameter larger than 20 nm can be estimated from the UV-Vis spectra [22].

2.2. TEM and DLS Measurements

The UV-Vis data were correlated with those obtained by dynamic light scattering (DLS) and transmission electron microscopy (TEM). During sample preparation for TEM, the protein was washed out, therefore the TEM images provide information only on the nanoparticle metallic core size.

Figure 3 shows the TEM images of AuNPs formed in BSA solution after seven days, evidencing a non-uniform size distribution and the presence of nanoparticles with various morphologies (triangular, rhombohedral, hexagonal; see also Figure S1 in the Supplementary Information file). Other studies [12,20] have also reported triangular and hexagonal morphologies for AuNPs obtained in the presence of BSA. Although the reduction process in the presence of BSA alone was slower compared with the similar processes reported by Bakshi [12] and Xia [20] that use additional physico-chemical factors like temperature, ionic strength and the presence of surfactants, in our case the average particle size of AuNPs was of 20 nm. The dimensions of AuNPs obtained in this slower process were smaller than those reported in the above-mentioned studies. The TEM images were collected at different time intervals and it was observed that, despite a small increase in particle size, the particles remained well-separated, indicative of the fact that aggregation does not occur (Figure S2). Another experiment in which an additional quantity of Au(III) was added to the colorless solution led to the formation of AuNPs with smaller size (less than 5 nm), as shown in Figure S3.

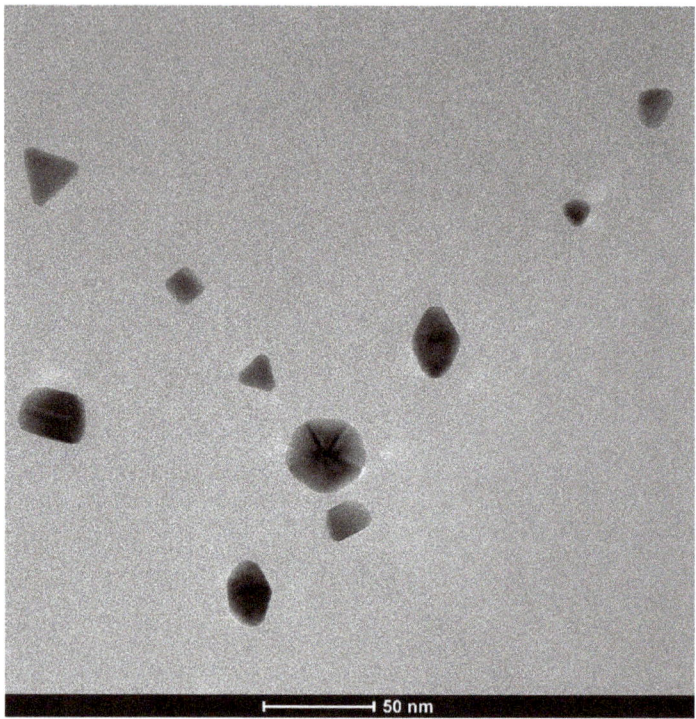

Figure 3. TEM image of gold nanoparticles formed in BSA.

DLS measurements evidence the changes in the dispersion size during the formation of AuNPs. The measurements were performed for a solution of BSA (2 mg/mL) prior to the addition of HAuCl$_4$, immediately after addition, and the evolution of the particle size was then followed over the course of one month. Literature DLS data report a size of BSA in solution less than 10 nm [23], a value lower than in our measurements (Table 1). This suggests the presence of BSA dimers in solution. It is important to notice that, in the presence of Au(III), prior to the formation of AuNPs, the size of BSA increases, which might indicate either a denaturation of BSA leading to a less compact protein conformation or the formation of small nanoparticles protected by BSA. DLS measurements performed after three weeks indicated an average particle diameter of 38.6 nm, while after 30 days this value increased to 56.9 nm (Table 1).

DLS measurements indicate also an increase of the nanoparticle size over time. It is possible that BSA plays a role in the growth process of AuNPs, which might be accompanied by conformational changes of BSA that can be revealed by spectroscopic methods.

Table 1. Evolution of the average particle diameter in a solution of BSA (2 mg/mL) in the absence and presence of HAuCl$_4$ (10^{-3} M).

Sample	Day	Diameter (nm) (Intensity Distribution)	Diameter (nm) (Volume Distribution)
BSA	Initial	20.4 ± 11.9	9.6 ± 4.5
BSA/Au^{3+}	Initial	31.1 ± 20.4	12.3 ± 6.3
BSA/Au^0	21	38.6 ± 19.4	21.5 ± 8.9
BSA/Au^0	30	73.3 ± 30.1	46.5 ± 17.7

2.3. Circular Dichroism and Raman Spectroscopy

We have shown that BSA acts as reducing agent, but it is also known that albumins can form protective layers for nanoparticles. To demonstrate that BSA is protecting AuNPs and to evidence the conformational changes associated with this process, we performed CD, Raman and EPR experiments.

As it was shown above, the reduction process induced by albumins is quite slow at room temperature. The CD spectra of BSA have been monitored for the entire period of time corresponding to the formation of AuNPs. The extent of BSA secondary structure alteration due to nanoparticle formation can be evidenced from the spectra.

The CD spectrum of BSA is characterized by the presence of two negative bands at 209 and 222 nm. These bands arise from π–π* and n–π* transitions in the amide groups, and are typical for proteins with predominantly helical content [24]. Prior to AuNPs formation, the intensity of this signal is only affected by time to a small extent. However, a significant decrease in the CD intensity of BSA occurs upon the growth of AuNPs up to a certain size, as can be seen from Figure 4. We consider that this is the effect of protein corona formation at the AuNPs surface.

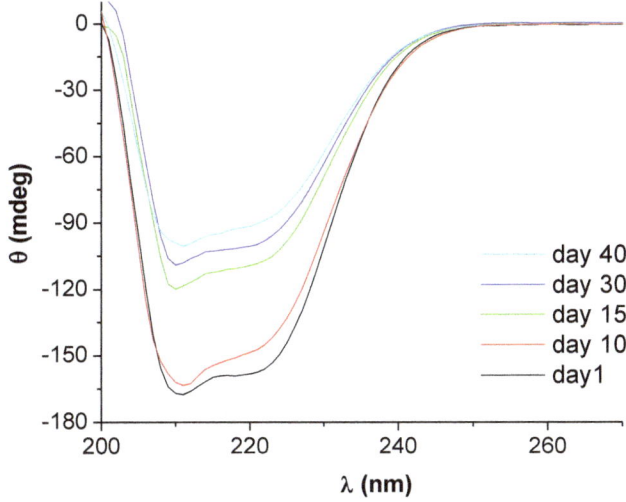

Figure 4. Circular dichroism spectra of BSA (2 mg/mL) recorded at various time intervals and evidencing the alteration of the BSA secondary structure upon gold nanoparticle formation and growth; [HAuCl$_4$] = 2 × 10^{-3} M.

The CD spectrum of BSA reflects changes in secondary structure occurring during the reduction of Au(III) as well as during nanoparticle growth. These changes indicate that the reduction step does induce conformational changes in the BSA structure to a lower extent as compared to the growth step, which most probably involves the formation of the BSA corona around nanoparticles with stabilizing effect. Both the reducing and protecting role of BSA induce conformational changes of the protein. As can be observed from Table 2, this consists in the increase of the β-sheet and unordered conformations contents on the expense of the α-helix content.

Table 2. Secondary structure content of BSA prior to and after gold nanoparticle formation and growth, as evidenced by circular dichroism.

Day	α-Helix	β-Sheet	Random Coil
1	0.60	0.07	0.33
10	0.59	0.08	0.33
15	0.40	0.21	0.39
30	0.36	0.18	0.46
40	0.30	0.18	0.52

Further information related to albumin denaturation and AuNPs/albumin interactions is provided by Raman spectroscopy, as will be discussed in the next section.

2.3.1. Modes of Adsorption of BSA onto Gold Nanoparticles

As stated above, the secondary structure of BSA is predominantly α-helical, with the helices being held together by disulfide bridges [25,26]. Protein unfolding, as evidenced by circular dichroism, is also associated to the breaking/reduction in the number of these S-S bonds. The disulfide bridges between Cys residues are a key structural parameter of BSA, as they stabilize the protein's folded state. Two spectral regions of the Raman spectra of BSA can be used to follow the change in protein conformation occurring upon AuNPs formation: the S-S (480–550 cm^{-1}) and C-S (630–720 cm^{-1}) stretching modes, sensitive to conformational changes and/or cleavage of the C_α–C_β–S–S'–C'_β–C'_α disulfide bridges in BSA [27,28], and the amide I and amide III Raman bands of BSA, indicative of the changes occurring in its secondary structure.

2.3.2. The Configuration of Disulfide Bridges

The Raman spectra of BSA prior to and after AuNPs formation are presented in Figure 5. In order to obtain Raman spectra with good signal-to-noise ratio, the concentrations of BSA and HAuCl$_4$ were increased compared with those used for other determinations.

The S-S Raman band of BSA includes three main contributions, indicating different configurations of the disulfide bridges: 500–510 cm^{-1} (gauche–gauche–gauche, *ggg*, rotamers), 515–525 cm^{-1} (gauche–gauche–trans, *ggt* or *tgg*) and 530–550 cm^{-1} (trans–gauche–trans, *tgt*) [29,30]. The position of these bands is altered when internal rotation about the S-S and C-S bonds occurs [31].

Band deconvolution in the S-S spectral range yielded six components for BSA (Figure 6A), with an additional component in the presence of AuNPs (Figure 6B). The S-S populations content was computed from the area of the individual components expressed as a fraction of the total area of the bands (Table 3). Initially, the disulfide bridges of BSA assume *ggg* (9%), *ggt* (34%) and *tgt* (57%) configurations. The contributions of *ggt* and *tgt* conformers to the Raman band corresponding to the S-S bond decrease as a result of AuNPs formation, while the bands corresponding to the *ggg* conformer become prevalent and shift towards lower wavenumbers. This indicates that BSA adsorption onto AuNPs favors the more energetically stable S-S conformation, *ggg* [32]. The localization of this band below 500 cm^{-1} indicates that some of these conformers are now found in a restricted medium, the S-S bonds being less flexible [33]. Such bands have been previously reported at 489 [34] and 463 cm^{-1} [35]. Thus, the *ggt* conformation of the disulfide bridges is converted to a *ggg* conformation as a result of protein corona forming onto AuNPs. A similar phenomenon was observed by Nakamura et al. [30] upon BSA conversion from N to F and E forms. The significant enhancement of this vibration in presence of nanoparticles may indicate that the respective disulfide bridges are located near the gold surface, yielding a surface enhanced Raman signal [36]. Differences in peak widths indicate different heterogeneities in the environment of the respective populations [37]. After adsorption, the disulfide bridges are 52% *ggg*, 22% *ggt* and 26% *tgt* configurations.

The reduction in the number of the different S–S bridge populations after AuNPs formation also correlates to changes observed in the features in the ν(C–S) spectral region (Figure 5).

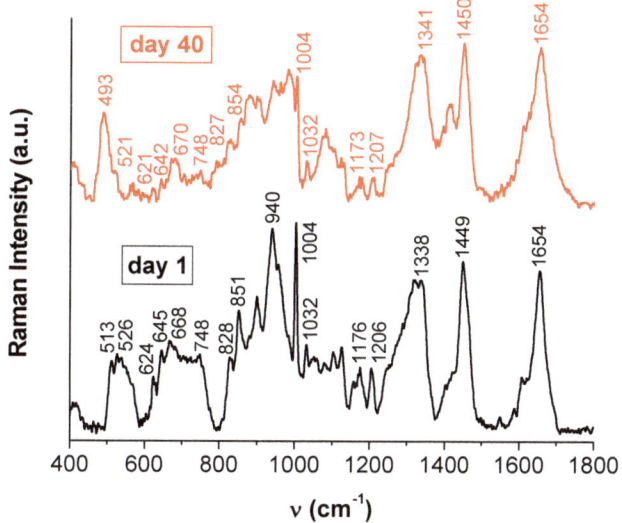

Figure 5. Raman spectra of BSA (20 mg/mL) prior to (day 1) and after (day 40) gold nanoparticle formation; peak positions are marked; [HAuCl$_4$] = 4 × 10^{-2} M.

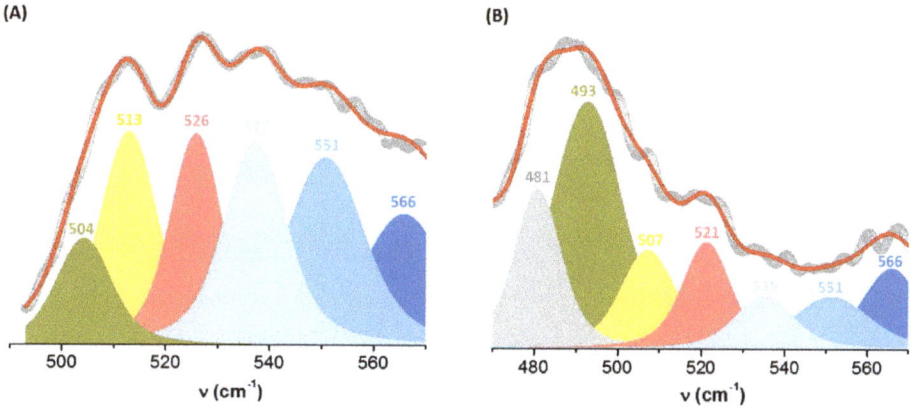

Figure 6. Deconvolution of the Raman bands in the spectral range corresponding to the ν(S–S) vibrations: (**A**) free BSA (day 1, R^2 = 0.995), (**B**) BSA adsorbed onto gold nanoparticles (day 40, R^2 = 0.994); the experimental spectra are depicted by grey dots, the convoluted spectra by red lines and the individual component functions are in color.

Table 3. The population of each conformer, calculated as the fraction of its band area from the total area of the bands.

Day 1		Day 40	
ν (cm^{-1})	Area (%)	ν (cm^{-1})	Area (%)
–	–	481	16
504	9	493	36
513	18	507	11
526	16	521	11
537	19	535	7
551	22	551	9
566	16	566	10

2.3.3. Estimating the Change in BSA Secondary Structure by Raman Spectroscopy

The analysis of the amide I band in the Raman spectrum provides information on the secondary structure of BSA that can be corroborated with those obtained from CD spectra. Figure 7 presents the deconvolution of the amide I band of BSA initially and after the AuNPs were formed, and the contributions of integrated peaks are listed in Table 4. The integrated peaks correspond to the α-helix (1650–1657 cm^{-1}), β-sheet (1612–1640 cm^{-1} and 1670–1690, antiparallel, and 1626–1640, parallel), β-turn (1655–1675 cm^{-1}, 1680–1696 cm^{-1}) and random coil (1640–1651 cm^{-1}) conformations [38]. One observes that both CD and Raman spectroscopies predict the same effect of AuNPs on the helical content of BSA. Moreover, the band at 940 cm^{-1}, the skeletal ν(C–C) vibration, another indicator of the helical structure, is intense in free BSA and much weaker when BSA is adsorbed onto the nanoparticles (Figure 5), thus evidencing as well the loss of helical content [30,39].

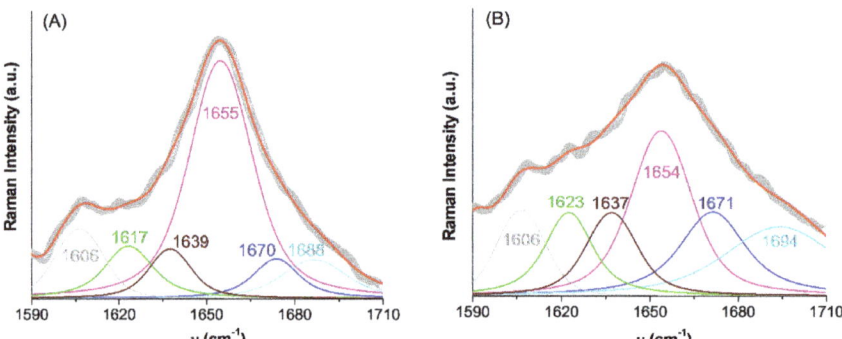

Figure 7. Deconvolution of the amide I region of the Raman spectra of free BSA (**A**, day 1) and BSA adsorbed onto gold nanoparticles (**B**, day 40). The secondary structure contents estimated from the respective peak areas are listed in Table 4.

Table 4. Secondary structure content of BSA prior to and after gold nanoparticle formation and growth, as evidenced by Raman spectroscopy.

Day 1		Day 40		Assignment
ν (cm^{-1})	Area (%)	ν (cm^{-1})	Area (%)	
1617	11	1623	13	β-sheet
1639	9	1637	13	random coil
1655	62	1654	31	α-helix
1670	8	1671	19	β-turn
1688	10	1694	24	β-turn

The Raman data allowed us to reveal some of the specific molecular groups involved in adsorption. Such local changes caused by AuNPs formation and growth can also be evidenced by spin probe method of EPR spectroscopy.

2.4. Electron Paramagnetic Resonance Measurements

All species involved in the formation of AuNPs are diamagnetic and, in order to monitor the processes of AuNPs formation and growth in the presence of BSA, we selected the spin probe method of EPR spectroscopy. Thiol derivatives have a high affinity for the Au surface and form protective layers that ensure the stability of AuNPs [40]. Therefore, we selected as a first spin probe the biradical DIS3 (Figure 1) with paramagnetic moieties linked through a disulfide bridge, which has been used for studying dynamic aspects of ligands protecting the Au surface, exchange processes and the stability of AuNPs [18,19,30]. Figure 8A shows the changes in the EPR spectrum of DIS3 during the formation of AuNPs.

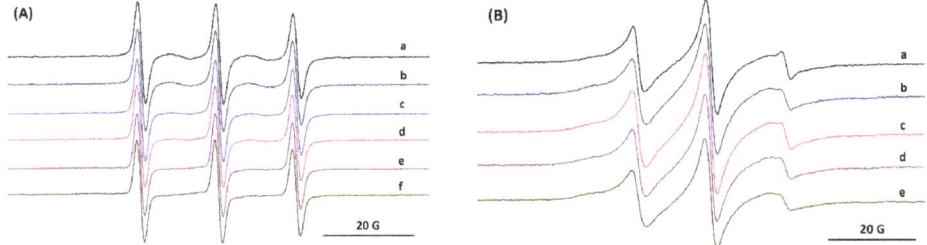

Figure 8. The evolution of the electron paramagnetic resonance (EPR) spectra of (**A**) DIS3: a) initial, b) 4 days, c) 6 days, d) 10 days, e) 13 days, f) 17 days, and (**B**) CAT16 in BSA/Au system: a) initial, b) 4 days, c) 10 days, d) 13 days, e) 17 days.

In water solution, the EPR spectrum of biradical DIS3 (Figure 8A, spectrum a) shows the additional lines corresponding to exchange interactions (second and forth lines in the spectrum), although their intensities are lower than in organic solvents [18]. The evolution of spectral changes for DIS3 indicates that the lines attributed to the exchange interactions gradually decreased during the reduction step and the formation of AuNPs. After 10 days, the EPR spectrum indicates only three lines, without evidencing a restricted motion (Figure 8A, spectrum d). As it was reported in literature [18,19], adsorption of DIS3 at the surface of AuNPs induces the breaking of the disulfide bond. The changes in the EPR spectrum occur on the same time scale as the changes in the CD spectrum.

The spin probe method of EPR spectroscopy can be used to prove small conformational changes in the BSA structure [15–17]. The spin probe CAT16 binds to the BSA sites exposed to water, due to the ionic character of the probe. This determines a relatively fast molecular motion of the nitroxide group in CAT16 as compared to that of 5-doxyl stearic acid, which is almost immobilized in the complex with BSA [16,41,42]. Considering this different behavior of spin probes with affinity for BSA binding sites, we selected CAT16 to analyze the evolution of the BSA/Au system. Figure 8B shows the EPR spectra of CAT16 in solution of BSA in the absence and in the presence of Au, at different time intervals. Spectrum a in Figure 8B presents a double component feature, one with restricted motion and another with fast motion. The formation of AuNPs slows down the motion of the initial faster component, which indicates the fact that the nitroxide group senses the increase in size of the BSA/AuNP assembly (Figure 8B, spectrum e). The simulated spectra of CAT16 in BSA and in BSA/AuNP systems are shown in Figure S4. The simulation of EPR spectra suggests a slower motion of both components once AuNPs are formed. The percentages of the two components corresponding to the EPR spectrum of CAT16 in BSA/AuNPs are approximately the same (Table S1).

A series of EPR studies reported in literature evidenced that some aerobic oxidations in the presence of AuNPs involve the generation of reactive species [43–45]. Starting from this, we tested if reactive radicals that can be trapped by the spin trap DMPO (5,5-dimethyl-1-pyrroline N-oxide) are generated in solution of BSA/AuNPs. In Figure 9, the blue spectrum corresponds to the DMPO adduct with the hydroxyl radical that can be generated in the presence of adsorbed oxygen at the AuNPs surface. The spectrum is a superposition of two components, one corresponding to the DMPO–HO adduct and another corresponding to a nitroxide degradation product of DMPO. The hyperfine coupling constants of the DMPO–HO adduct are a_N = 14.85 G and a_H = 14.78 G. Generation of hydroxyl radicals in BSA/AuNPs might be significant for the biomedical applications of such systems.

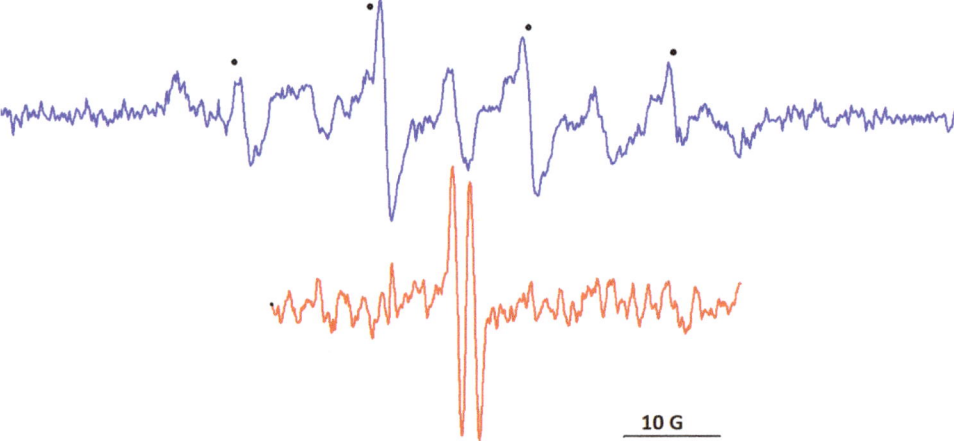

Figure 9. The EPR spectra of 5,5-dimethyl-1-pyrroline N-oxide (DMPO)–HO adduct (in blue) and ascorbyl free radical (in red) generated in the BSA/AuNPs system.

A well-known component of biofluids is ascorbate, a species that acts as radical scavenger. Oxidation of ascorbate is a two-step process that involves the ascorbyl free radical as intermediate. Hydroxyl radicals can generate the ascorbyl radical [46–48]. Addition of calcium ascorbate to the BSA/AuNPs solution led to the formation of the ascorbyl radical characterized by a_H = 1.78 G, as shown in Figure 9, spectrum in red.

The spin-trapping experiments performed in solution of BSA/AuNPs after purging argon did not lead to the formation of DMPO–HO adducts. The spin trapping measurements suggest that AuNPs might be involved in oxidative processes during their transport to the specific targets.

3. Materials and Methods

BSA fatty acid free fraction V was purchased from Fluka (Buchs, Switzerland). Chloroauric acid ($HAuCl_4$, >97%) and DMPO were purchased from Sigma Aldrich (St. Louis, MI, USA). These were used without further purification. The solutions were prepared in phosphate buffer at pH 7. The spin probe CAT16 was obtained from Molecular Probes (Leiden, The Netherlands), whereas the spin probe DIS3 was obtained as described in literature [18,19].

3.1. Sample Preparation

BSA was dissolved in a solution of $HAuCl_4$ (2×10^{-3} M) in phosphate buffer at pH 7. The resulting system was left at room temperature over a period of over one month. The concentration of albumin for measurements using different methods were as follows: 2 mg/mL for UV-Vis, DLS, circular dichroism, 10 mg/mL for EPR measurements and 20 mg/mL for Raman spectroscopy.

3.2. Instrumentation

3.2.1. UV-Vis Absorption

Absorption spectra were recorded on a Jasco V-560 UV-VIS spectrophotometer in the range 300-700 nm.

3.2.2. DLS Measurements

The particle size evolution in BSA and BSA/Au solutions was analyzed by dynamic light scattering (DLS) measurements in water on a Beckman Coulter particle size analyzer (Brea, CA, USA) using the Delsa Nano software (Beckman Coulter, Brea, CA, USA). The measurements were performed at room temperature.

3.2.3. Circular Dichroism Measurements

Circular dichroism spectra of BSA (2 mg/mL) in the absence and in the presence of $HAuCl_4$ (2×10^{-3} M) were recorded, at various time intervals, on a Jasco J-815 circular dichroism spectropolarimeter, in 0.05 cm path length cuvettes, at 4 s response time, 100 nm/min scan speed, 1 nm bandwidth and 1000 millidegrees (mdeg) sensitivity. The characteristic dichroic signal of BSA observed in the 200–300 nm spectral range was averaged over six scans with subtraction of the buffer blank. The results were expressed in ellipticity (θ, in mdeg). The BSA secondary structure contents prior to and after nanoparticle formation were determined with the DichroWeb online server [49,50]. The α-helix, β-sheet and unordered contributions were estimated employing the K2D deconvolution algorithm [51]. All fits led to normalized root mean square deviations lower than 0.2 [52,53].

3.2.4. Raman Spectroscopy

Raman spectra of BSA (20 mg/mL) in the absence and in the presence of $HAuCl_4$ (4×10^{-2} M) were collected on a Jasco NRS-3100 Raman spectrometer with laser excitation at 785 nm (laser power 150 mW). The spectra were averaged over 300 scans and the spectrum of the control ($HAuCl_4$ in phosphate buffer) was subtracted from each sample spectrum. Baseline correction and smoothing of spectra with a 15 point Savitsky–Golay function were performed with the Spectra Analysis program. The spectra were deconvoluted to the minimum number of components by fitting the Raman bands of BSA with a sum of pseudo-Voigt functions (Gaussian and Lorentzian, in equal contributions). Band position and width were constrained within reasonable limits.

3.2.5. Transmission Electron Microscopy

The TEM analysis was performed on Tecnai G2-F30 S-twin microscope (FEG-STEM) at an accelerating voltage of 300kV. Sample preparation was minimal, consisting in mounting a drop of AuNPs alcoholic suspension on holey carbon film copper grid allowing the solvent to evaporate at room temperature.

3.2.6. EPR Spectroscopy

EPR spectra of DIS3 and CAT16 were recorded on a JEOL FA 100 spectrometer equipped with a cylindrical type resonator TE011, with a frequency modulation of 100 kHz, microwave power of 0.998 mW, sweep time of 480 s, modulation amplitude of 1 G, time constant of 0.3 s, and magnetic field scan range of 100 G. For the spin trapping experiments, the following settings were used: sweep field 150 G, frequency 100 kHz, gain 800, sweep time 240 s, time constant 0.1 s, modulation width 1 G, microwave power 1 mW.

EPR spectral simulations of CAT16 spectra were performed using the program developed by Budil et al. [54] based on non-linear least–squares (NLLS) fits, allowing fitting of spectra encompassing two components with different dynamics.

4. Conclusions

In this study we monitored the formation and stabilization of gold nanoparticles in the presence of BSA by a combination of different physico-chemical techniques. The gold (III) reduction and the growth of gold nanoparticles are processes accompanied by conformational changes in the BSA. DLS measurements evidenced the presence in solution of assemblies with molecular size growing from the size of albumin up to the average size of 40 nm. TEM images evidenced different morphologies of the nanoparticles, with an average size of 20 nm after 10 days and a slow increase to approximately 50 nm over a one-month interval. The results of this study highlight that albumin, as a reducing agent of Au^{3+}, can be an initiator for the formation of gold nanoparticles. Moreover, by using EPR and Raman spectroscopies it was possible to prove that albumin covers the gold nanoparticles surface. Using EPR spectroscopy, we were able to monitor the formation of nanoparticles, to analyze the changes in the behavior of spin probes at the surface of albumin, and to evidence the formation of reactive species, intermediated by gold nanoparticles. This physico-chemical study thus provides relevant information for the application of such systems in the biomedical field of nanoparticle delivery.

Supplementary Materials: The following are available online at http://www.mdpi.com/1420-3049/24/18/3395/s1, Figure S1: TEM images of gold nanoparticles formed in BSA evidencing the presence of various shapes. Figure S2: TEM images of gold nanoparticles formed in BSA after three weeks. Figure S3: TEM image of AuNPs with smaller particle size (roughly 5 nm) formed after adding a new quantity of Au(III). Figure S4: Simulated EPR spectra of CAT16 in (A) BSA (10 mg/mL), (B) BSA/AuNPs after 10 days and (C) BSA/AuNPs after 17 days. Table S1: EPR parameters obtained by spectral simulation.

Author Contributions: Conceptualization, G.I.; methodology, G.I. and I.M.; EPR measurements, M.C.B. and I.M.T.; DLS measurements, D.C.; TEM images, C.M.; CD and Raman spectroscopy, I.M.; UV-Vis, I.M., I.M.T.; writing—original draft preparation, all authors.; writing—review and editing, I.M. and G.I.; supervision, G.I.; project administration, G.I.; funding acquisition, G.I.

Funding: This work was supported by the Romanian National Authority for Scientific Research, CNCS–UEFISCDI, project number PN-III-P4-ID-PCE-2016-0734.

Acknowledgments: Romanian Academy for the PhD fellowship of Maria Cristina Buta is acknowledged. Petre Ionita is acknowledged for the synthesis of the DIS3 biradical.

Conflicts of Interest: The authors declare no conflict of interest.

References

1. Sotnikov, D.V.; Berlina, A.N.; Ivanov, V.S.; Zherdev, A.V.; Dzantiev, B.B. Adsorption of proteins on gold nanoparticles: One or more layers? *Coll. Surf. B* **2019**, *173*, 557–563. [CrossRef] [PubMed]
2. Dominguez-Medina, S.; McDonough, S.; Swanglap, P.; Landes, C.F.; Link, S. In situ measurement of bovine serum albumin interaction with gold nanospheres. *Langmuir* **2012**, *28*, 9131–9139. [CrossRef] [PubMed]
3. Casals, E.; Pfaller, T.; Duschl, A.; Oostingh, G.J.; Puntes, V. Time evolution of the nanoparticle protein corona. *ACS Nano* **2010**, *4*, 3623–3632. [CrossRef] [PubMed]
4. Murawala, P.; Tirmale, A.; Shiras, A.; Prasad, B.L.V. In situ synthesized BSA capped gold nanoparticles: Effective carrier of anticancer drug Methotrexate to MCF-7 breast cancer cells. *Mater. Sci. Eng. C* **2014**, *34*, 158–167. [CrossRef] [PubMed]
5. Sardar, R.; Funston, A.M.; Mulvaney, P.; Murray, R.W. Gold nanoparticles: Past, present and future. *Langmuir* **2009**, *25*, 13840–13851. [CrossRef] [PubMed]
6. Turkevich, J.; Stevenson, P.C.; Hillier, J.A. Study of the nucleation and growth processes in the synthesis of colloidal gold. *Discuss. Faraday Soc.* **1951**, *11*, 55–59. [CrossRef]
7. Brewer, S.H.; Glomm, W.R.; Johnson, M.C.; Knag, M.K.; Franzen, S. Probing BSA binding to citrate-coated gold nanoparticles and surfaces. *Langmuir* **2005**, *21*, 9303–9307. [CrossRef]
8. Nakata, S.; Kido, N.; Hayashi, M.; Hara, M.; Sasabe, H.; Sugawara, T.; Matsuda, T. Chemisorption of proteins and their thiol derivatives onto gold surfaces: Characterization based on electrochemical nonlinearity. *Biophys. Chem.* **1996**, *62*, 63–72. [CrossRef]

9. Silin, V.; Weetall, H.; Vanderah, D.J. SPR studies of the nonspecific adsorption kinetics of human IgG and BSA on gold surfaces modified by self-assembled monolayers (SAMs). *J. Colloid Interface Sci.* **1997**, *185*, 94–103. [CrossRef]
10. De Paoli Lacerda, A.H.; Park, J.J.; Curt Meuse, C.; Pristinski, D.; Becker, M.L.; Karim, A.; Douglas, J.F. Interaction of gold nanoparticles with common human blood proteins. *ACS Nano* **2010**, *4*, 365–379. [CrossRef]
11. Khandelia, R.; Jaiswal, A.; Ghosh, S.S.; Chattopadhyay, A. Gold nanoparticle–protein agglomerates as versatile nanocarriers for drug delivery. *Small* **2013**, *9*, 3494–3505. [CrossRef] [PubMed]
12. Khullar, P.; Singh, V.; Mahal, A.; Dave, P.N.; Thakur, S.; Kaur, G.; Singh, J.; Kamboj, S.S.; Bakshi, M.S. Bovine serum albumin bioconjugated gold nanoparticles: Synthesis, Hemolysis, and cytotoxicity toward cancer cell lines. *J. Phys. Chem. C* **2012**, *116*, 8834–8843. [CrossRef]
13. Xie, J.; Lee, J.Y.; Wang, D.I.C. Synthesis of single-crystalline gold nanoplates in aqueous solutions through biomineralization by serum albumin protein. *J. Phys. Chem. C* **2007**, *111*, 10226–10232. [CrossRef]
14. Basu, N.; Bhattacharya, R.; Mukherjee, P. Protein-mediated autoreduction of gold salts to gold nanoparticles. *Biomed. Mater.* **2008**, *3*, 034105. [CrossRef] [PubMed]
15. Rogozea, A.; Matei, I.; Turcu, I.M.; Ionita, G.; Sahini, V.E.; Salifoglou, A. EPR and circular dichroism solution studies on the interactions of bovine serum albumin with ionic surfactants and β-cyclodextrin. *J. Phys. Chem. B* **2012**, *116*, 14245–14253. [CrossRef] [PubMed]
16. Neacsu, M.V.; Matei, I.; Micutz, M.; Staicu, T.; Precupas, A.; Popa, V.T.; Salifoglou, A.; Ionita, G. Interaction between albumin and Pluronic F127 block copolymer revealed by global and local physicochemical profiling. *J. Phys. Chem. B* **2016**, *120*, 4258–4267. [CrossRef] [PubMed]
17. Matei, I.; Ariciu, A.M.; Neacsu, M.V.; Collauto, A.; Salifoglou, A.; Ionita, G. Cationic spin probe reporting on thermal denaturation and complexation–decomplexation of BSA with SDS. Potential applications in protein purification processes. *J. Phys. Chem. B* **2014**, *118*, 11238–11252. [CrossRef]
18. Ionita, P.; Caragheorgheopol, A.; Gilbert, B.C.; Chechik, V. EPR study of a place-exchange reaction on Au nanoparticles: Two branches of a disulfide molecule do not adsorb adjacent to each other. *J. Am. Chem. Soc.* **2002**, *124*, 9048–9049. [CrossRef]
19. Chechik, V.; Wellsted, H.J.; Korte, A.; Gilbert, B.C.; Horia Caldararu, H.; Ionita, P.; Caragheorgheopol, A. Spin-labelled Au nanoparticles. *Faraday Discuss.* **2004**, *125*, 279–291. [CrossRef]
20. Leslie Au, L.; Lim, B.; Colletti, P.; Jun, Y.-S.; Xia, Y. Synthesis of gold microplates using bovine serum albumin as a reductant and a stabilizer. *Chem. Asian J.* **2010**, *5*, 123–129.
21. Hirayama, K.; Akashi, S.; Furuya, M.; Fukuhara, K.I. Rapid confirmation and revision of the primary structure of bovine serum albumin by ESIMS and Frit-FAB LC/MS. *Biochem. Biophys. Res. Commun.* **1990**, *173*, 639–646. [CrossRef]
22. Haiss, W.; Thanh, N.T.K.; Aveyard, J.; Fernig, D.G. Determination of size and concentration of gold nanoparticles from UV-Vis spectra. *Anal. Chem.* **2007**, *79*, 4215–4221. [CrossRef] [PubMed]
23. Das, T.; Ghosh, P.; Shanavas, M.S.; Maity, A.; Mondal, S.; Purkayastha, P. Protein-templated gold nanoclusters: Size dependent inversion of fluorescence emission in the presence of molecular oxygen. *Nanoscale* **2012**, *4*, 6018–6024. [CrossRef] [PubMed]
24. Greenfield, N.J. Applications of circular dichroism in protein and peptide analysis. *Trends Anal. Chem.* **1999**, *18*, 236–244. [CrossRef]
25. Peters, T. *All about Albumin. Biochemistry, Genetics and Medical Application*; Academic Press: San Diego, CA, USA, 1996.
26. Paris, G.; Kraszewski, S.; Ramseyer, C.; Enescu, M. About the structural role of disulfide bridges in serum albumins: Evidence from protein simulated unfolding. *Biopolymers* **2012**, *97*, 889–898. [CrossRef] [PubMed]
27. Tu, A.T. *Advances in Spectroscopy, Spectroscopy of Biological Systems*; Clark, R.J.H., Hester, R.E., Eds.; Wiley: Chichester, UK, 1986; Volume 13, pp. 47–112.
28. Hernandez, B.; Pfluger, F.; Lopez-Tobar, E.; Kruglik, S.G.; Garcia-Ramos, J.V.; Sanchez-Cortes, S.; Ghomi, M. Disulfide linkage Raman markers: A reconsideration attempt. *J. Raman Spectrosc.* **2014**, *45*, 657–664. [CrossRef]
29. Fleury, F.; Ianoul, A.; Berjot, M.; Feofanov, A.; Alix, A.J.P.; Nabiev, I. Camptothecin-binding site in human serum albumin and protein transformations induced by drug binding. *FEBS Lett.* **1997**, *411*, 215–220. [CrossRef]

30. Nakamura, K.; Era, S.; Ozaki, Y.; Sogami, M.; Hayashi, T.; Murakami, M. Conformational changes in seventeen cystine disulfide bridges of bovine serum albumin proved by Raman spectroscopy. *FEBS Lett.* **1997**, *417*, 375–378. [CrossRef]
31. Jurasekova, Z.; Marconi, G.; Sanchez-Cortes, S.; Torreggiani, A. Spectroscopic and molecular modeling studies on the binding of the flavonoid luteolin and human serum albumin. *Biopolymers* **2009**, *91*, 917–927. [CrossRef]
32. Navarra, G.; Tinti, A.; Di Foggia, M.; Leone, M.; Militello, V.; Torreggiani, A. Metal ions modulate thermal aggregation of beta-lactoglobulin: A joint chemical and physical characterization. *J. Inorg. Biochem.* **2014**, *137*, 64–73. [CrossRef]
33. Van Wart, H.E.; Scheraga, H.A. Raman spectra of strained disulfides. Effect of rotation about sulfur-sulfur bonds on sulfur-sulfur stretching frequencies. *J. Phys. Chem.* **1976**, *80*, 1823–1832. [CrossRef]
34. Liu, R.; Zi, X.; Kang, Y.; Si, M.; Wu, Y. Surface-enhanced Raman scattering study of human serum on PVA-Ag nanofilm prepared by using electrostatic self-assembly. *J. Raman Spectrosc.* **2011**, *42*, 137–144. [CrossRef]
35. Jurasekova, Z.; Tinti, A.; Torreggiani, A. Use of Raman spectroscopy for the identification of radical-mediated damages in human serum albumin. *Anal. Bioanal. Chem.* **2011**, *400*, 2921–2931. [CrossRef] [PubMed]
36. Podstawka, E.; Ozaki, Y.; Proniewicz, L.M. Adsorption of S–S containing proteins on a colloidal silver surface studied by surface-enhanced Raman spectroscopy. *Appl. Spectrosc.* **2004**, *58*, 1147–1156. [CrossRef] [PubMed]
37. David, C.; Foley, S.; Mavon, C.; Enescu, E. Reductive unfolding of serum albumins uncovered by Raman spectroscopy. *Biopolymers* **2008**, *89*, 623–634. [CrossRef] [PubMed]
38. Pelton, J.T.; McLean, L.R. Spectroscopic methods for analysis of protein secondary structure. *Anal. Biochem.* **2000**, *277*, 167–176. [CrossRef] [PubMed]
39. Ma, Y.; Chechik, V. Aging of Gold Nanoparticles: Ligand Exchange with Disulfides. *Langmuir* **2011**, *27*, 14432–14437. [CrossRef] [PubMed]
40. Brust, M.; Walker, M.; Bethell, D.; Schiffrin, D.J.; Whyman, R. Synthesis of thiol-derivatised gold nanoparticles in a two-phase liquid–liquid system. *Chem. Commun.* **1994**. [CrossRef]
41. Junk, M.J.N.; Spiess, H.W.; Hinderberger, D. The distribution of fatty acids reveals the functional structure of human serum albumin. *Angew. Chem. Int. Ed.* **2010**, *49*, 8755–8759. [CrossRef]
42. Muravsky, V.; Gurachevskaya, T.; Berezenko, S.; Schnurr, K.; Gurachevsky, A. Fatty acid binding sites of human and bovine albumins: Differences observed by spin probe ESR. *Spectrochim. Acta A* **2009**, *74*, 42–47. [CrossRef]
43. Ionita, P.; Conte, M.; Gilbert, B.C.; Chechik, V. Gold nanoparticle-initiated free radical oxidations and halogen abstractions. *Org. Biomol. Chem.* **2007**, *5*, 3504–3509. [CrossRef] [PubMed]
44. Conte, M.; Miyamura, H.; Kobayashi, S.; Chechik, V. Spin trapping of Au-H intermediate in the alcohol oxidation by supported and unsupported gold catalysts. *J. Am. Chem. Soc.* **2009**, *131*, 7189–7196. [CrossRef] [PubMed]
45. Conte, M.; Liu, X.; Murphy, D.M.; Whiston, K.; Hutchings, G.J. Cyclohexane oxidation using Au/MgO: An investigation of the reaction mechanism. *Phys. Chem. Chem. Phys.* **2012**, *14*, 16279–16285. [CrossRef] [PubMed]
46. Spasojević, I. Free radicals and antioxidants at a glance using EPR spectroscopy. *Crit. Rev. Clin. Lab. Sci.* **2011**, *48*, 114–142. [CrossRef] [PubMed]
47. Buettner, G.R.; Jurkiewicz, B.A. Catalytic metals, ascorbate and free radicals: Combinations to avoid. *Radiation Res.* **1996**, *145*, 532–541. [CrossRef] [PubMed]
48. Buettner, G.R.; Jurkiewicz, B.A. Ascorbate free radical as a marker of oxidative stress: An EPR study. *Free Radic. Biol. Med.* **1993**, *14*, 49–55. [CrossRef]
49. Dichroweb. On-line Analysis for Protein Circular Dichroism Spectra. Available online: http://dichroweb.cryst.bbk.ac.uk (accessed on 2 October 2018).
50. Whitmore, L.; Wallace, B.A. DICHROWEB: An online server for protein secondary structure analyses from circular dichroism spectroscopic data. *Nucleic Acids Res.* **2004**, *32*, W668–W673. [CrossRef] [PubMed]
51. Andrade, M.A.; Chacon, P.; Merelo, J.J.; Moran, F. Evaluation of secondary structure of proteins from UV circular dichroism using an unsupervised learning neural network. *Prot. Eng.* **1993**, *6*, 383–390. [CrossRef] [PubMed]
52. Mao, D.; Wachter, E.; Wallace, B.A. Folding of the mitochondrial proton adenosine triphosphatase proteolipid channel in phospholipid vesicles. *Biochemistry* **1982**, *21*, 4960–4968. [CrossRef]

53. Whitmore, L.; Wallace, B.A. Methods of Analysis for Circular Dichroism Spectroscopy of Proteins and the DichroWeb Server. In *Advances in Biomedical Spectroscopy, vol. 1, Modern Techniques for Circular Dichroism and Synchrotron Radiation Circular Dichroism Spectroscopy*; Wallace, B.A., Janes, R.W., Eds.; IOS Press: Amsterdam, The Netherlands, 2009; pp. 165–182.
54. Budil, D.E.; Lee, S.; Saxena, S.; Freed, J.H. Nonlinear-least squares analysis of slow-motion EPR spectra in one and two dimensions using a modified Levenberg–Marquardt algorithm. *J. Magn. Reson. Ser. A* **1996**, *120*, 155–189. [CrossRef]

Sample Availability: Samples of AuNPs/BSA are available from the authors.

© 2019 by the authors. Licensee MDPI, Basel, Switzerland. This article is an open access article distributed under the terms and conditions of the Creative Commons Attribution (CC BY) license (http://creativecommons.org/licenses/by/4.0/).

Article

Converging Fate of the Oxidation and Reduction of 8-Thioguanosine

Katarzyna Taras-Goslinska [1,†], Fabrizio Vetica [2,3,†], Sebastián Barata-Vallejo [3,4], Virginia Triantakostanti [3], Bronisław Marciniak [1,5] and Chryssostomos Chatgilialoglu [2,3,5,*]

1. Faculty of Chemistry, Adam Mickiewicz University, Wieniawskiego 1, 61-712 Poznań, Poland
2. R&D Laboratory, Lipinutragen srl, Via Piero Gobetti 101, 40129 Bologna, Italy
3. ISOF, Consiglio Nazionale delle Ricerche, Via Piero Gobetti 101, 40129 Bologna, Italy
4. Departamento de Quimíca Organíca, Facultad de Farmacia y Bioquímica, Universidad de Buenos Aires, Junin 954, RA-1113 Buenos Aires, Argentina
5. Center of Advanced Technologies, Adam Mickiewicz University, 61-712 Poznań, Poland
* Correspondence: chrys@isof.cnr.it; Tel.: +39-051-639-8309
† These authors contributed equally to the work.

Academic Editor: Derek J. McPhee
Received: 2 August 2019; Accepted: 28 August 2019; Published: 29 August 2019

Abstract: Thione-containing nucleobases have attracted the attention of the scientific community for their application in oncology, virology, and transplantology. The detailed understanding of the reactivity of the purine derivative 8-thioguanosine (8-TG) with reactive oxygen species (ROS) and free radicals is crucial for its biological relevance. An extensive investigation on the fate of 8-TG under both reductive and oxidative conditions is here reported, and it was tested by employing steady-state photooxidation, laser flash photolysis, as well as γ-radiolysis in aqueous solutions. The characterization of the 8-TG T_1 excited state by laser flash photolysis and the photooxidation experiments confirmed that singlet oxygen is a crucial intermediate in the formation of the unexpected reduced product guanosine, without the formation of the usual oxygenated sulfinic or sulfonic acids. Furthermore, a thorough screening of different radiolytic conditions upon γ-radiation afforded the reduced product. These results were rationalized by performing control experiments in the predominant presence of each reactive species formed by radiolysis of water, and the mechanistic pathway scenario was postulated on these bases.

Keywords: photolysis; laser flash photolysis; γ-radiolysis; singlet oxygen; nucleosides; free radicals; reaction mechanism

1. Introduction

The search for new purine and pyrimidine derivatives of biological significance has been conducted by many research groups [1,2]. The analogues of purine bases of nucleic acids containing a sulfur atom, such as mercaptopurine, 6-thioguanine, or azathioprine, have been applied in medicine for many years in the treatments of cancer, in particular leukemia in children, viral diseases, and in transplantology [3–5]. However, it has been recently evidenced that the skin of patients taking these drugs for a long time becomes much more sensitive to sunlight, and the chances of skin cancer development in these patients increase [6,7]. A small structural modification involving the attachment of a thione group to the purine ring system results in a decrease in energy of the electronic excited states, leading to absorption of long-wavelength light, including the UV-A range [8]. Moreover, thiocarbonyl compounds in excited states are highly reactive [9]. Therefore, thio-derivatives of guanine present in the chains of nucleic acids can take part in photochemical transformations or at least can initiate them. The photochemical processes upon excitation of thioguanosine chromophore in DNA include oxidation of DNA bases and/or proteins by $^1O_2$ or other reactive oxygen species (ROS), photo cross-linking of DNA strands, and generation of reactive radicals that can modify the components of nucleic acids or proteins. In the photochemical studies of 6-thioguanosine (6-TG), most attention has been paid to the oxidation reactions, as they are assumed to be mainly responsible for the adverse biological effects [10,11]. It has been reported that in aerated aqueous solutions, after absorption of UV-A, 6-TG undergoes Type I and Type II photooxidation processes and generates ROS, including singlet oxygen $^1O_2$ or radicals, harmful to biological systems [12,13]. After selective excitation, 6-TG built in DNA generates singlet oxygen $^1O_2$ through energy transfer from the excited electronic state of the 6-TG molecule to the ground state oxygen molecule. Other reactive individuals formed in the presence of oxygen include radicals generated in the process of electron transfer or charge–transfer complexes between the triplet state of 6-TG and $^3O_2$ [8,12].

Guanine substituted at the C8 position with either a bromine or thiol group is a potent adjuvant [14]. As an approach to reveal the photooxidation mechanism, we have started the investigations on the properties of 8-thioguanosine (8-TG, 1) excited states by steady-state and time-resolved techniques. The electronically excited 8-TG molecule in the presence of oxygen generates reactive species, including singlet oxygen $^1O_2$ or radicals.

On a structural point of view, the presence of a sulfur atom at the C8 position in 8-TG results in the establishment of a tautomeric equilibrium between the thione and the thiol form. Equilibrium of mercapto-azoles has attracted interest in both experimental and theoretical chemistry because of the relevance of this molecule in biological applications. Theoretical studies on 2-mercaptobenzimidazole, with a similar heterocyclic scaffold to 8-TG, revealed that the equilibrium favors the thione form (tautomerization energy thiol/thione is −34.2 kJ/mol) [15]. Thus, it is reasonable to consider that 8-TG would favor the thione form as well (Scheme 1).

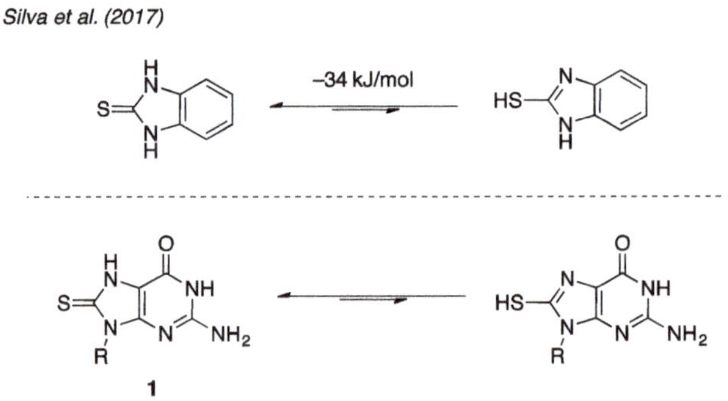

Scheme 1. Tautomeric equilibrium of 8-thioguanosine (8-TG).

Therefore, 8-thioguanosine could potentially have a similar reactivity as the thioureas, specifically cyclic thioureas. The thiourea scaffold has been widely investigated over the years for its applications in synthetic organic chemistry, industry [16], medicinal chemistry, and agrochemistry [17], and as an ROS scavenger [18,19]. Because of the highly applicability, several research groups have been interested in the redox chemistry of thioureas and have reported detailed mechanistic studies for these transformations. The reactivity of thioureas with hydrogen peroxide [20], common reducing agents such as K and NaBH$_4$ [21], singlet oxygen [22], and upon pulse radiolysis in both reducing and oxidizing conditions have been studied [23–29], outlining multiple mechanistic scenarios.

With these premises, in this study we report our investigation on the steady-state photooxidation and laser flash photolysis of 8-TG, as well as its γ-radiolysis in aqueous solution, under both reductive and oxidative conditions, highlighting an unusual and converging product formation.

2. Results and Discussion

2.1. Structural Characterization of 8-Thioguanosine (8-TG)

Initially, we decided to register NMR spectra in order to define which tautomeric form of 8-TG is favored. The proton NMR in DMSO-d_6 shows 2 different broad signals at 11.13 and 12.63 ppm, corresponding to NH from amide and thioamide moieties (Figures S1–S3). Moreover, both peaks exchange after adding D_2O. These results, combined with the previously reported theoretical studies (vide supra), suggest that the equilibrium favors the thione form.

Additionally, we decided to perform a UV titration of aqueous solutions of 8-TG to calculate the pK_a value. Using 12 different 0.1 mM solutions of 8-TG, each at a different pH (using NaOH/phosphoric acid buffers), we recorded the absorption values and we plotted the absorption versus pH of the solution (Figure 1). The resulting data were fit into sigmoid curves, and the inflection points were calculated. The titration curve showed 2 pK_a values: $pK_{a1} = 1.87$ and $pK_{a2} = 7.88$, which presumably correspond to NH_2 protonation and deprotonation of the amide moiety, respectively [30,31].

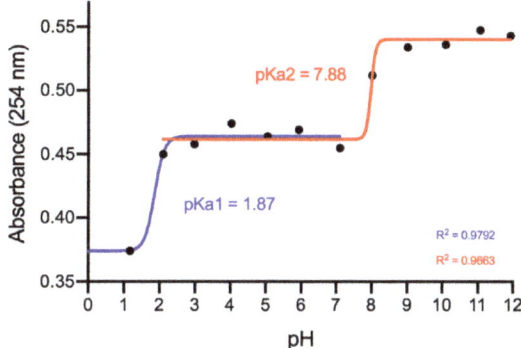

Figure 1. UV-pH titration of 8-TG.

2.2. Steady-State Photochemistry

Direct excitation of 8-TG in air-equilibrated aqueous solution with $\lambda > 300$ nm light or monochromatic radiation at 313 nm leads to disappearance of the intense absorption band ($\lambda_{max} = 284$ nm, $\varepsilon_{max} = 18{,}700$ $M^{-1} \cdot cm^{-1}$) of the substrate and increase in absorption in the range of 240–260 nm (Figure 2a). The spectral changes are compatible with the expectation that the thiocarbonyl group is engaged in the photoreaction. The HPLC analysis revealed 88.5% conversion of the substrate and a formation of guanosine (Guo, **2**) as the major photoproduct (Figure 2b). The product was formed with a chemical yield of 89% based on the substrate reacted (Scheme 2).

Guo (**2**) was found to be the major photoproduct under very low (<15%) conversion of 8-TG as well (Scheme 2). The UV absorption spectra recorded during the irradiation and chromatogram of the irradiated solution are presented in Figure 2. The product concentration profile showed that it was present at very low substrate conversion (Figure 3a). Therefore, it did not arise from secondary photochemistry.

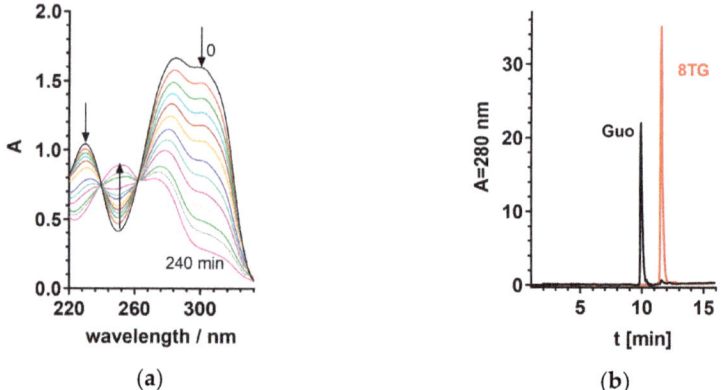

Figure 2. (**a**) UV spectra of 8-TG (c = 1.8 × 10^{-4} M) in air-equilibrated aqueous solution before (black) and after 240 min. Irradiation at λ = 313 nm, (**b**) Chromatogram of the solution of 8-TG before (red) and after irradiation (black).

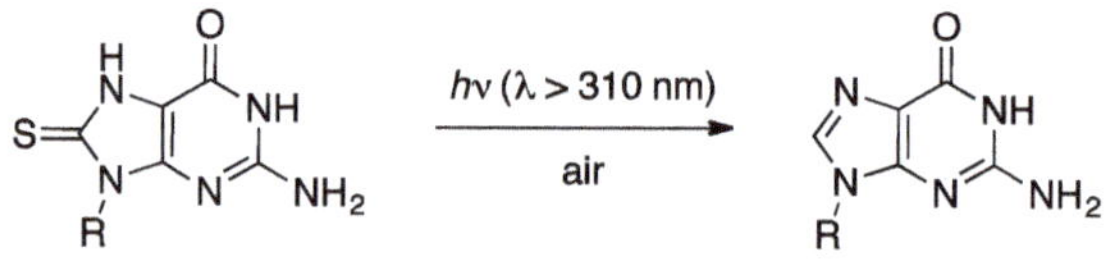

Scheme 2. The formation of Guo (**2**) upon excitation of 8-TG (**1**).

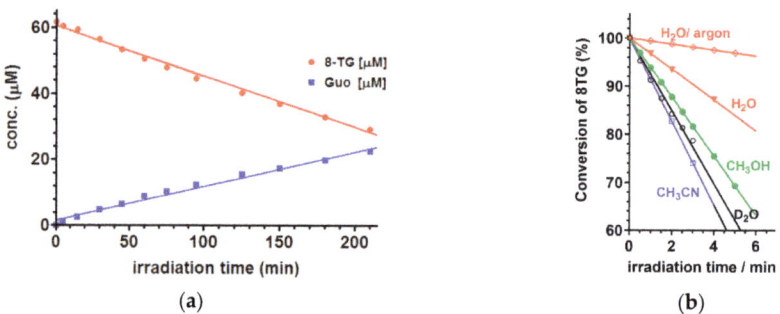

Figure 3. (**a**) The concentration of 8-TG (red) and Guo (blue) as a function of the irradiation time in air-equilibrated aqueous solution. (**b**) Rates of substrate disappearance upon radiation at λ > 300 nm of 8-TG in air-equilibrated solutions with different solvents.

In contrast to the efficient photochemistry in the presence of atmospheric oxygen, the disappearance of the substrate was slowed down by a factor of 10 when aqueous solution was purged by argon for 15 min prior to irradiation (Figure 3b). Several products were observed, but guanosine was not identified. Thus, the presence of oxygen is required for the photochemical transformation of 8-TG to Guo, suggesting that the observed transformation is a photooxidation reaction. The nature of the solvent has an impact on the rate of 8-TG disappearance (Figure 3b). A higher rate was observed in aerated organic solvents. Moreover, the photochemical reaction rate of 8-TG in D$_2$O increased twofold as compared to the rate determined in aqueous (H$_2$O) solution.

In general, two mechanisms of photochemical oxidation of organic compounds can be distinguished: a mechanism (type II) with the involvement of singlet oxygen (^{1}O$_2$), a highly reactive oxygen molecule in its excited electronic state, and a nonsinglet oxygen mechanism (type I) [7,32].

In order to establish which mechanism was operating in the case of 8-TG, the compound was subjected to a reaction with $^1O_2$. Methylene blue (MB) and rose bengal (RB) are very effective sensitizers of $^1O_2$ upon visible light irradiation in aerobic conditions [33]. Figure 4 shows the absorption spectra of a mixture of MB and 8-TG before and after irradiation (λ > 400 nm) and a chromatogram of the irradiated mixture. Guo was found to be a major product in both MB- and RB-sensitized reactions of 8-TG, with chemical yields of 57% and 54%, respectively.

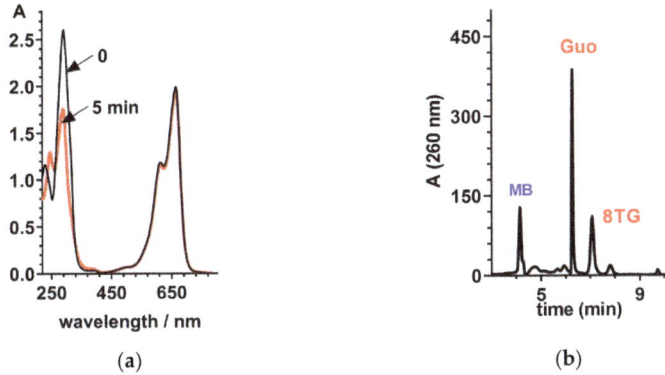

Figure 4. (a) UV absorption spectra recorded in the course of radiation of methylene blue (MB) and 8-TG with λ > 400 nm; (b) Chromatogram of the irradiated mixture.

The formation of Guo upon excitation of 8-TG with λ = 313 nm light and in a reaction sensitized by dyes (λ > 400 nm) suggests that the reactive oxygen species, $^1O_2$, is involved in the direct photooxidation of 8-TG (Scheme 3a). The generation of $^1O_2$ by 8-TG, like in the case of 6-TG, may occur via an energy transfer from the 8-TG triplet excited state (T_1) to the oxygen molecule in the electronic ground state.

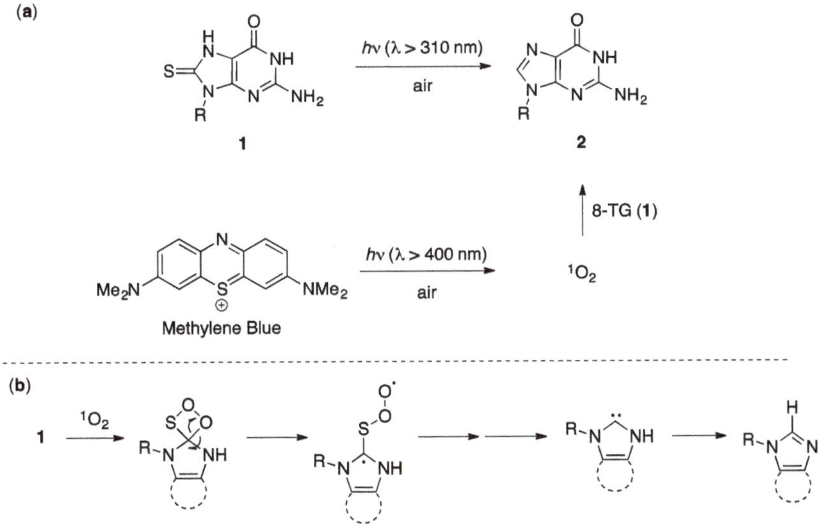

Scheme 3. (a) Photochemical oxidation of 8-TG and photosensitized generation of singlet oxygen; (b) Proposed mechanistic pathway of the $^1O_2$-promoted desulfurization of 8-TG.

The desulfurized substrate is an unexpected photoproduct obtained from photooxidation of heteroaromatic thiocarbonyl compounds [34,35]. The photoproduct molecules identified so far always contained oxygen atom(s). The major photoproducts were: ketones, sulfines, and, in the case of enolizable thiones, sulfinates and sulfonates. For example, the photoproducts obtained from 6-thioguanine, under experimental conditions identical to that applied for 8-TG, have been identified as guanine 6-sulfinate, guanine 6-sulfonate, and guanine as a minor photoproduct [36]. Comprehensive theoretical and experimental investigations of 6-TG photoxidation performed by Zou et al. have shown that all stable isolated photoproducts, most likely, derive from a common, unstable peroxy intermediate 6-(SOOH)-G [37]. 6-(SOOH)-G is suggested to be a primary product of the $^1O_2$ attack on sulfur atoms in the 6-TG molecule.

Oxygenated sulfur compounds, like sulfinic and sulfonic acids, as well as the carbonyl analog of 8-TG, have not been detected. Considering what has been previously described for 6-TG and the proposed mechanism hypothesized for thiourea in the presence of singlet oxygen [22], we propose that a similar [2 + 2] attack of $^1O_2$ on the thione moiety would lead to a thioperoxyl radical via 4-membered-ring opening. Subsequently, this unstable intermediate could, instead of proceeding to an oxygenated sulfur compound, evolve to form a carbene, leading afterwards to guanosine (**2**) (Scheme 3b).

2.3. Laser Flash Photolysis

It has been established that the excited triplet state plays a key role in the generation of singlet oxygen. That is why the characterization of this state, configuration, radiative and radiation-less pathways of deactivation, and yield of singlet oxygen generation is so important [38].

The properties of the lowest energy excited states of 8-TG have been studied by laser flash photolysis. In argon-saturated aqueous solution, only transient absorption of the excited triplet state (T_1) of 8-TG can be observed in nanosecond and microsecond time scales. Transient absorption spectra of 8-TG in Ar-saturated aqueous solution are shown in Figure 5a. Immediately after the laser, a broad spectrum extending over 330–750 nm was observed. From these kinetic traces, it was found that the decay of T_1 was independent of the monitoring wavelength, suggesting that the two bands can be assigned to a single excited species. Single exponential fits were made for the transient absorption decays measured for a series of 8-TG solutions having concentrations in the range of 0.05–0.5 mM. The determined rate constants were then used in the Stern–Volmer plot. Under these conditions, the T_1 state of 8-TG was relatively long-lived. The intrinsic (concentration-independent) T_1 state lifetime, τ^0 = 3.6 µs, was determined from the Stern–Volmer plot (Figure 5c). The T_1 state is quenched rapidly by the ground state O_2 molecules with a rate constant $k_Q = 5 \times 10^9$ M^{-1}·s^{-1} (in air-equilibrated aqueous solution). The kinetic traces of the T_1 state of 8-TG in argon-saturated and air-equilibrated aqueous solution, respectively, determined by monitoring the absorption decay at λ = 620 nm are presented in Figure 5b. The primary photochemical processes in aqueous solution are triplet state formations, as shown by flash photolysis.

Since 8-TG $(T_1)^*$ is quenched by dissolved molecular oxygen, a sensitization reaction between 8-TG* and molecular oxygen should occur [39]. In the presence of oxygen, radicals can also be formed in the process of electron transfer. In particular, evidence for presence of the 8-TG radical or radical cation was sought in transient absorption measurements. The transient absorption spectrum of 8-TG obtained under aerated conditions is identical to the spectrum recorded in the Ar-saturated solution.

However, the compared decay dynamics of the T_1 state at 340 nm shows the appearance of a new transient. We observed a long-lived transient (λ_{max} = 340 nm) formed in reaction of 8-TG with oxygen. Based on our experience and literature reports we can assume that the observed species is a radical or radical cation. (Figure 6). The transient was found to disappear 35.0 µs after the laser pulse, it was not quenched by oxygen, and the recorded spectrum (λ_{max} = 350 nm) was in the range where guanosine radicals were previously observed [31].

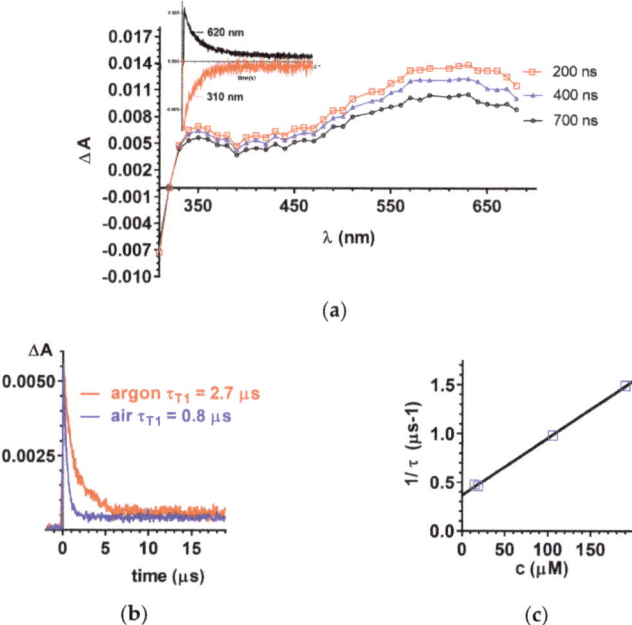

Figure 5. (**a**) Transient absorption spectrum of 8-TG (0.62 mM) in argon-saturated aqueous solution at various delay times (λ_{exc} = 266 nm, 3 mJ). Insert: kinetic traces of the 620 nm absorbance decay and 310 nm the ground state depopulation of 8-TG in argon-saturated aqueous solution; (**b**) Kinetic traces of the 620 nm absorbance decay in argon-saturated and air-equilibrated aqueous solution; (**c**) Stern–Volmer plot of reciprocal of triplet lifetime vs. concentration of 8-TG in argon-saturated solution.

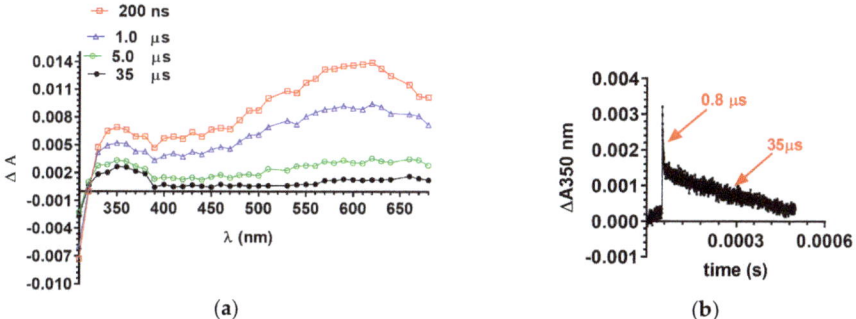

Figure 6. (**a**) Transient absorption spectrum of 8-TG (0.62 mM) in air-equilibrated aqueous solution at various delay times (λ_{exc} = 266 nm, 3 mJ). (**b**) Kinetic trace of the 340 nm absorbance decay in air-equilibrated aqueous solution.

For 8-TG, the radical processes have not been observed so far, but they are well known for guanine [40]. Molecular oxygen is known to act as an electron acceptor whether it is in its ground state or in its excited singlet state [41]. Both processes are presented below with the participation of 8-TG in its T_1 excited state or in its ground state (S_0).

$$8\text{-TG}(T_1) + {}^3O_2 \rightarrow 8\text{-TG}^{+\bullet} + O_2^{\bullet-}, \tag{1}$$

$$8\text{-TG}(S_0) + {}^1O_2 \rightarrow 8\text{-TG}^{+\bullet} + O_2^{\bullet-}, \tag{2}$$

2.4. γ-Radiolysis of 8-TG in Aqueous Solutions

Radiolysis of neutral water leads to the reactive species e_{aq}^-, $HO^\bullet$, and $H^\bullet$, together with H^+ and H_2O_2, as shown in Equation (3). The values in brackets represent the radiation chemical yield (G) in $\mu mol \cdot J^{-1}$ [42].

$$H_2O + \gamma\text{-irr.} \rightarrow e_{aq}^-(0.27), HO^\bullet(0.28), H^\bullet(0.06), H^+(0.06), H_2O_2(0.07) \tag{3}$$

Having discovered the unexpected reductive desulfurization of 8-TG in oxidative conditions, our investigation proceeded with the γ-radiolysis of **1** in the presence, initially, of all the reactive species generated by radiolytic decomposition of water: e_{aq}^-, $HO^\bullet$, and $H^\bullet$. Irradiations were performed at room temperature using a ^{60}Co-Gammacell at different doses. The exact absorbed radiation dose was determined by employing the Fricke chemical dosimeter, by taking $G(Fe^{3+}) = 1.61$ $\mu mol \cdot J^{-1}$ [43].

We investigated this reaction by detailed product studies in a dose-dependent experiment, irradiating 6 different independent samples (each 0.196 mM of **1**) for 100 Gy dose intervals, until achieving a total dose of 500 Gy. The reaction mixtures were monitored by HPLC-UV (254 nm) analysis, and the chromatograms were plotted in Figure 7a. In the dose course, we could observe consistent consumption of the starting material **1** (red peak, RT = 13.01 min) and simultaneous formation of the main product (blue peak, retention time (RT) = 9.8 min). Mass spectra of this product showed the mass of guanosine **2**, also confirmed by comparison with commercially available guanosine as well as spiking experiments. The product yield (mol·kg^{-1}) divided by the absorbed dose (1 Gy = 1 J·kg^{-1}) gives the reaction's chemical yield (G). The G values at each analyzed dose were calculated and plotted with the dependence of the dose (Figure 7b). The radiation chemical yields extrapolated to the zero dose are: $G(-1) = 0.17$ and $G(2) = 0.12$ $\mu mol \cdot J^{-1}$. Considering the sum of the G values of the reactive species (Equaction (3)) of 0.61 $\mu mol \cdot J^{-1}$ (0.68 if we consider H_2O_2 contribution), we could observe that 0.44 (or 0.51 respectively) of the reacting radicals returned to starting materials, while 0.17 reacted productively with **1**. Moreover, 70.6% of these radicals (G = 0.12 $\mu mol \cdot J^{-1}$), lead to the formation of product **2**. Unknown products formed and were visible in the formation of small peaks at RT = 8.6 and 14.45 min.

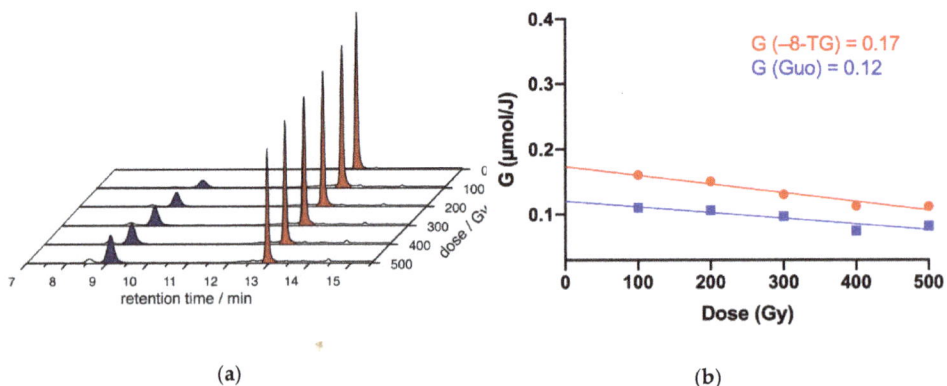

Figure 7. γ-radiolysis of **1** (0.196 mM) in N_2-purged water at a natural pH at a dose rate of 1.85 Gy·min^{-1}. (**a**) HPLC runs of the reactions. The HPLC peaks of **1** are highlighted in red, while the peaks of guanosine (**2**) are highlighted in blue. (**b**) The chemical irradiation yields $G(-1)$ (●) and $G(2)$ (■) as a function of the irradiation dose. The line extrapolated to the zero dose leads to the G values reported in the graph.

The results obtained in the above γ-radiolysis experiment and in the absence of oxidation products, such as 8-oxo-Guo or the disulfide dimer of 8-TG, underline an unexpected and more complex mechanistic scenario that is in accordance with the results obtained in the photooxidation experiments. Therefore, we decided to investigate some aspects of these reactions in more detail.

Before moving forward to the reactive species, we decided to first test the reactivity of **1** with H_2O_2, which is also formed by radiolysis of water. Indeed, the reaction of thiourea derivatives with hydrogen peroxide is a well-known desulfurization process, leading to the reduced product and SO_2 [20]. Therefore, we carried out an experiment outside the Gammacell using an aqueous solution of 8-TG (0.3 mM) in the presence of 10% H_2O_2, kept without stirring for 4.5 h, and a time comparable to 500 Gy if the reaction was done in Gammacell (Figure S12, Supporting Information). As expected, the formation of Guo (**2**) as a major product was observed.

2.4.1. γ-radiolysis of 8-TG in Aqueous Solutions under Reductive Conditions

Afterwards, we focused first on the reactivity of e_{aq}^- and $H^\bullet$ with **1**, employing N_2-saturated water solutions at a natural pH. tBuOH (0.25 M) was added to the reaction mixture in order to suppress the $HO^\bullet$ generated from the radiolytic decomposition of water (Equation (4)). In fact, efficient conversion of hydroxyl radicals into the less reactive *tert*-butyl radicals allows studying the reaction of the e_{aq}^- and $H^\bullet$ atoms because the presence of tBuOH also suppresses $H^\bullet$ radicals, but only slightly since the reaction rate is more than three orders of magnitude lower (Equation (5)) [42].

$$HO^\bullet + tBuOH \rightarrow (CH_3)_2(OH)CH_2^\bullet + H_2O \qquad (k = 6.0 \times 10^8 \ M^{-1} \cdot s^{-1}) \qquad (4)$$

$$H^\bullet + tBuOH \rightarrow (CH_3)_2(OH)CH_2^\bullet + H_2 \qquad (k = 1.7 \times 10^5 \ M^{-1} \cdot s^{-1}) \qquad (5)$$

From Figure 8a we can infer that the gradual disappearance of 8-TG occurred with the formation of Guo (**2**). Furthermore, no evidence of any side reaction was found, a part from some negligible peaks eluted after the starting material (RT between 14 and 15 min). The G values at each analyzed dose were calculated and plotted with the dependence of the dose (Figure 8b). The radiation chemical yields extrapolated to the zero dose are: $G(-1) = 0.15$ and $G(2) = 0.14$ µmol·J^{-1}. These values underline that all the reacted 8-TG was converted solely to product **2** or returned to the starting material and, in a similar scenario, to the previous experiments without tBuOH.

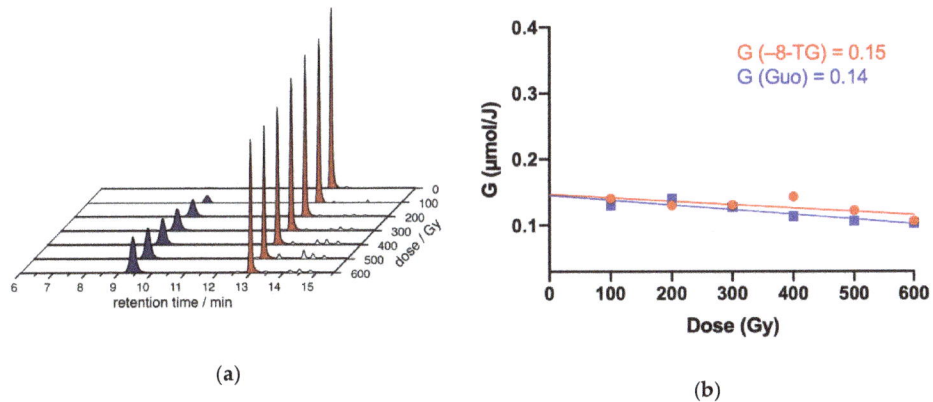

(a)

(b)

Figure 8. γ-radiolysis of **1** (0.25 mM) in N_2-purged water at a natural pH, containing tBuOH (0.25 M) at a dose rate of 1.85 Gy·min^{-1}. (**a**) HPLC runs of the reactions. The HPLC peaks of **1** are highlighted in red, while the peaks of guanosine (**2**) are highlighted in blue. (**b**) The chemical irradiation yields $G(-1)$ (●) and $G(2)$ (■) as function of the irradiation dose. The line extrapolated to the zero dose leads to the G values reported in the graph.

On the basis of these and the previous results, our mechanistic hypothesis is depicted in Scheme 4. We postulated that the contribution of H_2O_2 would proceed, as reported in literature [20], via formation of the oxygenated intermediate **C**, which eliminated SO_2 and generated a carbene, which then led to product **2**. On the other hand, H• radicals could add on the thione moiety of 8-TG, forming the radical **A**, as suggested by previously reported studies [25]. Subsequently, this species could eliminate HS• radicals, forming the same carbene **B** formed with H_2O_2 and achieving Guo (**2**). The combined G values of H• radicals and H_2O_2 correspond to 0.13 µmol·J^{-1} and the G (**2**) = 0.12 µmol·J^{-1}; hence, they were in accordance with the suggested pathway. However, the reaction of e_{aq}^-/H$^+$ should also potentially contribute to the formation of intermediate **A**.

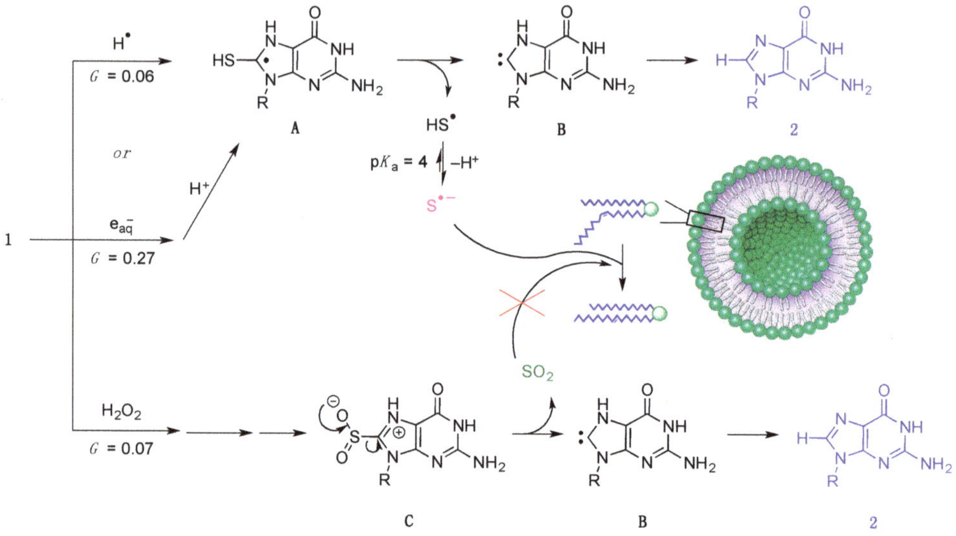

Scheme 4. Postulated mechanism of desulfurization promoted by H• atoms and H_2O_2.

Concerning the experiment suppressing HO• radicals, on the basis of the observed similar G values, we postulated that the reaction could involve H• atoms and H_2O_2 (Scheme 4) as well. Indeed, both these species are present in the reaction medium. However, we were interested to also understand the involvement of e_{aq}^- in the formation of Guo (**2**). The reaction of 8-TG with e_{aq}^- could generate a radical anion, which, upon protonation, could afford the previously discussed radical formed by the attack of H• atoms. Therefore, the contribution of both H• radicals and/or e_{aq}^-/H$^+$ would provide the formation of Guo (**2**) via elimination of HS• and formation of carbene **B**.

With these premises, to confirm whether the mechanistic pathway of this desulfurization proceeded via heterolytic bond cleavage or by thiyl radical formation, we performed the same three experiments in the presence of 1-palmitoyl-2-oleoyl-sn-glycero-3-phosphocholine (POPC) liposomes (2 mM) on the basis of our extensive experience in lipid cis/trans isomerization catalyzed by thiyl radicals (Scheme 4) [44–46]. In fact, if the mechanism involves the elimination of sulfhydryl radicals, we should observe isomerization, while no isomerization should occur if SO_2 is produced. After 4.5 h reaction time or 500 Gy irradiation, the phospholipids were extracted, transesterified to the methyl esters in methanolic KOH (0.5 M) solution, and analyzed via gas chromatography (GC) (for details see 3.6–3.8). GC analysis of the fatty acid methyl esters revealed that no isomerization occurred in the reaction with H_2O_2, supporting our hypothesis, which involves the formation of SO_2 (Figure S12). Conversely, in the γ-radiolysis experiment we observed 4.5% of isomerization, supporting the postulated elimination of HS• radicals (Figure S7). Moreover, the reaction performed under reductive conditions achieved the desulfurized product **2** while providing isomerization of the oleic double bonds up to 13.7% (Figure S8).

2.4.2. The Role of HO• Radicals and Other Oxidants

In a subsequent experiment, we deemed it proper to study the reaction of **1** with HO• radicals. In doing so, we performed radiolysis of **1** in N_2O-saturated water on the basis of the previously discussed transformation. In N_2O-saturated solution at a natural pH ($\approx$0.02 M of N_2O), e_{aq}^- are efficiently transformed into HO• radicals via Equation (6) ($k = 9.1 \times 10^9$ $M^{-1} \cdot s^{-1}$), affording G (HO•) = 0.55 µmol·J^{-1}; that is, HO• and H• account for 90% and 10%, respectively, of the reactive species [42].

$$e_{aq}^- + N_2O + H_2O \rightarrow HO^\bullet + N_2 + HO^- \qquad (k = 9.1 \times 10^9 \text{ M}^{-1}\cdot\text{s}^{-1}) \qquad (6)$$

Solutions of **1** (0.3 mM) were irradiated and stopped at intervals of 100 Gy, up to 500 Gy, and the results are plotted in Figure 9a. Also in this experiment, we observed a gradual consumption of **1** (red peaks) with concurrent formation of the reduction product (**2**, blue) despite the oxidative conditions. At a dose of 600 Gy the product yield was 14%, while the conversion was 54%. However, in this case, we observed the formation of 2 additional peaks, one with RT of $\approx$9 min and the other $\approx$14.5 min, which could explain the lower conversion measured. The radiation chemical yields extrapolated at zero dose from the plots in Figure 9b were: G (−1) = 0.16 and G (2) = 0.12 µmol·J^{-1}. Assuming that these results derive as before from the contribution of H• radicals and H_2O_2, since G (HO•) is 0.55 µmol·J^{-1}, the arising question is what happened to HO• radicals since we did not observe further consumption of the substrate **1**.

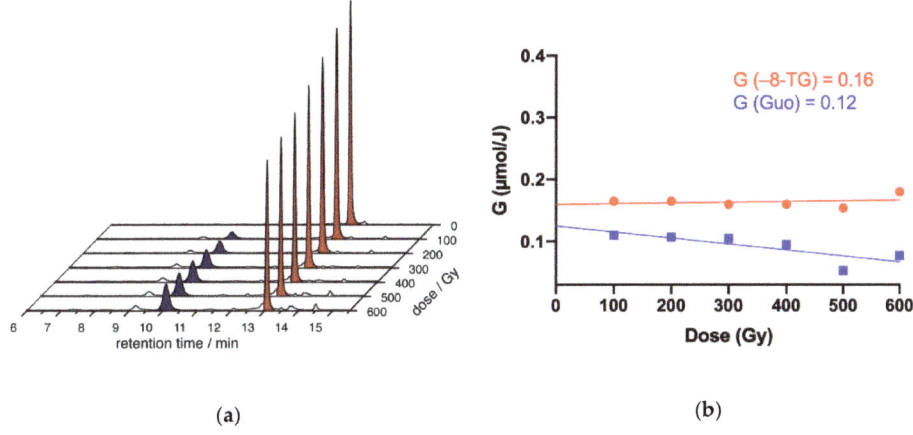

(a)   (b)

Figure 9. γ-radiolysis of **1** (0.3 mM) in N_2O-purged water at a natural pH at a dose rate of 1.85 Gy·min^{-1}. (**a**) HPLC runs of the reactions. The HPLC peaks of **1** are highlighted in red, while the peaks of guanosine (**2**) are highlighted in blue. (**b**) The chemical irradiation yields G (−1) (●) and G (2) (■) as function of the irradiation dose. The line extrapolated to a zero dose leads to the G values reported in the graph.

To answer this question, we decided to employ different oxidant species generated from the conversion of hydroxyl radicals (from Equation (6)), depending of the type of additive present in an N_2O-saturated solution. In particular, we decided to test the reactivity of **1** in the presence $Br_2^{\bullet-}$ and $N_3^\bullet$, formed via Equations (7) and (8) from KBr (0.1 M) and NaN_3 (0.1 M) (respectively G ($Br_2^{\bullet-}$) and G ($N_3^\bullet$) = 0.55 µmol·J^{-1}).

$$HO^\bullet + N_3^- \rightarrow N_3^\bullet + HO^- \qquad (k = 1.2 \times 10^{10} \text{ M}^{-1}\cdot\text{s}^{-1}) \qquad (7)$$

$$HO^\bullet + 2Br^- \rightarrow Br_2^{\bullet-} + HO^- \qquad (k = 1.1 \times 10^{10} \text{ M}^{-1}\cdot\text{s}^{-1}) \qquad (8)$$

In Figure S5 the HPLC-UV runs for the dose-dependent experiments in the presence of $N_3^\bullet$ are depicted. The main product formed was the reduced product guanosine (**2**) at a 33.6% yield and 95.7% conversion after 600 Gy irradiation. From the dose course studies, we could observe the gradual consumption of **1** while **2** consistently formed. The radiation chemical yields highlight that about half of $N_3^\bullet$ radicals reacted with **1** in a productive way, while 66% of these contributed to formation of the product. Interestingly, the same behavior was observed by using $Br_2^{\bullet-}$, which measured similar G values and an identical difference between G (−1) and G (**2**) of 0.07 µmol·J^{-1} (Figure S4).

With these results in hand, we were interested in understanding the involvement of these oxidant species in the overall desulfurization of 8-TG (**1**). In fact, in all the reactions we could observe similar results to the ones obtained in the previous experiments in terms of G values, while no further consumption of the substrate was achieved via side product formation. On the basis of our strong background of radical chemistry of guanosine derivatives [31,47–52], it was established that HO$^\bullet$ radicals could perform hydrogen abstraction predominantly on the exocyclic NH_2 group [48–50], leading to the radical intermediate **D** (Scheme 5). This species could undergo tautomerization to the more stable radical **E**. We proved that this tautomeric equilibrium was not affected by substitution at the 8-position of the guanine moiety with H, Br, or alkyl group, and it occurred at a constant of about 3×10^4 s^{-1} [50]. Hence, we could assume that this could be the case as well for 8-TG-derived radical **D**. Once the tautomeric radical **E** formed, this could lead back to the starting material **1**. In fact, this has been reported for similar intermediates for the other studied guanosine derivatives, which resulted in regeneration of the substrate with a still unclear mechanism [31,52]. This pathway could explain the absence of further consumption of 8-TG in the predominant presence of HO$^\bullet$ radicals. Additionally, the intermediate **E** could be directly formed by $Br_2^{\bullet-}$ and $N_3^\bullet$ and provide regeneration of **1** in the same fashion, explaining the comparable results in the presence of these two radical species [31,51,52].

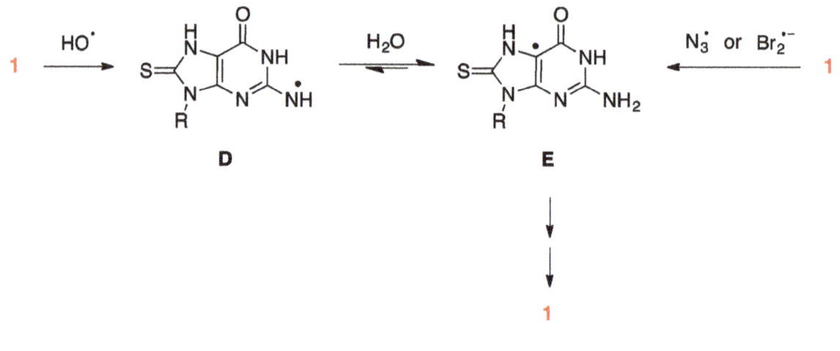

Scheme 5. Hypothesized mechanism of the reaction between HO$^\bullet$, $Br_2^{\bullet-}$ and $N_3^\bullet$, forming radical intermediates and leading back to the substrate.

Subsequently, we carried out all these experiments under oxidative conditions in the presence of POPC liposomes, to check whether the contribution of H$^\bullet$ atoms could lead to the formation of thiyl radicals in this case as well. In all the reactions, we could observe the promotion of *cis/trans* isomerization of the oleic moiety between 2.1–7.9% (Figures S9–S11), confirming that the contribution of the oxidant species leads to the regeneration of substrate **1**.

3. Materials and Methods

3.1. Materials

Unless otherwise noted, all commercially available compounds were used without further purification. Solvents were purchased at HPLC grade and used without further purification. Water was purified with a Millipore system. 1-Palmitoyl-2-oleoyl-sn-glycero-3-phosphocholine (POPC) was purchased from Larodan Inc. (Solna, Sweden); chloroform, methanol, diethyl ether, and n-hexane (HPLC grade) were purchased from Baker (Dover, NJ, USA). 8-TG was purchased from Santa Cruz

Biotechnology Inc. (Dallas, TX, USA). Guanosine, D_2O, and DMSO-d_6 were purchased from Sigma Aldrich (St. Louis, MO, USA).

3.2. UV-Vis Absorption Spectra

The UV absorptions for the determination of the pKa were recorded with a UV-vis spectrometer Agilent Carey 100 (Santa Clara, CA, USA). Using 12 different 0.1 mM solutions of 8-TG, each at a different pH (using NaOH/phosphoric acid buffers), we recorded the absorption values and we plotted the absorption versus pH of the solution. The resulting data were fit into sigmoid curves, and the inflection points were calculated. Figure 1 was prepared with Graphpad Prism 8.0 (San Diego, CA, USA).

All other UV spectra were recorded at room temperature with a Cary300Bio spectrophotometer.

3.3. HPLC-MS Analysis

HPLC-MS analyses were performed with an Agilent 1260 Infinity II HPLC system equipped with a Fortis 5 µm UniverSil C18 column (250 × 4.6 mm) using, as eluent, 0.01% formic acid in water with 0–55% gradient of acetonitrile, coupled with an InfinityLab single quadrupole liquid chromatography/mass selective detector (LC/MSD) and with a diode array detector set at 254 nm. Quantitative studies of the performed reactions were done by multiple point calibration curves (6 points) in HPLC-MS equipped with a Fortis 5 µm UniverSil C18 column (250 × 4.6 mm) using, as eluent, 0.01% formic acid in water with 0–55% gradient of acetonitrile (injection volume 5 µL, flow rate 0.7 mL/min, detection at 254 nm). Elution method: (A = H_2O + 0.01% formic acid, B = ACN) A:B = 95:5 1 min, 45:55 at 11 min and held for 5 min, 45:55 at 15 min, back at 95:5 at 18 min, and held for 5 min. R^2 = 0.99998 8-TG, 1.0 Guo. Retention times: 8-TG 13.02 min, Guo 9.8 min.

3.4. Steady-State Irradiation

Solutions (2.5 mL) of 8-TG (1.8×10^{-4} M) in ACN or H_2O in a quartz cell (l = 1 cm) were irradiated using a 200 W high-pressure mercury lamp equipped with a pyrex (λ > 300 nm) or an interference filter (λ = 313 nm). For irradiation in the absence of oxygen, solutions were deaerated by bubbling with argon; otherwise, the solutions were irradiated in an open-to-air cell (air-equilibrated). In dye-sensitized photoreactions, solutions containing methylene blue or rose bengal (c = 0.05 mM) and 8-TG (1.8×10^{-4} M) were irradiated in air-equilibrated ACN or H_2O solutions at >400 nm. The photoreactions were monitored by measuring the absorbance spectrum. The concentrations of substrate and photoproducts were determined by HPLC analysis. The photoproduct guanosine was identified by a comparison of the HPLC peak retention time, UV–VIS absorption, and MS spectra with a standard reference.

3.5. Laser Flash Photolysis (LFP) Experiments

The set up for the nanosecond laser flash photolysis (LFP) experiments and its data acquisition system have previously been described in detail [41]. LFP experiments employed a pulsed Nd:YAG laser (λ = 266 nm, 3 mJ, 3–9 ns) for excitation. Transient decays were recorded at individual wavelengths by the step-scan method with a step distance of 10 nm in the range of 320–800 nm as the mean of 10 xenon-lamp pulses.

3.6. Experiments in Gammacell

Irradiations were performed at room temperature (22 ± 2 °C) using a ^{60}Co-Gammacell at different dose rates. The exact absorbed radiation dose was determined with the Fricke chemical dosimeter, by taking $G(Fe^{3+})$ = 1.61 µmol·J^{-1} [43].

In the radiation experiments, 4 mL vials equipped with a rubber septum were used. Two milliliters of freshly prepared stock solutions (0.09 or 0.3 mM) of 8-TG were added together with specific amounts of additives, depending of the experiments (see above). Afterwards, the solution was degassed with either N_2 or N_2O for 10 min via a cannula, sealed with parafilm, and irradiated in a Gammacell. After the irradiation time, the reaction mixtures were injected in HPLC-MS without further workup.

For kinetic experiments, different reactions were prepared from a single mother solution, and then each one was stopped at a different dose. Figures were prepared with the software Graphpad Prism 8.0 and OriginLab.

For the experiments in the presence of POPC liposomes, the workup of the reactions was done as follows: 2:1 chloroform/methanol (5 × 4 mL) was added to the reaction mixture according to the Folch method [53]. The organic layer was collected and dried over anhydrous Na_2SO_4 then evaporated under vacuum to dryness. One milliliter of 0.5 M solution of KOH in methanol was added to each sample for transesterification of the fatty acid containing esters, which was performed by stirring for 30 min at room temperature. The methanolic reaction mixture was quenched with brine (1 mL), then the extraction was repeated with n-hexane (3 × 2 mL), and the organic phases were collected, dried over anhydrous Na_2SO_4, and evaporated to dryness. The fatty acid methyl ester (FAME) residue of each sample was dissolved in 50 µL of n-hexane, and 1 µL was injected for GC analysis.

3.7. Preparation of 1-Palmitoyl-2-oleoyl-sn-glycero-3-phosphocholine (POPC) Liposomes

Large unilamellar vesicles by extrusion technique (LUVETs) were prepared using the hydration-extrusion technique [54]. Briefly, POPC (76 mg; 0.1 mmol) was dissolved in $CHCl_3$/MeOH 2:1 until a clear lipid solution was obtained. The organic solvent was removed using a rotary evaporator to yield a homogeneous lipid film on the sides of a round-bottom flask. The lipid film was thoroughly dried to remove residual organic solvent by placing the vial or flask on a vacuum pump for 1 h. The dried lipid film was left to hydrate for 30 min in PBS and then vortexed for 10 min until a milky monophasic solution containing multilamellar vesicles (MLVs) was obtained. Once a stable suspension was formed, MLVs were downsized to LUVETs using a LiposoFast hand extrusion device (Avestin Inc., Ottawa, ON, Canada) equipped with 100 nm pore size polycarbonate filters through which the lipid suspension was passed 19 times, controlling the temperature with a heat block when required. The resulting suspension was used in the irradiation experiments.

3.8. GC Analysis of Fatty Acid Methyl Esters

Fatty acid methyl esters (FAMEs) were analyzed by GC (Agilent 6850, Milan, Italy), using the split mode (50:1), equipped with a 60 m × 0.25 mm × 0.25 µm (50% cyanopropyl)-methylpolysiloxane column (DB23, Agilent, USA), and a flame ionization detector with the following oven program: temperature started at 165 °C, held for 3 min, followed by an increase of 1 °C/min up to 195 °C, held for 40 min, followed by a second increase of 10 °C/min up to 240 °C and held for 10 min. A constant pressure mode (29 psi) was chosen with helium as the carrier gas. FAMEs were identified by comparison with authentic samples, and chromatograms were examined as described previously.

3.9. NMR Analysis of 8-TG

NMR spectra were recorded at ambient temperature on a Varian 500 MHz spectrometer. D_2O and DMSO-d_6, purchased from Sigma Aldrich (St. Louis, MO, USA), were used as solvents.

4. Conclusions

In conclusion, we reported a detailed study on the reactivity of biologically relevant 8-thioguanosine under oxidative and reductive conditions by using photooxidation, laser flash photolysis, and γ-radiolysis.

Direct excitation (λ > 310 nm) of 8-thioguanosine in aerated aqueous solution gave guanosine instead of oxidized sulfur derivatives (purine sulfonic and sulfinic acids). It appears reasonable to postulate that 8-TG transformation under direct irradiation occurs with intermediacy of $^1O_2$. The properties of the lowest energy excited state of 8-TG (T_1) have been studied by laser flash photolysis, and it was subsequently established that this excited T_1 state could act as a sensitizer, while in its ground state it could act as an acceptor of $^1O_2$. Guanosine, formally a reduction product of 8-TG, is a rather unexpected photoproduct obtained from the photooxidation reaction of thiocarbonyl compounds, and singlet oxygen is a key intermediate in the desulfurization of the compound studied. Subsequent investigations on the γ-radiolysis of 8-TG highlighted a complex mechanistic scenario, which was studied by evaluating the effects of the predominant presence of each reactive species derived by the radiolysis of water. The presence of oxidant species such as $HO^\bullet$ radicals, $Br_2^{\bullet-}$, and $N_3^\bullet$ resulted, supposedly, in the regeneration of substrate 1. Conversely, control experiments in the presence of POPC liposomes and reductive conditions displayed the possible involvement of $H^\bullet$ atoms, H_2O_2, and/or e_{aq}^-/H^+ in the desulfurization of 8-TG by the generation of thiyl radicals and the simultaneous formation of the desulfurized product 2.

Supplementary Materials: The following are available online at http://www.mdpi.com/1420-3049/24/17/3143/s1, Figure S1: ^1H NMR spectrum of 8-TG; Figure S2: Zoomed ^1H NMR spectrum of 8-TG; Figure S3: ^1H NMR spectrum of 8-TG after adding D_2O; Figure S4: Reaction of $Br_2^{\bullet-}$ with 8-TG; Figure S5: Reaction of $N_3^\bullet$ with 8-TG; Figure S6: Reaction of 8-TG in the presence of H_2O_2 outside Gammacell; Figure S7: Reaction of $HO^\bullet$, $H^\bullet$, and e_{aq}^- with 8-TG in the presence of POPC liposomes; Figure S8: Reaction of e_{aq}^- with 8-TG in the presence of POPC liposomes; Figure S9: Reaction of $HO^\bullet$ with 8-TG in the presence of POPC liposomes; Figure S10: Reaction of $Br_2^{\bullet-}$ with 8-TG in the presence of POPC liposomes; Figure S11: Reaction of $N_3^\bullet$ with 8-TG in the presence of POPC liposomes; Figure S12: Reaction of H_2O_2 with 8-TG in the presence of POPC liposomes.

Author Contributions: K.T.-G. and F.V. contributed equally to this work; K.T.-G., B.M., C.C., and F.V designed the study; C.C. and B.M. coordinated the work; C.C., F.V., and K.T.-G. wrote the paper; C.C., S.B.-V., F.V., B.M., and K.T.-G. discussed the project; F.V. and S.B.-V. performed the experiments in Gammacell and collected the worked-up samples; K.T.-G. performed the time steady-state photooxidation and laser flash photolysis; F.V. performed the HPLC-MS analyses of the experiments in Gammacell; C.C., S.B.-V., and F.V. discussed the HPLC-MS results; V.T. performed the GC analyses of the experiments with liposomes; C.C., S.B.-V., F.V., and K.T.-G. prepared all the figures. All the authors reviewed the manuscript.

Funding: This research received no external funding.

Acknowledgments: The authors thankfully acknowledge the contribution of Grażyna Wenska.

Conflicts of Interest: The authors declare no conflict of interest.

Abbreviations

8-TG	8-thioguanosine
6-TG	6-thioguanosine
ROS	reactive oxygen species
Guo/G	guanosine
FAME	fatty acid methyl ester
HPLC-MS	high performance liquid chromatography mass spectrometry
GC	gas chromatography
POPC	1-palmitoyl-2-oleoyl-sn-glycero-3-phosphocholine
MB	Methylene Blue
RB	Rose Bengal
8-oxo-G	8-oxo-guanosine

References

1. De Clercq, E. Antiviral drug discovery: Ten more compounds, and ten more stories (part B). *Med. Res. Rev.* **2009**, *29*, 571–610. [CrossRef] [PubMed]
2. Cantara, W.A.; Crain, P.F.; Rozenski, J.; McCloskey, J.A.; Harris, K.A.; Zhang, X.; Vendeix, F.A.P.; Fabris, D.; Agris, P.F. The RNA modification database, RNAMDB: 2011 update. *Nucleic Acids Res.* **2010**, *39*, D195–D201. [CrossRef] [PubMed]
3. Elion, G.B. The purine path to chemotherapy. *Vitro Cell. Dev. Biol.* **1989**, *25*, 321–330. [CrossRef]
4. Dervieux, T.; Blanco, J.G.; Krynetski, E.Y.; Vanin, E.F.; Roussel, M.F.; Relling, M.V. Differing contribution of thiopurine methyltransferase to mercaptopurine versus thioguanine effects in human leukemic cells. *Cancer Res.* **2001**, *61*, 5810. [PubMed]

5. Karran, P.; Attard, N. Thiopurines in current medical practice: Molecular mechanisms and contributions to therapy-related cancer. *Nat. Rev. Cancer* **2008**, *8*, 24. [CrossRef] [PubMed]
6. Massey, A.; Xu, Y.-Z.; Karran, P. Photoactivation of DNA thiobases as a potential novel therapeutic option. *Curr. Biol.* **2001**, *11*, 1142–1146. [CrossRef]
7. Attard, N.R.; Karran, P. UVA photosensitization of thiopurines and skin cancer in organ transplant recipients. *Photochem. Photobiol. Sci.* **2012**, *11*, 62–68. [CrossRef] [PubMed]
8. Ashwood, B.; Pollum, M.; Crespo-Hernández, C.E. Photochemical and photodynamical properties of sulfur-substituted nucleic acid bases. *Photochem. Photobiol.* **2019**, *95*, 33–58. [CrossRef] [PubMed]
9. Favre, A.; Saintomé, C.; Fourrey, J.L.; Clivio, P.; Laugâa, P. Thionucleobases as intrinsic photoaffinity probes of nucleic acid structure and nucleic acid-protein interactions. *J. Photochem. Photobiol. B Biol.* **1998**, *42*, 109–124. [CrossRef]
10. Brem, R.; Daehn, I.; Karran, P. Efficient DNA interstrand crosslinking by 6-thioguanine and UVA radiation. *DNA Repair* **2011**, *10*, 869–876. [CrossRef]
11. Brem, R.; Karran, P. Reactive oxygen-mediated damage to a human DNA replication and repair protein. *Cancer Res.* **2012**, *77*, 4787–4797. [CrossRef] [PubMed]
12. Pollum, M.; Ortiz-Rodrıguez, L.A.; Jockusch, S.; Crespo-Hernández, C.E. The triplet state of 6-thio-2′-deoxyguanosine: Intrinsic properties and reactivity toward molecular oxygen. *Photochem. Photobiol.* **2016**, *92*, 286–292. [CrossRef] [PubMed]
13. Brem, R.; Guven, M.; Karran, P. Oxidatively-generated damage to DNA and proteins mediated by photosensitized UVA. *Free Radic. Biol. Med.* **2017**, *107*, 101–109. [CrossRef] [PubMed]
14. Goodman, M.G. Mechanism of synergy between T cell signals and C8-substituted guanine nucleosides in humoral immunity: B lymphotropic cytokines induce responsiveness to 8-mercaptoguanosine. *J. Immunol.* **1986**, *136*, 3335. [PubMed]
15. Silva, A.L.R.; Ribeiro da Silva, M.D.M.C. Energetic, structural and tautomeric analysis of 2-mercaptobenzimidazole. *J. Therm. Anal. Calorim.* **2017**, *129*, 1679–1688. [CrossRef]
16. Sahu, S.; Rani Sahoo, P.; Patel, S.; Mishra, B.K. Oxidation of thiourea and substituted thioureas: A review. *J. Sulfur Chem.* **2011**, *32*, 171–197. [CrossRef]
17. Shakeel, A.; Altaf, A.A.; Qureshi, A.M.; Badshah, A. Thiourea derivatives in drug design and medicinal chemistry: A short review. *J. Drug Des. Med. Chem.* **2016**, *2*, 10. [CrossRef]
18. Pandey, M.; Srivastava, A.K.; D'Souza, S.F.; Penna, S. Thiourea, a ROS scavenger, regulates source-to-sink relationship to enhance crop yield and oil content in *Brassica juncea* (L.). *PLoS ONE* **2013**, *8*, e73921. [CrossRef]
19. Lou, E.K.; Cathro, P.; Marino, V.; Damiani, F.; Heithersay, G.S. Evaluation of hydroxyl radical diffusion and acidified thiourea as a scavenger during intracoronal bleaching. *J. Endod.* **2016**, *42*, 1126–1130. [CrossRef]
20. Grivas, S.; Ronne, E. Facile desulfurization of cyclic thioureas by hydrogen peroxide in acetic acid. *Acta Chem. Scand.* **1995**, *49*, 225–229. [CrossRef]
21. Kuhn, N.; Kratz, T. Synthesis of imidazol-2-ylidenes by reduction of imidazole-2(3*H*)-thiones. *Synthesis* **1993**, 561–562. [CrossRef]
22. Azarifar, D.; Golbaghi, M. Selective and facile oxidative desulfurization of thioureas and thiobarbituric acids with singlet molecular oxygen generated from trans-3,5-dihydroperoxy-3,5-dimethyl-1,2-dioxolane. *J. Sulfur Chem.* **2016**, *37*, 1–13. [CrossRef]
23. Dey, G.R.; Naik, D.B.; Kishore, K.; Moorthy, P.N. Nature of the transient species formed in the pulse radiolysis of some thiourea derivatives. *J. Chem. Soc. Perkin Trans. 2* **1994**, 1625–1629. [CrossRef]
24. Dey, G.R.; Naik, D.B.; Kishore, K.; Moorthy, P.N. Kinetic and spectral characteristics of transients formed in the pulse radiolysis of phenylthiourea in aqueous solution. *Radiat. Phys. Chem.* **1994**, *43*, 365–369. [CrossRef]
25. Dey, G.R.; Naik, D.B.; Kishore, K.; Moorthy, P.N. Kinetic and spectral properties of the intermediates formed in the pulse radiolysis of 2-mercaptobenzimidazole. *Res. Chem. Intermed.* **1995**, *21*, 47–58. [CrossRef]
26. Wang, W.-F.; Schuchmann, M.N.; Schuchmann, H.-P.; Knolle, W.; von Sonntag, J.; von Sonntag, C. Radical cations in the OH-radical-induced oxidation of thiourea and tetramethylthiourea in aqueous solution. *J. Am. Chem. Soc.* **1999**, *121*, 238–245. [CrossRef]
27. Dey, G. A comparative study of radical cations of thiourea, thiosemicarbazide, and diethylthiourea in aqueous sulphuric acid media employing pulse radiolysis technique. *J. Phys. Org. Chem.* **2013**, *26*, 927–932. [CrossRef]

28. Serobatse, K.R.N.; Kabanda, M.M. An appraisal of the hydrogen atom transfer mechanism for the reaction between thiourea derivatives and OH radical: A case-study of dimethylthiourea and diethylthiourea. *Comput. Theor. Chem.* **2017**, *1101*, 83–95. [CrossRef]
29. Gain, S.; Das, R.S.; Banerjee, R.; Mukhopadhyay, S. Kinetics and mechanism of oxidation of thiourea by a bridging superoxide in the presence of Ellman's reagent. *J. Coord. Chem.* **2016**, *69*, 2136–2147. [CrossRef]
30. Miyata, S.; Hoshino, M.; Isozaki, T.; Yamada, T.; Sugimura, H.; Xu, Y.-Z.; Suzuki, T. Acid dissociation equlibrium and singlet molecular oxygen quantum yield of acetylated 6,8-dithioguanosine in aqueous buffer solution. *J. Phys. Chem. B* **2018**, *122*, 2912–2921. [CrossRef]
31. Ioele, M.; Bazzanini, R.; Chatgilialoglu, C.; Mulazzani, Q.G. Chemical radiation studies of 8-bromoguanosine in aqueous solutions. *J. Am. Chem. Soc.* **2000**, *122*, 1900–1907. [CrossRef]
32. Baptista, M.S.; Cadet, J.; Di Mascio, P.; Ghogare, A.A.; Greer, A.; Hamblin, M.R.; Lorente, C.; Nunez, S.C.; Ribeiro, M.S.; Thomas, A.H.; et al. Type I and type II photosensitized oxidation reactions: Guidelines and mechanistic pathways. *Photochem. Photobiol.* **2017**, *93*, 912–919. [CrossRef] [PubMed]
33. Rao, V.J.; Muthuramu, K.; Ramamurthy, V. Oxidations of thio ketones by singlet and triplet oxygen. *J. Org. Chem.* **1982**, *47*, 127–131. [CrossRef]
34. Wenska, G.; Koput, J.; Burdziński, G.; Taras-Goslinska, K.; Maciejewski, A. Photophysical and photochemical properties of the T1 excited state of thioinosine. *J. Photochem. Photobiol. A Chem.* **2009**, *206*, 93–101. [CrossRef]
35. Hemmens, V.J.; Moore, D.E. Photo-oxidation of 6-mercaptopurine in aqueous solution. *J. Chem. Soc. Perkin Trans. 2* **1984**, 209–211. [CrossRef]
36. Ren, X.; Li, F.; Jeffs, G.; Zhang, X.; Xu, Y.-Z.; Karran, P. Guanine sulphinate is a major stable product of photochemical oxidation of DNA 6-thioguanine by UVA irradiation. *Nucleic Acids Res.* **2009**, *38*, 1832–1840. [CrossRef]
37. Zou, X.; Zhao, H.; Yu, Y.; Su, H. Formation of guanine-6-sulfonate from 6-thioguanine and singlet oxygen: A combined theoretical and experimental study. *J. Am. Chem. Soc.* **2013**, *135*, 4509–4515. [CrossRef] [PubMed]
38. Miyata, S.; Yamada, T.; Isozaki, T.; Sugimura, H.; Xu, Y.Z.; Suzuki, T. Absorption characteristics and quantum yields of singlet oxygen generation of thioguanosine derivatives. *Photochem. Photobiol.* **2018**, *94*, 677–684. [CrossRef]
39. Miyata, S.; Tanabe, S.; Isozaki, T.; Xu, Y.-Z.; Suzuki, T. Characteristics of the excited triplet states of thiolated guanosine derivatives and singlet oxygen generation. *Photochem. Photobiol. Sci.* **2018**, *17*, 1469–1476. [CrossRef]
40. Bensasson, R.V.; Land, E.J.; Truscott, T.G. *Excited States and Free Fadicals in Biology and Medicine*; Oxford University Press Inc.: New York, NY, USA, 1993; p. 150.
41. Pedzinski, T.; Markiewicz, A.; Marciniak, B. Photosensitized oxidation of methionine derivatives. Laser flash photolysis studies. *Res. Chem. Intermed.* **2009**, *35*, 497–506. [CrossRef]
42. Buxton, G.V.; Greenstock, C.L.; Helman, W.P.; Ross, A.B. Critical review of rate constants for reactions of hydrated electrons, hydrogen atoms and hydroxyl radicals (($^\bullet$OH)/$^\bullet$O−) in aqueous solution. *J. Phys. Chem. Ref. Data* **1988**, *17*, 513–886. [CrossRef]
43. Spinks, J.W.T.; Woods, R.J. *An Introduction to Radiation Chemistry*, 3rd ed.; John-Wiley and Sons, Inc.: New York, NY, USA, 1990; p. 100.
44. Lykakis, I.N.; Ferreri, C.; Chatgilialoglu, C. The sulfhydryl radical (HS$^\bullet$/S$^\bullet$−): A contender for the isomerization of double bonds in membrane lipids. *Angew. Chem. Int. Ed.* **2007**, *46*, 1914–1916. [CrossRef] [PubMed]
45. Chatgilialoglu, C.; Ferreri, C.; Melchiorre, M.; Sansone, A.; Torreggiani, A. Lipid geometrical isomerism: From chemistry to biology and diagnostics. *Chem. Rev.* **2014**, *114*, 255–284. [CrossRef] [PubMed]
46. Chatgilialoglu, C.; Ferreri, C.; Torreggiani, A.; Salzano, A.M.; Renzone, G.; Scaloni, A. Radiation-induced reductive modifications of sulfur-containing amino acids within peptides and proteins. *J. Proteom.* **2011**, *74*, 2264–2273. [CrossRef]
47. Chatgilialoglu, C.; Caminal, C.; Altieri, A.; Vougioukalakis, G.C.; Mulazzani, Q.G.; Gimisis, T.; Guerra, M. Tautomerism in the guanyl radical. *J. Am. Chem. Soc.* **2006**, *128*, 13796–13805. [CrossRef]
48. Chatgilialoglu, C.; Caminal, C.; Guerra, M.; Mulazzani, Q.G. Tautomers of one-electron-oxidized guanosine. *Angew. Chem. Int. Ed.* **2005**, *44*, 6030–6032. [CrossRef] [PubMed]
49. Chatgilialoglu, C.; D'Angelantonio, M.; Guerra, M.; Kaloudis, P.; Mulazzani, Q.G. A reevaluation of the ambident reactivity of the guanine moiety towards hydroxyl radicals. *Angew. Chem. Int. Ed.* **2009**, *48*, 2214–2217. [CrossRef]

50. Chatgilialoglu, C.; D'Angelantonio, M.; Kciuk, G.; Bobrowski, K. New insights into the reaction paths of hydroxyl radicals with 2′-deoxyguanosine. *Chem. Res. Toxicol.* **2011**, *24*, 2200–2206. [CrossRef]
51. D'Angelantonio, M.; Russo, M.; Kaloudis, P.; Mulazzani, Q.G.; Wardman, P.; Guerra, M.; Chatgilialoglu, C. Reaction of hydrated electrons with guanine derivatives: Tautomerism of intermediate species. *J. Phys. Chem. B* **2009**, *113*, 2170–2176. [CrossRef]
52. Kaloudis, P.; D'Angelantonio, M.; Guerra, M.; Spadafora, M.; Cismaş, C.; Gimisis, T.; Mulazzani, Q.G.; Chatgilialoglu, C. Comparison of isoelectronic 8-HO-G and 8-NH2-G derivatives in redox processes. *J. Am. Chem. Soc.* **2009**, *131*, 15895–15902. [CrossRef]
53. Folch, J.; Lees, M.; Sloane Stanley, G.H. A simple method for the isolation and purification of total lipides from animal tissues. *J. Biol. Chem.* **1957**, *226*, 497–509. [PubMed]
54. Olson, F.; Hunt, C.A.; Szoka, F.C.; Vail, W.J.; Papahadjopoulos, D. Preparation of liposomes of defined size distribution by extrusion through polycarbonate membranes. *Biochim. Biophys. Acta Biomembr.* **1979**, *557*, 9–23. [CrossRef]

Sample Availability: Not available.

© 2019 by the authors. Licensee MDPI, Basel, Switzerland. This article is an open access article distributed under the terms and conditions of the Creative Commons Attribution (CC BY) license (http://creativecommons.org/licenses/by/4.0/).

Article

Radiation Induced One-Electron Oxidation of 2-Thiouracil in Aqueous Solutions

Konrad Skotnicki [1,*], Katarzyna Taras-Goslinska [2], Ireneusz Janik [3,*] and Krzysztof Bobrowski [1]

1. Centre of Radiation Research and Technology, Institute of Nuclear Chemistry and Technology, 03-195 Warsaw, Poland; kris@ichtj.pl
2. Faculty of Chemistry, Adam Mickiewicz University, 61-614 Poznan, Poland; karem@amu.edu.pl
3. Notre Dame Radiation Laboratory, University of Notre Dame, Notre Dame, IN 46556, USA
* Correspondence: k.skotnicki@ichtj.waw.pl (K.S.); ijanik@nd.edu (I.J.); Tel.: +48-22-5041292 (K.S.)

Academic Editor: Chryssostomos Chatgilialoglu
Received: 6 October 2019; Accepted: 27 November 2019; Published: 2 December 2019

Abstract: Oxidative damage to 2-thiouracil (2-TU) by hydroxyl ($^\bullet$OH) and azide ($^\bullet$N$_3$) radicals produces various primary reactive intermediates. Their optical absorption spectra and kinetic characteristics were studied by pulse radiolysis with UV-vis spectrophotometric and conductivity detection and by time-dependent density functional theory (TD-DFT) method. The transient absorption spectra recorded in the reactions of $^\bullet$OH with 2-TU depend on the concentration of 2-TU, however, only slightly on pH. At low concentrations, they are characterized by a broad absorption band with a weakly pronounced maxima located at λ = 325, 340 and 385 nm, whereas for high concentrations, they are dominated by an absorption band with $\lambda_{max} \approx 425$ nm. Based on calculations using TD-DFT method, the transient absorption spectra at low concentration of 2-TU were assigned to the $^\bullet$OH-adducts to the double bond at C5 and C6 carbon atoms (3$^\bullet$, 4$^\bullet$) and 2c-3e bonded $^\bullet$OH adduct to sulfur atom (1 ... $^\bullet$OH) and at high concentration of 2-TU also to the dimeric 2c-3e S-S-bonded radical in neutral form (2$^\bullet$). The dimeric radical (2$^\bullet$) is formed in the reaction of thiyl-type radical (6$^\bullet$) with 2-TU and both radicals are in an equilibrium with $K_{eq} = 4.2 \times 10^3$ M^{-1}. Similar equilibrium (with $K_{eq} = 4.3 \times 10^3$ M^{-1}) was found for pH above the pK$_a$ of 2-TU which involves admittedly the same radical (6$^\bullet$) but with the dimeric 2c-3e S-S bonded radical in anionic form (2$^{\bullet-}$). In turn, $^\bullet$N$_3$-induced oxidation of 2-TU occurs via radical cation with maximum spin location on the sulfur atom which subsequently undergoes deprotonation at N1 atom leading again to thiyl-type radical (6$^\bullet$). This radical is a direct precursor of dimeric radical (2$^\bullet$).

Keywords: 2-thiouracil; radiosensitizers; $^\bullet$OH and $^\bullet$N$_3$ radicals; 2c-3e S∴S-bonded intermediates; pulse radiolysis; TD-DFT methods; thiobases; nucleobase derivatives

---

1. Introduction

In 1988 George H. Hitchings and Gertrude B. Elion were awarded Nobel Prize in Physiology and Medicine for their groundbreaking work which laid the foundations for development of new drugs against a variety of diseases. Interestingly, their research in the 1950-s on the therapeutic properties of sulfur-substituted nucleobases (thiobases) resulted in two new chemotherapeutic drugs, thioguanine and 6-mercaptopurine, which were approved by US Food and Drug Administration (FDA) for treatment of acute leukemia and are still used for this purpose. Their revolutionary work was based on the fact, that substitution of carbonyl oxygen atom in canonical nucleobase by sulfur atom affects cell metabolism and leads to their death [1].

The aforementioned studies were devoted mostly to purine based thiobases but thiopyrimidines also have their place on the wide spectrum of biologically active compounds. First, they naturally occur in bacterial tRNA, up to date 11 different thiopyrimidine based compounds have been identified in bacteria [2], where they play important role in cellular metabolism [3]. There is also some evidence that thiobases play important role in metabolism of higher organism, for example, lack of certain sulfur-substituted nucleobases in mitochondria has been proven to lead to development of diabetes in mice [4].

Since replacement of oxygen atom by sulfur atom in DNA/RNA bases induces a substantial red-shift in absorption spectrum of the ground state, sulfur substituted nucleic acid bases are known photosensitive probes. Regular nucleobases absorb light mostly in ultra violet C (UVC) region (100–280 nm), while thiobases, depending on the structure absorb UVB (280–315 nm) and UVA (315–400 nm) radiation [5]. For this reason, 2-thiotymine is used for specific targeting of selected sites in nucleic acids [6]. What is even more important, thiobases (in contrast to canonical nucleobases) populate long-lived, reactive triplet state in high yields [7]. These properties are believed to be utilized in photodynamic therapy of various cancers. In fact, 4-thio-2-deoxythymidine has been proven to treat cancerous cell lines in vitro upon UVA irradiation [8,9] as well as skin cancer cells in biopsies [10,11] and bladder cancer in mice [12]. This compound have been currently moved to clinical trials [13]. For these reasons, it is not surprising that the vast majority of studies have been devoted to the photophysical and photochemical properties of thiobases related with their medical applications. There are several comprehensive research articles and reviews addressing these issues [14–30].

Moreover, sulfur substituted nucleobases were suggested to be good candidates for radiation therapy since they can potentially develop—similar to their halogenated analogues—radiosensitization properties [31]. The mechanism by which the former radiosensitizers operate was attributed to their reduction by thermalized/prehydrated [32] or ballistic electrons [33]. This results in formation of reactive radicals derived from the respective nucleobases which are believed to be responsible for DNA damage [34]. Therefore, it was highly desirable to understand how the sulfur-substituted nucleobases can be altered by the secondary electrons. A series of papers was published in recent five years addressing the dissociative electron attachment to 2-thiouracil (2-TU) and 2-thiothymine in gas-phase and showing that the fragmentation of these molecules arises mainly at the sulfur site [35–40]. Some attention was also paid to electron paramagnetic resonance (EPR) studies [41–49] combined with DFT calculations [50–52] of radicals derived from γ-irradiated single crystals of sulfur-substituted nucleobases.

Surprisingly, much less attention was directed into studies of radicals derived from sulfur nucleobases exposed to reactions with one-electron reductants and oxidants in aqueous phase. Reactions of hydrated electrons e^-_{aq} with bis(1-substituted-uracil-4-yl) disulfide [53], 2-thio analogues of cytosine and uracil [54], 8-thioguanosine [55], 5-iodo-4-thio-2′-deoxyuridine [56] and 5-bromo-4-thio-2′-deoxyuridine [57], H atoms (H$^\bullet$) with 8-thioguanosine [55], formate ($CO_2^{\bullet-}$) and 2-hydroxypropan-2-yl ((CH_3)$_2{}^\bullet COH$) radicals with 2-thio analogues of cytosine and uracil [54], hydroxyl radicals ($^\bullet OH$), dichloride ($Cl_2^{\bullet-}$) and dibromide ($Br_2^{\bullet-}$) radical anions with 2-TU [58], 4-thiouracil [59] and 8-thioguanosine [55] and azide radicals ($^\bullet N_3$) with 4-thiouracil [59] and 8-thioguanosine [55] are all but a few examples.

The main conclusions derived from the above mentioned radiation chemical studies is that substitution by sulfur substantially changes pathway of radical reactions. Sulfur atom is the main target of one-electron oxidation by $^\bullet OH$, $Br_2^{\bullet-}$, $Cl_2^{\bullet-}$, leading to the formation of sulfur centered radical cation/thiyl radical, which instantaneously dimerize with excess of sulfur nucleobase, forming three-electron bonded, sulfur centered radical cation (see Scheme 1 for 2-TU) [58,59]).

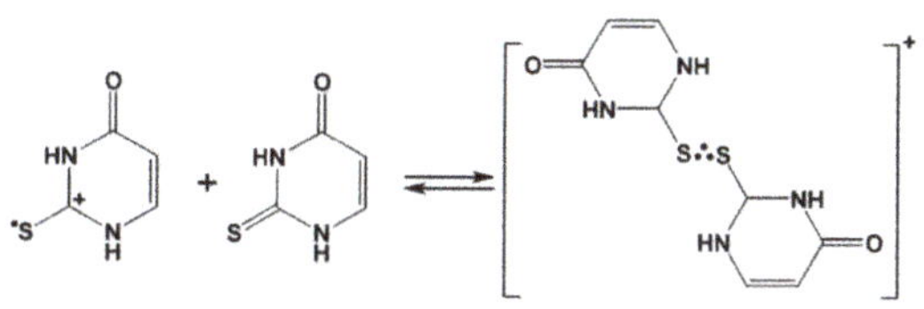

Scheme 1. Dimeric radical cation formation in 2-thiouracil [58].

On the other hand, in the presence of sulfur atom at the position C2 in 2-TU and 2-thiocytosine, one-electron reduction by e^-_{aq} leads only to radicals protonated on heteroatoms—an exocyclic O-atom in 2-TU and a ring N-atom in 2-thiocytosine [54].

Studies conducted so far have undoubtedly demonstrated a very complex mechanistic scenario of oxidation and reduction of sulfur substituted nucleobases. Moreover, the presence of a sulfur atom (instead of an oxygen atom) at position C2 in 2-TU may result in the establishment of an additional tautomeric equilibrium (in relation to uracil), namely between the thione and thiol form [60,61]. This is a key issue since the presence or lack of specific functional groups may drastically affect the reaction pathway. Theoretical [62–64] and experimental [65,66] studies revealed that the energetically preferred tautomer of the neutral form of 2-TU in solution is the oxo-thione (Scheme 2). On the other hand, it was also suggested that 2-TU exists as the equilibrium mixture of oxo-thione and oxo-thiol forms; however, the population ratio of tautomers were not evaluated [67]. The calculated relative free energies indicated that two out of several possible thiol tautomers (hydroxy-thiol/oxo-thiol) are energetically higher by about 50–60 kJ mol^{-1} compared to the oxo-thione tautomer [62–64]. In turn, anionic form of 2-TU (pK$_a$ = 7.75) [68] exists in equimolar mixture of two hydroxy-thione tautomers in the form of monoanions (Scheme 2) due to deprotonation either at N1 or N3 nitrogen atoms [65].

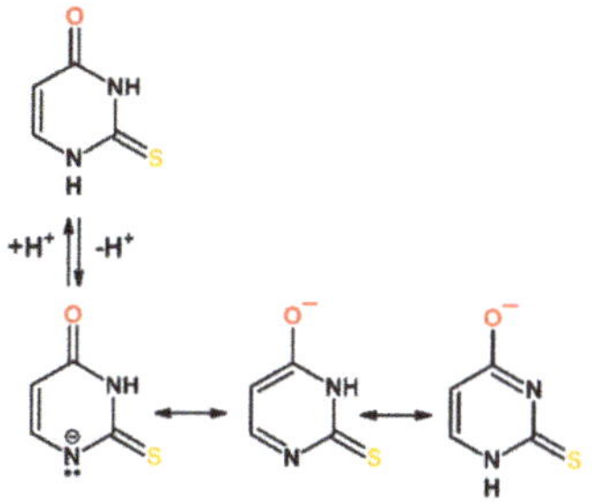

Scheme 2. Acid-base and tautomeric equilibria in 2-TU.

With this information and premises, we report in this paper our in-depth studies on the •OH and •N$_3$ radicals induced oxidation of 2-TU in aqueous solution. The current paper is dedicated to extension of previous studies on oxidation of 2-TU [58], addressing the influence of 2-TU concentration, pH and the character of one-electron oxidant on the transient absorption spectra and the kinetics of transients formed. The experimental transient absorption spectra observed in irradiated aqueous solutions containing 2-TU will be compared with the respective spectra of radicals and radical ions calculated using TD-DFT methods.

2. Results

2.1. Oxidation by •OH Radicals—Influence of 2-TU Concentration on Absorption Spectra

The transient absorption spectrum with λ_{max} = 430 nm, which resulted from the reaction of •OH radicals with 2-TU present in aqueous solution in 1 mM concentration, reported in the earlier work [58], was assigned to the formation of dimeric radical cation with 2c-3e sulfur-sulfur bond (Scheme 1). It was the only intermediate identified as the oxidation product derived from 2-TU. Therefore, we decided to record UV-vis transient absorption spectra at various 2-TU concentrations in order to check whether and to what extent concentration of 2-TU affects spectral characteristics. The spectral changes obtained from the pulse irradiation of N$_2$O-saturated unbuffered aqueous solutions containing 2-TU in the concentration range 0.1 mM to 1 mM are shown in Figure 1. The recording times are selected at the maximum intensity of the absorption signal after electron pulse for the specified concentration of 2-TU, which corresponds to the steady-state concentrations of transient species.

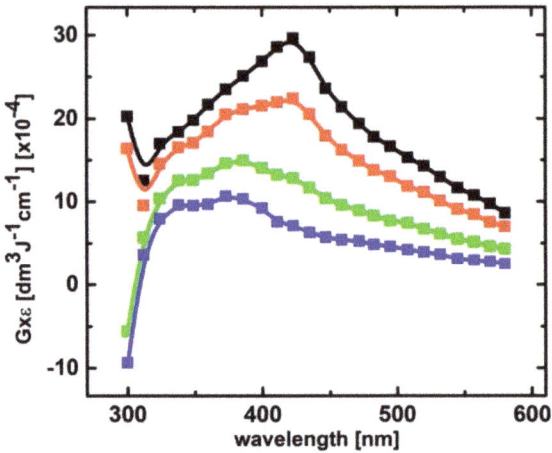

Figure 1. Transient absorption spectra recorded in N$_2$O-saturated unbuffered aqueous solution at pH 4 containing 0.1 mM (■), 0.2 mM (■), 0.5 mM (■) and 1 mM (■) of 2-TU, recorded 3 µs, 2 µs, 1 µs and 0.8 µs, respectively, after electron pulse.

It is clearly seen that the absorption spectra recorded for two low concentration of 2-TU are different in shape and position of the absorption maxima in comparison to the absorption spectra recorded for two high concentration of 2-TU. They are characterized by a broad absorption band with a weakly pronounced maximum at $\lambda \approx 385$ nm and a shoulder within 320–360 nm range. In turn, the latter two spectra are similar to the spectrum recorded in the earlier work and are characterized by $\lambda_{max} \approx 425$ nm [58]. This is a strong indication that some other products are formed and their spectral contribution at higher 2-TU concentration is probably hidden under the absorption assigned earlier to the dimeric radical cation [58]. More experimental evidence for their formation are obtained by studying time evolution of absorption spectra at low 2-TU concentration.

2.1.1. Time Evolution of Absorption Spectra—Low Concentration of 2-TU

The spectral changes observed after pulse irradiation of N$_2$O-saturated solutions containing 0.1 mM and 0.2 mM of 2-TU yielded complex series of spectral changes, however, their behavior in both cases is very similar (Figure 2A and Figure S1A in Supplementary Materials, respectively). A broad absorption spectrum with a weakly pronounced maximum at $\lambda \approx 325$ nm and a shoulder within 360–385 nm range began to form within 300 ns time domain. With the time elapsed, both absorptions underwent further growth, however, after 1 µs with the inversion of intensities. The intensity of absorption at $\lambda \approx 385$ nm became stronger and 2 µs (left insert in Figure S1A in Supplementary Materials) or 3 µs (left insert in Figure 2A) after the pulse the absorption spectra were fully developed and are characterized by a broad absorption band with $\lambda_{max} \approx 385$ nm and a pronounced shoulder within 340–360 nm range.

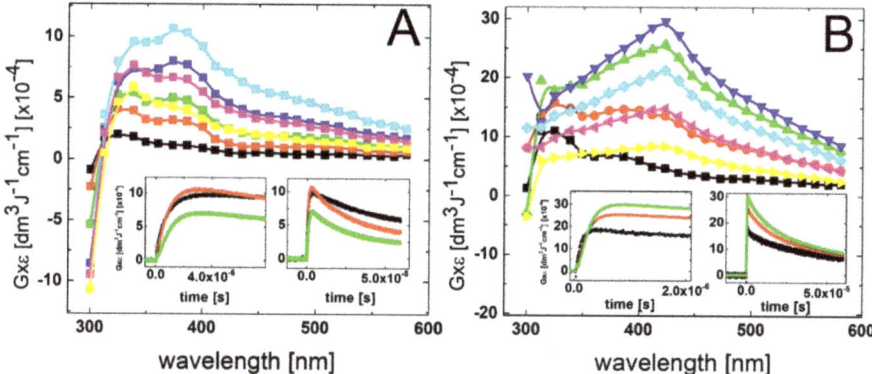

Figure 2. Transient absorption spectra recorded in N_2O-saturated unbuffered aqueous solution at pH 4 containing (**A**) 0.1 mM of 2-TU 100 ns (■), 300 ns (■), 500 ns (■), 1 µs (■), 3 µs (■), 25 µs (■) and 60 µs (■) after electron pulse, (**B**) 1 mM of 2-TU 100 ns (■), 200 ns (■), 400 ns (■), 1 µs (■), 10 µs (■), 25 µs (■) and 60 µs (■) after electron pulse. Inserts: time profiles representing growth (left) and decay (right) of transient absorptions at λ = 338 nm (■), 386 nm (■) and 426 nm (■).

With the time further elapsed, the absorption spectra observed at longer times (25 µs and 60 µs) after the pulse are characterized by a broad absorption bands, however, with the inversed position of λ_{max} (340 nm) and a shoulder (360–385 nm). Interestingly, no formation of the absorption band with a clear maximum at λ = 425 nm was observed at any time domain up to 60 µs, however, its presence this time might be probably hidden under the absorption band with λ_{max} = 385 nm. Nonetheless, the kinetic behavior followed at the two selected wavelengths (338 nm and 386 nm) indicates the presence of two transient species characterized by different lifetimes (right inserts in Figure 2A and Figure S1A in Supplementary Materials).

2.1.2. Time Evolution of Absorption Spectra—High Concentration of 2-TU

The spectral changes observed after pulse irradiation of N_2O-saturated solutions containing 0.5 mM and 1 mM of 2-TU yielded even more complex series of spectral changes in comparison to solutions with lower concentration of 2-TU. However, the time evolution of spectra observed within first 200 ns time domain is very similar. A broad absorption spectrum with weakly pronounced maxima at λ ≈ 325 nm and 385 nm was also observed (Figure S1B in Supplementary Materials and Figure 2B, respectively) as for low concentrations of 2-TU. However, with the time elapsed, the absorption spectra underwent further growth and also a substantial spectral shift. The absorption spectra were fully developed within 1 µs and are characterized by a broad absorption band with λ_{max} ≈ 425 nm (Figure S1B in Supplementary Materials and Figure 2B). Moreover, a pronounced shoulder within 360–385 nm is also seen for lower concentration of 2-TU (0.5 mM), (Figure S1B in Supplementary Materials).

With the time further elapsed, the absorption band with λ_{max} ≈ 425 nm started to decay and 10 µs after the pulse still dominated the spectra. However, the spectra observed at longer times (25 µs and 60 µs) after the pulse are characterized by a very broad and flat absorption without a clearly pronounced maximum (Figure S1B in Supplementary Materials and Figure 2B). The kinetic behavior followed at the three selected wavelengths (338 nm, 386 nm and 426 nm) (right inserts in Figure S1B in Supplementary Materials and Figure 2B) also indicates the presence of three different transient species.

2.1.3. Influence of pH on Time Evolution of Absorption Spectra at Low and High Concentration of 2-TU

The subsequent chemical systems subjected to irradiation were the basic aqueous solutions containing low (0.1 mM) and high (2 mM) concentration of 2-TU at pH = 10. At this pH, 2-TU exists in equimolar mixture of two hydroxy-thione tautomers in the form of monoanions (*vide* Scheme 2). In basic solutions, the spectral changes observed for the lowest concentration of 2-TU are different from those observed at pH = 4 as follows—(i) at short time domain up to 500 ns, a broad absorption spectrum with a weakly pronounced maximum at $\lambda \approx 350$ nm (not at $\lambda \approx 325$ nm) and a shoulder within 380–430 nm range (not within 360–385 nm range) began to form (Figure 3A). The first change in spectral features can be easily rationalized by higher absorption of 2-TU in the ground state < $\lambda \approx$ 350 nm at pH = 10 in comparison to pH = 4 (Figure S2 in Supplementary Materials) causing a stronger decrease in absorption due to the consumption of 2-TU. In turn, an appearance of a shoulder finds its justification in spectral features observed at 6 µs showing an absorption band with $\lambda_{max} \approx 430$ nm. This absorption band was not observed at pH = 4 at low concentration of 2-TU (Figure 2A and Figure S1 in Supplementary Materials).

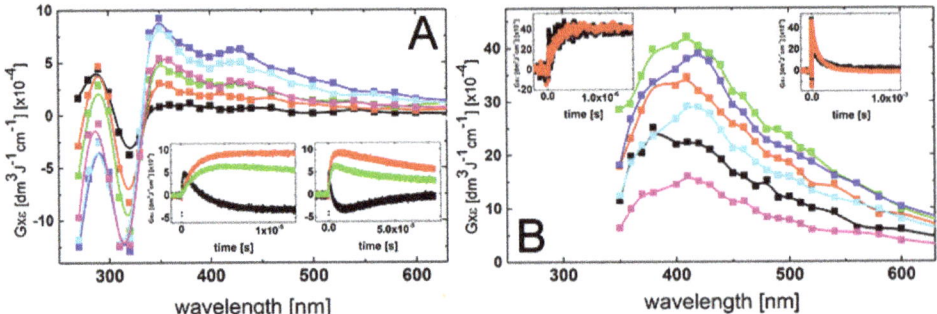

Figure 3. Transient absorption spectra recorded in N$_2$O-saturated unbuffered aqueous solution at pH = 10 containing (**A**) 0.1 mM of 2-TU 200 ns (■), 500 ns (■), 1 µs (■), 6 µs (■), 20 µs (■), and 80 µs (■) after electron pulse. Inserts: time profiles representing growth (left) and decay (right) of transient absorptions at λ = 290 nm (■), 350 nm (■) and 420 nm (■); (**B**) 2 mM of 2-TU 100 ns (■), 240 ns (■), 1 µs (■), 6 µs (■), 25 µs (■), and 80 µs (■) after electron pulse. Inserts: time profiles representing growth (left) and decay (right) of transient absorptions at λ = 380 nm (■) and 420 nm (■).

Meanwhile, the spectral changes observed for the highest concentration of 2-TU are very similar to those observed at pH = 4. At short time (100 ns), a broad absorption spectrum with a weakly pronounced maximum at $\lambda \approx 380$ nm and a shoulder within 380–430 nm range was observed which with the further time elapsed is transformed into absorption spectrum with a clearly pronounced maximum at $\lambda \approx 425$ nm. Moreover, due to the strong absorption of 2-TU in the ground state, measurements for λ < 350 nm were not possible (Figure S2 in Supporting Materials).

2.2. Oxidation by •OH Radicals—Kinetics

2.2.1. The Rate Constant of the •OH Reaction with 2-TU

In order to directly determine the bimolecular rate constant of the •OH reaction with 2-TU, the build-up kinetics at various concentrations of 2-TU were recorded at two wavelengths—324 nm and 386 nm. The rate of formation, followed at these wavelengths fits to a single exponential (inserts in Figure 4). These wavelengths correspond to absorption maxima of two bands which were formed initially and were the only ones fully developed at low concentration of 2-TU (Figure 2A and Figure S1A in Supplementary Materials). These results support the tentative hypothesis that these absorption bands cannot be assigned to dimeric radical ions which are formed in a secondary process and are characterized by the absorption band with $\lambda_{max} \approx 425$ nm [58]. With this information in hand, the pseudo-first-order rate constants of the formation of the 324-nm and 386-nm absorption were plotted as a function of 2-TU concentration (Figure 4). It is clearly seen that the pseudo-first-order rate constants measured at λ = 324 nm show a linear dependence on the concentration of 2-TU in the full range of concentration studied (Figure 4A) with the slope representing the second-order rate constant for the formation of transient(s) resulting from the reaction of •OH radicals with 2-TU. The calculated second-order rate constant $k_{324} = 1.3 \times 10^{10}$ M^{-1}s^{-1} is similar to the rate constants determined previously for 2-TU (9.6 × 10^9 M^{-1}s^{-1}) [58] and thiourea (1.2 × 10^{10} M^{-1}s^{-1}) [69] by competition kinetics using 2-propanol.

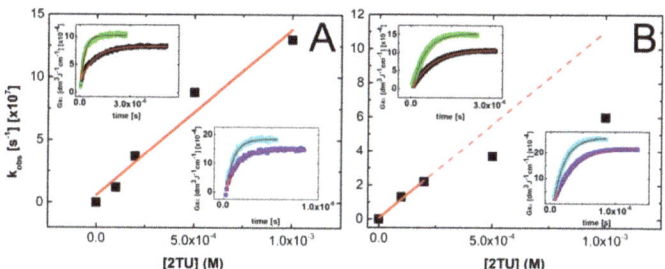

Figure 4. Plots of the observed pseudo-first-order rate constants of the formation of the 324-nm absorption (**A**) and the 386-nm absorption (**B**) as a function of 2-TU in N$_2$O-saturated unbuffered aqueous solution at pH = 4. Inserts in A: time profiles representing growth of transient absorption at λ = 324 nm for the concentration of 2-TU (■) 0.1 mM and (■) 0.2 mM (top) and (■) 0.5 mM and (■) 1 mM (bottom). Inserts in B: time profiles representing growth of transient absorption at λ = 386 nm for the concentration of 2-TU (■) 0.1 mM and (■) 0.2 mM (top) and (■) 0.5 mM and (■) 1 mM (bottom).

On the other hand, the pseudo-first-order rate constants measured at λ = 386 nm show a departure from linearity for high concentration of 2-TU (Figure 4B). This is not surprising taking into account the following facts—high efficiency of dimeric radical ions formation at high concentration of 2-TU and significant contribution of their absorption band at this wavelength. In order to suppress this inconvenience, the pseudo-first-order rate constants measured for two lowest concentration of 2-TU were taken into account in the linear fit. The calculated second-order rate constant $k_{386} = 1.1 \times 10^{10}$ M^{-1}s^{-1} fits very well to the expected rate constant value [58].

2.2.2. Equilibrium Constant and Rate Constants of Reactions Involved in Equilibrium

For both pHs, the maximum value of the 426-nm absorbance is dependent on the 2-TU concentration. When this is increased from 0.047 mM to 3 mM, $G \times \varepsilon$ increases from 6.2 × 10^{-4} dm^3 J^{-1} cm^{-1} to 25.6 × 10^{-4} dm^3 J^{-1} cm^{-1} and from 5.6 × 10^{-4} dm^3 J^{-1} cm^{-1} to 20.2 × 10^{-4} dm^3 J^{-1} cm^{-1} (values corrected for the loss by bimolecular decay) for pH = 4 and 10, respectively (Figure S3, panels C and D). This increase cannot be accounted for by the increase in G(•OH) due to the higher 2-TU concentration [70]. It rather points to the existence of an equilibrium situation presented in Scheme 1 where a dimeric radical ion is formed which is responsible for the strong absorption at $\lambda_{max} \approx 425$ nm.

The equilibrium constant $K = k_{forward}/k_{backward}$ can be obtained from Equation (1), where A_0 is the absorbance at λ = 426 nm in 2-TU solution of 3 mM and A is the absorbance at a given concentration of 2-TU.

$$A_0/A - 1 = K^{-1} [2\text{-TU}]^{-1} \qquad (1)$$

In Figure 5 (panels A and B), the term $A_0/A - 1$ is plotted against the reciprocal of the 2-TU concentration for pH = 4 and pH = 10, respectively. From the reciprocal values of the slopes of these linear plots, K values 3.8(5) × 10^3 M^{-1} and 4.6(5) × 10^3 M^{-1} were obtained for pH = 4 and pH = 10, respectively. These values are very close to that reported for 2-TU at pH = 6.5 (4.7 × 10^3 M^{-1}) [58], however, substantially lower to that reported for 4-TU at pH = 7 (1.8×10^4 M^{-1}) [59] by using the same experimental approach.

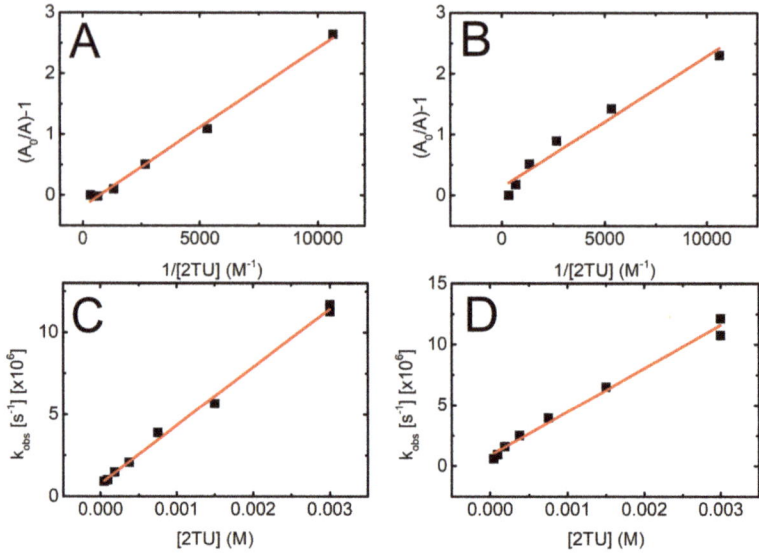

Figure 5. Dependence of the term (A_0/A) − 1 at λ = 426 nm on the reciprocal of 2-TU in the pulse radiolysis of N$_2$O-saturated unbuffered aqueous solution at pH = 4 (panel **A**) and pH 10 (panel **B**) containing 2-TU in the range of concentration 0.047 mm – 3 mM. Plots of the observed pseudo-first order rate constants of the formation of the 426-nm absorption as a function of 2-TU in N$_2$O-saturated unbuffered aqueous solution at pH = 4 (panel **C**) and pH = 10 (panel **D**).

A kinetic treatment of the equilibration process shown in Scheme 1 can also be represented by Equation (2), where k_{obs} is the experimental pseudo first-order rate constant for the formation of dimeric radical ion (Figure S3A,B in Supplementary Materials).

$$k_{obs} = k_{forward} [\text{2-TU}] + k_{backward} \qquad (2)$$

This approach allows measurement of not only equilibrium constant (K) but also rate constants involved in the equilibrium that is, ($k_{forward}$ and $k_{backward}$). Figure 5 (panels C and D) shows plots based on Equation (2) for pH = 4 and pH = 10, respectively. From the slopes and intercepts of the linear plots, we obtained $k_{forward}$ = 3.6×10^9 M^{-1}s^{-1} and $k_{backward}$ = 7.8×10^5 s^{-1} for pH 4 and $k_{forward}$ = 3.6×10^9 M^{-1}s^{-1} and $k_{backward}$ = 9.0×10^5 s^{-1} for pH = 10 and hence $K = k_{forward}/k_{backward}$ = 4.6×10^3 M^{-1} and 4.0×10^3 M^{-1} for pH = 4 and 10, respectively. Considering the errors in the measurement of absorbencies and rate constants based on relatively weak signals, the agreement in the respective K values obtained by the two approaches is very good. From these two independent measurements the average values for K are 4.2×10^3 M^{-1} and 4.3×10^3 M^{-1} for pH = 4 and pH = 10, respectively.

The obtained values of $k_{forward}$ for 2-TU are comparable to that reported (5.0×10^9 M^{-1}s^{-1}) in analogous equilibrium for thiourea. On the other hand, the obtained values of $k_{backward}$ for 2-TU are significantly higher to that reported (9.1×10^3 s^{-1}) for thiourea [69]. These facts explain lower K values for 2-TU and indicates lower stability of its dimeric radical ion in comparison with thiourea.

2.3. Oxidation by •OH Radicals—Time-Resolved Conductivity

It has been reported in a number of pulse radiolytic studies that time-resolved conductivity detection can help untangle mechanistic nuances encountered in analysis of kinetic changes of optical absorption especially when either multiple transients are formed in the similar spectral range [71,72] or the transients have optical absorption outside the available detection range [73]. Upon one-electron •OH radical induced oxidation, hydroxide anion (OH^-) is produced and instantaneously changes pH of irradiated solutions which manifests itself in change of apparent conductivity. In acidic conditions it results in the decrease of apparent conductivity as a consequence of neutralization reaction with protons ($H^+ + OH^- \rightarrow H_2O$). In turn, in basic solutions it causes increase of apparent conductivity due to increase of concentration of conducting OH^- anions. From the amplitude of conductivity change the yield of •OH radicals involved in one electron oxidation process can be easily deducted.

In both acidic and basic N_2O-saturated solutions of thiourea, a simpler compound with the same S=C < moiety like in 2-TU, transient conductivity changes after electron pulse were assigned to 100% yield of one-electron oxidation of thiourea to thiourea dimeric sulfur radical cation by •OH radicals [69]. Since analogous dimeric radical cation formation was invoked in the previous pulse radiolytic studies of 2-TU [58], we wanted to estimate what extent of one-electron oxidation process, that is, the part of •OH radicals yield, is involved in formation of radical cation $1^{•+}$, a potential precursor of dimeric radical cation $2^{•+}$ (Scheme 3).

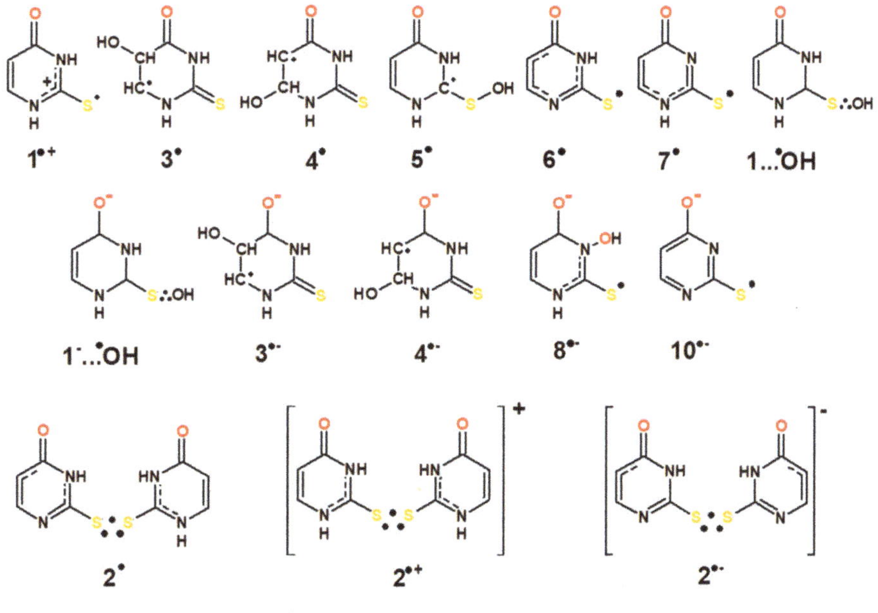

Scheme 3. Structures of potential transients produced in the reaction of •OH radicals with 2-TU at various pHs. The same symbols of transients were used as in the previous work [58] and expanded for the species which were not considered earlier.

Results of our studies are shown on Figure 6. After electron pulse in N_2O-saturated 1 mM 2-TU acidic solutions (pH = 4.1) we observed an instantaneous growth of conductivity followed by its fast decrease almost to the level detected prior to the pulse (Figure 6, red line). The initial transient conductivity spike is a result of net increase of conductivity due to production of conducting species of water radiolysis (hydrated electrons (e_{aq}^-) and protons (H^+). In N_2O-saturated aqueous solutions e_{aq}^- are quickly converted into •OH radicals with the side product of hydroxide anions (OH^-) being released as well within just few nanoseconds after the electron pulse. Therefore, the fast decrease of conductivity recorded within 1 µs after the pulse occurs through stoichiometric neutralization reaction ($H^+ + OH^- \rightarrow H_2O$) with k = 1.4 × 10^{11} $M^{-1}s^{-1}$ [74], of highly conducting protons by OH^-. Upon completion of this reaction stable conductivity level is reached and remains unchanged at the level of ~30 ± 10 S cm^2/100eV for the next 80 microseconds (Figure 6 shows just first 40 µs).

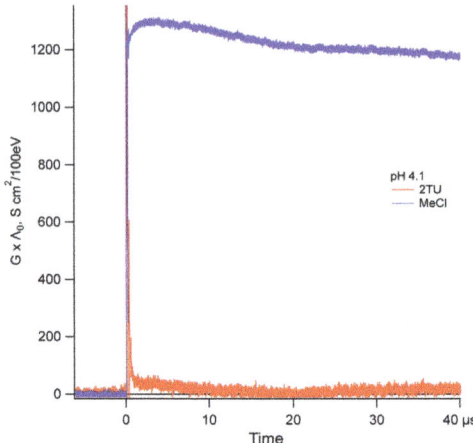

Figure 6. Comparison of equivalent transient conductivity changes represented as G × Λ_0 vs. time after the pulse of electrons in N_2O-saturated solution containing 1 mM of 2-TU at pH = 4.1 (red line) to CH_3Cl-saturated aqueous solutions at the same pH (blue line).

2.4. Oxidation by •N_3 Radicals and $CH_3CN^{•+}$ Radical Cations

The subsequent chemical systems subjected to irradiation were the N_2O-saturated aqueous solutions containing NaN_3 and two various concentrations of 2-TU (0.1 mM and 2 mM) at pH = 7. Taking into account the fact that oxidation potential of 2-TU at pH = 7 measured versus Ag/AgCl electrode falls in the vicinity of +0.5 V which is equivalent to ≈ +0.7 V versus SHE (the standard hydrogen electrode) [75] and the reduction potential of azide radicals (•N_3) (+1.33 V vs. SHE) [76], one would reasonably expect that in such designed systems, oxidation of 2-TU can be initiated by •N_3 radicals. This was, indeed, observed. A transient absorption spectrum recorded 4 µs after pulse irradiation of N_2O-saturated solutions containing 0.1 mM of 2-TU and 50 mM NaN_3 exhibits a narrow and distinct absorption band with λ_{max} = 320 nm (Figure 7A) which is developed with k = 1.5 × 10^6 s^{-1} (top right insert, Figure 7A). One has to note that similar absorption band was observed in O_2-saturated acetonitrile solution containing 0.1 mM of 2-TU, where strongly oxidizing radical cations ($CH_3CN^{•+}$) are formed [77] and can also initiate oxidation of 2-TU (left top insert, Figure 7A). With the time further elapsed, the absorption spectra observed 12 µs and 60 µs after the pulse are still characterized by a similar 320-nm absorption band (Figure 7A). However, the decay kinetics of the 320-nm absorption band reached a plateau within 150 µs with k_{320} = 4.7 × 10^4 s^{-1} (right bottom insert, Figure 7A). Interestingly, a kinetic trace recorded at λ = 420 nm represents a growth with k_{420} = 4.7 × 10^4 s^{-1} (right bottom insert, Figure 7A). Comparison of the pseudo-first order rate constants measured at these two wavelengths clearly indicates that formation of the 420-nm absorption occurs at the expense of the decay of the 320-nm absorption.

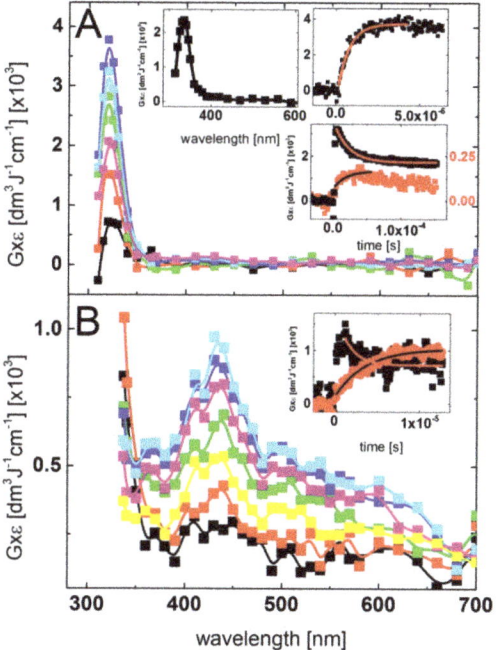

Figure 7. Transient absorption spectra recorded in N$_2$O-saturated unbuffered aqueous solution at pH = 7 containing 50 mM NaN$_3$ and (**A**) 0.1 mM of 2-TU 240 ns (■), 480 ns (■), 1 µs (■), 4 µs (■), 12 µs (■) and 60 µs (■) after electron pulse. <u>Inserts</u>: top left: transient absorption spectrum recorded 1 µs after the pulse in O$_2$-saturated solution of acetonitrile containing 0.1 mM of 2-TU; top right: time profile representing growth of transient absorption at λ = 320 nm; bottom: time profiles representing growth of transient absorption at λ = 420 nm (■) and decay of transient absorption at λ = 320 nm (■); (**B**) 2 mM of 2-TU 600 ns (■), 1.2 µs (■), 3.2 µs (■), 8 µs (■), 12 µs (■), 24 µs (■) and 48 µs (■) after electron pulse. <u>Inset</u>: time profiles representing growth of transient absorption at λ = 420 nm (■) and decay of transient absorption at λ = 337nm (■).

The spectral changes observed after pulse irradiation of N$_2$O-saturated solutions containing 2 mM of 2-TU are more visible in comparison to solutions with lower concentration of 2-TU. A transient absorption spectrum recorded 1.2 µs after the pulse is characterized by a strong nondescript absorption band with no distinct λ$_{max}$ at wavelengths in the range 337-400 nm and a weaker, however pronounced, absorption band with λ$_{max}$ ≈ 430 nm (Figure 7B). Due to the strong absorption of 2-TU in the ground state, measurements for λ < 337 nm were not possible (*vide* Figure S2 in Supporting Materials). The intensity of this uncharacteristic absorption band decreased in intensity (as shown by a decay of absorption measured at λ = 337 nm) and reached a plateau within 10 µs with $k_{337} = 1.5 \times 10^5$ s^{-1}. At the same time absorption band with λ$_{max}$ ≈ 430 nm was fully developed with $k_{420} = 1.4 \times 10^5$ s^{-1} (insert in Figure 7B). Comparison of the pseudo-first order rate constants measured at these two wavelengths clearly indicates that formation of the 420-nm absorption occurs at the expense of the decay of the 337-nm absorption. Excellent agreement in k_{337} and k_{420} values is a strong support that the species absorbing in the region < 340 nm is a direct precursor of the species absorbing at λ = 420 nm.

2.5. Theoretical Calculations

We expanded previous computational work on transient products formed in •OH reaction with neutral form of 2-TU in aqueous solutions [58] to intermediates which can be produced with singly deprotonated 2-TU. The former should be observable in acidic solutions at pH = 4, the latter in basic solutions at pH = 10 (*vide* Scheme 2). The main intention of current work was to confront TD-DFT computations of potential transients (*vide* Scheme 3) with the experimental transient absorption spectra recorded at various concentrations of neutral and deprotonated 2-TU (*vide* Figure 2, Figure 3 and Figure S1 in Supplementary Materials). Upon encounter of •OH with neutral or deprotonated 2-TU one-electron oxidation, •OH addition (at positions C5, C6, S8) and H abstraction (at positions N1 and N3) can occur simultaneously. In order to help narrow down to the most efficient channels of •OH-induced reactivity we performed detailed analysis of activation barriers of •OH addition and H abstraction reactions.

Earlier work suggested [58] that transients produced with the highest yields have 2c-3e character. Therefore, we applied range separated hybrid functional ωB97x/aug-cc-pVTZ based computational methodology (*vide* Section 4.4) which proven quite successful in structural, spectral and thermochemical characterization of this group of open shell species [78,79]. Detailed solution phase optimized geometries of all transients with bond lengths as well as maximum spin populations are presented in Figures S4–S6 (Supplementary Materials). Our solution phase optimized geometries of monomer type transients ($1^{•+}$, $1 \ldots {}^{•}OH_{(1)}$, $1 \ldots {}^{•}OH_{(2)}$, $3^{•}$, $4^{•}$, $6^{•}$, $7^{•}$) compare very well with the solution phase optimized geometries obtained previously at BH and HLYP/6-311+G(d,p) level [58].

Interestingly, the intermediates denoted as $1 \ldots {}^{•}OH_{(1)}$, $1 \ldots {}^{•}OH_{(2)}$ were found to be hemibonded •OH adducts to sulfur atom. Typical shape expected for σ* SOMO orbitals in these type of intermediates is presented on Figure S7 (Supplementary Materials).

The computed harmonic stretch frequencies of SO bonds in 2c-3e OH adducts are 329 cm^{-1} and 338 cm^{-1}, respectively. These computed values match almost exactly the experimental stretching frequency of 338 cm^{-1} recorded for 2c-3e SO bond in DMS-OH radical by the time-resolved Raman spectroscopy [80] assuring further our assignment. It has to be noted that $1 \ldots {}^{•}OH_{(1)}$, $1 \ldots {}^{•}OH_{(2)}$ were suggested as possible precursors of radicals $6^{•}$ and $7^{•}$ in the process of direct H-atom abstraction from positions N1 and N3, respectively [58]. However, they were not considered as hemibonded •OH adducts to sulfur atom (*vide* Scheme 3) but as hydrogen-bonded adducts to H7 and H9 at N1 and N3 atoms in the ring, respectively. We performed detailed studies of energy profiles for H atom abstraction from positions N1 and N3 in aqueous phase. Results of these studies are summarized on Figures S8, S9A and Table S2 in Supplementary Materials. Contrary to the previous work we found that both channels of H atom abstraction at positions N1 and N3 with transition states (TS6• and TS7•), geometrically close to $1 \ldots {}^{•}OH_{(1)}$ and $1 \ldots {}^{•}OH_{(2)}$, are essentially barrierless in aqueous phase (Figure S8 in Supplementary Materials). There is a small 0.25 kcal mol^{-1} barrier in TS7• formation (Figure S8B in Supplementary Materials) which should contribute to some regioselectivity of H atom abstraction at position N1. The products resulting from H atom abstraction at positions N1 and N3 are hemibonded radicals with 2c-3e bonds between sulfur and oxygen of water molecules produced out of original OH moieties of respective hemibonded OH adducts upon H addition to oxygen (denoted as 6•OH$_2$ and 7•OH$_2$ (*vide* Figure S7 in Supplementary Materials), illustrating shapes of their σ* orbitals). These S-O hemibonded intermediates are energetically about 6–7 kcal mol^{-1} above free energy of hydrated thiyl radicals. Hence, one can only assume that at room temperature these intermediates will be in equilibrium with radicals $6^{•}$ and $7^{•}$ (Figure S8 in Supplementary Materials). We believe that these two channels represent paths of proton coupled electron transfer (PCET) to oxygens of 2c-3e adducts $1 \ldots {}^{•}OH_{(1)}$, $1 \ldots {}^{•}OH_{(2)}$ in production of thiyl radicals $6^{•}$ and $7^{•}$. We also found the second H abstraction pathway at position N3 with 14.6 kcal mol^{-1} barrier at transition state TS7$_2$• and pre-reactive complex RC7• in which •OH radical is hydrogen bonded to oxygen atom of 2-TU. This channel even though kinetically not feasible also leads to thiyl radical $7^{•}$ (vide Figure S9A in Supplementary Materials). The group of hemibonded •OH adducts to sulfur were characterized experimentally by optical absorption band with λ_{max} in the range of 330 nm–390 nm [81–83]. Hence, we can expect that $1 \ldots {}^{•}OH_{(1)}$ and $1 \ldots {}^{•}OH_{(2)}$ should have similar spectral features. However, in our experiment they may be too short lived to be observed due to the mentioned above conversion to SO hemibonded water radicals (6•OH$_2$ and 7•OH$_2$) in PCET process. These kind of water hemibonded intermediates have been characterized in the past to have characteristic optical absorption band due to charge transfer from solvent transition [84,85]. Even though their lifetime can be short, they can be populated in reaction with neighboring water molecules and be constantly reformed due to large abundance of neighboring water molecules. Therefore, we decided to compute their absorption spectra as well (Figure 8).

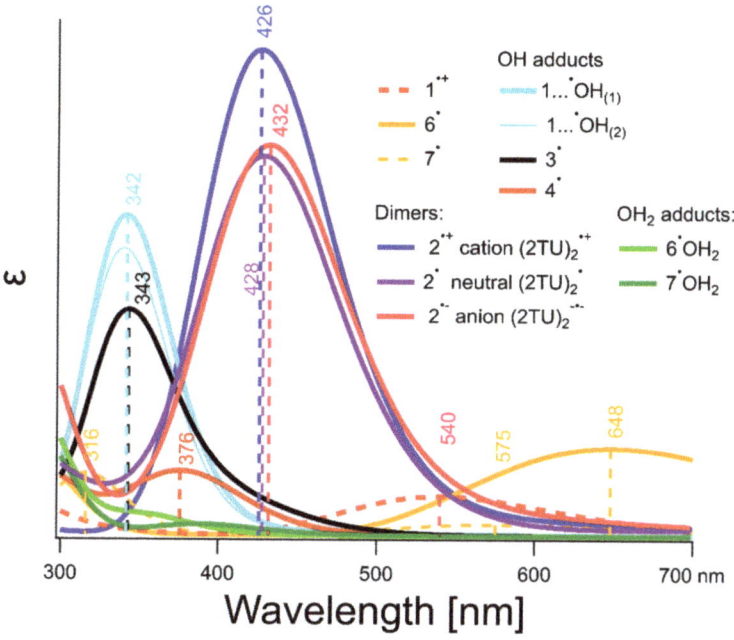

Figure 8. TD-DFT calculated absorption spectra of potential transients (see legend for symbols and Figures S4–S6 for geometries) produced in •OH–induced oxidation of 2-TU in water at pH = 4.

Contrary to the previous studies [58], we found that •OH radical addition to the most nucleophilic C5 and C6 sites of 2-TU ring and formation of the respective radicals 3• and 4• have energetic barriers (*vide* Figure S9B and Table S2 in Supplementary Materials). Even though the most stable conformer produced in the course of these reactions, radical 4• (radical with H atom pointing inward ring, *vide* Figure S4 in Supplementary Materials), is thermodynamically more stable than the most stable conformer of 3• (equatorial conformer with H atom of OH group pointing outward ring, *vide* Figure S4 in Supplementary Materials) by almost 1.9 kcal mol^{-1}. In spite of the fact that it is even more stable than the most stable •OH adduct to sulfur by almost 13 kcal mol^{-1}, chances of its formation are very small. Formation of the most thermodynamically stable conformers of •OH adducts to 2-TU ring proceed through 3.5 and 5.6 kcal mol^{-1} barriers at C5 and C6, respectively (*vide* Figure S9B and Table S2 in Supplementary Materials). As a result, analogously to reactions of •OH addition to uracil ring [86,87] one can expect regioselectivity in addition to position C5 over C6 but still kinetically formation of 2c-3e OH adducts to sulfur should dominate based on computed energy profiles for •OH addition reactions to neutral 2-TU.

Our calculated UV-vis spectra monomer type transients derived from 2-TU (2TU•+, 2TU• and •OH adducts, hemibonded water adducts and thiyl radicals) are shown in Figure 8. Figure 8 shows only computed spectra of intermediates from 300 nm up since it was not possible to reliably measure anything < 300 nm. Aqueous solutions of 2-TU attenuate analyzing light in that region due to its light absorption (Figure S2 in Supplementary Materials). This causes mixing of bleaching and absorbance signals which cannot be resolved without knowledge of individual contributions of all possible intermediates.

Among all monomer type transients derived from 2-TU expected to be produced at pH = 4, the strongest absorbing species are 1 ... •OH$_{(1)}$, 1 ... •OH$_{(2)}$ with their respective σ → σ* electronic transitions located near 342 nm–343 nm. The •OH adducts to the most nucleophilic C5 and C6 sites of 2-TU ring (radicals 3• and 4•, respectively) absorb light in the similar spectral range according to our computations. Their absorption bands are characterized by absorption maxima located at λ = 343 nm and 376 nm, respectively (Figure 8).

In the process of geometry optimization of dimeric radical cation 2•+ we noticed that this intermediate can exist in three different rotamers produced by 180° torsional rotation around C–S bonds. All isomers should be observable since the energy difference between them is less than 0.5 kcal mol^{-1}. The optimized geometry of the isomer of 2•+ with the lowest energy has the same structure as presented before [58] and is shown in Figure S6 (Supplementary Materials). All of the three isomers have almost identical absorption spectrum with λ$_{max}$ located at 426 nm (vide 2•+ in Figure 8) and slightly varying oscillator strength owing to the fact that σ and σ* orbitals are orthogonal and apparently are not much affected by the slightly varying geometries of both pendant molecular rings. Considering that either radical cation 1•+ can deprotonate in position N1 giving more thermodynamically stable radical 6• (rather than radical 7• formed upon deprotonation in position N3) or H-atom abstraction by •OH radicals can also preferably happen at position N$_1$ (vide supra and Figure S8 in Supplementary Materials), we can anticipate formation of neutral dimeric radical 2• resulting from dimerization of radical 6• with neutral 2-TU. Similar scenario can be expected upon dimerization of radical 7• with neutral 2-TU. We considered both possibilities computationally and found that dimer resulting from dimerization of radical 6• with neutral 2-TU is thermodynamically more stable. Stability of geometry of radical 2• depends on internal hydrogen bonding between N atoms and H atoms of the nearest protonated nitrogen. Rotation of the fully protonated 2-TU around C–S bond by 180° leads to its isomeric form. Two most stable isomers differ less than 0.7 kcal mol^{-1} in their respective energies. Optimized geometry of the most stable isomer of radical 2• is presented on Figure S6 (Supplementary Material). Calculated UV–Vis spectrum of this form of radical 2• is shown in Figure 8. Optical spectral parameters of radical 2• differ slightly as far as position of λ$_{max}$ is concerned and shows only 20% smaller oscillator strength. Isosurfaces of computed spin density of all 2c-3e intermediates considered in our studies are shown in Figure S7 (Supplementary Materials).

At pHs higher than the first pK$_a$ of 2-TU pronounced formation of anionic form of 2c-3e SS radical anion 2•− can be readily expected in reaction of radical 6• with deprotonated 2-thiouracil (2-TU$^-$) at N1 position. Calculated UV–Vis spectrum of 2•− is shown on Figure 8 and Figure S11 (Supplementary Materials). Absorption maximum of the anion radical 2•− is slightly red-shifted in comparison to absorption maxima of radicals 2• and 2•+ and has smaller oscillator strength than dimeric radical cation 2•+ but a bit larger than neutral dimeric radical 2•.

Additional forms of deprotonated monomer-type transients (1$_a$$^-$ ... •OH$_{(1)}$, 1$_a$$^-$ ... •OH$_{(2)}$, 1$_b$$^-$ ... •OH$_{(1)}$, 1$_b$$^-$ ... •OH$_{(2)}$, 3$_a$•−, 4$_a$•−, 3$_b$•−, 4$_b$•−, 8•−, 10•−) can be observed at pH = 10 due to •OH reaction with singly deprotonated 2-TU (i.e., 2-TU$_a$$^-$ or 2-TU$_b$$^-$, where subscripts 'a' or 'b' indicates the site of deprotonation at N1 or N3, respectively). Their optimized geometries are presented on Figure S5 (Supplementary Materials).

The energy profiles of deprotonated monomer-type transients formation are quite similar to energy profiles of their protonated counterparts generated at pH ~ 4. •OH addition to sulfur has no activation barrier and should dominate kinetically regioselectivity of all •OH addition reactions (vide Figure S10 in Supplementary Materials). Hemibonded negatively charged OH adducts are geometrically very close to transition states leading to almost barrierless H atom abstraction from positions N1 and N3. Unlike acidic conditions H abstraction at N3 has now small energy barrier of 0.17 kcal mol^{-1}. Either of two paths of a PCET in deprotonated 2-TU will lead first to a hemibonded SO water intermediate 10$_a$•−OH$_2$ or 10$_b$•−OH$_2$ (where 'a' or 'b' specifies deprotonation site in 2-TU molecule at N1 or N3, respectively) which then rearranges to thiyl radical anion $10^{\bullet-}$. In turn this radical can react with hydrating water molecule giving back water hemibonded $10_a{}^{\bullet-}$–OH_2 or $10_b{}^{\bullet-}$–OH_2. An alternate route for N3 H atom abstraction, also resulting in thiyl radical anion $10^{\bullet-}$, has a barrier of 13.6 kcal mol^{-1} and proceeds from pre-reaction complex $RC10_{a2}{}^{\bullet-}$ which consists of $^\bullet OH$ radical H-bonded to oxygen atom of 2-TU$_a{}^-$ (Figure S10A in Supplementary Materials). It is worth noting that, unlike in neutral form, in singly deprotonated thiobase a PCET process can proceed only on one side of a sulfur atom, where a nitrogen atom is protonated. This means that the opposite side of the sulfur is free to form a 2c3e OH adduct, which should have a longer lifetime due to hydrogen bonding between OH and deprotonated nitrogen site (for structures see Figure S5 and for spin density isosurfaces Figure S7). Since $1_a{}^- \ldots {}^\bullet OH_{(2)}$ and $1_b{}^- \ldots {}^\bullet OH_{(1)}$ are involved in a PCET, it means that in solutions at lowest thiobase concentration $1_b{}^- \ldots {}^\bullet OH_{(2)}$ and $1_a{}^- \ldots {}^\bullet OH_{(1)}$ may be longer observable than hemibonded OH adducts produced in analogous acidic conditions. The presence of tautomeric forms of 2-TU becomes relevant at pHs couple units below first pK_a of thiobase and therefore one can expect spectral contribution of intermediates $1_b{}^- \ldots {}^\bullet OH_{(2)}$ and $1_a{}^- \ldots {}^\bullet OH_{(1)}$ already at nearly neutral pH in sub-millimolar thiobase solutions, where hemibonded SS dimers formation is inhibited.

Among all 2-TU anionic monomer-type intermediates the strongest absorbing species are again hemibonded OH adducts to sulfur atom ($1_a{}^- \ldots {}^\bullet OH_{(1)}$, $1_a{}^- \ldots {}^\bullet OH_{(2)}$, $1_b{}^- \ldots {}^\bullet OH_{(1)}$, $1_b{}^- \ldots {}^\bullet OH_{(2)}$). Their absorption spectra are characterized by comparable intensities to their protonated counterparts observed at pH 4 (Figure 8), however, with λ_{max} locations shifted to longer wavelengths and varying in the range between 353 nm and 367 nm (Figure S11 in Supplementary Materials). Upon deprotonation of 2-TU also absorption spectra of anionic $^\bullet OH$ adducts to the most nucleophilic C5 and C6 sites of 2-TU ring (labeled as radicals $3_a{}^{\bullet-}/3_b{}^{\bullet-}$ and $4_a{}^{\bullet-}/4_b{}^{\bullet-}$, respectively) are different in comparison to their protonated forms. Their absorption maxima are varying in the range of λ_{max} between 314 nm and 477 nm, respectively (Figure S11 in Supplementary Materials).

Additionally, we considered formation of $^\bullet OH$ adduct at N3 position, since in basic solutions 2-TU can form double C=N bond. We labeled this $^\bullet OH$ adduct $8^{\bullet-}$. However since its formation is endothermic by 24 kcal mol^{-1} and its absorption spectrum is located <300 nm, this intermediate is not relevant to our discussion (for structure, thermochemistry and geometry of all discussed intermediates see Supplementary Materials Figures S5/S6 and Tables S1 and S3, respectively).

3. Discussion

3.1. Assignment of the Absorption Spectra to Respective Intermediates

The first key issue which has to be clarified before an attempt to propose the mechanism of one-electron oxidation of 2-TU by $^\bullet OH$ and $^\bullet N_3$ radicals is an assignment of the experimental transient spectra observed at low and high concentration of 2-TU to appropriate intermediates based on theoretical calculations and possible specific reactions of $^\bullet OH$ and $^\bullet N_3$ radicals. Earlier studies on reactions of $^\bullet OH$ radicals and dihalogen radical anions ($Cl_2^{\bullet-}$, $Br_2^{\bullet-}$) with 2-TU reported the formation of only one intermediate, namely, dimeric radical cation $(2\text{-}TU)_2{}^{\bullet+}$ characterized by an absorption band with λ_{max} = 430 nm and which exists in an equilibrium shown on Scheme 2 [58]. However, except for theoretical calculations, there were no other proofs that its precursor and dimeric radical itself have cationic form. Moreover, no experimental transient spectra which can be assigned to other possible transients resulted from 2-TU oxidation were reported. One of the reason was that the earlier spectral studies were limited to only one and relatively high concentration of 2-TU [58].

Low concentration of 2-TU. Thus, what is the identity of the transient species absorbing in the wavelength range 320–400 nm and characterized by weakly pronounced absorption maxima located at λ_{max} = 325, 340 and 385 nm (*vide* Figure 2A and Figure S1A in Supplementary Materials) in solutions containing low concentration of 2-TU at pH = 4? Taking into account the molecular structure of the most energetically favorable tautomer of 2-TU (Scheme 2) and plausible primary reactions of $^\bullet OH$ radicals, it is reasonable to assume the following primary sites of $^\bullet OH$-attack—(i) a double C=C bond at C5 and C6 position and (ii) a double C=S bond at S8 position. The •OH adducts at C5 and C6 positions labelled as 3• and 4• (Scheme 3) are characterized by absorption bands with calculated maxima located at λ = 343 nm and 376 nm, respectively (Figure 8). By analogy to uracil, C5 position is of the higher electron density than C6 position and should be more preferred site to electrophilic •OH attack [88]. Moreover, their relative free energies in solution (ΔG_{rel}), with respect to the sum of substrates (•OH and 2-TU) are lower (−6.85 kcal mol^{-1} and −10,53 kcal mol^{-1}, respectively) and therefore their formation is thermodynamically possible [58]. Even though the most stable conformer produced in the course of these reactions, radical 4•, is thermodynamically more stable than the most stable conformer of radical 3• (*vide* Figure S9B and Table S2 in Supplementary Materials) its formation proceeds through 5.6 kcal mol^{-1} barrier in comparison to 3.5 kcal mol^{-1} calculated for 3•. Therefore, we can expect higher contribution of 3• in experimental absorption spectra presented in Figure 2A and Figure S1A in Supplementary Materials.

An interesting case is an addition of •OH to S8 in a double C=S bond. This reaction can potentially lead to the C2-centered radical labelled as 5• (Scheme 3) characterized by absorption band with calculated maximum at λ = 340 nm. However, its formation is thermodynamically unfavorable [58] and 5• is not considered as a species contributing to experimental absorption spectra observed in this work. However, this kind of radical was considered as a primary species during oxidation of thiourea by •OH radicals [69].

Numerous theoretical studies about the character of bond in •OH radicals adducts to sulfur atom has been performed over the years and in organic sulfides, in particular, 2c-3e character of the S-O bond was established [89,90]. This fact motivated us to consider similar intermediate, namely •OH adduct to sulfur atom in a double C=S bond, having 2c-3e bond character. The OH adducts at S8 position labelled as 1 ... •OH$_{(1)}$, 1 ... •OH$_{(2)}$ (Scheme 3) meet all requirements for hemibonded •OH adducts to sulfur atom. The S–O bond length is ~2.33 Å and maximum spin population is almost evenly distributed over S and O atoms (*vide* Figure S4 in Supporting Materials). What is equally important both 1 ... •OH$_{(1)}$, 1 ... •OH$_{(2)}$ radicals are characterized by absorption bands with calculated maxima located at λ in the range 340–342 nm (Figure 8) and can also contribute to experimental absorption spectra presented in Figure 2A and Figure S1A in Supplementary Materials. However, one can reasonably assume that 1 ... •OH$_{(1)}$, 1 ... •OH$_{(2)}$ radicals are precursors of 6• and 7• radicals which can be formed in the concerted electron/proton transfer where formation of 1•+ occurs in concert with its deprotonation at positions N1 and N3, respectively. Alternatively, it would not be surprising, that the both hemibonded •OH adducts to sulfur can be precursors of radical cation 1•+ (Scheme 3) via analogous spontaneous and/or proton catalyzed dissociation processes which were observed in •OH-induced oxidation of aliphatic sulfides [91] and methionine containing peptides [71,72,92]. However, based only on changes of absorption spectra, it was not possible to determine which reaction pathway takes place. However, these reaction pathways can be distinguished based on net changes of the apparent conductivity (*vide* Section 3.2).

The absorption spectra recorded at pH = 10 (Figure 3A), taking into account the molecular structure of the most energetically favorable tautomers of 2-TU (Scheme 2), can be in principle assigned to the similar type of species as for pH = 4, that is, 1$^-$ ... •OH$_{(1)}$, 1$^-$ ... •OH$_{(2)}$, 3•$^-$ and 4•$^-$ (Scheme 3). The •OH adducts at S8 position meet also all requirements for hemibonded •OH adducts to sulfur atom. The S–O bond length is ~2.38 Å and maximum spin population is almost evenly distributed over S and O atoms (*vide* Figure S5 in Supplementary Materials). Interestingly, the absorption band with λ_{max} = 425 nm was also observed, which can be assigned to dimeric radical anion 2•$^-$ based on calculated λ_{max} = 432 nm (Figure S11 in Supplementary Materials).

High concentration of 2-TU. The transient absorption spectra observed at high concentration of 2-TU are similar (Figure S1B in Supplementary Materials and Figure 2B). to that observed in the earlier work and were assigned to the formation of dimeric radical cation with 2c-3e sulfur-sulfur bond 2•+ [58]. As a consequence, monomeric radical cation 1•+ was taken as a direct precursor of 2•+ which is presented in the form of equilibrium (*vide* Scheme 1).

In our calculations, we considered $2^{\bullet+}$ as a possible transient responsible for the absorption band with λ_{max} = 425 nm, however, we considered also possibility of formation of dimeric radicals in neutral ($2^{\bullet}$) and anionic ($2^{\bullet-}$) forms. Since their calculated spectral parameters are very similar to that calculated for $2^{\bullet+}$ (Figure 8) their character and as well of their direct precursors cannot be decided on the basis of absorption spectra. The observed net change of apparent conductivity should dispel doubts about the character of the dimeric radical formed (*vide* Section 3.2).

There is no reason to exclude contribution of the $^{\bullet}$OH adducts at C5 and C6 positions labelled as $3^{\bullet}$ and $4^{\bullet}$ formed at low concentration of 2-TU to the observed transient spectra. Their presence is clearly manifested in absorption spectra at short time domains (Figure 2B and Figure S1B in Supporting Materials), however, at long time domains are hidden under the absorption assigned to one of the dimeric radical forms.

3.2. Justification of the Reaction Pathway Involving Hemibonded $^{\bullet}$OH Adducts to Sulfur Atom

Contrary to our expectations, based on previous results with thiourea [69], decrease of the net conductivity below zero was not observed which would imply consumption of H$^+$ through neutralization reaction by OH$^-$ (*vide* Section 2.3) which are released with simultaneous formation of radical cation $1^{\bullet+}$, in another words replacement of highly conducting protons by the resulting low conducting radical cations $1^{\bullet+}$. Hence, formation of $1^{\bullet+}$ followed by formation of $2^{\bullet+}$ does not seem to be the reaction pathway during $^{\bullet}$OH induced oxidation of neutral form of 2-TU.

Instead, there is a couple of possible scenarios which can be suggested. The first one, in which $^{\bullet}$OH radicals react mainly by processes of addition and H-atom abstraction upon which no new conducting species are being formed. This process was suggested earlier, however, with activation energies in the range 2.53 kcal mol^{-1}–3.79 kcal mol^{-1} [58]. Taking also into account an additional well known fact that H-abstraction reactions by $^{\bullet}$OH radicals occur with lower rate constants than their addition to sulfur atom, this possibility can be rather discarded. The second one seems to be more reasonable. The $^{\bullet}$OH radicals form the hemibonded adduct to sulfur (1 ... $^{\bullet}$OH) which decays by separated coupled electron proton transfer. The OH$^-$ generated in the inner-sphere electron transfer, (that leads simultaneously to radical $1^{\bullet+}$) is neutralized by the proton released from either N1 or N3 atoms of $1^{\bullet+}$ and therefore formation of $6^{\bullet}$ or $7^{\bullet}$ is not associated with any net change of conductivity. Moreover, the lack of decrease of the net conductivity below zero excludes the process which operates in case of methionine containing peptides [71,72,92]. In this processes, the hemibonded S∴OH radical undergoes OH$^-$ elimination, however, by protons from the bulk of solution, leading simultaneously to monomeric radical cation (S$^{\bullet+}$) and further to dimeric radical cation with 2c-3e S–S bonds (S∴S)$^+$. This process is accompanied by decrease of net conductivity due to substitution of highly conducting protons by the resulting low conducting radical cations S$^{\bullet+}$ and (S∴S)$^+$. This observation is crucial in determining the nature of both the dimer radical and its precursor (*vide* Section 3.3).

3.3. Mechanisms of $^{\bullet}$OH and $^{\bullet}$N$_3$-Induced Oxidation of 2-Thiouracil

On the basis of the identification of transients and complementary theoretical calculations, the mechanisms presented in Scheme 4 to Scheme 5 are proposed for the $^{\bullet}$OH and $^{\bullet}$N$_3$-induced oxidation of 2-TU in aqueous solutions.

The initial steps are the additions of $^{\bullet}$OH radical to C5=C6 double bond and to S8 atom yielding $3^{\bullet}$ and $4^{\bullet}$ radicals and hemibonded adduct to sulfur (1 ... $^{\bullet}$OH), respectively. Based on our calculations we can assume regioselectivity in addition to position C5 over C6 and in a consequence more efficient formation of radical $3^{\bullet}$. Furthermore formation of 2c-3e OH adducts to sulfur should dominate based on computed energy profiles for $^{\bullet}$OH addition reactions to neutral 2-TU.

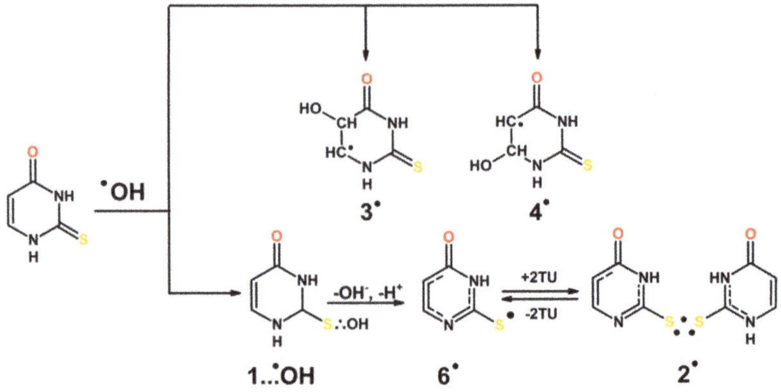

Scheme 4. •OH-induced oxidation of 2-TU at pH below its pK$_a$.

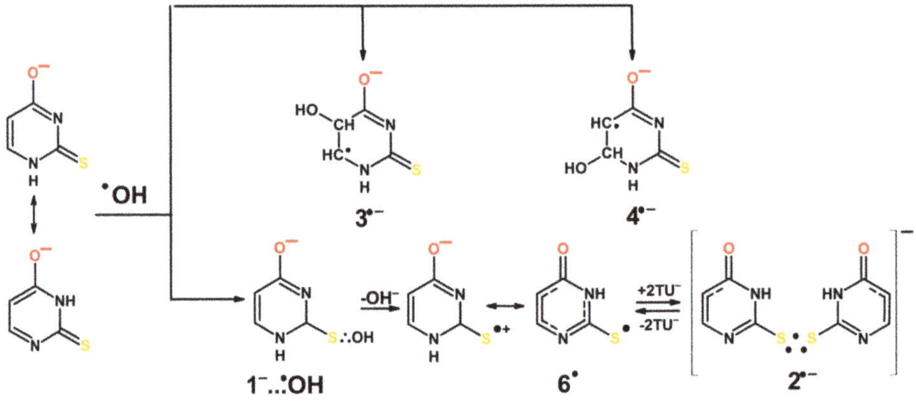

Scheme 5. •OH-induced oxidation of 2-TU at pH above its pK$_a$.

The hemibonded adduct to sulfur (1 ... •OH) decays by separated coupled electron proton transfer leading to formation of 6• or 7• radicals. Since formation of radical 6• is thermodynamically preferred (*vide* Table S2 in Supplementary Materials) this form is presented in the scheme. At higher 2-TU concentration 6• radicals are converted into dimeric radical 2•, however, both radicals exist in an equilibrium (Scheme 4). The dimeric radical exists in neutral form since both substrates (6• and 2-TU) are present in neutral forms, too.

At higher pH, the initial steps of •OH reactions with 2-TU are practically the same and lead to the following species—3•−, 4•− and 1− ... •OH. The hemibonded adduct to sulfur is then converted to 6•. Similarly, as for pH = 4, at higher concentration of 2-TU radical 6• is in an equilibrium with the dimeric radical anion (2•−). The anionic character of dimeric radical is due to the fact that at pH = 10, 2-TU exists in anionic forms (Scheme 5). Formation of 3•− is even more preferable than formation of 4•−, in comparison to 3• and 4•. In this case both lower activation barrier and lower ΔG favor its formation (Figure S10B,C; Table S2 in Supplementary Materials).

Since azide radicals (•N$_3$) are commonly considered as one-electron oxidants the initial step leads to the radical cation (1•+). Formation of this radical cannot be observed directly because its absorption spectrum is located in inconvenient observation area < 300 nm. In principle, the radical cation (1•+) could be a precursor of dimeric radical cation (2•+). Its spectral features are very similar to those showing by 2• and 2•− (Figure 8). However, one can assume another scenario. Radical cation (1•+) can also undergo fast deprotonation to radical 6• which further undergo transformation to dimeric radical

$2^{\bullet}$. One important observation strongly supports this reaction pathway. At low concentration of 2-TU, a transient absorption spectrum with λ_{max} = 320 nm is observed which can be tentatively assigned to radical $6^{\bullet}$ (Figure 7A) Moreover, the transient absorption in the vicinity of λ = 350 nm observed at high concentration of 2-TU (Figure 7B) can be also tentatively assigned to radical $6^{\bullet}$. Since the pseudo-first order rate constant of $6^{\bullet}$ radical decay matches the pseudo-first order rate constant growth measured at the λ_{max} assigned to the absorption bands of dimeric radicals, this is a strong support that radical $6^{\bullet}$ is a direct precursor of $2^{\bullet}$ (Scheme 6).

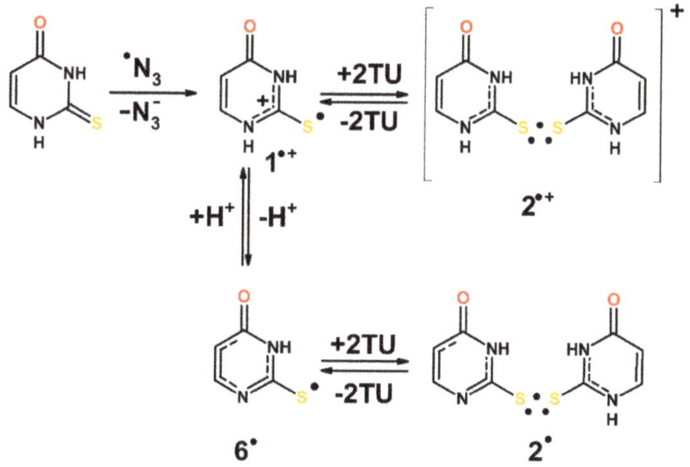

Scheme 6. $^{\bullet}N_3$-induced oxidation of 2-TU at pH below its pK_a.

4. Materials and Methods

4.1. Chemicals

2-Thiouracil (2-TU) (≥99% purity) and sodium azide (NaN$_3$) (≥99.5% purity) were purchased from Sigma-Aldrich (St. Louis, MO, USA) and used without further purification. Nitrous oxide (N$_2$O) > 98% was from Messer (Warsaw, Poland).

4.2. Preparation of Solutions

All solutions were made with water triply distilled provided by a Millipore Direct-Q 3-UV system. The pH was adjusted by the addition of NaOH (≥99% purity) or HClO$_4$ (70%, 99% purity), both form Sigma-Aldrich (St. Louis, MO, USA). Prior to irradiation, the samples were purged gently with N$_2$O for 30 min. per 200 mL volume.

4.3. Pulse Radiolysis

Pulse radiolysis experiments with time-resolved UV-vis optical absorption detection were carried out at the Institute of Nuclear Chemistry and Technology (INCT) in Warsaw, Poland and the Notre Dame Radiation Laboratory (NDRL), Notre Dame, IN, USA. In the INCT, the linear electron accelerator (INCT LAE 10) delivering 10 ns pulses with electron energy about 10 MeV was applied as a source of irradiation. The 150 W xenon arc lamp E7536 (Hamamatsu Photonics K.K) was used as a monitoring light source. The respective wavelengths were selected by MSH 301 (Lot Oriel Gruppe) motorized monochromator/spectrograph with two optical output ports. The time dependent intensity of the analyzing light was measured by means of photomultiplier (PMT) R955 (Hamamatsu). A signal from detector was digitized using a WaveSurfer 104MXs-B (1 GHz, 10 GS/s, LeCroy) oscilloscope. A detailed description of the experimental setup has been given elsewhere along with the basic details of the equipment and the data collection system [93,94]. Absorbed doses per pulse were on the order of 10–20 Gy (1 Gy = 1 J·kg^{-1}). Experiments were performed with a continuous flow of sample solutions at room temperature (~22 °C). In order to avoid photodecomposition and/or photobleaching effects in the samples, the UV or VIS cut-off filters were used. However, no evidence of such effects was found within the time domains monitored. Water filter was used to eliminate near IR wavelengths. Absorption intensities are presented in $G \times \varepsilon$, where G is radiation-chemical yield (μM J^{-1}) of given species and ε represents molar absorption coefficient. This value is directly proportional to the absorbance.

At the NDRL, pulse radiolysis experiments were carried out using the 8-MeV linear accelerator (LINAC). Following 8 ns, ~15 Gy electron pulse, transient optical absorption signals were recorded in UV−visible range using two different detection systems. First, single channel PMT/monochromator detection system was described elsewhere [95]. Second, multichannel system with nanosecond response recorded array of 24 monochromatic kinetic signals on all input channels of 6 synchronously triggered Tektronix oscilloscopes. The oscilloscopes were connected to an array of equivalent 24 silicon photodiode/amplifier detectors. Each of the detectors was optically coupled with different monochromatic line generated at the exit focal plane of SpectraPro 2150 spectrograph (Princeton Instruments) using a custom bundle of 24 fused silica fiber optics. The probe light source for both systems was a 1000 W xenon lamp pulsed to high current for 2 milliseconds. Variation of the solute concentration was accomplished using two HPLC pumps, one of which was pumping N$_2$O saturated 2mM stock solution of 2-TU and the other was filled with N$_2$O saturated water at the same pH as the 2-TU solution. The solutions from the pumps were mixed at a tee connection before the inlet of the flow cell. By varying the flow rates of the two pumps, with constant total flow of 3 mL/min, the concentration of 2-TU could be easily remotely changed/controlled.

At both pulse radiolysis set-ups, the dosimetry was based on N$_2$O-saturated solutions containing 10 mM KSCN, taking a radiation chemical yield of G = 0.635 μmol J^{-1} and a molar absorption coefficient of 7580 M^{-1}·cm^{-1} at λ = 472 nm for the (SCN)$_2^{\bullet-}$ radical [96].

For time-resolved conductivity measurements, the conductivity apparatus was used which allows high-precision conductometric measurements over a pH range from 3 to 6. In the current experiments, pH was restricted to 4.1. A detailed description of the conductivity apparatus along with the measuring cell has been given elsewhere [97]. The dosimetry was achieved using acidic (pH = 4.1), aqueous solution saturated with methyl chloride (CH$_3$Cl). In this dosimeter system, pulse irradiation yields H$^+$ and Cl$^-$ with G(H$^+$) = G(Cl$^-$) = 0.285 μmol J^{-1}. The respective equivalent conductivities at 18 °C were taken as Λ(H$^+$) = 315 S cm^2 equiv^{-1} and Λ(Cl$^-$) = 65 S cm^2 equiv^{-1} [98].

4.4. Theoretical Procedures

The theoretical calculations were performed using the Gaussian 09 and Gaussian 16 program package [99]. One computational method have been used in this work for the geometry optimizations, ground state reactivities and excited state calculations based on the range separated hybrid (RSH) functionals ωB97x [100] and correlation consistent basis sets of triple ζ type, augmented with diffuse functions, denoted aug-cc-pVTZ [101,102]. The hydration effects were taken into consideration using a polarized continuum model (IEFPCM) [103]. The local minima were verified by frequency calculations. The Mulliken scheme was used to obtain spin density distribution. The transition states (TS) were located by the Synchronous Transit-Guided Quasi-Newton approach (QST3 method). The genuineness of the transition states in each case was ensured by the presence of one imaginary frequency related to either the stretching of the C-O bond (for OH addition) or H-O bond (for H abstraction) that connects the $^{\bullet}$OH and 2-TU neutral or anionic reactants. Intrinsic reaction coordinate (IRC) calculations have been carried out from the TS, leading to OH adducts (at C5 and C6 positions of neutral 2-TU) and the pre-reactive complex (RC$^{\bullet}$). IRC calculations were also carried for H abstraction reactions from N3 and N1 positions in neutral and anionic forms of 2-TU. In case of anionic 2-TU no effort was taken to find optimized structures of pre-reactive complexes for OH addition as preliminary potential energy surface scans showed lack of such a complex and its eventual existence did not seem relevant to most of important findings of the work presented. Electronic transition energies and oscillator strengths were calculated by the time-dependent DFT (TD-DFT) method. DFT ωB97x method was proven to be very satisfactory in characterization of ground state geometries, harmonic vibrational frequencies, dissociation energies and absorption maxima of 2c-3e intermediates of $(SCN)_2^{\bullet-}$ and $(SeCN)_2^{\bullet-}$ [78,79]. Similar good performance of ω B97x based TD-DFT method was documented earlier [104,105]. It compared well with higher levels of theory in modeling ground-state reactivity and activation barriers for •OH addition to double bonds in vacuum [106]. However, it could differ from them in obtaining proper activation barriers in certain cases in PCM solvent and/or explicit water molecules [107]. Since •OH addition leading to formation of 2c-3e SO bonds is essentially barrierless [90] application of this relatively economic computational method serves as a good compromise in quantitative estimation of relative pathways in molecular systems where H abstraction, OH addition to double bonds and SO hemibond formation can occur simultaneously. We performed verification of accuracy of ωB97x functional in determining vertical excitation energies of OH adducts to C5 and C6 positions in 2-TU. In addition to that we compared CASPT2 level transitions obtained previously for •OH adducts to uracil with transitions calculated using method applied in our studies of 2-TU. Using the starting geometries of OH adducts to uracil taken from the Supporting Information in Reference [108] we re-optimized them at the level of ωB97x/aug-cc-pvtz (PCM) and computed their absorption maxima. For most stable ground state structures—adduct U5OH (Hext)—we obtained a second transition at 4.13 eV versus CASPT 4.39 eV (−0.26 eV or +18 nm) and for adduct U6OH (Hint) we obtained transition at 3.56 eV versus CASPT 3.05 eV (+0.51eV or −58 nm). This may seem like a big difference but one needs to point out that both calculations should be referred to the experimental value at 3.26 eV. U6OH adduct CASPT computed transition was shifted from the experimental value by +27 nm and TD-DFT wB97x/aug-cc-pvtz computed transition was shifted −31 nm from the experimental value of 380 nm. For a broad spectrum with half width of ~1 eV we would consider both values as quite satisfactory. Our TD-DFT calculations predicted transition for U5OH to occur at ~299.9 nm versus 282 nm obtained by CASPT. These values are qualitatively quite comparable to previous findings obtained at much higher computational cost [108]. It has been documented that DFT and TD-DFT results always display some spin contamination, namely, the mean value of the S^2 spin operator for the ground and excited states differs from the theoretical result $<S^2> = S(S+1) = 0.75$. This is due to the single determinant form of the wave function in DFT [109]. Consequently, these states may no more be called spin doublet states but spin-contaminated states where only the quantum number Ms = ±1/2 is ensured. These contaminated states ψ_C may be considered a combination of the doublet state ψ_D and quartet state ψ_Q, $\psi_C = \psi_D + \varepsilon\psi_Q$ (with Ms = ±1/2). Analyzing results of our TD-DFT computations we noticed that most relevant transitions (with the highest oscillator strengths) in all considered 2c-3e intermediates never had $<S^2>$ value higher than 0.77, hence we considered them not spin contaminated. On the other hand, thiyl radicals as well as OH adducts to C5 and C6, which were found to have much lower oscillator strengths in our experimental spectral range of interest have shown $<S^2>$ values extending up to 1.0 for the strongest transitions and >1 for the transitions with oscillator strengths 1–2 orders of magnitude lower than these strongest transitions. Therefore, results of computations of excited states in these intermediates are affected by spin contamination and in order to get more accurate predictions of their absorption spectra their ground and excited states should be treated using higher level computational methods which was beyond interest in our current studies. This was justified by the fact that contribution of these intermediates to the observed spectra was relatively minor. We still decided to include these spectra to all computed 2c3e intermediates for the qualitative comparison. All computed UV-Vis spectra were generated using the default setting of GaussView 6 in which peaks assume a Gaussian band shape with half widths of 0.4eV [110].

5. Conclusions

In the current paper, we provided an experimental proof supported by the TD-DFT calculations that character of primary and secondary reactive intermediates depends on the concentration of

2-thiouracil and character of the oxidant ($^{\bullet}$OH vs. $^{\bullet}$N$_3$). At low concentration of 2-TU, $^{\bullet}$OH-adducts to the double bond at C5 position (3$^{\bullet}$) and 2c-3e bonded $^{\bullet}$OH adducts to sulfur (1 ... $^{\bullet}$OH) are dominant species. Their relative contribution seems to depend on the individual rate constants of $^{\bullet}$OH addition to the double bonds and sulfur atom, respectively. In turn, at high concentration of 2-TU, beyond contribution of 3$^{\bullet}$ radicals, contribution of dimeric 2c-3e S-S-bonded radicals in neutral form (2$^{\bullet}$) has to be taken into account. Their direct precursors are thiyl-type radicals (6$^{\bullet}$) which are formed from 1 ... $^{\bullet}$OH radicals via separated coupled electron proton transfer. Dimeric 2c-3e S-S-bonded radicals are also formed at pH above the pK_a of 2-TU but they have anionic character (2$^{\bullet-}$) owing to the anionic form of 2-TU. In turn, $^{\bullet}$N$_3$-induced oxidation of 2-TU occurs via radical cations with maximum spin location on the sulfur atom (1$^{\bullet+}$) which subsequently undergo deprotonation at N1 atom leading to thiyl-type radicals (6$^{\bullet}$), direct precursors of dimeric radicals (2$^{\bullet}$).

Supplementary Materials: The following are available online at http://www.mdpi.com/1420-3049/24/23/4402/s1, Figure S1: Transient absorption spectra recorded in N$_2$O-saturated unbuffered aqueous solution at pH = 4 containing 0.2 and 0.5 mM of 2-TU, Figure S2: Absorption spectra recorded in aqueous solutions containing 2-TU at pH = 4 and pH = 10, Figure S3: Time profiles representing growth and decay of transient absorptions at λ = 420 nm at pH = 4 and at pH = 10 at various concentration of 2-TU, Figure S4: Solution phase (PCM) optimized geometries of monomeric type transients formed at pH lower than pK_a of 2-TU, Figure S5: Solution phase (PCM) optimized geometries of monomeric type transients formed at pH higher than pK_a of 2-TU, Figure S6: Solution phase (PCM) optimized geometries of 2c-3e SS dimers. Figure S7: Computed SCF spin density isosurfaces of various 2c-3e intermediates. Figure S8: Relative energy profile in aqueous phase (PCM) for the H abstraction from neutral 2-TU via 2c-3e OH adducts. Figure S9: Relative energy profiles in aqueous phase (PCM) for the H abstraction and $^{\bullet}$OH addition from/to 2-TU. Figure S10: Relative energy profiles in aqueous phase (PCM) for the $^{\bullet}$OH addition (and H abstraction) reactions to/from 2-TU monoanions. Figure S11: TD-DFT calculated absorption spectra of potential transients produced in $^{\bullet}$OH-induced oxidation of 2-thiouracil (2-TU) in water at pH higher than its pK_a. Table S1: Thermochemistry values for the reactants and products optimized structures. Table S2: Free energies of reactions* of $^{\bullet}$OH addition to 2-TU at pH 4 and 10. Table S3: Cartesian coordinates of the studied structures.

Author Contributions: Conceptualization, K.S. and K.B.; methodology, K.S., K.B., I.J.; experiments in pulse radiolysis, K.S., I.J.; theoretical calculations, I.J.; data analysis, K.S., K.B., I.J., K.T.-G.; writing—original draft preparation, K.S., K.B., I.J., K.T.-G.; writing—review and editing, K.S., K.B., I.J.; funding acquisition, K.S.

Funding: This research was funded by the Polish National Science Centre under grant 2018/28/C/ST4/00479 (K.S.) and by the U.S. Department of Energy, Office of Science, Office of Basic Energy Sciences under Award Number DE-FC02-04ER15533 (I.J.) This is document number NDRL-5259 from the Notre Dame Radiation Laboratory.

Acknowledgments: We would like to express our thanks to the supporting staff in NDRL for their help in implementing improvements to the transient conductivity setup over the last few years. In particular, help from our glassblower Kiva Ford and our machinist Joseph Admave is greatly appreciated. Thanks are also due to Ian Carmichael for helpful discussions and suggestions on the computational approach in understanding the nature of 2c-3e dimers studied in this project.

Conflicts of Interest: The authors declare no conflict of interest.

References

1. Elion, G.B.; Hitchings, G.H. Antagonists of nucleic acid derivatives. 6. Purines. *J. Biol. Chem.* **1951**, *192*, 505–518. [PubMed]
2. Cantara, W.A.; Crain, P.F.; Rozenski, J.; McCloskey, J.A.; Harris, K.A.; Zhang, X.; Vendeix, F.A.P.; Fabris, D.; Agris, P.F. The RNA modification database, RNAMDB:2011 update. *Nucleic Acid Res.* **2011**, *39*, D195–D201. [CrossRef] [PubMed]
3. Ajitkumar, P.; Cherayil, J.D. Thionucleosides in transfer ribonucleic acid -diversity, structure, biosynthesis, and function. *Microbiol. Rev.* **1988**, *52*, 103–113.
4. Wei, F.Y.; Suzuki, T.; Watanabe, S.; Kimura, S.; Kaitsuka, T.; Fujimura, A.; Matsui, H.; Atta, M.; Michiue, H.; Fontecave, M.; et al. Deficit of tRNA(Lys) modification by Cdkal1 causes the development of type 2 diabetes in mice. *J. Clin. Investig.* **2011**, *121*, 3598–3608. [CrossRef] [PubMed]
5. Pollum, M.; Martinez-Fernandez, L.; Crespo-Hernandez, C.E. Photochemistry of Nucleic Acid Bases and Their Thio- and Aza-Analogues in Solution. In *Photoinduced Phenomena in Nucleic Acids I: Nucleobases in the Gas Phase and in Solvents*; Barbatti, M.B., Borin, A.C., Ullrich, S., Eds.; Springer: Cham, Switzerland, 2015; Volume 355, pp. 245–327.

6. Kutyavin, L.V.; Rhinehart, R.L.; Lukhtanov, E.A.; Gorn, V.V.; Meyer, R.B.; Gamper, H.B. Oligonucleotides containing 2-aminoadenine and 2-thiothymine act as selectively binding complementary agents. *Biochemistry-US* **1996**, *35*, 11170–11176. [CrossRef] [PubMed]
7. Martinez-Fernandez, L.; Gonzalez, L.; Corral, I. An ab initio mechanism for efficient population of triplet states in cytotoxic sulfur substituted DNA bases: The case of 6-thioguanine. *Chem. Commun.* **2012**, *48*, 2134–2136. [CrossRef]
8. Reelfs, O.; Karran, P.; Young, A.R. 4-thiothymidine sensitization of DNA to UVA offers potential for a novel photochemotherapy. *Photochem. Photobiol. Sci.* **2012**, *11*, 148–154. [CrossRef]
9. Brem, R.; Zhang, X.; Xu, Y.-Z.; Karran, P. UVA photoactivation of DNA containing halogenated thiopyrimidines induces DNA lesions. *J. Photochem. Photobiol. B Biol.* **2015**, *145*, 1–10. [CrossRef]
10. Gemenetzidis, E.; Shavorskaya, O.; Xu, Y.-Z.; Trigiante, G. Topical 4-thiothymidine is a viable photosensitiser for the photodynamic therapy of skin malignancies. *J. Dermatolog. Treat.* **2013**, *24*, 209–214. [CrossRef]
11. Reelfs, O.; Macpherson, P.; Ren, X.; Xu, Y.-Z.; Young, A.R. Identification of potentially cytotoxic lesions induced by UVA photoactivation of DNA 4-thithymidine in human cells. *Nucleic Acid Res.* **2011**, *39*, 9620–9632. [CrossRef]
12. Pridgeon, S.W.; Heer, R.; Taylor, G.A.; Newell, D.R.; O'Toole, K.; Robinson, M.; Xu, Y.-Z.; Karran, P.; Boddy, A.V. Thiothymidine combined with UVA as a potential novel therapy for bladder cancer. *Br. J. Cancer* **2011**, *104*, 1869–1876. [CrossRef] [PubMed]
13. Trigiante, G.; Xu, Y.-Z. 4-thiothymidine and its analogues as UVA-activated photosensitizers. In *Photodynamic Therapy: Fundamentals, Applications and Health Outcomes*; Hugo, A.G., Ed.; Nova Science Publishers: Hauppauge, NY, USA, 2015; pp. 193–206.
14. Ashwood, B.; Pollum, M.; Crespo-Hernandez, C.E. Photochemical and Photodynamical Properties of Sulfur-Substituted Nucleic Acid Bases. *Photochem. Photobiol.* **2019**, *95*, 33–58. [CrossRef] [PubMed]
15. Mai, S.; Pollum, M.; Martinez-Fernandez, L.; Dunn, N.; Marquetand, P.; Corral, I.; Crespo-Hernandez, C.E.; Gonzalez, L. The origin of efficient triplet state population in sulfur-substituted nucleobases. *Nat. Commun.* **2016**, *7*, 13077. [CrossRef] [PubMed]
16. Favre, A.; Saintome, C.; Fourrey, J.L.; Clivio, P.; Laugaa, P. Thionucleobases as intrinsic photoaffinity probes of nucleic acid structure and nucleic acid protein interactions. *J. Photochem. Photobiol. B Biol.* **1998**, *42*, 109–124. [CrossRef]
17. Taras-Goslinska, K.; Wenska, G.; Skalski, B.; Maciejewski, A.; Burdzinski, G.; Karolczak, J. Intra- and intermolecular electronic relaxation of the second excited singlet and the lowest excited triplet states of 1,3-dimethyl-4-thiouracil in solution. *Photochem. Photobiol.* **2002**, *75*, 448–456. [CrossRef]
18. Taras-Goslinska, K.; Wenska, G.; Skalski, B.; Maciejewski, A.; Burdzinski, G.; Karolczak, J. Spectral and photophysical properties of the lowest excited triplet state of 4-thiouridine and its 5-halogeno derivatives. *J. Photochem. Photobiol. A Chem.* **2004**, *168*, 227–233. [CrossRef]
19. Wenska, G.; Taras-Goslinska, K.; Skalski, B.; Maciejewski, A.; Burdzinski, G.; Karolczak, J. Putative phototautomerization of 4-thiouridine in the S$_2$ excited state revealed by fluorescence study using picosecond laser spectroscopy. *J. Photochem. Photobiol. A Chem.* **2006**, *181*, 12–18. [CrossRef]
20. Wenska, G.; Taras-Goslinska, K.; Filipiak, P.; Hug, G.L.; Marciniak, B. Photochemical reactions of 4-thiouridine disulfide and 4-benzylthiouridine the involvement of the 4-pyrimidinylthiyl radical. *Photochem. Photobiol. Sci.* **2008**, *7*, 250–256. [CrossRef]
21. Wenska, G.; Koput, J.; Burdzinski, G.; Taras-Goslinska, K.; Maciejewski, A. Photophysical and photochemical properties of the T$_1$ excited state of thioinosine. *J. Photochem. Photobiol. A Chem.* **2009**, *206*, 93–101. [CrossRef]
22. Wenska, G.; Filipiak, P.; Taras-Goslinska, K.; Sobierajska, A.; Gdaniec, Z. Orientation-dependent quenching of the triplet excited 6-thiopurine by nucleobases. *J. Photochem. Photobiol. A Chem.* **2011**, *217*, 55–61. [CrossRef]
23. Wenska, G.; Taras-Goslinska, K.; Lukaszewicz, A.; Burdzinski, G.; Koput, J.; Maciejewski, A. Mechanism and dynamics of intramolecular triplet state decay of 1-propyl-4-thiouracil and its alpha-methyl-substituted derivatives studied in perfluoro-1,3-dimethylcyclohexane. *Photochem. Photobiol. Sci.* **2011**, *10*, 1294–1302. [CrossRef] [PubMed]
24. Taras-Goslinska, K.; Burdzinski, G.; Wenska, G. Relaxation of the T$_1$ excited state of 2-thiothymine, its riboside and deoxyriboside-enhanced nonradiative decay rate induced by sugar substituent. *J. Photochem. Photobiol. A Chem.* **2014**, *275*, 89–95. [CrossRef]

25. Pollum, M.; Crespo-Hernandez, C.E. Communication: The dark singlet state as a doorway state in the ultrafast and efficient intersystem crossing dynamics in 2-thiothymine and 2-thiouracil. *J. Chem. Phys.* **2014**, *140*, 071101. [CrossRef] [PubMed]
26. Pollum, M.; Jockusch, S.; Crespo-Hernandez, C.E. 2,4-Dithiothymine as a Potent UVA Chemotherapeutic Agent. *J. Am. Chem. Soc.* **2014**, *136*, 17930–17933. [CrossRef] [PubMed]
27. Crespo-Hernandez, C.E.; Martinez-Fernandez, L.; Rauer, C.; Reichardt, C.; Mai, S.; Pollum, M.; Marquetand, P.; Gonzalez, L.; Corral, I. Electronic and Structural Elements That Regulate the Excited-State Dynamics in Purine Nucleobase Derivatives. *J. Am. Chem. Soc.* **2015**, *137*, 4368–4381. [CrossRef] [PubMed]
28. Pollum, M.; Jockusch, S.; Crespo-Hernandez, C.E. Increase in the photoreactivity of uracil derivatives by doubling thionation. *Phys. Chem. Chem. Phys.* **2015**, *17*, 27851–27861. [CrossRef] [PubMed]
29. Sanchez-Rodriguez, J.A.; Mohamadzade, A.; Mai, S.; Ashwood, B.; Pollum, M.; Marquetand, P.; Gonzalez, L.; Crespo-Hernandez, C.E.; Ullrich, S. 2-Thiouracil intersystem crossing photodynamics studied by wavelength-dependent photoelectron and transient absorption spectroscopies. *Phys. Chem. Chem. Phys.* **2017**, *19*, 19756–19766. [CrossRef]
30. Karran, P.; Attard, N. Thiopurines in current medical practice: Molecular mechanisms and contributions to therapy-related cancer. *Nat. Rev. Cancer* **2008**, *8*, 24–36. [CrossRef]
31. Herak, J.N.; Sankovic, K.; Hutterman, J. Thiocytosine as a radiation energy trap in a single crystal of cytosine hydrochloride. *Int. J. Radiat. Biol.* **1994**, *66*, 3–9. [CrossRef]
32. Ling, L.; Ward, J.F. Radiosensitization of Chinese Hamster V79 Cells by Bromodeoxyuridine Substitution of Thymidine: Enhancement of Radiation-Induced Toxicity and DNA Strand Break Production by Monofilar and Bifilar Substitution. *Radiat. Res.* **1990**, *121*, 76–83. [CrossRef]
33. Abdoul-Carime, H.; Dugal, P.-C.; Sanche, L. Damage Induced by 1-30 eV Electrons on Thymine- and Bromouracil-Substituted Oligonucleotides. *Radiat. Res.* **2000**, *153*, 23–28. [CrossRef]
34. Abdoul-Carime, H.; Huels, M.A.; Illenberger, E.; Sanche, L. Sensitizing DNA to secondary electron damage: Resonant oxidative radicals from 5-halouracils. *J. Am. Chem. Soc.* **2001**, *123*, 5354–5355. [CrossRef] [PubMed]
35. Kopyra, J.; Freza, S.; Abdoul-Carime, H.; Marchaj, M.; Skurski, P. Dissociative electron attachment to gas phase thiothymine: Experimental and theoretical approaches. *Phys. Chem. Chem. Phys.* **2014**, *16*, 5342–5348. [CrossRef] [PubMed]
36. Kopyra, J.; Abdoul-Carime, H.; Kossoski, F.; Varela, M.T.d.N. Electron driven reactions in sulphur containing analogues of uracil: The case of 2-thiouracil. *Phys. Chem. Chem. Phys.* **2014**, *16*, 25054–25061. [CrossRef] [PubMed]
37. Kopyra, J.; Abdoul-Carime, H. Dissociative electron attachment to gas phase nucleobases: Comparision of thymine and thiothymine. *J. Phys. Conf. Ser.* **2015**, *635*, 072066. [CrossRef]
38. Kopyra, J.; Abdoul-Carime, H.; Skurski, P. Temperature Dependence of the Dissociative Electron Attachment to 2-Thiothymine. *J. Phys. Chem. A* **2016**, *120*, 7130–7136. [CrossRef]
39. Kopyra, J.; Abdoul-Carime, H. Unusual temperature dependence of the dissociative electron attachment cross section of 2-thiouracil. *J. Chem. Phys.* **2016**, *144*, 034306. [CrossRef]
40. Kopyra, J.; Kopyra, K.K.; Abdoul-Carime, H.; Branowska, D. Insights into the dehydrogenation of 2-thiouracil induced by slow electrons: Comparison of 2-thiouracil and 1-methyl-2-thiouracil. *J. Chem. Phys.* **2018**, *148*, 234301. [CrossRef]
41. Besic, E. EPR study of the free radicals in the single crystals of 2-thithymine g-irradiated at 300 K. *J. Mol. Struct.* **2009**, *917*, 71–75. [CrossRef]
42. Besic, E.; Gomzi, V. EPR study of sulfur-centered p-radical in g-irradiated single crystal of 2-thiothymine. *J. Mol. Struct.* **2008**, *876*, 234–239. [CrossRef]
43. Besic, E.; Sankovic, K.; Gomzi, V.; Herak, J.N. Sigma radicals in gamma-irradiated single crystals of 2-thiothymine. *Phys. Chem. Chem. Phys.* **2001**, 2723–2725. [CrossRef]
44. Herak, J.N.; Sankovic, K.; Hole, E.O.; Sagstuen, E. ENDOR study of a thiocytosine oxidation product in cytosine monohydrate crystals doped with 2-thiocytosine, X-irradiated at 15 K. *Phys. Chem. Chem. Phys.* **2000**, *2*, 4971–4975. [CrossRef]
45. Herak, J.N.; Sankovic, K.; Krilov, D.; Hutterman, J. An EPR Study of the Transfer and Trapping of Holes Produced by Radiation in Guanine(Thioguanine) Hydrochloride Single Crystals. *Radiat. Res.* **1999**, *151*, 319–324. [CrossRef]

46. Herak, J.N.; Hutterman, J. Long-range hole migration in irradiated crystals of nucleic acid bases. An EPR study. *Int J. Radiat. Biol.* **1991**, *60*, 423–432. [CrossRef] [PubMed]
47. Sankovic, K.; Herak, J.N.; Krilov, D. EPR spectroscopy of the sulphur-centered radicals derived from thiocytosine. *J. Mol. Struct.* **1988**, *190*, 277–286. [CrossRef]
48. Jorgensen, J.-P.; Sagstuen, E. ESR of Irradiated 2-Thiouracil Single Crystals. A 3a-Hydrogen Radical. *Radiat. Res.* **1981**, *88*, 29–36. [CrossRef]
49. Claesson, O.; Lund, A.; Jorgensen, J.-P.; Sagstuen, E. Electron spin resonance of irradiated crystals of 2-thiouracil; hyperfine coupling tensor for nuclei with I = 1/2. *J. Magn. Reson.* **1980**, *41*, 229–239. [CrossRef]
50. Kumar, A.; Sevilla, M.D. SOMO-HOMO Level Inversion in Biologically Important Radicals. *J. Phys. Chem. B* **2018**, *122*, 98–105. [CrossRef]
51. Kumar, A.; Sevilla, M.D. p vs. s-Radical States of One-Electron-Oxidized DNA/RNA Bases: A Density Functional Theory Study. *J. Phys. Chem. B* **2013**, *117*, 11623–11632. [CrossRef]
52. Gomzi, V. DFT study of radicals formed in 2-thithymine single crystals at 77 K: 1- and 2-molecule models revised. *Comput. Theor. Chem.* **2011**, *963*, 497–502. [CrossRef]
53. Wenska, G.; Filipiak, P.; Asmus, K.-D.; Bobrowski, K.; Koput, J.; Marciniak, B. Formation of a Sandwich-Structure Assisted, Relatively Long-Lived Sulfur-Centered Three-electron Bonded Radical Anion in the Reduction of a Bis(1-substituted-uracilyl) Disulfide in Aqueous Solution. *J. Phys. Chem. B* **2008**, *112*, 100045–100053. [CrossRef] [PubMed]
54. Prasanthkumar, K.P.; Alvarez-Idaboy, J.R.; Kumar, P.V.; Singh, B.G.; Priyadarsini, K.I. Contrasting reactions of hydrated electron and formate radical with 2-thioanalogues of cytosine and uracil. *Phys. Chem. Chem. Phys.* **2016**, *18*, 28781–28790. [CrossRef] [PubMed]
55. Taras-Goslinska, K.; Vetica, F.; Barata-Vallejo, S.; Triantakostanti, V.; Marciniak, B.; Chatgilialoglu, C. Converging Fate of the Oxidation and Reduction of 8-Thioguanosine. *Molecules* **2019**, *24*, 3143. [CrossRef] [PubMed]
56. Makurat, S.; Spisz, P.; Kozak, W.; Rak, J.; Zdrowowicz, M. 5-iodo-4-thio-2′-deoxyuridine as a sensitizer of X-ray induced cancer cell killing. *Int. J. Mol. Sci.* **2019**, *20*, 1308. [CrossRef]
57. Spisz, P.; Zdrowowicz, M.; Makurat, S.; Kozak, W.; Skotnicki, K.; Bobrowski, K.; Rak, J. Why Does the Type of Halogen Atom Matter for the Radiosensitizing Properties of 5-Halogen Substituted 4-Thio-2′-Deoxyuridines? *Molecules* **2019**, *24*, 2819. [CrossRef]
58. Prasanthkumar, K.P.; Suresh, C.H.; Aravindakumar, C.T. Oxidation Reactions of 2-Thiouracil: A Theoretical and Pulse Radiolysis Study. *J. Phys. Chem. A* **2012**, *116*, 10712–10720. [CrossRef]
59. Prasanthkumar, K.P.; Suresh, C.H.; Aravindakumar, C.T. Dimer radical cation of 4-thiouracil: A pulse radiolysis and theoretical study. *J. Phys. Org. Chem.* **2013**, *26*, 510–516. [CrossRef]
60. Giuliano, B.M.; Feyer, V.; Prince, K.C.; Coreno, M.; Evangelisti, L.; Melandri, S.; Caminati, W. Tautomerism in 4-Hydroxypyrimidine, S-Methyl-2-thiouracil, and 2-Thiouracil. *J. Phys. Chem. A* **2010**, *114*, 12725–12730. [CrossRef]
61. Katrizky, A.R.; Szafran, M.; Stevens, J. The Tautomeric Equilibria of Thio Analogues of Nucleic Acid Bases. Part 2. AM1 and ab initio Calculations of 2-Thiouracil and its Methyl Derivatives. *J. Chem. Soc. Perkin Trans. II* **1989**, 1507–1511. [CrossRef]
62. Yekeler, H. Ab initio study on tautomerism of 2-thiouracil in the gas phase and in solution. *J. Comput. -Aided Mol. Des.* **2000**, *14*, 243–250. [CrossRef]
63. Babu, N.S. Theoretical Study of Stability, Tautomerism, Equilibrium Constants (pK_T) of 2-Thiouracil in Gas Phase and Different Solvents (Water and Acetonitrile) by the Density Functional Theory Method. *Amer. Chem. Sci. J.* **2013**, *3*, 137–150. [CrossRef]
64. Marino, T.; Russo, N.; Sicilia, E.; Toscano, M. Tautomeric Equilibria of 2- and 4-Thiouracil in Gas Phase and in Solvent: A Density Functional Study. *Int. J. Quantum Chem.* **2001**, *82*, 44–52. [CrossRef]
65. Psoda, A.; Shugar, D. Structure and Tautomerism of The Neutral and Monoanionic Forms of 2-Thiouracil, 2,4-Dithiouracil, Their Nucleosides, and Some Related Derivatives. *Acta Biochim. Pol.* **1979**, *26*, 55–72.
66. Ghomi, M.; Letellier, R.; Taillandier, E.; Chinsky, L.; Laigle, A.; Turpin, P. Interpretation of the vibrational modes of uracil and its ^{18}O-substituted and thio derivatives studied by resonance Raman spectroscopy. *J. Raman Spectrosc.* **1986**, *17*, 249–255. [CrossRef]

67. Igarashi-Yamamoto, N.; Tajiri, A.; Hatano, M.; Shibuya, S.; Ueda, T. Ultraviolet absorption, circular dichroism studies of sulfur-containing nucleic acid bases and their nucleosides. *Biochim. Biophys. Acta* **1981**, *656*, 1–15. [CrossRef]
68. Christensen, H.N. Ultrafiltrability of Thiouracil in Human Serum; Determination of thiouracil. *J. Biol. Chem.* **1945**, *160*, 425–433.
69. Wang, W.; Schuchmann, M.N.; Schuchmann, H.-P.; Knolle, W.; Von Sonntag, J.; Von Sonntag, C. Radical Cations in the OH-Radical-Induced Oxidation of Thiourea and Tetramethylthiourea in Aqueous Solution. *J. Am. Chem. Soc.* **1999**, *121*, 238–245. [CrossRef]
70. Schuler, R.H.; Hartzell, A.L.; Behar, B. Track effects in radiation chemistry. Concentration dependence for the scavenging of OH by ferrocyanide in N_2O-saturated aqueous solutions. *J. Phys. Chem.* **1981**, *85*, 192–199. [CrossRef]
71. Schöneich, C.; Pogocki, D.; Hug, G.L.; Bobrowski, K. Free radical reactions of methionine in peptides: Mechanisms relevant to b-amyloid oxidation and Alzheimer's disease. *J. Am. Chem. Soc.* **2003**, *125*, 13700–13713. [CrossRef]
72. Bobrowski, K.; Hug, G.L.; Pogocki, D.; Marciniak, B.; Schoneich, C. Stabilization of sulfide radical cations through complexation with the peptide bond: Mechanisms relevant to oxidation of proteins containing multiple methionine residues. *J. Phys. Chem. B* **2007**, *111*, 9608–9620. [CrossRef]
73. Asmus, K.-D. Some conductivity experiments in pulse radiolysis. *Int. J. Radiat. Phys. Chem.* **1972**, *4*, 417–437. [CrossRef]
74. Eyring, E.M. Fast Reactions in Solutions. *Surv. Prog. Chem.* **1964**, *2*, 57–89.
75. Goyal, R.N.; Singh, U.P.; Abdullah, A.A. Electrochemical oxidation of 2-thiouracil at pyrolitic graphite electrode. *Bioelectrochemistry* **2005**, *67*, 7–13. [CrossRef] [PubMed]
76. Wardman, P. The reduction potentials of one-electron couples involving free radicals in aqueous solution. *J. Phys. Chem. Ref. Data* **1989**, *18*, 1637–1753. [CrossRef]
77. Skotnicki, K.; De la Fuente, J.R.; Cañete, A.; Berrios, E.; Bobrowski, K. Radical Ions of 3-Styryl-quinoxalin-2-one Derivatives Studied by Pulse Radiolysis in Organic Solvents. *J. Phys. Chem. B* **2018**, *122*, 4051–4066. [CrossRef] [PubMed]
78. Janik, I.; Carmichael, I.; Tripathi, G.N.R. Transient Raman spectra, structure, and thermochemistry of the thiocyanate dimer radical anion in water. *J. Chem. Phys.* **2017**, *146*, 214305. [CrossRef]
79. Janik, I.; Tripathi, G.N.R. The selenocyanate dimer radical anion in water: Transient Raman spectra, structure, and reaction dynamics. *J. Chem. Phys.* **2019**, *150*, 094304. [CrossRef]
80. Janik, I.; Tripathi, G.N.R. The early events in the OH radical oxidation of dimethyl sulfide in water. *J. Chem. Phys.* **2013**, *138*, 044506. [CrossRef]
81. Schöneich, C.; Bobrowski, K. Intramolecular hydrogen transfer as the key step in the dissociation of hydroxyl radical adducts of (alkylthio)ethanol derivatives. *J. Am. Chem. Soc.* **1993**, *115*, 6538–6547. [CrossRef]
82. Wang, F.; Pernot, P.; Marignier, J.-L.; Archirel, P.; Mostafavi, M. Mechanism of (SCN)2·− Formation and Decay in Neutral and Basic KSCN Solution under Irradiation from a Pico- to Microsecond Range. *J. Phys. Chem. B* **2019**, *123*, 6599–6608. [CrossRef]
83. Bobrowski, K.; Schöneich, C. Hydroxyl radical adduct at sulfur in substituted organic sulfides stabilized by internal hydrogen bond. *J. Chem. Soc., Chem. Commun.* **1993**, 795–797. [CrossRef]
84. Chipman, D.M. Hemibonding between Hydroxyl Radical and Water. *J. Phys. Chem. A* **2011**, *115*, 1161–1171. [CrossRef] [PubMed]
85. Treinin, A.; Hayon, E. Charge transfer spectra of halogen atoms in water. Correlation of the electronic transition energies of iodine, bromine, chlorine, hydroxyl, and hydrogen radicals with their electron affinities. *J. Am. Chem. Soc.* **1975**, *97*, 1716–1721. [CrossRef]
86. Prasanthkumar, K.P.; Suresh, C.H.; Aravindakumar, C.T. Theoretical study of the addition and abstraction reactions of hydroxyl radical with uracil. *Radiat. Phys. Chem.* **2012**, *81*, 267–272. [CrossRef]
87. Francés-Monerris, A.; Merchán, M.; Roca-Sanjuán, D. Theoretical Study of the Hydroxyl Radical Addition to Uracil and Photochemistry of the Formed U6OH• Adduct. *J. Phys. Chem. B* **2014**, *118*, 2932–2939. [CrossRef]

88. Fujita, S.; Steenken, S. Pattern of OH radical addition to uracil and methyl-, and carboxyl-substituted uracils. Electron transfer of OH adduct with N,N,N',N'-tetramethyl-p-phenylene diamine and tetranitromethane. *J. Am. Chem. Soc.* **1981**, *103*, 2540–2545. [CrossRef]
89. Domin, D.; Braida, B.; Berges, J. Influence of Water on the Oxidation of Dimethyl Sulfide by the •OH Radical. *J. Phys. Chem. B* **2017**, *121*, 9321–9330. [CrossRef]
90. Fourré, I.; Bergés, J. Structural and topological characterization of three-electron bond: The S\O radicals. *J. Phys. Chem. A* **2004**, *108*, 898–906. [CrossRef]
91. Bobrowski, K.; Pogocki, D.; Schöneich, C. Mechanism of the OH radical-induced decarboxylation of 2-(alkylthio)ethanoic acid derivatives. *J. Phys. Chem.* **1993**, *97*, 13677–13684. [CrossRef]
92. Schöneich, C.; Pogocki, D.; Wisniowski, P.; Hug, G.L.; Bobrowski, K. Intramolecular Sulfur-Oxygen Bond Formation in Radical Cations of N-Acetylmethionine Amide. *J. Am. Chem. Soc.* **2000**, *122*, 10224–10225. [CrossRef]
93. Bobrowski, K. Free radicals in chemistry, biology and medicine: Contribution of radiation chemistry. *Nukleonika* **2005**, *50* (Supplement 3), S67–S76.
94. Mirkowski, J.; Wisniowski, P.; Bobrowski, K. *INCT Annual Report 2000*; INCT: Warsaw, Poland, 2001; pp. 31–33.
95. Hug, G.L.; Wang, Y.; Schöneich, C.; Jiang, P.-Y.; Fessenden, R.W. Multiple time scales in pulse radiolysis. Application to bromide solutions and dipeptides. *Radiat. Phys. Chem.* **1999**, *54*, 559–566. [CrossRef]
96. Schuler, R.H.; Patterson, L.K.; Janata, E. Yield for the scavenging of hydroxyl radicals in the radiolysis of nitrous oxide-saturated aqueous solutions. *J. Phys. Chem.* **1980**, *84*, 2088–2090. [CrossRef]
97. Janata, E. Pulse radiolysis conductivity measurements in aqueous solutions with nanosecond time resolution. *Radiat. Phys. Chem.* **1982**, *19*, 17–21. [CrossRef]
98. Veltwisch, D.; Janata, E.; Amus, K.-D. Primary processes in the reaction of OH-radicals with sulphoxides. *J.C.S. Perkin Trans. II* **1980**, 146–153. [CrossRef]
99. Frisch, M.J.; Trucks, G.W.; Schlegel, H.B.; Scuseria, G.E.; Robb, M.A.; Cheeseman, J.R.; Scalmani, G.; Barone, V.; Mennucci, B.; Petersson, G.A.; et al. *Gaussian 09, Revision E.01*; Gaussian Inc.: Wallingford, CT, USA, 2009.
100. Vydrov, A.; Scuseria, G.E. Assessment of a long-range corrected hybrid functional. *J. Chem. Phys.* **2006**, *125*, 234109. [CrossRef]
101. Dunning, T.H. Gaussian basis sets for use in correlated molecular calculations. I. The atoms boron through neon and hydrogen. *J. Chem. Phys.* **1989**, *90*, 1007. [CrossRef]
102. Woon, D.E.; Dunning, T.H. Gaussian basis sets for use in correlated molecular calculations. III. The atoms aluminum through argon. *J. Chem. Phys.* **1993**, *98*, 1358–1371. [CrossRef]
103. Tomasi, J.; Mennuci, B.; Cammi, R. Quantum Mechanical Continuum Solvation Models. *Chem. Rev.* **2005**, *105*, 2999–3094. [CrossRef]
104. Laurent, A.D.; Jacquemin, D. TD-DFT benchmarks: A review. *Int. J. Quantum Chem.* **2013**, *113*, 2019–2039. [CrossRef]
105. Dupont, C.; Dumont, É.; Jacquemin, D. Superior Performance of Range-Separated Hybrid Functionals for Describing σ* ← σ UV–Vis Signatures of Three-Electron Two-Center Anions. *J. Phys. Chem. A* **2012**, *116*, 3237–3246. [CrossRef] [PubMed]
106. Milhøj, B.O.; Sauer, S.P.A. Kinetics and Thermodynamics of the Reaction between the •OH Radical and Adenine: A Theoretical Investigation. *J. Phys. Chem. A* **2015**, *119*, 6516–6527. [CrossRef] [PubMed]
107. Francés-Monerris, A.; Merchán, M.; Roca-Sanjuán, D. Mechanism of the OH Radical Addition to Adenine from Quantum-Chemistry Determinations of Reaction Paths and Spectroscopic Tracking of the Intermediates. *J. Org. Chem.* **2017**, *82*, 276–288. [CrossRef] [PubMed]
108. Frances-Monerris, A.; Merchan, M.; Roca-Sanjuan, D. Communication: Electronic UV-Vis transient spectra of the center dot OH reaction products of uracil, thymine, cytosine, and 5,6-dihydrouracil by using the complete active space self-consistent field second-order perturbation (CASPT2//CASSCF) theory. *J. Chem. Phys.* **2013**, *139*. [CrossRef]
109. Ipatov, A.; Cordova, F.; Doriol, L.C.; Casid, M.E. Excited-state spin-contamination in time-dependent density-functional theory for molecules with open-shell ground states. *J. Mol. Struc.-Theochem.* **2009**, 60–73. [CrossRef]

110. Creating UV/Visible Plots from the Results of Excited States Calculations. Available online: http://gaussian.com/uvvisplot/ (accessed on 29 November 2019).

Sample Availability: Samples of the 2-thiouracil are available from the authors.

© 2019 by the authors. Licensee MDPI, Basel, Switzerland. This article is an open access article distributed under the terms and conditions of the Creative Commons Attribution (CC BY) license (http://creativecommons.org/licenses/by/4.0/).

Article

Anomeric Spironucleosides of β-D-Glucopyranosyl Uracil as Potential Inhibitors of Glycogen Phosphorylase

Aggeliki Stathi [1], Michael Mamais [1,2], Evangelia D. Chrysina [2] and Thanasis Gimisis [1,*]

[1] Organic Chemistry Laboratory, Department of Chemistry, National and Kapodistrian University of Athens, 10571 Athens, Greece; iamaggeliki@yahoo.gr (A.S.); mmamais@chem.uoa.gr (M.M.)
[2] Institute of Chemical Biology, National Hellenic Research Foundation, 11635 Athens, Greece; echrysina@eie.gr
* Correspondence: gimisis@chem.uoa.gr; Tel.: +30-210-727-4928

Received: 13 May 2019; Accepted: 17 June 2019; Published: 25 June 2019

Abstract: In the case of type 2 diabetes, inhibitors of glycogen phosphorylase (GP) may prevent unwanted glycogenolysis under high glucose conditions and thus aim at the reduction of excessive glucose production by the liver. Anomeric spironucleosides, such as hydantocidin, present a rich synthetic chemistry and important biological function (e.g., inhibition of GP). For this study, the Suárez radical methodology was successfully applied to synthesize the first example of a 1,6-dioxa-4-azaspiro[4.5]decane system, not previously constructed via a radical pathway, starting from 6-hydroxymethyl-β-D-glucopyranosyluracil. It was shown that, in the rigid pyranosyl conformation, the required [1,5]-radical translocation was a minor process. The stereochemistry of the spirocycles obtained was unequivocally determined based on the chemical shifts of key sugar protons in the ^{1}H-NMR spectra. The two spirocycles were found to be modest inhibitors of RMGPb.

Keywords: type 2 diabetes; glycogen phosphorylase; anomeric spironucleosides; 1,6-dioxa-4-azaspiro[4.5]decane; [1,5]-radical translocation

1. Introduction

Despite the prevalence of type 2 diabetes worldwide, no sufficient treatment has been identified; therefore, a molecular approach based on the three-dimensional structure of enzymes directly involved in glycogen metabolism has received increasing attention. Glycogen phosphorylase (GP) is an allosteric enzyme with a regulatory role in glycogen breakdown to glucose [1]. Since glucose is the physiological substrate of GP, it promotes the inactive form of the enzyme acting synergistically with insulin towards reducing the rate of glycogen degradation and shifting the equilibrium towards glycogen synthesis. GP three-dimensional structure in the T state (GPb) has been exploited as a target for the design of glucose-based compounds that inhibit enzymic action preventing glycogenolyis and acting as regulators of glucose levels in the bloodstream [2,3]. The glucose specificity for the catalytic site of the enzyme has been utilized to drive glucose derivatives to the active site of the enzyme, exploiting the catalytic channel by adding a variety of structural features to these compounds in terms of rigidity and functional groups [4]. One of the early lead inhibitors of rabbit muscle glycogen phosphorylase b (RMGPb) was pyranosyl spironucleosides [5] (**2a,b**, Figure 1), the structure of which was inspired from hydantocidin (**1**), a natural spiro compound with herbicidal and plant growth regulatory activity. To this end, a number of attempts to synthesize more potent inhibitors were made leading to other spiro-heterocycles that exhibited stronger affinity for RMGPb [6]. Similar studies in our laboratory led to N^4-aryl-N^1-(β-D-glucopyranosyl)cytidine nucleosides which exhibit RMGPb inhibition in the nanomolar range [7].

We present here a methodology, that involves a key [1,5]-radical translocation step for the synthesis of anomeric spironucleosides **4a,b** (Figure 1), which were found to be modest inhibitors of rabbit muscle glycogen phosphorylase (RMGP*b*). Spirocyclic nucleosides present a rich synthetic chemistry and important biological function [8,9]. The targets in this paper contain a rare 1,6-dioxa-4-azaspiro[4.5]decane structure [10], which has not been previously constructed via a radical pathway [11]. In the similar 1,6-dioxa-4-azaspiro[4.4]nonane system (**3**, Figure 1), we reported an efficient protocol, using a 6-lithiation strategy for generating a 6-hydroxymethyluridine intermediate followed by oxidative cyclization through a [1,5]-radical translocation strategy [12,13]. We were interested in applying this protocol to the related "decane" system.

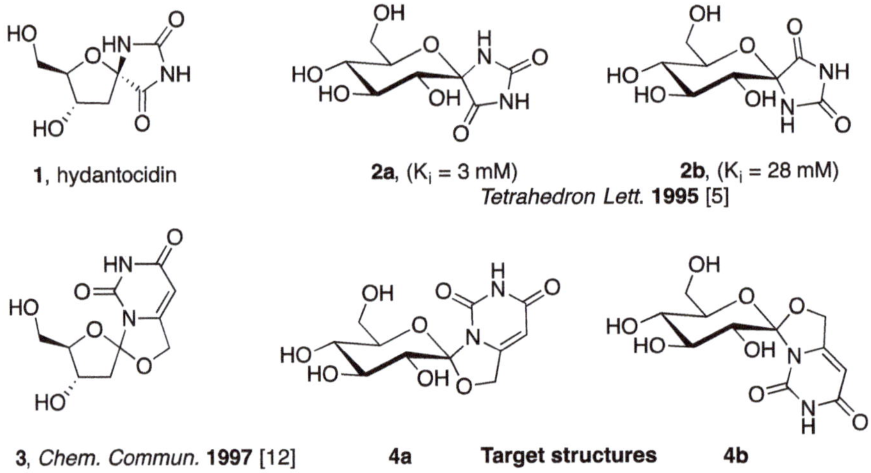

Figure 1. Natural and synthetic anomeric spironucleosides.

2. Results

2.1. Synthesis

Initially, we attempted to access compound **7** (Scheme 1) by direct lithiation of the known 2,3,4,6-tetra-*O*-benzyl-β-D-glucopyranosyluracil [14]. The protocol was based on the previous well-established 6-lithiation of protected 2-deoxy- and ribouridines, followed by the reaction with dimethylformamide or ethyl formate to generate the corresponding 6-formyluridines [12,15,16]. Although a major product formed under these conditions, spectral analysis revealed that it was the product of a selective mono-debenzylation and did not contain a formyl group. Although we could not unequivocally determine the position of debenzylation, we hypothesized that this had occurred in position-3, proximal to the 6-position where the initial lithiation is expected to occur (data not shown). This reductive debenzylation could be reminiscent of a previous method employing lithium naphthalenide [17].

The above result prompted us to change our strategy and include a previously reported [18] N^1-(β-D-glucopyranosyl)-6-methyluracil (**6a**) intermediate in our synthesis, as exemplified in Scheme 1. Under optimized conditions, the *N*-glycosylation reaction of persilylated 6-methyluracil, in the presence of an excess of TMSOTf in DCE, led to the formation of three products, namely, **6a–c**, isolated after column chromatography in 67%, 30%, and 3% yield, respectively. As determined by ESI-MS and NMR, the major product was the expected N^1-glycosylated 6-methyluridine **6a**, whereas the N^3-glycosylated analogue **6b** was isolated in 30% yield, together with a small amount of N^1,N^3-bisglycosylated isomer **6c**.

The main feature that differentiated **6b** from **6a** in the ^{1}H-NMR spectrum was a substantial downfield shift of H-2' ($\delta\Delta$ = 0.8 ppm) in **6b**, induced by the magnetic anisotropy effect of the second vicinal amidic 4-carbonyl. A similar effect was observed in the more complex spectra of the bis-substituted analogue **6c**, and the [M + H]$^+$ peak at 786.3 amu (ESI-MS) clearly differentiated **6c** from the other two isomers ([M + H]$^+$ at 457.2 amu).

The allylic methyl group of **6a** was oxidized to the corresponding aldehyde **7** in the presence of selenium dioxide (3 equivs) in dioxane: acetic acid [19], in 67% yield. Apart from the aldehyde **7**, isolated in 67% yield and recognized in ^{1}H-NMR by its characteristic aldehydic proton at 9.90 ppm, a second more polar product was isolated in low yield, identified as the allylic alcohol **8**. Its formation is expected by the mechanism of the reaction which follows an electrophilic allylic addition to selenium followed by a [2,3]-sigmatropic rearrangement [20,21]. Application of stoichiometric amounts of SeO$_2$ also led to aldehyde **7** as the major product, although full conversion was not observed even after prolonged reaction times. When the reaction was performed with 3 equivs of SeO$_2$, under strictly anhydrous conditions, the yield of aldehyde **7** was maximized. Reduction of aldehyde **7** with NaBH$_4$ in CHCl$_3$/isopropanol, in the presence of silica gel, at 0 °C [22] led to partial removal of acetate groups. By lowering the reaction temperature to −30 °C, exclusive formation of the allylic alcohol **8** was observed, and the product was isolated in 90% yield.

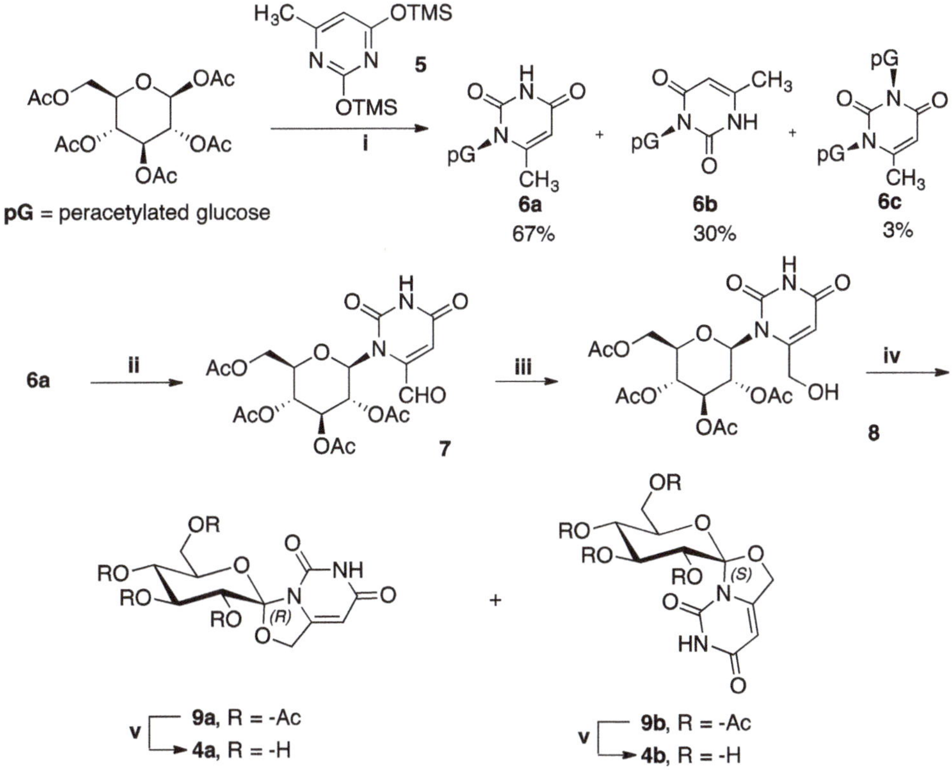

Scheme 1. (i) TMSOTf, DCE, reflux, 1 h, 67%, (ii) SeO$_2$, dioxane, AcOH, reflux, 5 h, 67%, (iii) NaBH$_4$, silica gel, CHCl$_3$, propanol, −30 °C, 1 h, 90%, (iv) DIB, I$_2$, CH$_2$Cl$_2$, hv, r.t., 2.5 h, 18% (9a:9b = 1.25:1), and 50% (**7**), (v) NH$_3$ (7N in MeOH), r.t., 16 h, 100%.

The key step photolysis of **8**, under the standard optimized Suárez conditions [23], utilized for alkoxy radical generation in hydrogen atom transfer (HAT) reactions [11], in the presence of DIB and I$_2$, in DCM and under visible light (150 W) irradiation, led to the isolation of three products in 50%, 10%, and 8% yield. The major product was, surprisingly, the aldehyde **7**, whereas the two minor products corresponded to the expected isomeric spironucleosides **9a,b**. The reaction is expected to proceed through the generation of an alkoxyl radical intermediate, followed by a [1,5]-radical translocation [24] to generate a C-1' radical intermediate which, after oxidation and ionic cyclization, provides the spirocycles **9a,b** [12,13]. The main formation of aldehyde **7** can be explained by a possible disproportionation reaction of the above alkoxy radical intermediate to aldehyde **7** and alcohol **8**, with the latter re-entering the reaction cycle (see discussion below). The application of the same conditions as in the similar 1,6-dioxa-4-azaspiro[4.4]nonane system [12,13] aids in the comparison of the two systems and indicates that the rigidity of the β-D-glucopyranosyl ring renders the [1,5]-hydrogen atom transfer less favorable in this system than in the previously observed flexible ribosyl system. It should be noted that when the same reaction was attempted with a similar photolysis in the presence of Pb(OAc)$_4$, I$_2$, and CaCO$_3$ [25], a complex mixture of products was obtained that could not be further characterized. Final removal of the acetate protection of **9a,b** with ammonia in methanol led to the isolation of the target spirocycles **4a,b** in quantitative yield.

2.2. Kinetic Experiments

RMGP*b* was isolated, purified, and recrystallized according to previously established protocols [26]. Compounds **4a,b** were assayed in the direction of glycogen synthesis for their inhibitory effect on RMGP*b* as described before [26,27]. They both exhibited competitive inhibition with respect to the substrate glucose 1-phosphate (Glc-1-P), at constant concentrations of glycogen (0.2% *w/v*) and AMP (1 mM). Compound **4b** was found to be a stronger inhibitor of RMGP*b* (35% inhibition at 1 mM) than **4a** (26% inhibition at 1 mM).

3. Discussion

The stereochemistry of the two spirocyclic products could be inferred unequivocally, from the ^{1}H-NMR spectra, as can be seen in Figure 2. Both spectra contain features that can be explained by the magnetic anisotropy induced by the 2-C=O group onto the sugar α- or β- hydrogens depending on the stereochemistry of the new spiro-center. As the new spirocycle locks the configuration of the pyrimidine ring with respect to the sugar ring, the 2-C=O is spaced in the vicinity of H-2' in the *R*-anomer **9a** and on the other hand, in the vicinity of H-3' and H-5', in the *S*-anomer **9b**. This results in significant downfield shifts of the corresponding sugar Hs in the ^{1}H-NMR spectra. Specifically, there is a 0.55 ppm shift of H-2' going from the *S*- to *R*-anomer (5.62 to 6.17 ppm), whereas there is a 0.48 and 0.84 ppm shift for H-5' (4.18 to 4.65 ppm) and H-3' (5.47 to 6.31 ppm), respectively, going from the *R*- to the *S*-anomer (Figure 2). The remaining Hs (H-4', H-6', H-5, and H-7) have similar chemical shifts in the two spectra, with the difference that the pairs of H-6' and H-7 protons appear as AB quartets in the case of the more congested *S*-anomer **9b**, whereas they collapse to singlets in the case of the *R*-anomer **9a**.

The same trends reported above for the protected derivatives were also observed in the case of the final compounds **4a,b**. Specifically, there was a 0.61 ppm shift of H-2' going from the *S*- (**4b**) to *R*- (**4a**) anomer (3.95 to 4.56 ppm), whereas there is a 0.37 and 0.79 ppm shift for H-5' (3.90 to 4.27 ppm) and H-3' (3.83 to 4.62 ppm), respectively, going from the *R*- (**4a**) to the *S*- (**4b**) anomer. The remaining Hs (H-4', H-6', H-5, and H-7) had similar chemical shifts in the two spectra, with the difference, again in this case, being that the pairs of H-6' and H-7 protons appear as AB quartets in the case of the more congested *S*-anomer **4b**, whereas they collapse to singlets in the case of the *R*-anomer **4a**. It should be noted that 2D NOESY spectroscopy did not provide any additional information for the above systems, as the only correlations that were observed were those between either protons H-2' and H-4' or H-3' and H-5', above and below the plane of the glucopyranosyl ring, respectively.

The above analysis allowed us to better interpret the ^{1}H-NMR spectrum of compound **8** (in DMSO-d6) which appeared as a mixture of rotamers, indicating a slow, on the NMR time scale, rotation around the glycosidic $C^{1'}$-N^1 bond, due to the new 6-hydroxymethyl substituent. The existence of rotamers was indicated by the presence of two amidic hydrogens at 11.48 and 11.32 ppm in a ≈1:2 ratio. Two more characteristic low field signals were a doublet at 6.43 ppm and a triplet at 6.08 ppm exhibiting the same ≈1:2 ratio. By applying the analysis above, performed for the final spirocyclic products, one can assign the doublet to the H-2' of the rotamer with the 6-CH$_2$OH group in α-position and the triplet to the H-3' of the conformer with the 6-CH$_2$OH group in β-position. When the ^{1}H-NMR spectrum of **8** was obtained at a higher temperature (75 °C), the above signals collapsed to broad singlets, confirming the above hypothesis. The observed ratio of the two rotamers at equilibrium is significant in the next key step, as explained below.

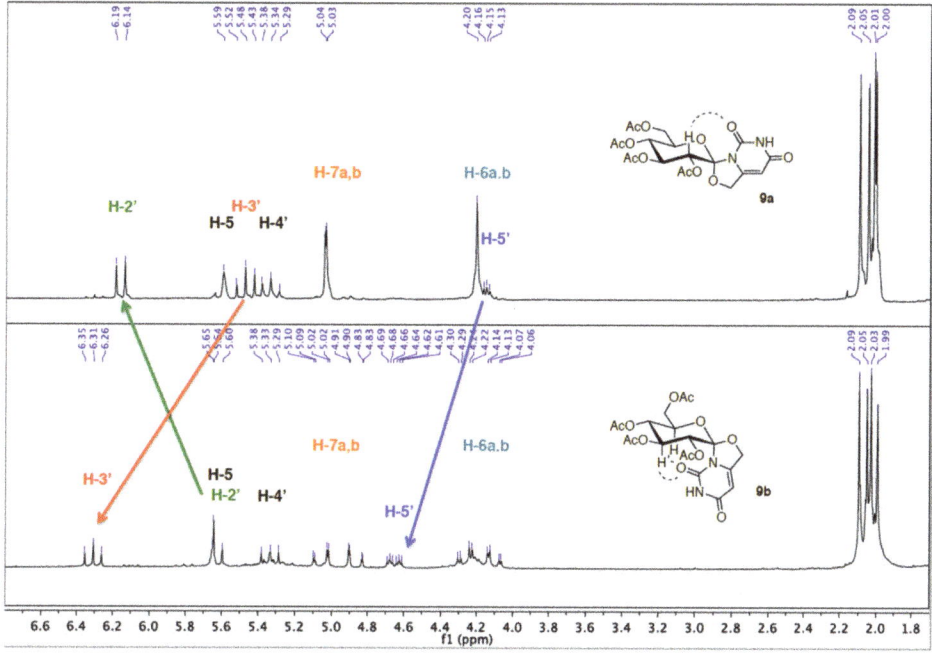

Figure 2. Comparison of ^{1}H-NMR spectral shifts in compounds **9a,b**.

Regarding the mechanism of the key step, it is expected that, under the Suárez conditions, an alkoxy radical intermediate is produced that may exist in two possible conformations (**10-syn**, **10-anti**, Scheme 2). These two conformers are similar to those observed above for alcohol **8** and are expected to be formed in a similar ≈1:2 ratio (**10-syn:10-anti**). The rigidity of the pyranosyl chair conformation may not allow a fast interconversion between the conformers for steric reasons, and the product distribution may be affected directly by these two conformer populations and their corresponding reactivity. Specifically, **10-anti** cannot undergo a [1,5]-hydrogen shift and an alternative [1,6]-hydrogen atom transfer from the 2-position of the sugar is known to be disfavored in the presence of acetyl protection [23]. The only available pathway for conformer **10-anti** is a radical disproportionation reaction leading to **7** and **8**, of which the latter re-enters the reaction cycle, while the former accumulates (Scheme 2).

On the other hand, conformer **10-syn** possesses a suitable conformation for a [1,5]-radical translocation leading to **11-syn** intermediate. After oxidation of **11-syn**, in the presence of I$_2$, the **12-syn** oxonium ion may exist in equilibrium with conformer **12-anti**, through rotation of the $C^{1'}$-N^1 bond, and also with a possible Vorbrüggen-type intermediate, formed through the anchimeric assistance of the 2'-acetyl group. The Vorbrüggen intermediate is the main species that determines the stereochemistry of the final product in *N*-glycosylation reactions [28], and if the same was true in our system, exclusive formation of the *S*-anomer **9b** would be expected. Nevertheless, in our system, a 1:1.25 *S:R* mixture of anomers **9b:9a** was obtained. This result allows us to draw two major conclusions regarding the mechanism. First, rotation around the $C^{1'}$-N^1 bond in the oxonium ion **12-syn** to produce **12-anti** has to be faster than cyclization in order to allow the formation of the second, prior to cyclization, and there must be no major thermodynamic difference between the two conformers. Second, the formation of the Vorbrüggen intermediate must not be favored in this system, for steric reasons, and even if it is formed, through conformer **12-anti**, the process rate is comparable to that of the cyclization of **12-syn** conformer to the observed *S*-anomer **9a** (Scheme 2).

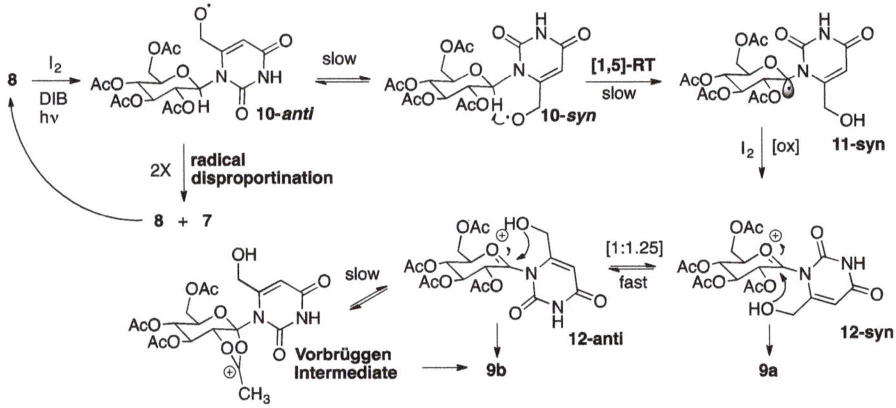

Scheme 2. Proposed mechanism for the formation of **9a,b**.

Kinetics determined that the *S* anomeric spirocycle **4b** exhibited 1.25 times higher inhibition than the *R* anomer **4a**. The difference could be associated with the locked *syn* conformation of the pyrimidine ring with regard to the β-D-glucose moiety and possible unfavorable interactions of **4a** within the catalytic site between the backbone CO of His377 with the uracil 2-C=O, as has been observed previously with protein crystallography (unpublished results). Although we attempted to obtain X-ray crystallographic data by soaking crystals of RMGP*b* with either **4a** or **4b**, the rather low affinity of both spirocycles did not provide sufficient data for establishing their binding in the catalytic site and studying their interactions. Both spironucleosides are stronger binders than the natural inhibitors of GP, β- and α-D-glucose [29]. For example, **4b** is about 7 and 1.5 times stronger than β- and α-D-glucose, respectively. The new compounds, nevertheless, exhibit a rather low inhibition profile compared with the known spirohydantoin derivative of glucopyranose [5] and other known strong catalytic site inhibitors of RMGP*b* [7,30]. We established previously that *anti* is the desirable conformation of the pyrimidine ring at the anomeric position of β-D-glucose leading to strong inhibition [7], and the current results confirm this finding. Anomeric spironucleosides are rigid structures and, given that they possess the correct conformation, are expected to bind strongly to the catalytic site of GP. Our current studies are therefore directed towards anomeric spironucleosides with locked *anti* conformations, and these results will be reported in due course.

4. Materials and Methods

All reagents and solvents were purchased from commercial sources (Sigma–Aldrich, Merck, NJ, USA; Alfa-Aesar, Fisher Scientific, MA, USA) and used without further purification, unless otherwise stated. All reactions were carried out under an argon atmosphere on a magnetic stirrer (IKA®-Werke GmbH & Co. KG, Staufen, Germany) and monitored by thin-layer chromatography. Compounds were purified by flash chromatography on silica gel 40–60 μm, 60 Å. NMR measurements were performed with a Varian Mercury 200 Nuclear Magnetic Resonance Spectrometer (Varian Inc., Agilent Technologies, Palo Alto, CA, USA) at 200 MHz for ^{1}H and at 50 MHz for ^{13}C, respectively. The deuterated solvents used for NMR spectroscopy were CDCl$_3$ and D$_2$O. Chemical shifts are given in ppm and were referenced on residual solvent peaks for CDCl$_3$ (δ 7.26 ppm for ^{1}H-NMR and 77.16 ppm for ^{13}C-NMR), whereas for D$_2$O an external reference of 3-(trimethylsilyl)-1-propanesulfonic acid sodium salt was used. Coupling constants were measured in Hz. Hydrogen atom assignments, when given, are based on COSY spectra. Melting points were obtained by using a Gallenkamp Sanyo apparatus (Fisher Scientific, MA, USA) and are uncorrected. Mass spectrometry experiments were carried out in a Thermo Finnigan Surveyor MSQ plus Mass Spectrometer (ThermoFisher Scientific, MA, USA), using the Electron Spray Ionization technique (ESI-MS). High-resolution mass spectrometry experiments were carried out in a Q-TOF Bruker MaXis Impact HR-Mass Spectrometer (Bruker, MA, USA). 1,2,3,4,6-Penta-O-acetyl-β-D-glucopyranose was synthesized using standard synthetic protocols [31]. AMP, Glc-1-P (dipotassium salt), and oyster glycogen were obtained from Sigma-Aldrich (Merck, NJ, USA) and used without further purification. Oyster glycogen was freed of AMP according to Helmreich and Cori [32]. ^{1}H and ^{13}C-NMR spectra of new compounds are available in the Supplementary Materials.

2,4-Di-(trimethylsilyloxy)-(6-methylpyrimidine) (**5**). A suspension of 6-methyluracil (1 g, 7.93 mmol) and well-grinded ammonium sulphate (80 mg, 0.60 mmol 0.076 eq) in HMDS (8.4 mL, 39.7 mmol, 5 eq) was heated to 120 °C under anhydrous conditions until full dissolution occurred. Upon completion, the excess of HMDS was removed through distillation, toluene was added twice (5 mL) followed by distillation to remove all traces of excess HMDS to yield 2.1 g (7.8 mmol, 97%) of the title compound which was characterized without further purification. ^{1}H-NMR (200 MHz, CDCl$_3$): δ = 0.34 (s, 18H), 1.95 (3H, s), 5.81 ppm (1H, s). ^{13}C-NMR (50 MHz, CDCl$_3$): δ = 0.00 (3C), 0.03 (3C), 23.4, 102.6, 162.6, 169.8, 170.0 ppm.

N-Glycosylation of 6-methyluridine. To a solution of **5** (1.5 g, 5.6 mmol, 1.5 eq) in dry 1,2-dichloroethane (7 mL) at r.t., a solution of TMSOTf (1.61 mL, 8.33 mmol, 2.25 eq) and 2,3,4,6-tetra-O-acetyl-β-D-glucopyranose[31] (1.44 g, 3.7 mmol) in dry DCE (3.5 mL) was added. The reaction mixture was heated at reflux until full consumption of the sugar (≈1 h). The mixture was then cooled, diluted with DCM, and washed successively twice with saturated aq. NaHCO$_3$ solution, water, and brine. The organic layer was dried over anhydrous Na$_2$SO$_4$, filtered, the solvents evaporated, and the crude product was purified by column chromatography (30–70% Et$_2$O in EtOAc) to give, in order of elution, **6c** as a white solid (86 mg, 0.11 mmol, 3%), **6a** as a white solid (1.13 g, 2.48 mmol, 67%) and **6b** as a white solid (0.51 g, 1.11 mmol, 30%).

1-(Tetra-O-acetyl-β-D-glucopyranosyl)-6-methyluracil (**6a**): R$_f$ = 0.40 (70:30 Et$_2$O:EtOAc). ^{1}H-NMR (200 MHz, CDCl$_3$): δ = 2.00 (s, 3H), 2.02 (s, 3H), 2.06 (s, 3H), 2.08 (s, 3H), 2.55 (s, 3H), 3.90 (bd, *J* = 9.5 Hz, 1H), 4.16 (dd, 1H, *J* = 12.2, 1.7 Hz), 4.27 (dd, 1H, *J* = 12.6, 4.2 Hz), 5.14 (t, 1H, *J* = 9.5 Hz), 5.35 (t, 1H, *J* = 9.5 Hz), 5.46 (t, 1H, *J* = 9.4 Hz), 5.57 (bs, 2H), 6.27 (d, 1H, *J* = 9.1 Hz), 8.62 ppm (bs, 1H). ^{13}C-NMR (50 MHz, CDCl$_3$): δ = 20.60, 20.78, 20.80 (2C), 20.97, 61.8, 67.9, 69.5, 72.8, 75.1, 80.5, 104.1, 139.3, 150.5, 162.7, 169.7, 169.8, 170.0, 170.7 ppm. HRMS (ESI): calcd. for C$_{19}$H$_{25}$N$_2$O$_{11}$$^+$ [M + H]$^+$ 457.1458 found 457.1465.

3-(Tetra-O-acetyl-β-D-glucopyranosyl)-6-methyluracil (**6b**): R$_f$ = 0.35 (70:30 Et$_2$O:EtOAc). ^{1}H-NMR (200 MHz, CDCl$_3$): δ = 1.95 (s, 3H), 2.01 (s, 3H), 2.04 (s, 3H), 2.06 (s, 3H), 2.17 (s, 3H), 3.84 (ddd, *J* = 10.1, 5.3, 2.5 Hz, 1H), 4.30–4.10 (m, 2H), 5.15 (t, *J* = 9.7 Hz, 1H), 5.30 (dd, *J* = 9.4, 8.6 Hz, 1H), 5.52 (s, 1H), 6.08 (d, *J* = 9.4 Hz, 1H), 6.18 (dd, *J* = 9.4, 8.4 Hz, 1H), 9.70 ppm (s, 1H). ^{13}C-NMR (50 MHz, CDCl$_3$): δ = 18.6, 20.4, 20.5 (2C), 20.6, 62.0, 67.9, 68.0, 73.7, 74.5, 78.3, 98.9, 151.5, 152.1, 162.5, 169.4, 169.6, 169.9, 170.5 ppm. ESI-MS: 457.2 [M + H]$^+$.

1,3-Bis-(tetra-O-acetyl-β-D-glucopyranosyl)-6-methyluracil (**6c**): R$_f$ = 0.50 (70:30 Et$_2$O:EtOAc). ^{1}H-NMR (200 MHz, CDCl$_3$): δ = 2.01 (s, 6H), 2.02 (s, 6H), 2.04 (s, 6H), 2.06 (s, 6H), 2.57 (s, 1H), 3.96–3.74 (m, 2H), 4.37–4.10 (m, 4H), 5.52–5.02 (m, 5H), 5.55 (s, 1H), 6.02 (dd, *J* = 9.3, 9.3 Hz, 1H), 6.10 (d, *J* = 9.3 Hz, 1H), 6.34 ppm (d, *J* = 9.9 Hz, 1H). ^{13}C-NMR (50 MHz CDCl$_3$): δ = 20.0, 20.3, 20.6 (4C), 20.7 (3C), 61.4,

61.9, 67.5, 67.8, 68.1, 69.2, 73.3, 73.5, 74.7, 75.0, 79.2, 81.3, 103.4, 150.5, 152.9, 160.8, 169.40, 169.49, 169.62, 169.96, 170.20, 170.40, 170.60, 170.78 ppm. ESI-MS: 786.3 [M + H]$^+$.

1-(Tetra-O-acetyl-β-D-glucopyranosyl)-6-formyluracil (**7**). To a solution of **6a** (2.1 g, 4.63 mmol) in dry dioxane (40 mL), selenium oxide was added (1.5 g, 13.9 mmol, 3 eq) and acetic acid (1.32 mL, 23.1 mmol). The reaction mixture was heated at reflux until full consumption of the starting material (≈5 h). The mixture was then cooled, diluted with ethyl acetate, and washed successively with saturated aq. NaHCO$_3$ solution, water, and brine. The organic layer was dried over anhydrous Na$_2$SO$_4$, filtered, the solvents evaporated, and the crude product was purified by column chromatography (97:3, EtOAc:Et$_2$O) to yield the title compound as a white solid (1.5 g, 3.1 mmol, 67%). R$_f$ = 0.50 (100% EtOAc). ^{1}H-NMR (200 MHz, CDCl$_3$): δ = 2.01 (s, 3H), 2.04 (s, 3H), 2.06 (s, 3H), 2.10 (s, 3H), 3.92 (ddd, *J* = 9.9, 3.3, 3.0 Hz, 1H), 4.19 (m, 2H), 5.24 (t, *J* = 9.7 Hz, 1H), 5.41 (t, *J* = 9.4 Hz, 1H), 5.61 (t, *J* = 9.3 Hz, 1H), 6.11 (d, *J* = 9.4 Hz, 1H), 6.28 (d, *J* = 2.3 Hz, 1H), 8.53 (s, 1H), 9.90 ppm (s, 1H). ^{13}C-NMR (50 MHz, CDCl$_3$): δ = 20.2, 20.4 (2C), 20.5, 61.0, 67.2, 70.3, 72.2, 74.9, 81.3, 110.0, 147.0, 150.2, 161.6, 169.4, 169.7, 170.0, 170.4, 183.4 ppm. HRMS (ESI): calcd. for C$_{19}$H$_{23}$N$_2$O$_{12}$$^+$ [M + H]$^+$ 471.1251 found 471.1255.

1-(Tetra-O-acetyl-β-D-glucopyranosyl)-6-hydroxymethyluracil (**8**). To a solution of **7** (0.2 g, 0.425 mmol) in 2-propanol (2.5 mL) and chloroform (0.6 mL), dry silica gel was added (43 mg) and the suspension was cooled to −30 °C. Then, NaBH$_4$ (0.161 g, 4.25 mmol, 10 eq) was added and the reaction mixture was stirred until full consumption of the starting material (≈1 h). Then, the mixture was diluted with DCM, filtered through Celite®, and the filtrate washed successively with saturated aq. NaHCO$_3$ solution, water, and saturated sodium chloride solution. The organic layer was dried over anhydrous Na$_2$SO$_4$, filtered, the solvents were evaporated, and the crude product was purified by column chromatography (EtOAc) to give the title compound as a white solid (146 mg, 0,38 mmol, 90%). R$_f$ = 0.40 (70:30, Et$_2$O:EtOAc). ^{1}H-NMR (200 MHz, DMSO-d6): (mixture of tautomers) δ = 1.92 (s, 3H), 1.96 (s, 3H), 2.01 (s, 6H), 4.30 (m, 2H), 4.92 (t, *J* = 9.6 Hz, 1H), 5.26–5.62 (m, 2H), 5.69 (s, 1H), 5.84 (s, 2H), 6.08 (t, *J* = 9.0 Hz, 1H), 6.42 (d, *J* = 9.4 Hz, 1H), 11.32 ppm (s, 1H), 11.48 ppm (s, 1H). ^{13}C-NMR (50 MHz, CDCl$_3$): δ = 20.4, 20.5 (2C), 20.7, 60.1, 61.2, 67.5, 69.8, 72.6, 75.2, 81.5, 102.9, 151.3, 157.2, 162.8, 169.5, 169.8, 170.4, 170.7 ppm. HRMS (ESI): calcd. for C$_{19}$H$_{25}$N$_2$O$_{12}$$^+$ [M + H]$^+$ 473.1407 found 473.1410.

Spirocyclization of **8**. A solution of **8** (150 mg, 0.32 mmol) in dichloromethane (16 mL) was degassed by argon gas bubbling for 10 min. Then, diacetoxyiodobenzene (155 mg, 0.48 mmol, 1.5 eq) and iodine (91 mg, 0.32 mmol, 1 eq) were added. Photolysis was carried out at r.t., with two 75 W Philips Standard 230 V visible light lamps, for 2.5 h. Afterwards, the reaction was quenched by 10% aq. Na$_2$S$_2$O$_3$ solution and then extracted with dichloromethane. The organic layer was collected, dried over anhydrous Na$_2$SO$_4$, and then filtered, the solvent was evaporated, and the crude product was purified by column chromatography (EtOAc: Et$_2$O gradient) to give, in order of elution, compound **7** (75 mg, 0.16 mmol, 50%), **9a** as a white solid, (12 mg, 0.026 mmol, 8%), and **9b** as a white solid (15 mg, 0.032 mmol, 10%).

(3*R*,3′*R*,4′*S*,5′*R*,6′*R*)-6′-(Acetoxymethyl)-5,7-dioxo-1,3′,4′,5′,6′,7-octahydrospiro[oxazolo[3,4-c] pyrimidine-3,2′-pyran]-3′,4′,5′-triyl triacetate (**9a**): R$_f$ = 0.60 (70:30 Et$_2$O:EtOAc). ^{1}H-NMR (200 MHz, CDCl$_3$): δ = 2.00 (s, 3H), 2.01 (s, 3H), 2.05 (s, 3H), 2.09 (s, 3H), 4.17–4.10 (m, 1H), 4.20 (s, 2H), 5.04 (m, 2H), 5.34 (t, *J* = 9.5 Hz, 1H), 5.48 (t, *J* = 10.0 Hz, 1H), 5.59 (s, 1H), 6.16 ppm (d, *J* = 9.8 Hz, 1H). ^{13}C-NMR (50 MHz, CDCl$_3$): δ = 20.52, 20.60, 20.62, 20.72, 61.2, 67.4, 68.2, 68.6, 71.3, 71.4, 93.8, 112.1, 145.7, 151.3, 163.1, 168.7, 169.2, 170.1, 170.7 ppm. HRMS (ESI): calcd. for C$_{19}$H$_{23}$N$_2$O$_{12}$$^+$ [M + H]$^+$ 471.1246 found 471.1239.

(3*S*,3′*R*,4′*S*,5′*R*,6′*R*)-6′-(Acetoxymethyl)-5,7-dioxo-1,3′,4′,5′,6′,7-octahydrospiro[oxazolo[3,4-c] pyrimidine-3,2′-pyran]-3′,4′,5′-triyl triacetate (**9b**): R$_f$ = 0.40 (70:30 Et$_2$O:EtOAc). ^{1}H-NMR (200 MHz, CDCl$_3$): δ = 1.99 (s, 3H), 2.03 (s, 3H), 2.05 (s, 3H), 2.09 (s, 3H), 4.10 (dd, *J* = 12.7, 2.5 Hz, 1H), 4.26 (dd, *J* = 12.7, 3.6 Hz, 1H), 4.65 (ddd, *J* = 9.6, 3.6, 2.5 Hz, 1H), 4.87 (dd, *J* = 14.5, 1.0 Hz, 1H), 5.06 (dd, *J* = 14.5, 1.6 Hz, 1H), 5.33 (dd, *J* = 10.1, 8.9 Hz, 1H), 5.63 (d, *J* = 9.3 Hz, 1H), 5.65 (s, 1H), 6.31 (t, *J* = 9.2 Hz,

1H), 8.61 ppm (s, 1H). ^{13}C-NMR (50 MHz, CDCl$_3$): δ = 20.52, 20.60, 20.62, 20.70, 61.4, 67.2, 67.3, 71.0, 72.4, 72.6, 94.2, 114.5, 148.3, 152.7, 163.3, 169.1, 169.6, 170.0, 170.6 ppm. HRMS (ESI): calcd. for C$_{19}$H$_{23}$N$_2$O$_{12}$$^+$ [M + H]$^+$ 471.1246 found 471.1241.

(3*R*,3'*R*,4'*S*,5'*S*,6'*R*)-3',4',5'-Trihydroxy-6'-(hydroxymethyl)-3',4',5',6'-tetrahydrospiro[oxazolo[3,4-*c*]pyrimidine-3,2'-pyran]-5,7(1H,6H)-dione (**4a**). A solution of **9a** (20 mg, 0.043 mmol) in methanolic ammonia (7 N, 0.35 mL) was stirred for 12 h at r.t., until full conversion to a single compound. Then, the solvent was evaporated and the compound was dried under high vacuum to yield the title compound as a white solid (13 mg, 0.043 mmol, 100%). ^{1}H-NMR (200 MHz, D$_2$O): δ = 3.64 (t, *J* = 9.1 Hz, 1H), 3.97–3.75 (m, 4H), 4.56 (d, *J* = 9.7 Hz, 1H), 5.20 (s, 2H), 5.88 ppm (s, 1H). ^{13}C-NMR (50 MHz, D$_2$O): δ = 63.0, 70.8, 71.6, 72.5, 76.5, 78.3, 96.7, 116.8, 142.5, 157.1, 169.6 ppm. HRMS (ESI): calcd. for C$_{11}$H$_{15}$N$_2$O$_8$$^+$ [M + H]$^+$ 303.0823 found 303.0830.

(3*S*,3'*R*,4'*S*,5'*S*,6'*R*)-3',4',5'-Trihydroxy-6'-(hydroxymethyl)-3',4',5',6'-tetrahydrospiro[oxazolo[3,4-*c*]pyrimidine-3,2'-pyran]-5,7(1H,6H)-dione (**4b**). A solution of **9b** (29 mg, 0.062 mmol) in methanolic ammonia (7 N, 0.51 mL) was stirred for 12 h at r.t., until full conversion to a single compound. Then, the solvent was evaporated and the compound was dried under high vacuum to yield the title compound as a white solid (19 mg, 0.062 mmol, 100%). ^{1}H-NMR (200 MHz, D$_2$O): δ = 3.58 (t, *J* = 9.6 Hz, 1H), 3.70 (dd, *J* = 12.5, 5.5 Hz, 2H), 3.85 (dd, *J* = 12.4, 2.1 Hz, 1H), 3.95 (d, *J* = 9.5 Hz, 2H), 4.26 (ddd, *J* = 10.1, 5.3, 2.2 Hz, 1H), 4.62 (t, *J* = 9.2 Hz, 2H), 5.08 (d, *J* = 14.9 Hz, 1H), 5.18 (d, *J* = 15.0 Hz, 1H), 5.81 ppm (s, 1H). ^{13}C-NMR (50 MHz, D$_2$O): δ = 169.5, 158.2, 149.4, 119.6, 96.4, 79.5, 77.3, 76.2, 71.6, 70.2, 63.4 ppm. HRMS (ESI): calcd. for C$_{11}$H$_{15}$N$_2$O$_8$$^+$ [M + H]$^+$ 303.0823 found 303.0828.

5. Conclusions

In conclusion, we successfully applied the Suárez radical methodology to synthesize the first example of a 1,6-dioxa-4-azaspiro[4.5]decane system starting from 6-hydroxymethyl-β-D-glucopyranosyluracil. We showed that, in the rigid pyranosyl conformation, the required [1,5]-radical translocation is a minor process. The stereochemistry of the spirocycles obtained was unequivocally determined by the chemical shifts of key sugar protons in the ^{1}H-NMR spectra. Finally, the two spirocycles were found to be modest inhibitors of RMGP*b*, corroborating the finding that *anti* should be the desired conformation of the pyrimidine ring of future anomeric spironucleosides, which may lead to strong inhibition of GP.

Supplementary Materials: The following are available online. ^{1}H and ^{13}C-NMR spectra of new compounds (Figures S1–10). Tables of kinetic measurements (Tables S1 and S2).

Author Contributions: Conceptualization, T.G.; methodology, T.G. and E.D.C.; investigation, A.S. and M.M.; writing—original draft preparation, T.G. and M.M.; writing—review and editing, T.G.; supervision, T.G. and E.D.C.; project administration, T.G.; funding acquisition, T.G. and E.D.C.

Funding: This work was funded by Heracleitus II (M.M.) and the Special Account of N.K.U.A. (T.G.).

Conflicts of Interest: The authors declare no conflict of interest.

References

1. Rines, A.K.; Sharabi, K.; Tavares, C.D.J.; Puigserver, P. Targeting hepatic glucose metabolism in the treatment of type 2 diabetes. *Nat. Rev. Drug Discov.* **2016**, *15*, 786–804. [CrossRef] [PubMed]
2. Gimisis, T. Synthesis of N-Glucopyranosidic Derivatives as Potential Inhibitors that Bind at the Catalytic Site of Glycogen Phosphorylase. *Mini-Rev. Med. Chem.* **2010**, *10*, 1127–1138. [CrossRef] [PubMed]
3. Donnier-Maréchal, M.; Vidal, S. Glycogen phosphorylase inhibitors: A patent review (2013–2015). *Expert Opin. Ther. Pat.* **2016**, *26*, 199–212. [CrossRef] [PubMed]
4. Chrysina, E.D.; Chajistamatiou, A.; Chegkazi, M. From structure-based to knowledge-based drug design through X-ray protein crystallography: Sketching glycogen phosphorylase binding sites. *Curr. Med. Chem.* **2011**, *18*, 2620–2629. [CrossRef] [PubMed]

5. Bichard, C.J.F.; Mitchell, E.P.; Wormald, M.R.; Watson, K.A.; Johnson, L.N.; Zographos, S.E.; Koutra, D.D.; Oikonomakos, N.G.; Fleet, G.W.J. Potent inhibition of glycogen phosphorylase by a spirohydantoin of glucopyranose: First pyranose analogues of hydantocidin. *Tetrahedron Lett.* **1995**, *36*, 2145–2148. [CrossRef]
6. Goyard, D.; Kónya, B.; Chajistamatiou, A.S.; Chrysina, E.D.; Leroy, J.; Balzarin, S.; Tournier, M.; Tousch, D.; Petit, P.; Duret, C.; et al. Glucose-derived spiro-isoxazolines are anti-hyperglycemic agents against type 2 diabetes through glycogen phosphorylase inhibition. *Eur. J. Med. Chem.* **2016**, *108*, 444–454. [CrossRef] [PubMed]
7. Mamais, M.; Degli Esposti, A.; Kouloumoundra, V.; Gustavsson, T.; Monti, F.; Venturini, A.; Chrysina, E.D.; Markovitsi, D.; Gimisis, T. A New Potent Inhibitor of Glycogen Phosphorylase Reveals the Basicity of the Catalytic Site. *Chem. A Eur. J.* **2017**, *23*, 8800–8805. [CrossRef] [PubMed]
8. Soto, M.; Rodríguez-Solla, H.; Soengas, R. Recent Advances in the Chemistry and Biology of Spirocyclic Nucleosides. In *Topics in Heterocyclic Chemistry*; Springer: Berlin, Heidelberg, Germany, 2019; pp. 1–43.
9. Chatgilialoglu, C.; Ferreri, C.; Gimisis, T.; Roberti, M.; Balzarini, J.; De Clercq, E. Synthesis and Biological Evaluation of Novel 1′-Branched and Spiro-Nucleoside Analogues. *Nucleosides Nucleotides Nucleic Acids* **2004**, *23*, 1565–1581. [CrossRef] [PubMed]
10. Gómez-García, O.; Gómez, E.; Toscano, R.; Salgado-Zamora, H.; Álvarez-Toledano, C. One-Pot Synthesis of Spirotetrahydrooxino [3,4-c] pyridines and Spirotetrahydrofuro [3,2-b] pyridin-2-ones via Lactonization from Activated Pyridyldihydrooxazoles and Bis(trimethylsilyl)ketene Acetals. *Synthesis* **2016**, *48*, 1371–1380. [CrossRef]
11. Martín, A.; Suárez, E. Carbohydrate Spiro-heterocycles via Radical Chemistry. In *Topics in Heterocyclic Chemistry*; Springer: Berlin/Heidelberg, Germany, 2019; pp. 1–54.
12. Chatgilialoglu, C.; Gimisis, T.; Spada, G.P. C-1′ Radical-Based Approaches for the Synthesis of Anomeric Spironucleosides. *Chem. A Eur. J.* **1999**, *5*, 2866–2876. [CrossRef]
13. Gimisis, T.; Chatgilialoglu, C.; Gimisis, T.; Castellari, C. A new class of anomeric spironucleosides. *Chem. Commun.* **1997**, 2089–2090. [CrossRef]
14. Liao, J.; Sun, J.; Yu, B. Effective synthesis of nucleosides with glycosyl trifluoroacetimidates as donors. *Tetrahedron Lett.* **2008**, *49*, 5036–5038. [CrossRef]
15. Groziak, M.P.; Koohang, A. Facile addition of hydroxylic nucleophiles to the formyl group of uridine-6-carboxaldehydes. *J. Org. Chem.* **1992**, *57*, 940–944. [CrossRef]
16. Tanaka, H.; Hayakawa, H.; Miyasaka, T. "Umpulong" of reactivity at the C-6 position of uridine: A simple and general method for 6-substituted uridines. *Tetrahedron* **1982**, *38*, 2635–2642. [CrossRef]
17. Liu, H.-J.; Yip, J.; Shia, K.-S. Reductive cleavage of benzyl ethers with lithium naphthalenide. A convenient method for debenzylation. *Tetrahedron Lett.* **1997**, *38*, 2253–2256. [CrossRef]
18. Wittenburg, E. Nucleoside und verwandte Verbindungen. VII. Alkylierung und Glykosidierung der Silyl-derivate 6-substituierter Uracile. *Collect. Czechoslov. Chem. Commun.* **1971**, *36*, 246–261. [CrossRef]
19. Felczak, K.; Drabikowska, A.K.; Vilpo, J.A.; Kulikowski, T.; Shugar, D. 6-Substituted and 5,6-Disubstituted Derivatives of Uridine: Stereoselective Synthesis, Interaction with Uridine Phosphorylase, and In Vitro Antitumor Activity. *J. Med. Chem.* **1996**, *39*, 1720–1728. [CrossRef] [PubMed]
20. Warpehoski, M.A.; Chabaud, B.; Sharpless, K.B. Selenium dioxide oxidation of endocyclic olefins. Evidence for a dissociation-recombination pathway. *J. Org. Chem.* **1982**, *47*, 2897–2900. [CrossRef]
21. Młochowski, J.; Brząszcz, M.; Giurg, M.; Palus, J.; Wójtowicz, H. Selenium-Promoted Oxidation of Organic Compounds: Reactions and Mechanisms. *Eur. J. Org. Chem.* **2003**, *2003*, 4329–4339. [CrossRef]
22. Florent, J.-C.; Dong, X.; Gaudel, G.; Mitaku, S.; Monneret, C.; Gesson, J.-P.; Jacquesy, J.-C.; Mondon, M.; Renoux, B.; Andrianomenjanahary, S.; et al. Prodrugs of Anthracyclines for Use in Antibody-Directed Enzyme Prodrug Therapy. *J. Med. Chem.* **1998**, *41*, 3572–3581. [CrossRef]
23. Francisco, C.G.; Freire, R.; Herrera, A.J.; Pérez-Martín, I.; Suárez, E. Intramolecular 1,5-versus 1,6-hydrogen abstraction reaction promoted by alkoxyl radicals in pyranose and furanose models. *Tetrahedron* **2007**, *63*, 8910–8920. [CrossRef]
24. Barton, D.H.R.; Beaton, J.M.; Geller, L.E.; Pechet, M.M. A New Photochemical Reaction 1. *J. Am. Chem. Soc.* **1961**, *83*, 4076–4083. [CrossRef]
25. Kittaka, A.; Kato, H.; Tanaka, H.; Nonaka, Y.; Amano, M.; Nakamura, K.T.; Miyasaka, T. Face selective 6,1′-(1-oxo) ethano bridge formation of uracil nucleosides under hypoiodite reaction conditions. *Tetrahedron* **1999**, *55*, 5319–5344. [CrossRef]

26. Oikonomakos, N.G.; Kontou, M.; Zographos, S.E.; Watson, K.A.; Johnson, L.N.; Bichard, C.J.F.; Fleet, G.W.J.; Acharya, K.R. N-acetyl-β-D-glucopyranosylamine: A potent T-state inhibitor of glycogen phosphorylase. A comparison with α-D-glucose. *Protein Sci.* **1995**, *4*, 2469–2477. [CrossRef] [PubMed]
27. Saheki, S.; Takeda, A.; Shimazu, T. Assay of inorganic phosphate in the mild pH range, suitable for measurement of glycogen phosphorylase activity. *Anal. Biochem.* **1985**, *148*, 277–281. [CrossRef]
28. Vorbrüggen, H.; Krolikiewicz, K.; Bennua, B. Nucleoside syntheses, XXII1) Nucleoside synthesis with trimethylsilyl triflate and perchlorate as catalysts. *Chem. Ber.* **1981**, *114*, 1234–1255. [CrossRef]
29. Martin, J.L.; Veluraja, K.; Ross, K.; Johnson, L.N.; Fleet, G.W.J.; Ramsden, N.G.; Bruce, I.; Orchard, M.G.; Oikonomakos, N.G. Glucose analog inhibitors of glycogen phosphorylase: The design of potential drugs for diabetes. *Biochemistry* **1991**, *30*, 10101–10116. [CrossRef] [PubMed]
30. Bokor, É.; Kun, S.; Docsa, T.; Gergely, P.; Somsák, L. 4(5)-Aryl-2-C-glucopyranosyl-imidazoles as New Nanomolar Glucose Analogue Inhibitors of Glycogen Phosphorylase. *ACS Med. Chem. Lett.* **2015**, *6*, 1215–1219. [CrossRef] [PubMed]
31. Grugel, H.; Minuth, T.; Boysen, M. Novel Olefin-Phosphorus Hybrid and Diene Ligands Derived from Carbohydrates. *Synthesis* **2010**, *19*, 3248–3258.
32. Helmreich, E.; Cori, C.F. The role of adenylic acid in the activation of phsphorylase. *Proc. Natl. Acad. Sci. USA* **1964**, *51*, 131–138. [CrossRef]

Sample Availability: Samples of compounds **4**, **6–9** are available from the authors.

© 2019 by the authors. Licensee MDPI, Basel, Switzerland. This article is an open access article distributed under the terms and conditions of the Creative Commons Attribution (CC BY) license (http://creativecommons.org/licenses/by/4.0/).

Article

Two-Step Azidoalkenylation of Terminal Alkenes Using Iodomethyl Sulfones

Nicolas Millius, Guillaume Lapointe and Philippe Renaud *

Department of Chemistry and Biochemistry, University of Bern, Freiestrasse 3, CH-3012 Bern, Switzerland; Nicolas.Millius@dsm.com (N.M.); guillaume.lapointe@novartis.com (G.L.)
* Correspondence: philippe.renaud@dcb.unibe.ch

Academic Editor: Chryssostomos Chatgilialoglu
Received: 31 October 2019; Accepted: 11 November 2019; Published: 18 November 2019

Abstract: The radical azidoalkylation of alkenes that was initially developed with α-iodoesters and α-iodoketones was extended to other activated iodomethyl derivatives. By using iodomethyl aryl sulfones, the preparation of γ-azidosulfones was easily achieved. Facile conversion of these azidosulfones to homoallylic azides using a Julia–Kocienski olefination reaction is reported, making the whole process equivalent to the azidoalkenylation of terminal alkenes.

Keywords: radical reaction; azidoalkylation; carboazidation; sulfones; azides; Julia–Kocienski olefination

1. Introduction

Organic alkyl azides are highly versatile compounds for synthesis [1–4]. They are unreactive towards a broad range of reaction conditions but, under dedicated conditions, they may become nitrene [5] and aminyl radical precursors [6–11] as well as suitable substrates for Schmidt reaction [12–14], aza-Wittig [15] reaction, and 1,3-dipolar cycloaddition [16,17].

They are commonly prepared via nucleophilic substitution of halides and related electrophiles using inorganic azides [18]. For tertiary alkyl azides, the nucleophilic substitution approach is often difficult, and free radical processes have proven to be a very convenient alternative. Radical azidation reactions are run under mild conditions, and they are compatible with a broad range of functional groups [19–21]. The carboazidation of alkenes represents a particularly attractive method to transform terminal alkenes into functionalized alkyl azides [22]. It is performed under chain transfer conditions and has been employed as a key step in several alkaloids syntheses [20,23–29]. Alternatively, carboazidation using under transition metal catalysis has also been reported [30–34]. Except for one reaction involving CCl₃Br [22], the reaction has mainly been used with α-iodoesters and α-iodoketones (Scheme 1I) under ditin [35,36] or triethylborane [37,38] mediation. More recently, a very efficient desulfitative approach was reported starting from α-azidosulfonyl esters [26]. This approach is the best in terms of atom economy and efficiency, but it is less convenient to test the applicability of the method with a broad range of substituted radicals since every starting azide has to be prepared separately. The iodide approach remains very attractive in terms of availability of the starting material (the starting iodide and the azidating agents are either commercially available or easily prepared), and it can be easily used to introduce of variety of functional groups. Here, we report the extension of the carboazidation process for the preparation of azido-nitriles, -phosphonates, -phthalimides, and sulfones according to Scheme 1II. The reaction with sulfones is particularly attractive since it allows one to prepare homoallylic azides.

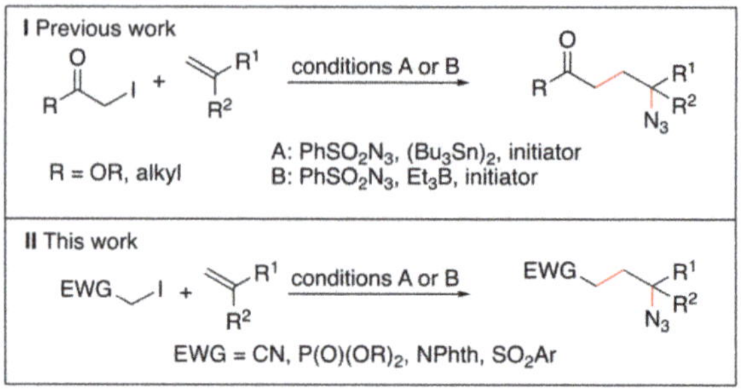

Scheme 1. The radical carboazidation reaction.

2. Results

Iodoacetonitrile **1**, *N*-iodomethylphthalimide **2**, diethyl iodomethanephosphonate **3**, and iodomethyl phenyl sulfones **4** are either commercially available or easily prepared (see supporting information). They were tested for the carboazidation of terminal alkenes **5** under ditin (A) or triethylborane (B) conditions (Scheme 2). Under ditin-mediated conditions A, reactions of **1–4** with methylenecyclohexane **5a** worked fine and provided the desired tertiary azide **6a–9a** in good yields. Azidonitrile **6a** is a potential precursor for 1,4-diamines, and azidophthalimide **7a** is a bis-protected 1,3-diamine. γ-Azidophosphonates such as **8a** are interesting precursors of γ-aminophosphonic acids, a well-established class of biologically active compounds [39]. Finally, the rich chemistry of sulfones renders γ-azidosulfones such as **9a** as potential precursors for a broad range of functionalized amines. The reaction of iodomethylsulfone **4** with **5a** mediated by Et$_3$B (method B) provided the azidosulfones **9a** in an increased 92% yield. Sulfone **4** was also employed for the carboazidation of methylenecycloheptene **5b** and the two substituted methylenecyclohexanes **5c** and **5d** as well as the monosubstituted terminal alkene **5e** under conditions B. The tertiary azides **9b–9d** were obtained in moderate to good yields, and the level of stereoselectivity observed for **9c** and **9d** (2–3:1) corresponded to expectations [40]. The secondary azidosulfone **9e** was obtained in 45% yield under conditions B. The crude product was contaminated with the iodide **10e** (9%) and the alcohol **11e** (13%). The alcohol **13e** presumably resulted from a sulfone assisted hydrolysis of the iodide **10e**, but reaction of the intermediate radical with oxygen could not be excluded. When the reaction was run at a higher temperature according to method A, no azide **9e** was obtained, and the iodine atom transfer product **10e** (34% yield) was the only isolated product.

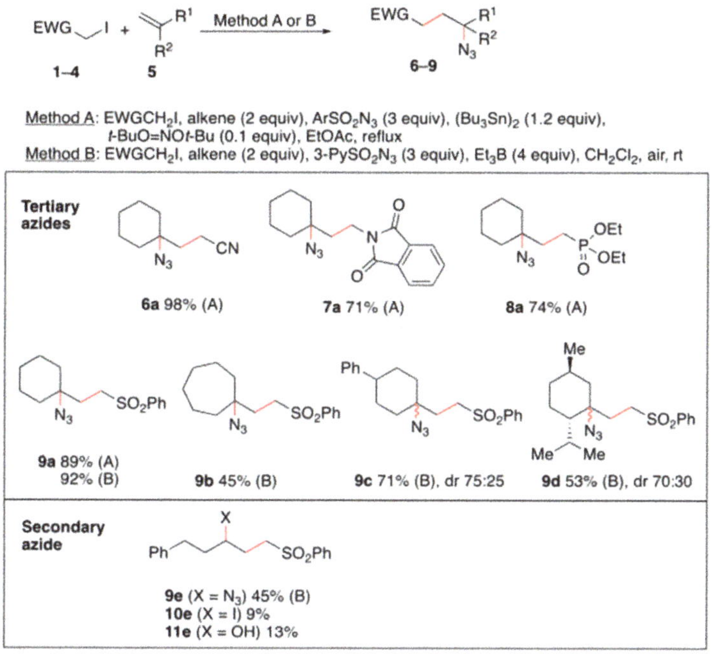

Scheme 2. Radical carboazidation with cyano-, phthalimido-, diethoxylphosphonyl-, and benzenesulfonyl-substituted radicals.

To illustrate the utility of γ-azidosulfones, compound **9c** was sulfurized to **12** by treatment with lithium hexamethyldisilazane (LiHMDS) and diphenyl disulfide (PhSSPh). The sulfide **12** was easily converted to the unsaturated γ-azido vinyl sulfone **13**, an attractive and versatile building block for synthesis, upon oxidation to the sulfoxide and standing in CDCl$_3$ (Scheme 3). The whole reaction sequence allowed us to convert a terminal 2,2-disubsituted alkene into a tertiary 1-sufonylated allylic azide. Attempts to convert **12** into a β-azido ester upon treatment successively with *meta*-chloroperbenzoic acid (*m*-CPBA) and trifluoroacetic acid (TFA) to promote a Pummerer rearrangement according to a procedure reported by Barton and co-workers failed to give the desired product [41].

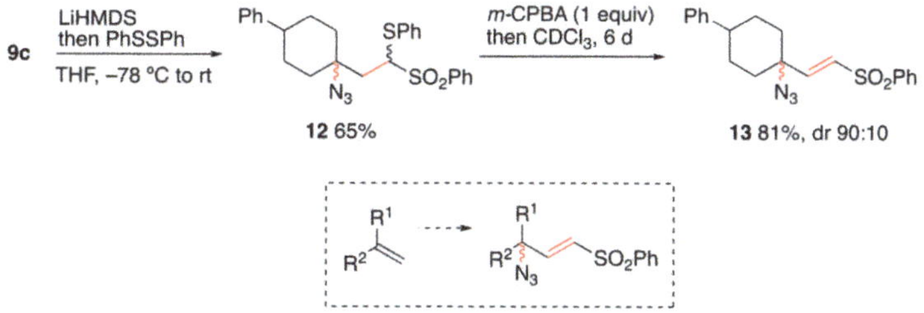

Scheme 3. Preparation of unsaturated γ-azido vinyl sulfone **13** from the azidosulfone **9c**.

The carboazidation with sulfones also offers a potential approach for the preparation of homoallylic azides [42] by taking advantage of the Julia–Kocienski olefination process [43,44]. For this purpose, 1-phenyl-1*H*-tetrazole-5-yl iodomethyl sulfone **14** was prepared from the commercially available 1-phenyl-1*H*-tetrazole-5-thiol [45,46]. Carboazidation was then tested with methylene cyclohexene **5a** and 2-butyl-1-hexene **5f** using the ditin procedure (Scheme 4). With **5a**, the tertiary azide **15a** was obtained in high yield. The reaction with 2-butyl-1-hexene **5f** proved to be more challenging. The desired azide **15f** was isolated in 31% yield together with a side product identified as being **16f** in 20% yield. Compound **16f** most likely resulted from the ipso attack of a tin radical to the 1H-tetrazole-5-yl sulfone followed by reaction of the primary alkanesulfonyl radical with 3-pyridinesulfonyl azide. A related intermolecular ipso substitution was recently reported by Kamijo and co-workers [47].

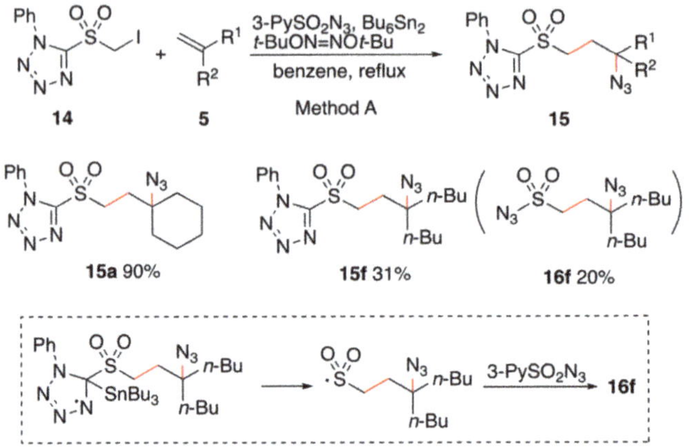

Scheme 4. Tin mediated azidoalkylation with 1-phenyl-1H-tetrazole-5-yl iodomethyl sulfone **14**.

Following this observation, all carboazidation reactions involving **14** and different alkenes **5** were using the Et$_3$B method B. Results are summarized in Scheme 5, and moderate to good yields were observed for the formation of γ-azidosulfones **15** with a broad range of 2,2-substituted alkenes. No side product resulting from an ipso substitution at the tetrazole could be detected in those reactions. The radical nature of the process was demonstrated by formation of the ring-opening reaction product **15j** from (−)-β-pinene.

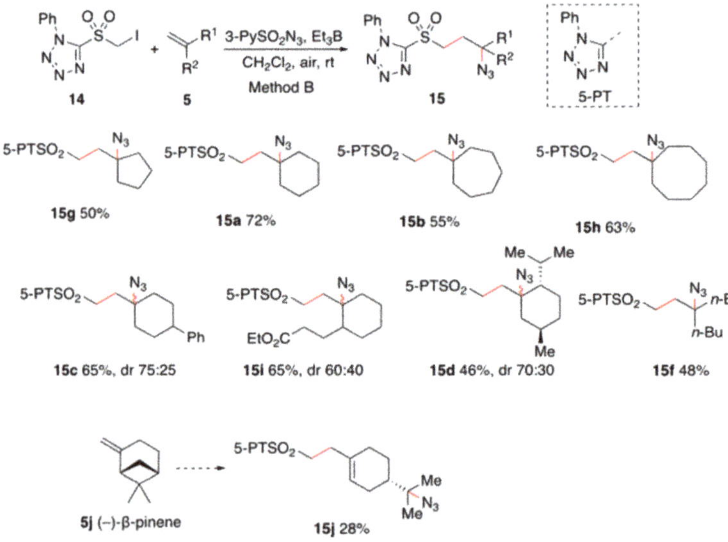

Scheme 5. *Cont.*

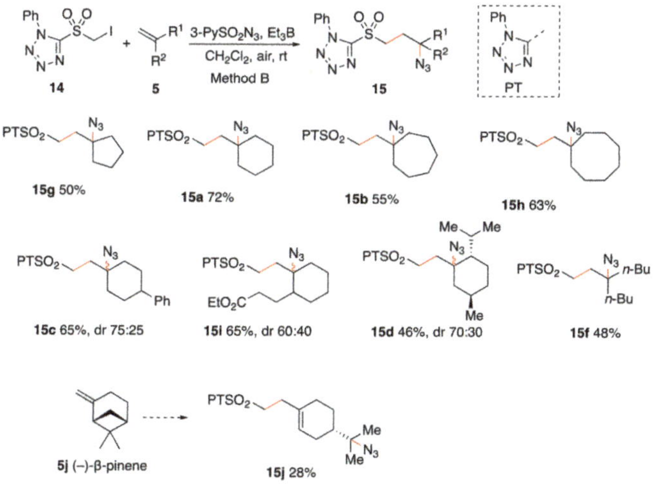

Scheme 5. Et$_3$B mediated azidoalkylation with 1-phenyl-1H-tetrazole-5-yl iodomethyl sulfone **14**.

Finally, the 1-phenyl-1H-tetrazole sulfones **15** were submitted to the Julia–Kocienski olefination. Deprotonation of the sulfones **15** with LiHMDS followed by treatment with aldehydes afforded the homoallylic tertiary azides **17–21**. Moderate to good yields and high E selectivity were obtained with aromatic (**17**), aliphatic (**18, 19**), and α,β-unaturated (**20, 21**) aldehydes (Scheme 6). Interestingly, the homoallylic tertiary azides **17–21** were found to be stable and easily purified by column chromatography on silicagel.

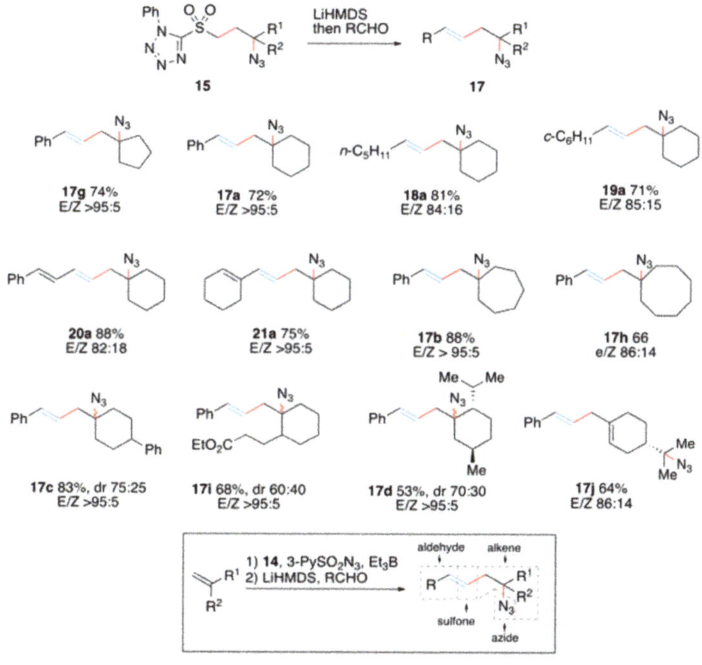

Scheme 6. Julia–Kocienski olefination of γ-azidosulfones **15** with aldehydes, a formal 4-component azidovinylation of alkenes.

3. Experimental Procedures

3.1. General Methods

All glassware was oven-dried at 160 °C and assembled hot or flame dried under vacuum, and allowed to cool under a nitrogen atmosphere. Unless otherwise stated, all the reactions were performed under a nitrogen atmosphere. For flash chromatography (FC) silica gel P60 (40–63 μm) (Silicycle, Basel, Switzerland) was used. Thin layer chromatography (TLC) was performed on silica gel F-254 plates (Silicycle, Basel, Switzerland) visualisation under UV (254 nm) or by staining. Staining solutions: (1) $KMnO_4$ (1.5 g), K_2CO_3 (10 g) and NaOH 10% (1.25 mL) in H_2O (200 mL); (2) ammonium molybdate tetrahydrate (50 g), $CeSO_4$ (2 g) and conc. H_2SO_4 (100 mL) in H_2O (900 mL); (3) p-anisaldehyde (3.7 mL), acetic acid (1.5 mL) and conc. H_2SO_4 (5 mL) in EtOH (135 mL). ^1H and ^{13}C NMR spectra were recorded on a Bruker Advance 300 (^1H: 300.18 MHz, ^{13}C: 75.48 MHz) (Bruker BioSpin AG, Fällanden, Switzerland). Chemical shifts (d) were reported in parts per million (ppm) with the residue solvent peak used as internal standard ($CHCl_3$: d = 7.26 ppm, C_6H_6: d = 7.16 ppm and THF: d = 1.72 ppm for ^1H NMR spectra and $CHCl_3$: d = 77.00 ppm, C_6H_6: d = 128.00 ppm and THF: d = 67.21 ppm for ^{13}C NMR spectra). Multiplicities were abbreviated as follows: s (singlet), d (doublet), t (triplet), q (quadruplet), m (multiplet) and br (broad). Coupling constants (J), are reported in Hz. ^{13}C NMR measurements were run using a proton-decoupled pulse sequence. The number of carbon atoms for each signal is indicated only when more than one. High-resolution mass spectrometry (HRMS) analyses were measured on an Applied Biosystems Sciex QSTAR Pulsar (hybrid quadrupole time-of-flight mass spectrometer using electrospray ionisation (ESI) (Sciex, Baden, Switzelrand). Low resolution mass-spectrometry (LRMS) analyses were performed Finnigan Trace GC-MS (Thermo Scientific, Schlieren, Switzerland) (EI mode at 70 eV); GC column: Optima Delta 3 0.25 μm, 20 m, 0.25 mm (Macherey-Nagel, Oensingen, Switzerland). The infrared measurements were performed on a Jasco FTIR-460 Plus spectrometer equipped with a Specac MKII Golden Gate Single Reflection Diamond ATR System and are reported in wave numbers (cm^{-1}). All reagents were obtained from commercial sources and used without further purification, unless otherwise mentioned. All reactions solvents (distilled THF, distilled Et_2O, distilled dichloromethane, commercial toluene and benzene) were filtered over columns of activated alumina under a positive pressure of argon. Solvents for flash chromatography and extractions were of technical grade and were distilled prior to use. Hexamethyldisilazane (HMDS) was fractionally distilled under a nitrogen atmosphere before use. 1,2-Dichloroethane (DCE) was distilled over CaH_2 under a nitrogen atmosphere.

3.2. General Procedures

Hexabutylditin-mediated carboazidation (procedure A)

Di-tert-butyl hyponitrite (DTBHN) [48] (0.1 equiv) was added in one portion to a solution of alkene (2–4 equiv), iodomethyl derivative (1 equiv), $(Bu_3Sn)_2$ (1.2 equiv), and $ArSO_2N_3$ [49] (3 equiv.) in benzene (0.5 M). The solution was stirred at 70 °C for 3 h. The crude mixture was directly purified by flash chromatography (FC) using KF/silica [50].

Et_3B-mediated carboazidation (procedure B)

A 1 M solution of Et_3B (3–4 equiv) was added at room temperature (rt) over 2 h via syringe pump to an open flask and then charged with a vigorously stirred mixture of alkene (2–4 equiv), iodomethyl derivative (1 equiv), and 3-$PySO_2N_3$ [49] (3 equiv) in solvent (0.66 M). Caution: the needle should be immersed into the reaction mixture in order to avoid direct contact of Et_3B drops with air. The reaction vessel should be protected from direct light exposure by aluminum foil. After 1 h stirring, H_2O and CH_2Cl_2 were added, and the layers were separated. The aqueous layer was extracted with CH_2Cl_2 (3×). The combined organic layers were washed with brine and dried over Na_2SO_4. The solvent was removed under reduced pressure, and the crude product was purified by FC.

Julia–Kocienski olefination

The phenyltetrazole sulfone derivative (1 equiv) was dissolved/diluted in THF (0.15 M) and cooled to −78 °C. A freshly prepared LiHMDS solution in THF (1.5 equiv) was added slowly and stirred for a further 30 min at −78 °C. Aldehyde (2 equiv) was added neat and stirred for a further 3 h at −78 °C. The reaction mixture was allowed to reach rt and was further stirred at rt overnight. H_2O and Et_2O were added to the reaction suspension, and the layers were separated. The aqueous phase was extracted with Et_2O (3×). The combined organic layers were washed with brine and dried over Na_2SO_4. The solvent was removed under reduced pressure, and the crude product was purified by FC.

5-((2-(1-Azidocyclohexyl)ethyl)sulfonyl)-1-phenyl-1H-tetrazole (**15a**)

According to the procedure A from di-*tert*-butylhyponitrite (17 mg, 0.10 mmol), methylenecyclohexane **5a** (0.24 mL, 2.00 mmol), 5-((iodomethyl)sulfonyl)-1-phenyl-1H-tetrazole **14** (350 mg, 1.00 mmol), hexabutylditin (0.61 mL, 1.20 mmol), and 3-PySO$_2$N$_3$ (552 mg, 3.00 mmol) in benzene (2.0 mL). The crude mixture was directly purified by FC using KF/silica gel (cyclohexane/EtOAc, 95:5) to afford **15a** (325 mg, 90%).

According to the procedure B from a 1 M solution of Et$_3$B in CH$_2$Cl$_2$ (4.00 mL, 4.00 mmol), methylenecyclohexane **5a** (0.24 mL, 2.00 mmol), 5-((iodomethyl)sulfonyl)-1-phenyl-1H-tetrazole **14**, (350 mg, 1.00 mmol), 3-PySO$_2$N$_3$ (552 mg, 3.00 mmol), and CH$_2$Cl$_2$ (0.50 mL). Purification by FC (cyclohexane/EtOAc, 95:5) afforded **15a** (260 mg, 72%). The NMR spectra of some compounds are in the Supplementary Materials.

Colorless crystals: m.p. 90.9–93.6 °C. ^{1}H NMR (300 MHz, CDCl$_3$): δ = 7.76–7.72 (m, 2H), 7.69–7.62 (m, 3H), 3.90–3.84 (m, 2H), 2.24–2.18 (m, 2H), 1.81–1.73 (m, 2H), 1.69–1.29 (m, 8H). ^{13}C NMR (75 MHz, CDCl$_3$): δ = 153.31, 132.96, 131.50, 129.75 (2C), 125.00 (2C), 62.45, 51.61, 34.41 (2C), 31.72, 25.07, 21.96 (2C). IR (neat): 2933, 2856, 2098, 1497, 1337, 1253, 1150. HRMS (ESI): calcd. for [M + H]$^+$: C$_{15}$H$_{20}$N$_7$O$_2$S calcd 362.1394; found: 362.1400.

(3-(1-Azidocyclohexyl)prop-1-en-1-yl)benzene (**17a**)

According to the Julia–Kocienski procedure from **15a** (260 mg, 0.72 mmol), LiHMDS in THF (1.66 mL, 1.08 mmol), benzaldehyde (0.15 mL, 1.44 mmol), and THF (3.00 mL). Purification by FC (cyclohexane/EtOAc, 98:2) afforded the alkene **17a** as an inseparable mixture of isomers (141 mg, E/Z > 95:5, 81%). Colorless oil.

(E)-**17a** (major): ^{1}H NMR (300 MHz, CDCl$_3$): δ = 7.39–7.20 (m, 5H), 6.47 (d, J = 15.8 Hz, 1H), 6.24 (dt, J = 15.8, 7.4 Hz, 1H), 2.46 (dd, J = 7.4, 1.2 Hz, 2H), 1.72 (d, J = 13.1 Hz, 2H), 1.65–1.39 (m, 7H), 1.32–1.21 (m, 1H). ^{13}C NMR (75 MHz, CDCl$_3$): = 137.24, 133.74, 128.51 (2C), 127.29, 126.17 (2C), 124.36, 64.22, 43.93, 34.52 (2C), 25.33, 22.07 (2C).

Characteristic signals for (Z)-**17a** (minor): ^{1}H NMR (300 MHz, CDCl$_3$): δ = 2.55 (d, J = 5.8 Hz, 2H). IR (neat): 3027, 2931, 2858, 2096, 1495, 1447, 1254, 1138, 1102, 1029. EI-MS m/z (%): M−N$_2$: 213.3 (21), 198.3 (7), 170.3 (20), 156.3 (16), 128.3 (10), 117.3 (100), 115.3 (73), 96.3 (63), 91.3 (40), 69.3 (34), 55.3 (39). HRMS (ESI): calcd. for [M + H]$^+$: C$_{15}$H$_{20}$N$_3$: 242.1652; found: 242.1655.

4. Conclusions

In conclusion, we demonstrated that the azidoalkylation of terminal alkenes is not limited to α-iodoester and α-iodoketones. The reaction also works well with nitriles, phosphonates, phthalimides, and aryl sulfones. This last class of compounds is particularly interesting in terms of potential synthetic applications. This point was illustrated by the preparation of homoallylic azides by merging the azidoalkylation process with a Julia–Kocienski olefination reaction. Recently, 1-phenyl-1H-tetrazole sulfones have also been shown to be privileged substrates for reductive cross-coupling processes, opening new opportunities for further functionalization [51,52].

Supplementary Materials: Detailed experimental procedures and NMR spectra of all compounds are available online at http://www.mdpi.com/1420-3049/24/22/4184/s1.

Author Contributions: Conceptualization and methodology, N.M., G.L. and P.R.; Experimental work N.M., G.L.; writing—original draft preparation, N.M.; Writing—review and editing, P.R.; Supervision, project administration and funding acquisition, P.R.

Funding: This research was funded by the Swiss National Science Foundation (Project 200020_172621).

Conflicts of Interest: The authors declare no conflict of interest.

References

1. Bräse, S.; Banert, K. *Organic Azides: Syntheses and Applications*; John Wiley & Sons: Chichester, UK, 2010; ISBN 978-0-470-51998-1.
2. Tanimoto, H.; Kakiuchi, K. Recent applications and developments of organic azides in total synthesis of natural products. *Nat. Prod. Commun.* **2013**, *8*, 1021–1034. [CrossRef]
3. Chiba, S. Application of organic azides for the synthesis of nitrogen-containing molecules. *Synlett* **2012**, *23*, 21–44. [CrossRef]
4. Bräse, S.; Gil, C.; Knepper, K.; Zimmermann, V. Organic azides. An exploding diversity of a unique class of compounds. *Angew. Chem. Int. Ed.* **2005**, *44*, 5188–5240. [CrossRef]
5. Gritsan, N.; Platz, M. Photochemistry of azides: The azide/nitrene interface. In *Organic Azides—Syntheses and Applications*; Bräse, S., Banert, K., Eds.; John Wiley & Sons: Chichester, UK, 2010; pp. 311–372. ISBN 978-0-470-68251-7.
6. Kim, S. Radical cyclizations involving the evolution of nitrogen. *Pure Appl. Chem.* **1996**, *68*, 623–626. [CrossRef]
7. Kim, S.; Joe, G.H.; Do, J.Y. Novel radical cyclizations of alkyl azides. A new route to N-Heterocycles. *J. Am. Chem. Soc.* **1994**, *116*, 5521–5522. [CrossRef]
8. Montevecchi, P.C.; Navacchia, M.L.; Spagnolo, P. A study of vinyl radical cyclization onto the azido group by addition of sulfanyl, stannyl, and silyl radicals to alkynyl azides. *Eur. J. Org. Chem.* **1998**, 1219–1226. [CrossRef]
9. Wyler, B.; Brucelle, F.; Renaud, P. Preparation of the core structure of aspidosperma and strychnos alkaloids from aryl azides by a cascade radical cyclization. *Org. Lett.* **2016**, *18*, 1370–1373. [CrossRef] [PubMed]
10. Brucelle, F.; Renaud, P. Synthesis of a leucomitosane via a diastereoselective radical cascade. *J. Org. Chem.* **2013**, *78*, 6245–6252. [CrossRef] [PubMed]
11. Minozzi, M.; Nanni, D.; Spagnolo, P. From azides to nitrogen-centered radicals: Applications of azide radical chemistry to organic synthesis. *Chem. Eur. J.* **2009**, *15*, 7830–7840. [CrossRef] [PubMed]
12. Wrobleski, A.; Coombs, T.C.; Huh, C.W.; Li, S.-W.; Aubé, J. The Schmidt reaction. *Org. React.* **2012**, *78*, 1–320.
13. Aubé, J.; Fehl, C.; Liu, R.; McLeod, M.C.; Motiwala, H.F. Hofmann, Curtius, Schmidt, Lossen, and related reactions. *Compr. Org. Synth.* **2014**, *6*, 598–635.
14. Nyfeler, E.; Renaud, P. Intramolecular Schmidt reaction: Applications in natural product synthesis. *CHIMIA Int. J. Chem.* **2006**, *60*, 276–284. [CrossRef]
15. Palacios, F.; Alonso, C.; Aparicio, D.; Rubiales, G.; de los Santos, J.M. Aza-Wittig reaction in natural product syntheses. In *Organic Azides*; Wiley: Chichester, UK, 2010; pp. 437–467. ISBN 978-0-470-68251-7.
16. Binder, W.H.; Kluger, C. Azide/alkyne-"click" reactions: Applications in material science and organic synthesis. *Curr. Org. Chem.* **2006**, *10*, 1791–1815. [CrossRef]
17. Schilling, C.; Jung, N.; Bräse, S. Cycloaddition Reactions with azides: An overview. In *Organic Azides—Syntheses and Applications*; Bräse, S., Banert, K., Eds.; John Wiley & Sons: Chichester, UK, 2010; pp. 269–284. ISBN 978-0-470-68251-7.
18. Pinho e Melo, T.M.V.D. Synthesis of azides. In *Organic Azides—Syntheses and Applications*; Bräse, S., Banert, K., Eds.; John Wiley & Sons: Chichester, UK, 2010; pp. 53–94. ISBN 978-0-470-68251-7.
19. Jimeno, C.; Renaud, P. Radical chemistry with azides. In *Organic Azides—Syntheses and Applications*; Bräse, S., Banert, K., Eds.; John Wiley & Sons: Chichester, UK, 2010; pp. 239–267. ISBN 978-0-470-68251-7.
20. Lapointe, G.; Kapat, A.; Weidner, K.; Renaud, P. Radical azidation reactions and their application in the synthesis of alkaloids. *Pure Appl. Chem.* **2012**, *84*, 1633–1641. [CrossRef]

21. Panchaud, P.; Chabaud, L.; Landais, Y.; Ollivier, C.; Renaud, P.; Zigmantas, S. Radical amination with sulfonyl azides: A powerful method for the formation of CN bonds. *Chem. Eur. J.* **2004**, *10*, 3606–3614. [CrossRef]
22. Renaud, P.; Ollivier, C.; Panchaud, P. Radical carboazidation of alkenes: An efficient tool for the preparation of pyrrolidinone derivatives. *Angew. Chem. Int. Edit.* **2002**, *41*, 3460–3462. [CrossRef]
23. Panchaud, P.; Ollivier, C.; Renaud, P.; Zigmantas, S. Radical carboazidation: Expedient assembly of the core structure of various alkaloid families. *J. Org. Chem.* **2004**, *69*, 2755–2759. [CrossRef]
24. Chabaud, L.; Landais, Y.; Renaud, P. Total synthesis of hyacinthacine A(1) and 3-epi-hyacinthacine A(1). *Org. Lett.* **2005**, *7*, 2587–2590. [CrossRef]
25. Schär, P.; Renaud, P. Total synthesis of the marine alkaloid (+/−)-lepadiformine via a radical carboazidation. *Org. Lett.* **2006**, *8*, 1569–1571. [CrossRef]
26. Weidner, K.; Giroult, A.; Panchaud, P.; Renaud, P. Efficient carboazidation of alkenes using a radical desulfonylative azide transfer process. *J. Am. Chem. Soc.* **2010**, *132*, 17511–17515. [CrossRef]
27. Lapointe, G.; Schenk, K.; Renaud, P. Concise synthesis of pyrrolidine and indolizidine alkaloids by a highly convergent three-component reaction. *Chem. Eur. J.* **2011**, *17*, 3207–3212. [CrossRef]
28. Lapointe, G.; Schenk, K.; Renaud, P. Total synthesis of (±)-cylindricine C. *Org. Lett.* **2011**, *13*, 4774–4777. [CrossRef]
29. Gonçalves-Martin, M.G.; Zigmantas, S.; Renaud, P. Formal synthesis of (−)-cephalotaxine. *Helv. Chim. Acta* **2012**, *95*, 2502–2514. [CrossRef]
30. Huang, W.-Y.; Lü, L. The reaction of perfluoroalkanesulfinates VII. Fenton reagent-initiated addition of sodium perfluoroalkanesulfinates to alkenes. *Chin. J. Chem.* **1992**, *10*, 365–372. [CrossRef]
31. Wei, X.-H.; Li, Y.-M.; Zhou, A.-X.; Yang, T.-T.; Yang, S.-D. Silver-catalyzed carboazidation of arylacrylamides. *Org. Lett.* **2013**, *15*, 4158–4161. [CrossRef]
32. Bunescu, A.; Ha, T.M.; Wang, Q.; Zhu, J. Copper-catalyzed three-component carboazidation of clkenes with acetonitrile and sodium azide. *Angew. Chem. Int. Ed.* **2017**, *56*, 10555–10558. [CrossRef] [PubMed]
33. Li, W.-Y.; Wu, C.-S.; Wang, Z.; Luo, Y. Fe-Catalyzed three-component carboazidation of alkenes with alkanes and trimethylsilyl azide. *Chem. Commun.* **2018**, *54*, 11013–11016. [CrossRef] [PubMed]
34. Xiong, H.; Ramkumar, N.; Chiou, M.-F.; Jian, W.; Li, Y.; Su, J.-H.; Zhang, X.; Bao, H. Iron-catalyzed carboazidation of alkenes and alkynes. *Nat. Commun.* **2019**, *10*, 1–7. [CrossRef] [PubMed]
35. Ollivier, C.; Renaud, P. Formation of carbon-nitrogen bonds via a novel radical azidation process. *J. Am. Chem. Soc.* **2000**, *122*, 6496–6497. [CrossRef]
36. Ollivier, C.; Renaud, P. A novel approach for the formation of carbon - nitrogen bonds: Azidation of alkyl radicals with sulfonyl azides. *J. Am. Chem. Soc.* **2001**, *123*, 4717–4727. [CrossRef]
37. Panchaud, P.; Renaud, P. A convenient tin-free procedure for radical carboazidation and azidation. *J. Org. Chem.* **2004**, *69*, 3205–3207. [CrossRef] [PubMed]
38. Panchaud, P.; Renaud, P. 3-Pyridinesulfonyl azide. In *Encyclopedia of Reagents for Organic Synthesis*; John Wiley Sons, Ltd.: Chichester, UK, 2006; ISBN 978-0-470-84289-8.
39. *Aminophosphonic and Aminophosphinic Acids: Chemistry and Biological Activity*; Kukhar, V.P.; Hudson, H.R. (Eds.) John Wiley Sons, Ltd.: Chichester, UK, 2000; ISBN 978-0-471-89149-9.
40. Cren, S.; Schär, P.; Renaud, P.; Schenk, K. Diastereoselectivity control of the radical carboazidation of substituted methylenecyclohexanes. *J. Org. Chem.* **2009**, *74*, 2942–2946. [CrossRef] [PubMed]
41. Barton, D.H.R.; Chern, C.Y.; Jaszberenyi, J.C. The Invention of radical reactions. XXXIII. Homologation reactions of carboxylic acids by radical chain chemistry. *Aust. J. Chem.* **1995**, *48*, 407–425. [CrossRef]
42. Vita, M.V.; Caramenti, P.; Waser, J. Enantioselective synthesis of homoallylic azides and nitriles via palladium-catalyzed decarboxylative allylation. *Org. Lett.* **2015**, *17*, 5832–5835. [CrossRef] [PubMed]
43. Blakemore, P.R.; Cole, W.J.; Kocieński, P.J.; Morley, A. A stereoselective synthesis of trans-1,2-sisubstituted alkenes based on the condensation of aldehydes with metallated 1-phenyl-1H-tetrazol-5-yl sulfones. *Synlett* **1998**, *1998*, 26–28. [CrossRef]
44. Marko, I.; Pospisil, J. Julia, Julia–Kocienski, and related sulfur-based alkenations. In *Category 6, Compounds with All-Carbon Functions, Science of Synthesis*; Neier, R., Bellus, D., Eds.; G. Thieme Verlag: Stuttgart, Germany, 2010; Volume 47a, ISBN 978-3-13-119011-6.
45. Lebrun, M.-E.; Le Marquand, P.; Berthelette, C. Stereoselective synthesis of Z alkenyl halides via Julia olefination. *J. Org. Chem.* **2006**, *71*, 2009–2013. [CrossRef] [PubMed]

46. Wang, X.; Bowman, E.J.; Bowman, B.J.; Porco, J.A. Total synthesis of the salicylate enamide macrolide oximidine III: Application of relay ring-closing metathesis. *Angew. Chem. Int. Ed.* **2004**, *43*, 3601–3605. [CrossRef]
47. Kamijo, S.; Kamijo, K.; Murafuji, T. Aryl ketone mediated photoinduced radical coupling for the alkylation of benzazoles employing saturated heterocyclic compounds. *Synthesis* **2019**, *51*, 3859–3864.
48. David Mendenhall, G. The Lewis acid catalyzed reaction of trans-hyponitrite ion with alkyl halides. *Tetrahedron Lett.* **1983**, *24*, 451–452. [CrossRef]
49. Panchaud, P.; Renaud, P. 3-Pyridinesulfonyl azide: A useful reagent for radical azidation. *Adv. Synth. Catal.* **2004**, *346*, 925–928. [CrossRef]
50. Harrowven, D.C.; Guy, I.L. KF–Silica as a stationary phase for the chromatographic removal of tin residues from organic compounds. *Chem. Commun.* **2004**, 1968–1969. [CrossRef] [PubMed]
51. Merchant, R.R.; Edwards, J.T.; Qin, T.; Kruszyk, M.M.; Bi, C.; Che, G.; Bao, D.-H.; Qiao, W.; Sun, L.; Collins, M.R.; et al. Modular radical cross-coupling with sulfones enables access to sp3-rich (fluoro)alkylated scaffolds. *Science* **2018**, *360*, 75–80. [CrossRef] [PubMed]
52. Hughes, J.M.E.; Fier, P.S. Desulfonylative arylation of redox-active alkyl sulfones with aryl bromides. *Org. Lett.* **2019**, *21*, 5650–5654. [CrossRef] [PubMed]

Sample Availability: Samples are not available from the authors.

© 2019 by the authors. Licensee MDPI, Basel, Switzerland. This article is an open access article distributed under the terms and conditions of the Creative Commons Attribution (CC BY) license (http://creativecommons.org/licenses/by/4.0/).

Article

Axially Ligated Mesohemins as Bio-Mimicking Catalysts for Atom Transfer Radical Polymerization

Liye Fu [†], Antonina Simakova [†], Sangwoo Park, Yi Wang, Marco Fantin and Krzysztof Matyjaszewski *

Department of Chemistry, Carnegie Mellon University, Pittsburgh, PA 15213, USA; liyef@andrew.cmu.edu (L.F.); simakova@cmu.edu (A.S.); sangwoo82@gmail.com (S.P.); ywang4@andrew.cmu.edu (Y.W.); mfantin@andrew.cmu.edu (M.F.)
* Correspondence: matyjaszewski@cmu.edu
† Liye Fu and Antonina Simakova contributed equally.

Academic Editor: Chryssostomos Chatgilialoglu
Received: 22 September 2019; Accepted: 30 October 2019; Published: 2 November 2019

Abstract: Copper is the most common metal catalyst used in atom transfer radical polymerization (ATRP), but iron is an excellent alternative due to its natural abundance and low toxicity compared to copper. In this work, two new iron-porphyrin-based catalysts inspired by naturally occurring proteins, such as horseradish peroxidase, hemoglobin, and cytochrome P450, were synthesized and tested for ATRP. Natural protein structures were mimicked by attaching imidazole or thioether groups to the porphyrin, leading to increased rates of polymerization, as well as providing polymers with low dispersity, even in the presence of ppm amounts of catalysts.

Keywords: iron porphyrin; heme; ATRPase; iron-mediated ATRP; bio-mimicking catalyst

---

Atom transfer radical polymerization (ATRP) is one of the most widely used techniques in the field of reversible deactivation radical polymerization (RDRP) procedures, and it can provide well-defined polymers with predetermined molecular weight, low dispersity, and precisely controlled architecture [1–4]. ATRP catalysts are predominantly copper-based complexes, due to their extraordinary performance for the synthesis of a broad range of well-defined polymers [5,6]. Nevertheless, developing catalysts with transition metals other than copper is still of great interest [7]. Iron-mediated ATRP has also been extensively studied due to the biocompatibility and low toxicity of iron, which is especially important for biologically relevant systems [8–12]. Although iron-based catalysts offer these potential benefits, their use in ATRP has been limited due to their lower activity and selectivity. Therefore, the design and development of novel iron-based catalysts, which are comparable in activity to conventional copper-based catalysts, and capable of polymerizing a wider range of monomers, are critical to further advancements in this field.

ATRP is typically performed in organic solvents, but performing ATRP in aqueous media provides several advantages. Water is an environmentally benign solvent, enabling direct polymerization of water-soluble monomers, faster reactions, and polymerization in the presence of biomolecules [13–17]. Several methods for well-controlled Cu-based ATRP in water have been developed, but in the majority of reports a limited number of catalytic systems and a narrow range of monomers have been used [18–20]. Control of an ATRP in aqueous media is difficult due to some side reactions including catalyst and chain-end instabilities, as well as the creation of a large equilibrium constant that significantly increases the rate of the polymerization reaction [21–24].

We have previously reported the synthesis of protein–polymer hybrids using ATRP under biologically relevant conditions, which were designed to sustain the structure of a protein during polymerization while continuing to provide good control of the grafted polymer [25]. In these systems, proteins, appropriately modified with bromoesters or bromoamides, served as macroinitiators for the "grafting from" reaction [26]. Recent publications by Bruns [27,28] and diLena [29,30] have shown that certain proteins/enzymes, such as horseradish peroxidase (HRP), catalase or hemoglobin (Hb), could also serve directly as catalysts for ATRP. These protein-based ATRP catalysts, or "ATRPases", are proteins comprising heme centers that are able to produce high molecular weight (MW) polymers with dispersity around 1.5~1.6. The relatively high dispersity indicated limited control, plausibly due to insufficient deactivation provided by the bulky protein structures. Nevertheless, these catalytic systems can potentially expand the range of polymerizable monomers because of their different catalyst structure and tolerance to pH variation. However, a major drawback of using proteins as catalysts for ATRP is their sensitivity to reaction conditions and high molecular weight [31]. Therefore, it is necessary to pursue the development of synthetic analogues of natural ATRP enzymes that have enhanced properties, such as the ability to accommodate more stringent reaction conditions with increased mass-to-efficiency ratios of the catalyst complexes, that would allow for a wider range of applications for these biocatalytic systems.

Previously, a successful ATRP of neutral monomers with mesohemin-based catalysts was reported [32]. The hemin was modified with methoxy poly(ethylene glycol) (MPEG) chains to enhance water solubility, and additionally, the vinyl groups were hydrogenated to prevent catalyst copolymerization and consequent incorporation into the polymer chains. Oligo(ethylene oxide) methyl ether methacrylate (OEOMA, M_n = 475) was polymerized under benign aqueous conditions, generating polymers with well-defined molecular weight and low dispersity (<1.2) via activators regenerated by electron transfer (ARGET) ATRP. Acidic monomers, such as methacrylic acid, were directly polymerized with the same catalyst preparing polymers with predetermined MW and acceptable dispersity ~1.5 [33]. Thus, the design and discovery of a novel bio-catalytic systems is still of interest.

In this paper, two additional mesohemin catalysts were prepared with different ligands, each of which was selected to imitate the axial ligation from amino acid residues present in proteins. The iron center in heme, present in proteins, is often additionally ligated by residues of amino acids such as histidine, cysteine, methionine, or tyrosine [34,35]. Therefore, we chose two types of modification: one with an imidazole moiety to mimic complexation by histidine (Mesohemin-MPEG$_{550}$-Imidazole or MH-MPEG-N), and the other with a thioether moiety to mimic complexation by methionine (Mesohemin-MPEG$_{550}$-Thioether or MH-MPEG-S). The imidazole group has very high complexation affinity towards iron, and thus forms well-defined iron porphyrin complexes. The iron porphyrin complex with thiol has been extensively studied, but we chose to incorporate a thioether to prevent the strong radical transfer property of thiols (Figure 1a).

A series of axially ligated mesohemin complexes were synthesized (Scheme S1), to expand the scope of heme-based catalysts. In this series, one carboxyl group was modified with a poly(ethylene glycol) (PEG) tail and the second carboxyl group was modified with either an imidazole or a thioether via an amidation reaction. Hemin was selected as the starting material for the synthesis of modified iron porphyrins, because hemin is less photosensitive than protoporphyrin IX (hemin without iron), and this strategy did not require an additional step of metal insertion [36]. However, protoporphyrin IX could also have been used for synthesis of modified heme complexes, as it typically provides easier purification and analysis.

The modified mesohemin complexes were characterized by mass spectroscopy, Ultraviolet–visible (UV-Vis) spectroscopy, Infrared (IR) spectroscopy (Figure S1–S8) and cyclic voltammetry (CV; Figure 2a). According to CV measurements, the iron porphyrins formed complexes with varied redox potential $E_{1/2}$, indicating different reactivity. Two new complexes were characterized by less negative $E_{1/2}$ values when compared to the fully PEGylated mesohemin, but formed only a single catalytic species, even in the absence of excess bromide [20,37]. Imidazole-modified mesohemin was not significantly affected by the addition of excess bromide ions, but the CV of the thioether-modified mesohemin showed a shift towards a more negative potential (Figure 2b).

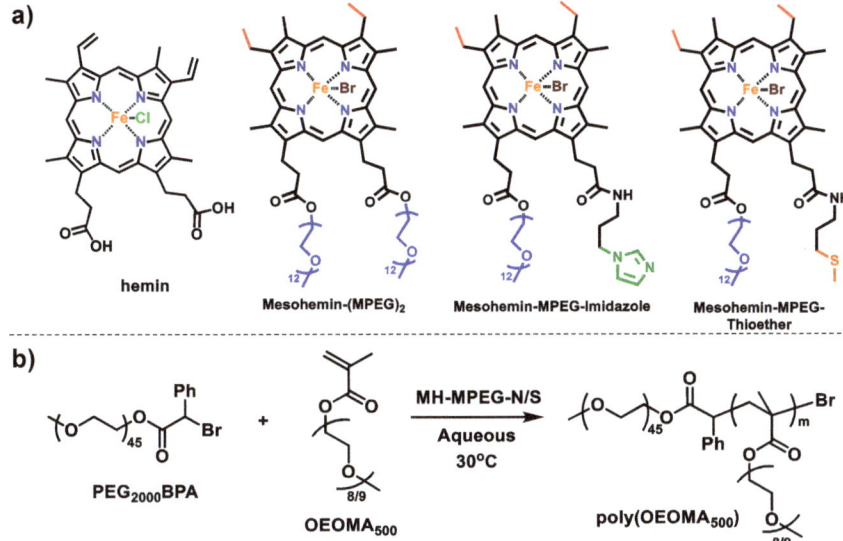

Figure 1. (**a**) Iron porphyrin derivatives used for catalysis of atom transfer radical polymerization (ATRP); (**b**) Scheme of Activator Generated by Electron Transfer (AGET) ATRP of oligo(ethylene oxide) methyl ether methacrylate (OEOMA)$_{500}$.

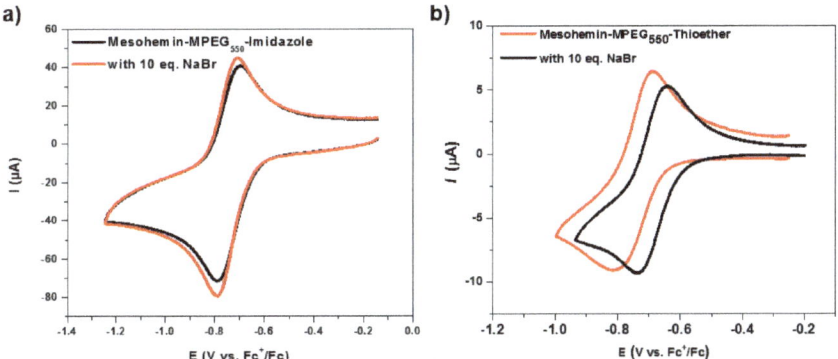

Figure 2. Cyclic voltammogram of (**a**) Mesohemin-methoxy poly(ethylene glycol) (MPEG)$_{550}$-N and (**b**) Mesohemin-MPEG$_{550}$-S, scan rate = 100 mV/s, supporting electrolyte = tetrabutylammonium hexafluorophosphate (TBAPF$_6$, 0.1 M in DMF).

To evaluate the feasibility of the new mesohemin catalysts, AGET ATRP's of OEOMA$_{500}$ were conducted in the presence of MH-MPEG-N/S with the initial polymerization conditions identical to those previously used for MH-(MPEG$_{550}$)$_2$ [32]. Polymerization with MH-MPEG-N was more than two times faster than that with MH-(MPEG$_{550}$)$_2$, with monomer conversion reaching 76% in only 2h. The final polymer possessed a relatively low dispersity of 1.27, which is slightly higher than that obtained with MH-(MPEG)$_2$. One plausible explanation for the increased activity is that it is mainly due to the lower E$_{1/2}$ based on electron donation from the attached imidazole. Polymerization in the presence of thioether-ligated mesohemin (MH-MPEG-S) did not proceed to high conversion, but the final polymer displayed the results of good control, with M_w/M_n < 1.3 (Figure 3). These reactions suggested that the modifications of mesohemin with axial ligands did provide complexes that could catalyze an ATRP, but additional optimization of reaction conditions needed to be addressed.

Table 1. Experimental conditions and results of ATRP catalyzed by axially ligated mesohemins [a].

Entry	M/I/RA/Cat	Catalyst	Conv./%, (Time, h)	$M_{n,th} \times 10^{-3}$ [b]	$M_{n,GPC} \times 10^{-3}$ [c]	M_w/M_n
1	216/1/1/1	MH-(MPEG)$_2$	60 (6)	61	62	1.19
2	216/1/1/1	MH-MPEG-N	76 (2.5)	84	76	1.27
3	216/1/1/1	MH-MPEG-S	25 (3.5)	27	40	1.28
4	227:1:0.3 × 2:1	MH-MPEG-N	75 (5)	83	108	1.16
5	227:1:0.3 × 2:1	MH-MPEG-S	41 (5)	43	43	1.07
6	216/1/1/1	MH-MPEG$_2$ + imidazole	33 (2)	37	78	1.91
7	216:1:0.3 × 2:0.1	MH-(MPEG)$_2$	70 (5)	72	98	1.17
8	216:1:0.3 × 2:0.1	MH-MPEG-N	60 (5)	69	190	1.42
9	216/1/0.3 × 2:0.1	MH-MPEG-S	61 (8.7)	61	57	1.18

[a] T = 30 °C; solvent: H$_2$O/DMF = 9/1; [NaBr] = 100 mM; RA = ascorbic acid; I = PEG$_{2000}$BPA; [I] = 2 mM; M = OEOMA$_{500}$; [M] = 20% (v/v); [b] $M_{n\,th}$ = ([M]$_0$/[I]$_0$)×conversion×M$_{monomer}$; [c] $M_{n,GPC}$ measured by Gel permeation chromatography (GPC) using universal PMMA standards with tetrahydrofuran (THF) as eluent.

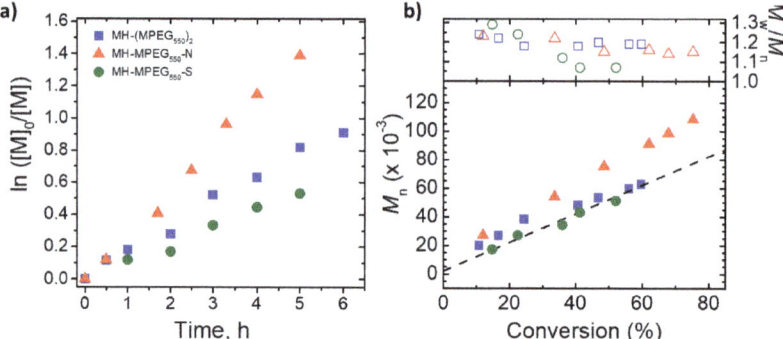

Figure 3. First-order kinetic plots (**a**), evolution of M_n and M_w/M_n with conversion (**b**) for entry 1-3 in Table 1.

In one such optimization, the addition of less reducing agent in a polymerization catalyzed by MH-MPEG-N resulted in a linear first-order kinetics and linear increase of MW with conversion, with values of MW close to theoretical values. This polymerization resulted in the formation of polymers with dispersity values lower than previously obtained with MH-(MPEG)$_{2,}$ indicating that conditions had been selected that provided a better controlled polymerization. Additionally, the reaction was faster despite decreased amounts of reducing agent.

In order to verify that the covalent attachment of the imidazole moiety was necessary for the formation of a 1/1 iron porphyrin/imidazole complex, an ATRP with fully PEGylated mesohemin was performed in the presence of free imidazole with a ratio of 1:1 to the iron porphyrin (Table 1, entry 6). This reaction resulted in a slow and poorly controlled polymerization. The final MW of the polymer formed under these conditions was double the theoretically predicted value, indicating inefficient initiation, and M_w/M_n was as high as 1.91. This poorly controlled polymerization could be explained by the fact that two imidazole molecules can complex to the iron porphyrin creating a situation in which a fraction of the catalyst is a hexa-coordinated mesohemin, and another fraction of the catalyst has no imidazole ligands. Since deactivation of a propagating radical cannot occur without the presence of Fe-Br species, the hexacoordinated species consequently results in the loss of deactivation efficiency, thereby providing a poorly controlled polymerization. Therefore, covalent attachment of an imidazole moiety forces preferential formation of a clean 1/1 complex of iron porphyrin and imidazole retaining the Fe-Br bond, which is necessary for performing a well-controlled ATRP.

Because iron porphyrins are highly colored compounds, reaction with a lower concentration of the catalyst would be beneficial for simplification of any desired purification procedures. In the next set of experiments, polymerizations were performed with a 10-fold lower concentration of catalyst. The MH-(MPEG)$_2$ did not provide polymers with well-defined M_n when concentration was reduced by a factor of 10 (Table 1, entry 7). However, the axially ligated mesohemins efficiently catalyzed ATRP when their concentrations were decreased by a factor of 10 (Table 1, entries 8, 9). The reaction catalyzed by imidazole-modified mesohemin resulted in formation of a polymer with higher MW than theoretically predicted and relatively high dispersity, reaching a value of 1.5. Nevertheless, the uniform shift in GPC traces toward higher MW indicated that some level of control over the polymerization was attained. Polymerizations conducted in the presence of thioether ligated mesohemin reached higher monomer conversions, over 60%, which was significantly higher than when the catalyst complex was used at higher concentrations. MWs were in good agreement with theoretically predicted values, and the final dispersity of the polymer was less than 1.2 (Figure 4). These results demonstrate that it is possible to use modified hemin complexes as ATRP catalysts at lower concentrations, but further optimization of the amount, and mode of addition, of the reducing agent is required.

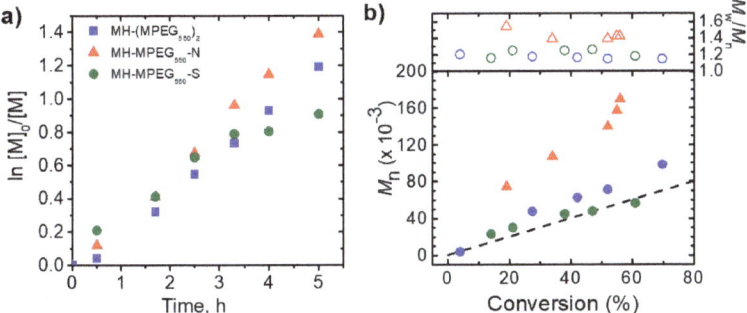

Figure 4. First-order kinetic plots (**a**), evolution of MW and dispersity with conversion (**b**) for entry 7-9 in Table 1.

In conclusion, a series of bioinspired iron porphyrin-based complexes were designed and successfully utilized as ATRP catalysts. Mesohemin-(MPEG$_{550}$)$_2$, prepared from naturally occurring hemin, performed significantly better than hemin itself, or the previously reported hematin-based complex. This can be attributed to the increased solubility of the catalysts in the reaction medium due to the presence of PEG tails. The hydrogenated vinyl bonds prevented copolymerization of the hemin and allowed for faster deactivation in the presence of excess bromide salt. Since this new environmentally benign class of ATRP catalysts showed promise, further modification of mesohemin-based catalysts with different axial ligands were studied. It was found that mesohemin-MPEG$_{550}$, additionally modified with imidazole and thioether units, efficiently catalyzed polymerizations in water at low catalyst concentrations. These bio-mimicking catalyst complexes will be further investigated in polymerization of acidic monomers and for other methods of low ppm ATRP.

Supplementary Materials: The following are available online at http://www.mdpi.com/1420-3049/24/21/3969/s1. Characterization of aqueous phase catalysts and polymerization results are supplied in the supporting information.

Author Contributions: A.S. and K.M. conceived and designed the project; L.F. designed experiments; Y.W, S.P. and M.F. collected and analyzed the CV data; L.F., A.S. and K.M. wrote the paper; K.M. was responsible for project administration.

Funding: This research was funded by NIH R01 DE020843.

Conflicts of Interest: The authors declare no conflict of interest.

References

1. Kamigaito, M.; Ando, T.; Sawamoto, M. Metal-catalyzed living radical polymerization. *Chem. Rev.* **2001**, *101*, 3689–3745. [CrossRef] [PubMed]
2. Matyjaszewski, K. Atom Transfer Radical Polymerization (ATRP): Current Status and Future Perspectives. *Macromolecules* **2012**, *45*, 4015–4039. [CrossRef]
3. Matyjaszewski, K.; Tsarevsky, N.V. Nanostructured functional materials prepared by atom transfer radical polymerization. *Nat. Chem.* **2009**, *1*, 276–288. [CrossRef] [PubMed]
4. Matyjaszewski, K.; Xia, J.H. Atom transfer radical polymerization. *Chem. Rev.* **2001**, *101*, 2921–2990. [CrossRef] [PubMed]
5. Braunecker, W.A.; Matyjaszewski, K. Controlled/living radical polymerization: Features, developments, and perspectives. *Prog. Polym. Sci.* **2007**, *32*, 93–146. [CrossRef]
6. Ribelli, T.G.; Lorandi, F.; Fantin, M.; Matyjaszewski, K. Atom Transfer Radical Polymerization: Billion Times More Active Catalysts and New Initiation Systems. *Macromol. Rapid Commun.* **2019**, *40*, 1800616. [CrossRef]
7. di Lena, F.; Matyjaszewski, K. Transition metal catalysts for controlled radical polymerization. *Prog. Polym. Sci.* **2010**, *35*, 959–1021. [CrossRef]
8. He, W.W.; Zhang, L.F.; Miao, J.; Cheng, Z.P.; Zhu, X.L. Facile Iron-Mediated AGET ATRP for Water-Soluble Poly(ethylene glycol) Monomethyl Ether Methacrylate in Water. *Macromol. Rapid Commun.* **2012**, *33*, 1067–1073. [CrossRef]
9. Schroeder, H.; Yalalov, D.; Buback, M.; Matyjaszewski, K. Activation-Deactivation Equilibrium Associated With Iron-Mediated Atom-Transfer Radical Polymerization up to High Pressure. *Macromol. Chem. Phys.* **2012**, *213*, 2019–2026. [CrossRef]
10. Wang, Y.; Zhang, Y.Z.; Parker, B.; Matyjaszewski, K. ATRP of MMA with ppm Levels of Iron Catalyst. *Macromolecules* **2011**, *44*, 4022–4025. [CrossRef]
11. Allan, L.E.N.; Perry, M.R.; Shaver, M.P. Organometallic mediated radical polymerization. *Prog. Polym. Sci.* **2012**, *37*, 127–156. [CrossRef]
12. Poli, R.; Shaver, M.P. ATRP/OMRP/CCT Interplay in Styrene Polymerization Mediated by Iron(II) Complexes: A DFT Study of the alpha-Diimine System. *Chem. Eur. J.* **2014**, *20*, 17530–17540. [CrossRef] [PubMed]
13. Gauthier, M.A.; Klok, H.A. Peptide/protein-polymer conjugates: Synthetic strategies and design concepts. *Chem. Commun.* **2008**, 2591–2611. [CrossRef] [PubMed]
14. Heredia, K.L.; Bontempo, D.; Ly, T.; Byers, J.T.; Halstenberg, S.; Maynard, H.D. In situ preparation of protein—"Smart" polymer conjugates with retention of bioactivity. *J. Am. Chem. Soc.* **2005**, *127*, 16955–16960. [CrossRef]
15. Peeler, J.C.; Woodman, B.F.; Averick, S.; Miyake-Stoner, S.J.; Stokes, A.L.; Hess, K.R.; Matyjaszewski, K.; Mehl, R.A. Genetically Encoded Initiator for Polymer Growth from Proteins. *J. Am. Chem. Soc.* **2010**, *132*, 13575–13577. [CrossRef] [PubMed]
16. Wang, X.S.; Armes, S.P. Facile atom transfer radical polymerization of methoxy-capped oligo(ethylene glycol) methacrylate in aqueous media at ambient temperature. *Macromolecules* **2000**, *33*, 6640–6647. [CrossRef]
17. Fantin, M.; Isse, A.A.; Matyjaszewski, K.; Gennaro, A. ATRP in Water: Kinetic Analysis of Active and Super-Active Catalysts for Enhanced Polymerization Control. *Macromolecules* **2017**, *50*, 2696–2705. [CrossRef]
18. Konkolewicz, D.; Magenau, A.J.D.; Averick, S.E.; Simakova, A.; He, H.K.; Matyjaszewski, K. ICAR ATRP with ppm Cu Catalyst in Water. *Macromolecules* **2012**, *45*, 4461–4468. [CrossRef]
19. Mougin, N.C.; van Rijn, P.; Park, H.; Muller, A.H.E.; Boker, A. Hybrid Capsules via Self-Assembly of Thermoresponsive and Interfacially Active Bionanoparticle-Polymer Conjugates. *Adv. Funct. Mater.* **2011**, *21*, 2470–2476. [CrossRef]
20. Simakova, A.; Averick, S.E.; Konkolewicz, D.; Matyjaszewski, K. Aqueous ARGET ATRP. *Macromolecules* **2012**, *45*, 6371–6379. [CrossRef]
21. Bortolamei, N.; Isse, A.A.; Magenau, A.J.D.; Gennaro, A.; Matyjaszewski, K. Controlled Aqueous Atom Transfer Radical Polymerization with Electrochemical Generation of the Active Catalyst. *Angew. Chem. Int. Ed.* **2011**, *50*, 11391–11394. [CrossRef] [PubMed]
22. Tsarevsky, N.V.; Pintauer, T.; Matyjaszewski, K. Deactivation efficiency and degree of control over polymerization in ATRP in protic solvents. *Macromolecules* **2004**, *37*, 9768–9778. [CrossRef]

23. Zhang, Q.; Wilson, P.; Li, Z.D.; McHale, R.; Godfrey, J.; Anastasaki, A.; Waldron, C.; Haddleton, D.M. Aqueous Copper-Mediated Living Polymerization: Exploiting Rapid Disproportionation of CuBr with Me6TREN. *J. Am. Chem. Soc.* **2013**, *135*, 7355–7363. [CrossRef]
24. Fantin, M.; Isse, A.A.; Gennaro, A.; Matyjaszewski, K. Understanding the Fundamentals of Aqueous ATRP and Defining Conditions for Better Control. *Macromolecules* **2015**, *48*, 6862–6875. [CrossRef]
25. Averick, S.; Simakova, A.; Park, S.; Konkolewicz, D.; Magenau, A.J.D.; Mehl, R.A.; Matyjaszewski, K. ATRP under Biologically Relevant Conditions: Grafting from a Protein. *ACS Macro Lett.* **2012**, *1*, 6–10. [CrossRef]
26. Russell, A.J.; Baker, S.L.; Colina, C.M.; Figg, C.A.; Kaar, J.L.; Matyjaszewski, K.; Simakova, A.; Sumerlin, B.S. Next generation protein-polymer conjugates. *AIChE J.* **2018**, *64*, 3230–3245. [CrossRef]
27. Sigg, S.J.; Seidi, F.; Renggli, K.; Silva, T.B.; Kali, G.; Bruns, N. Horseradish Peroxidase as a Catalyst for Atom Transfer Radical Polymerization. *Macromol. Rapid Commun.* **2011**, *32*, 1710–1715. [CrossRef]
28. Silva, T.B.; Spulber, M.; Kocik, M.K.; Seidi, F.; Charan, H.; Rother, M.; Sigg, S.J.; Renggli, K.; Kali, G.; Bruns, N. Hemoglobin and Red Blood Cells Catalyze Atom Transfer Radical Polymerization. *Biomacromolecules* **2013**, *14*, 2703–2712. [CrossRef]
29. Ng, Y.H.; di Lena, F.; Chai, C.L.L. Metalloenzymatic radical polymerization using alkyl halides as initiators. *Polym. Chem.* **2011**, *2*, 589–594. [CrossRef]
30. Ng, Y.H.; di Lena, F.; Chai, C.L.L. PolyPEGA with predetermined molecular weights from enzyme-mediated radical polymerization in water. *Chem. Commun.* **2011**, *47*, 6464–6466. [CrossRef]
31. Polizzi, K.M.; Bommarius, A.S.; Broering, J.M.; Chaparro-Riggers, J.F. Stability of biocatalysts. *Curr. Opin. Chem. Biol.* **2007**, *11*, 220–225. [CrossRef] [PubMed]
32. Simakova, A.; Mackenzie, M.; Averick, S.E.; Park, S.; Matyjaszewski, K. Bioinspired Iron-Based Catalyst for Atom Transfer Radical Polymerization. *Angew. Chem. Int. Ed.* **2013**, *52*, 12148–12151. [CrossRef] [PubMed]
33. Fu, L.Y.; Simakova, A.; Fantin, M.; Wang, Y.; Matyjaszewski, K. Direct ATRP of Methacrylic Acid with Iron-Porphyrin Based Catalysts. *ACS Macro Lett.* **2018**, *7*, 26–30. [CrossRef] [PubMed]
34. Marques, H.M.; Munro, O.Q.; Munro, T.; de Wet, M.; Vashi, P.R. Coordination of N-donor ligands by the monomeric ferric porphyrin N-acetylmicroperoxidase-8. *Inorg. Chem.* **1999**, *38*, 2312–2319. [CrossRef]
35. Tezcan, F.A.; Winkler, J.R.; Gray, H.B. Effects of ligation and folding on reduction potentials of heme proteins. *J. Am. Chem. Soc.* **1998**, *120*, 13383–13388. [CrossRef]
36. Chang, C.K.; Dolphin, D. Ferrous Porphyrin Mercaptide Complexes—Models for Reduced Cytochrome-P-450. *J. Am. Chem. Soc.* **1975**, *97*, 5948–5950. [CrossRef]
37. Fantin, M.; Isse, A.A.; Venzo, A.; Gennaro, A.; Matyjaszewski, K. Atom Transfer Radical Polymerization of Methacrylic Acid: A Won Challenge. *J. Am. Chem. Soc.* **2016**, *138*, 7216–7219. [CrossRef]

Sample Availability: Samples of the compounds are not available from the authors.

© 2019 by the authors. Licensee MDPI, Basel, Switzerland. This article is an open access article distributed under the terms and conditions of the Creative Commons Attribution (CC BY) license (http://creativecommons.org/licenses/by/4.0/).

MDPI
St. Alban-Anlage 66
4052 Basel
Switzerland
Tel. +41 61 683 77 34
Fax +41 61 302 89 18
www.mdpi.com

Molecules Editorial Office
E-mail: molecules@mdpi.com
www.mdpi.com/journal/molecules

www.ingramcontent.com/pod-product-compliance
Lightning Source LLC
LaVergne TN
LVHW071938080526
838202LV00064B/6633